Eclogite Facies Rocks

ECLOGITE FACIES ROCKS

edited by

D.A. CARSWELL
Reader in Metamorphic Petrology
Department of Geology
University of Sheffield

Springer Science+Business Media, LLC

ISBN 978-94-010-9265-4 ISBN 978-94-010-9263-0 (eBook)
DOI 10.1007/978-94-010-9263-0

Originally published by Blackie and Son Ltd in 1990.
First published 1990

British Library Cataloguing in Publication Data

Eclogite facies rocks.
1. Metamorphic rocks
I. Carswell, D.A.
552'.4

Library of Congress Cataloging-in-Publication Data

Eclogite facies rocks/edited by D.A. Carswell.
p. cm.
Bibliography: p.
Includes index.

1. Eclogite. I. Carswell, D.A.
QE475.E25E25 1989 89-15860
552'.4—dc20 CIP

Typesetting by Thomson Press (India) Ltd., New Delhi

Preface

The high pressures necessary for the stabilisation of eclogites in metabasic rocks and garnet peridotites in ultrabasic rocks have been long recognised and experimentally established. Xenoliths of such rocks brought up in volatile charged alkaline magmas, such as kimberlites, are widely accepted to be mostly of upper mantle derivation (Chapter 13). Eclogites are predicted to be thermodynamically stable also in the lower crust beneath cratonic regions. However, xenolith suite studies indicate that kinetic and/or compositional factors limit their distribution in the lower continental crust relative to granulite facies assemblages (Chapter 12).

Occurrences of eclogites and garnet peridotites in exposed crustal metamorphic terrains have been interpreted in the past as exotic tectonic blocks of deeper (largely mantle) origin, because of their apparent difference in metamorphic grade compared with the encompassing rocks. Only in recent years have metamorphic petrologists begun to recognise that such crustal terrains sometimes preserve co-facial (eclogite facies), high pressure mineral parageneses in other spatially associated lithologies such as metapelites and metagranitoids.

Placed in a modern, global geotectonic context, it is now apparent not only that eclogites can be expected to be stabilised in oceanic crust subducted at continental plate margins (Chapter 9), but also that eclogite facies mineral parageneses may be stabilised in a wider range of continental crust lithologies, where substantial tectonic thickening has occurred in continental plate collision zones (Chapters 8–10). Recent exciting evidence from the Western Alps (Chapter 10) suggests that continental crust may be subducted to depths approaching 100 km and yet exhumed during subsequent orogenic uplift. Exposures of eclogite facies assemblages in metamorphic terrains are therefore important indicators of palaeo-plate sutures and provide valuable constraints on tectonic and thermal models for the evolution of plate collision zones.

This book provides the first coherent and comprehensive review on the mineralogy (Chapter 2) and stability conditions (Chapters 3 and 4) of eclogites (*sensu stricto*) and eclogite facies assemblages in general. A temperature-based classification scheme which correlates equilibration temperature with particular environments for eclogite formation is followed (Chapter 1). The characterisation of co-facial mineral parageneses in lithologies other than basites and ultrabasites is also discussed, in far greater detail than previously available.

Emphasis is placed on the documentation and interpretation of eclogite facies rocks in European metamorphic belts: the Caledonides, Variscides and Alps (Tethysides). Pressure–temperature–time paths deduced from the observed sequences of prograde and retrograde mineral assemblages, in lithologies which have at least in part retained high pressure eclogite facies mineralogies, are demonstrated to be important in interpretative models of the dynamo-thermal evolution of these belts. Moreover, tectonic intercalation of peridotite bodies with upper mantle geochemical signatures and relict high pressure, garnet-bearing assemblages (Chapter 11) appears to be a particularly characteristic feature of these belts, and bears witness to profound crust-mantle interaction during orogenesis. Exposures of Grenvillian-age crustal eclogites have recently been proven, but it can be speculated that further back in geological history, the continental crust may not have been sufficiently cool and rigid for the operation of the contemporary-style subduction processes necessary for the stabilisation and successful exhumation of eclogite facies rocks. However, as discussed in Chapter 6, there are several inherent difficulties in the dating of eclogite facies mineral assemblages, and reaction kinetics (Chapter 7) undoubtedly play a crucial role in both their formation and preservation.

The enthusiastic support and endeavours of the contributors, the publisher, and Pat Mellor (secretary and typist) during the production of this book have been invaluable and much appreciated. Financial support from the University of Sheffield, the Natural Environment Research Council and the Royal Society, for my research work on eclogite facies rocks is also gratefully acknowledged. Even more important has been the love and unfailing support of my wife, Gill.

D.A.C.

Contributors

D.A. Carswell	Department of Geology, University of Sheffield, Mappin Street, Sheffield, S1 3JD, UK
C. Chopin	Départment de Géologie, Ecole Normale Supérieure, 24 Rue Lhomond, 75231 Paris, France
S.J. Cuthbert	Department of Geology, University of Glasgow, Kelvinside, Glasgow, G12 8QQ, UK
J.B. Dawson	Department of Geology, University of Sheffield, Mappin Street, Sheffield, S1 3JD, UK
G.T.R. Droop	Department of Geology, University of Manchester, Oxford Road, Manchester, M13 9PL, UK
D. Gebauer	Laboratory for Isotope Geochemistry and Mass Spectrometry, Swiss Federal Institute of Technology, CH-8092, Zurich, Switzerland
W.L. Griffin	Division of Mineral Physics and Mineralogy, CSIRO, PO Box 136, North Ryde, NSW 2113, Australia
S.L. Harley	Department of Geology and Geophysics, University of Edinburgh, Kings Buildings, West Mains Road, Edinburgh, EH9 3JW, UK
B. Lombardo	Centro di Studio sui Problemi dell'Orogeno delle Alpi Occidentali, C.N.R., via Accademia delle Scienze 5, 10123 Torino, Italy
L.G. Medaris Jr.	Department of Geology and Geophysics, University of Wisconsin, Madison, WI-53706, USA
A. Mottana	Dipartimento di Scienze della Terra, Sezione Mineralogico-Cristallografica, Università di Roma "La Sapienza", Piazzale Aldo Moro 5, 00185 Roma, Itlay
R. Oberhänsli	Labor für Mikroröntgenspektroskopie, Philosophisch Naturwissenschaftlichen Fakultat, Universität Bern, CH-3012, Bern, Switzerland.
P.J. O'Brien	Department of Geology, University of Sheffield, Mappin Street, Sheffield, S1 3JD, UK

S.Y. O'Reilly School of Earth Sciences, Macquarie University, North Ryde, NSW 2113, Australia

N.J. Pearson School of Earth Sciences, Macquarie University, North Ryde, NSW 2113, Australia

U. Pognante Dipartimento di Scienze della Terra dell' Università, via Valperga Caluso 37, 10125 Torino, Italy

D.C. Rubie Bayerisches Geoinstitut, Universität Bayreuth, Postfach 101251, D8580 Bayreuth, Federal Republic of Germany

M. Schliestedt Institut für Mineralogie, Universität Hannover, Welfengarten 1, D-3000 Hannover 1, Federal Republic of Germany

Contents

Glossary of Symbols and Abbreviations

(a) General

a_c^b	activity of chemical component *c* in phase *b*
C_p	heat capacity
D	depth (km = kilometres)
f	fugacity
G	Gibbs free energy
H	enthalpy
K	equilibrium constant for a balanced reaction
K_D	distribution coefficient
$K_{DFe^{2+}-Mg^{2+}}^{Grt-Cpx}$	distribution coefficient for Fe^{2+}/Mg^{2+} in garnet relative to clinopyroxene
LP, MP, HP	low, medium, and high pressure respectively
LT, MT, HT	low, medium, and high temperature respectively
P	total effective pressure (kbar = kilobars)
P_s	lithostatic confining pressure on solid aggregate of mineral grains
P_f	pore fluid pressure
P_{H_2O}	partial equilibrium vapour pressure of water in pore fluid phase
R	gas constant
S	entropy
s.s.	sensu stricto
s.l.	sensu lato
ss.	solid solution
T	temperature (°C = degree celsius; K = degree kelvin)
t	time (Ma = million years; Ga = 10^9 years)
V	volume
W	regular solution interaction parameter
X_c^b	mole fraction of chemical component *c* in phase *b*
α	coefficient of thermal expansion
β	coefficient of isothermal compressibility
γ	activity coefficient

(b) Chemical system abbreviations

A = Al_2O_3
C = CaO
Cr = Cr_2O_3
F = FeO
H = H_2O
K = K_2O
M = MgO
N = Na_2O
S = SiO_2
e.g. NCMASH = Na_2O–CaO–MgO–Al_2O_3–SiO_2–H_2O

(c) Chemical parameters

Mg* = 100 Mg/(Mg + Fe)
Cr* = 100 Cr/(Cr + Al)

(d) Mineral name abbreviations

Ab	albite
Acm	acmite
Act	actinolite
Adr	andradite
Agt	aegirine-augite
Alm	almandine
Aln	allanite
Amp	amphibole
An	anorthite
And	andalusite
Ank	ankerite
Ap	apatite
Ard	ardennite
Arg	aragonite
Atg	antigorite
Aug	augite
Bar	barroisite
Brn	braunite
Bt	biotite
CaEs	Ca-Eskola
Cal	calcite
Car	carpholite
CaTs	Ca-Tschermaks
Chl	chlorite
Chr	chromite
Chu	clinohumite
Cld	chloritoid
Cli	clinochlore
Clm	chloromelanite
Coe	coesite
Cor	corundum
Cpx	clinopyroxene
Crd	cordierite
Crs	crossite
Cum	cummingtonite
Czo	clinozoisite
Dan	dannemorite
Der	deerite
Di	diopside
Dmd	diamond
Dol	dolomite
Ed	edenite
Ell	ellenbergerite
En	enstatite
Ep	epidote
Fa	fayalite
Fl	fluorite
Fo	forsterite
Fs	ferrosilite
Fsp	feldspar
Gh	gehlenite
Gln	glaucophane
Gr	graphite
Grs	grossular
Grt	garnet
Gru	grunerite
Hbl	hornblende
Hc	hercynite
Hd	hedenbergite
Hem	haematite
Hol	hollandite
Hus	hausmannite
Hy	hypersthene
Ill	illite
Ilm	ilmenite
Jd	jadeite
Kfs	K-feldspar
Kln	kaolinite
Ky	kyanite
L	liquid, melt
La	larnite
Lab	labradorite
Lws	lawsonite
Mag	magnetite
Mc	microcline
Mer	merwinite
Mgs	magnesite
MgTs	Mg-Tschermaks
Mrg	margarite
Ms	muscovite
Mtc	monticellite
Mul	mullite
Ne	nepheline
Ol	olivine
Olg	oligoclase
Oln	oligonite
Omp	omphacite
Opx	orthopyroxene
Per	periclase
Pg	paragonite
Phe	phengite
Phl	phlogopite
Pie	piemontite
Pl	plagioclase
Pmp	pumpellyite
Po	pyrrhotite
Prg	paragasite
Prl	pyrophyllite
Prp	pyrope
Prs	pyrolusite
Pxm	pyroxmangite
Pyr	pyrophanite
Pyx	pyroxene
Py	pyrite
Qtz	quartz
Rbk	riebeckite
Rdn	rhodonite
Rds	rhodochrosite
Rt	rutile
Scp	scapolite
Sil	sillimanite
Spl	spinel
Spn	sphene
Spr	sapphirine
Sps	spessartine
St	staurolite
Stp	stilpnomelane
Thu	thulite
Tlc	talc
Tr	tremolite
Ts	Tschermaks-pyroxene
Ttn	titanite
Tur	tourmaline
Uvr	uvarovite
V	vapour
Ves	vesuvianite
Wnc	winchite
Wo	wollastonite
Yo	yoderite
Zo	zoisite
Zrn	zircon
Zus	zussmanite

1 Eclogites and the eclogite facies: definitions and classification

D.A. CARSWELL

1.1 Background

Since the classic early work of Eskola (1920, 1921), the name eclogite has been used both for a specific litho-type and to characterise a particular metamorphic facies. This has resulted in classificatory and interpretative problems, and these have been exacerbated by the general failure to recognise eclogite facies parageneses in other than basic and ultrabasic rock compositions.

The High Pressure Rocks Study Group of the IUGS (International Union of Geological Societies) Subcommission on the Systematics of Metamorphic Rocks continues to debate the definition of the petrographic rock name *eclogite* and of the *eclogite facies*. In order to clarify rather than add to current confusion and controversies (*see* reviews by Smith *et al* 1982, and Smith 1983), it is necessary to anticipate their conclusions and define these terms here.

1.2 Definition of eclogite as a litho-type

There appears to be broad acceptance of the original definition of Haüy (1822) that an eclogite is a rock composed mainly of garnet and clinopyroxene. However, general agreement more or less ends there. Most petrologists recognise eclogites to be characterised by omphacite or chloromelanite clinopyroxenes with high Al^{VI} and Na contents (e.g. jadeite/Tschermaks ratio $> 1/2$; White, 1964), and by relatively magnesian (pyrope rich) garnets (Eskola, 1921; Church, 1968; Miyashiro, 1973; Smith *et al* 1982; Smulikowski 1964, 1980). Others (e.g. Davidson, 1943; Lovering, 1964) have extended the usage to include garnet-clinopyroxene rocks with mineral compositions which are more typical of those in meta-basic granulites.

In accordance with the nature of the original eclogites described by Haüy (1822) and subsequent descriptions by Eskola (1921), most petrologists do *not* accept plagioclase as a primary phase in eclogites. Nonetheless, descriptions of so-called plagioclase eclogites have appeared in the literature (e.g. Church 1964, 1968; Kappel 1967; Kozlowski 1958; Subramanian 1956). In other instances, garnetiferous rocks with apparently coexisting omphacitic clinopyroxene and plagioclase have been referred to as eclogite-granulite or omphacite granulites (Smulikowski and Bakun-Czubarow, 1973; Smulikowski, 1980; Pouba *et al* 1985).

It is recognised that eclogites have a roughly basaltic composition (Yoder and Tilley 1962; Green and Ringwood 1967a; Banno 1970; Turner 1981). However, some petrologists have described rocks of non-basaltic compositions as eclogites (Kozlowski

1958; Forbes 1965; Miyashiro 1973). Indeed Ringwood (1975) has advocated the use of the term 'plagioclase eclogite' for relatively aluminous rock compositions (high-Al basalt, gabbroic anorthosite, andesite) which, based on experimental evidence, retain plagioclase in equilibrium with garnet + omphacitic clinopyroxene ± quartz under *P–T* conditions where the common basaltic rock compositions equilibrate to true plagioclase-free eclogites.

In order to avoid ambiguity and confusion, such usage is rejected here, where the term eclogite as a petrographic rock name is restricted to rocks of broadly basaltic composition which lack primary plagioclase and have a predominant assemblage of jadeite-bearing clinopyroxene (omphacite or chloromelanite[1]) and garnet.

Smaller amounts of certain other anhydrous minerals are widely recognised in eclogites. Quartz, kyanite, orthopyroxene and rutile are the most common of these. Coesite, diamond and aragonite are rarer, but notably are indicative of high confining pressures. The status of accessory hydrous silicates such as glaucophanitic, barroisitic or pargasitic amphibole, phengite, paragonite, phlogopite, talc, zoisite and clinozoisite is more debatable (see Smith *et al* 1982). Where textural and chemical considerations indicate equilibration of these phases with the primary clinopyroxene and garnet in eclogites, their recognition as minor primary phases is acceptable. But what modal content of the additional primary phases is permissible in eclogites? Here it is proposed that use of the specific litho-type name eclogite should be restricted to rocks with > 70% garnet + jadeite-bearing clinopyroxene, compatible with broadly basaltic rock compositions. Rocks with different modal mineralogies (< 70% garnet + clinopyroxene), stabilised in other rock compositions under equivalent (eclogite facies) *P–T* conditions, should be characterised by different specific rock names, such as garnet peridotite, mucronite and whiteschist, or by use of diagnostic mineralogical adjectives (e.g. omphacite-phengite schist).

The requirement that eclogites (*s.s.*) contain a primary omphacitic or chloromelanitic clinopyroxene needs to be stressed. In particular, plagioclase-free garnet + fassaitic clinopyroxene rocks developed in certain silica undersaturated rock compositions in association with granulite facies assemblages (Davidson, 1943; Lovering and White, 1969) should be termed *garnet pyroxenites* (Miyashiro, 1973) and *not* eclogites. As a consequence of low Na_2O whole rock contents, the clinopyroxenes in eclogite facies meta-peridotites are not omphacites, but rather chrome diopsides—albeit with high Jd/CaTs ratios. Hence the high pressure (HP) assemblages in such meta-peridotites should be termed garnet lherzolites, garnet harzburgites, garnet wehrlites or garnet dunites, depending on the actual modal mineralogy. Also, genetically associated meta-pyroxenite and rodingitised meta-gabbros are rarely true eclogites but rather are garnet websterites, garnet clinopyroxenites or grospydites.

1.3 Definition of the eclogite facies

The concept of a metamorphic facies, encompassing diagnostic mineral assemblages equilibrated under a certain combination of *P–T–X* conditions in a particular

1 The current widely used pyroxene name *chloromelanite* (Essene and Fyfe, 1967, fig. 1) is retained in this book for certain sodic clinopyroxenes with higher Acm/Jd ratios than omphacites. However, the recent report of the IMA Subcommittee on Pyroxenes (Morimoto *et al*, 1988) has recommended that it be discarded and the composition field of omphacite expanded to include essentially all the primary clinopyroxene compositions encountered in eclogites (sensu stricto).

metamorphic regime (recognised repeatedly both in space and time), is particularly valuable in a modern geotectonic context. However, it is now widely appreciated (see discussions by Miyashiro 1973 and Turner 1981) that it is necessary to adopt a less rigid and fixed approach to the definition of individual metamorphic facies than was originally intended by Eskola (1920, 1939). This is especially the case for the eclogite facies which, whilst universally recognised to be indicative of high confining pressures, in fact includes metamorphic mineral assemblages formed over an exceptionally wide range of *P–T–X* conditions. Moreover, as discussed below, the boundary conditions for this facies are complex and critically dependent upon the chemical potential of a mobile component, water.

The eclogite facies was first defined by Eskola (1920, 1921), largely on the basis of his study of eclogite occurrences in western Norway. The assemblage of omphacitic clinopyroxene with garnet (±kyanite, quartz, orthopyroxene, rutile) in rocks of broadly basaltic composition was considered to be diagnostic. Pyrope-rich garnet bearing assemblages in spatially associated magnesian meta-peridotite bodies were effectively inferred to be isofacial with the eclogites (*s.s.*), and this was later confirmed experimentally by O'Hara (1967a, 1975), O'Hara and Mercy (1963), O'Hara *et al* (1971) and others. More recently, it has come to be appreciated that other markedly different mineral assemblages, such as phengite + garnet ± omphacite ± zoisite + quartz in metapelites, are stabilised in other lithologies under eclogite facies conditions (Compagnoni 1977; Koons and Thompson 1985). These various assemblages are summarised in Table 1.1 and detailed in Chapter 2 . The eclogite facies is defined here as constituting a metamorphic regime of high confining pressure within which these assemblages, as well as the more traditionally recognised diagnostic assemblages, are stabilised in the various rock compositions.

It is now clear that the confining pressures required for plagioclase elimination in basaltic and non-basaltic rock compositions are markedly different, and indeed differ even within the observed spectrum of basaltic compositions (see Figures 1.1 and 3.2 in particular). Although plagioclase is not an acceptable primary phase in eclogites (meta-basaltic litho-type) it can be expected to persist as a stable phase in certain other co-facial mineral parageneses, most notably in meta-anorthosites and meta-andesites. Accordingly, plagioclase and omphacitic clinopyroxene may coexist for a limited range of *P–T–X* conditions in the vicinity of the granulite facies-eclogite facies boundary (Smulikowski, 1980). It is therefore important to emphasise that the absence of plagioclase, implicit in the definition of the litho-type eclogite is *not* essential to *all* mineral assemblages equilibrated within the eclogite facies *P–T* regime.

Eclogite facies assemblages are notably susceptible to retrogressive mineral changes, especially in response to fluid influx during the dynamic processes responsible for their exhumation. Consequently, the earlier HP history of metamorphic belts often has to be deciphered from just sporadic preservations of eclogite relicts. In practice, it is readily identified that metamorphic rock sequences have been subjected to eclogite facies metamorphic conditions only where the plagioclase-free assemblage of garnet + clinopyroxene + quartz in meta-basaltic rocks, or of garnet + olivine ± clinopyroxene ± orthopyroxene in magnesian meta-peridotites, is preserved. It is therefore appropriate to continue to define and recognise the eclogite facies essentially on the basis of the distinctive mineralogies developed in such rocks, especially in view of their better survival rates relative to other eclogite facies mineral parageneses.

Table 1.1 Principal eclogite facies litho-types and constituent mineral assemblages

Rock composition	Litho-type	Important mineral phases	Possible additional mineral phases
Basic igneous rocks			
	Fe-rich type	Fe-Grt, Clm-Cpx	Qtz/Coe, Ky, Zo/Czo, Lws, Mag, Amp(esp.Gln-Crs), Phe, Pg, Gr/Dmd, Rt, Py grading to
Basalts/gabbros etc.	ECLOGITE Mg-rich type	Mg-Grt, Omp-Cpx	Opx, Ky, Zo/Czo, Lws, Amp, Qtz/Coe, Tlc, Phl, Mg-Cld, Mg-Chl, Cor, Mgs, Gr/Dmd
Ultrabasic igneous rocks			
Peridotites	GARNET PERIDOTITE	Mg-Grt, Mg-Ol	Cpx, Opx, Cr-Spl, Mg-Chl, Mg-Chr, Phl, Amp(Prg), Mgs, Tlc, Dmd/Gr
Mg, Al-rich cumulates	ALKREMITE	Mg-Grt, Spl	Cor, Ol
Ca, Al-rich cumulates (rodingites?)	GROSPYDITE	Grt(> Grs_{50}), Cpx, Ky	Cor, Dmd/Gr, Coe/Qtz
Acid/intermediate igneous rocks			
Granites/diorites etc.	MUCRONITE	Jd or Omp-Cpx, Qtz/Coe, Phe	Grt, Pg, Ky, Zo/Czo, Kfs, Amp(esp. Gln-Crs), Rt
Fe-rich pelites	PHENGITE GARNET OMPHACITE SCHIST	Grt, Jd or Omp-Cpx, Phe	Qtz/Coe, Ky, Zo/Czo, Amp (esp. Gln-Grs), Pg, Cld, Rt, Py, Gr/Dmd grading to
Mg-rich pelites	WHITESCHIST	Mg-Grt, Tlc, Ky	Qtz/Coe, Jd-Cpx, Phe, Mg-Cld, Mg-Chl, Mg-St, Mg-Car, Ell, Yo
Psammites	JADEITE QUARTIZITE etc.	Jd or Omp-Cpx, Qtz/Coe	Kfs, Phe, Grt, Ky, Zo/Czo, Lws, Rt
Calcarenites	GARNET ARAGONITE MARBLE etc.	Arg/Cal, Ca-Grt	Mgs, Dol, Ank, Cpx, Zo/Czo, Lws, Ky, Phl, Qtz/Coe, Mg-Chl, Phe, Tr, Ttn, Gr/Dmd
Fe(Mn) rich sediments	GARNET PIEMONITE Mn-OMPHACITE META-CHERT etc.	Fe or Mn-Grt, Qtz/Coe Agt, Mn-Omp or Clm-Cpx	Fe or Mn-Ep, Phe Amp (Rbk-Crs), Mag, Py, Ttn, Der, Rdn, Brn, etc.

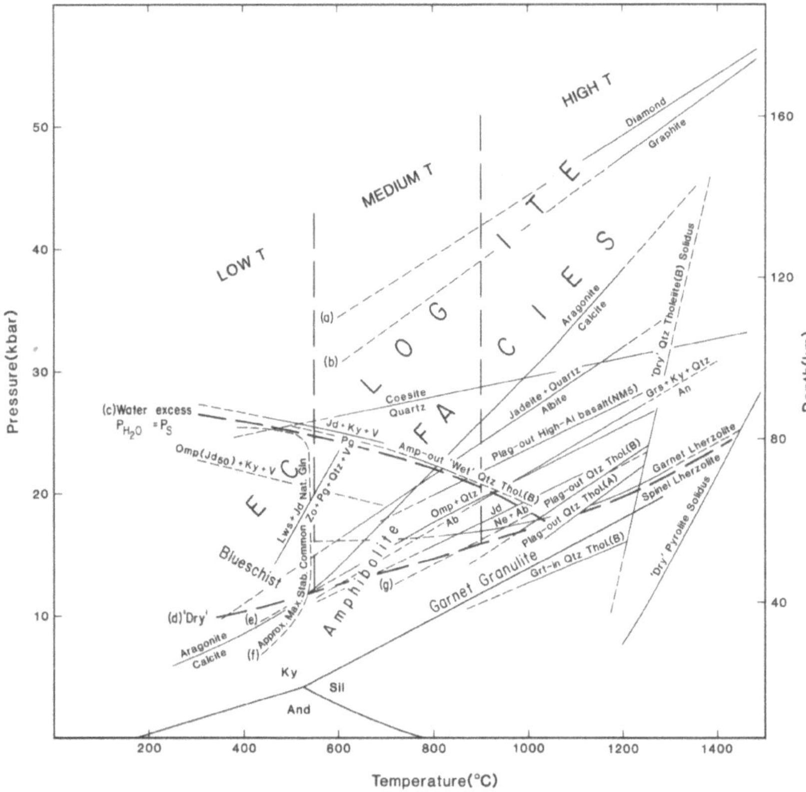

Figure 1.1 P_S–T petrogenetic grid showing conditions for the stability of eclogite facies assemblages in general and of eclogites and garnet peridotites in particular. This diagram includes experimentally determined reaction equilibria for both 'wet' and 'dry' rock systems. The boundary between the eclogite facies and the adjacent amphibole bearing facies (blueschist and amphibolite) should be viewed as a surface in P_S–P_{H_2O}–T space. Hence in the P_S–T section presented, two contrasting projected positions (heavy dashed curves *c* and *d* corresponding to water saturated and 'dry' rock systems respectively) are indicated for the low pressure boundary of the eclogite facies at $T < 1050°C$ (i.e. within the potential stability field of amphiboles). References to reaction curves shown, but not detailed in the general text, are as follows: Calcite⇌Aragonite—Carlson (1980); Lawsonite + Jadeite⇌Zoisite + Paragonite + Quartz + Vapour—Heinrich and Althaus (1980); Albite⇌Jadeite + Quartz—Holland (1980); thermodynamically calculated stability of omphacite ($Jd_{50}Di_{50}$) + quartz assemblage (curve *e*)—after Holland (1980, 1983); 3 Anorthite⇌Grossular + 2 Kyanite + Quartz—Koziol and Newton (1988); Albite + Nepheline⇌2 Jadeite—Robertson *et al* (1957); Plagioclase-out in high-Al basalt NM5—Ito and Kennedy (1971); Plagioclase-out and garnet-in reaction curves and solidus for quartz tholeiite compositions *A* and *B*—Green and Ringwood (1967a); 'dry' pyrolite solidus—Green and Ringwood (1967b); alumino-silicate polymorph stability fields—Salje (1986).

1.4 Eclogite facies P–T conditions and boundary relationships

Field associations indicate there is substantial overlap in the P–T stability fields for the diagnostic eclogite and garnet peridotite litho-types, and this has been confirmed by the experimental data reviewed in Chapter 3. However, the low pressure (LP) boundaries to their respective stability fields can be expected to coincide closely for only a limited range of P–T–X conditions (Figure 1.1, and see Figures 3.1 and 3.2).

The sliding nature of the mineral equilibria which define the granulite-eclogite facies boundary means that, like other metamorphic facies boundaries, it cannot be defined sharply in P–T space.

Experimental data indicate the minimum pressures required for eclogite or garnet peridotite stability in specific rock compositions for $T > 850$–$1000°C$, but P–T gradients appropriate to the lower temperature extrapolation of these LP stability field boundaries are less well known. Linear extrapolations of the plagioclase-out reaction curves of Green and Ringwood (1967a) for quartz tholeiite compositions suggest eclogite stability at very low pressures in cool, dry, crust (Figure 1.1), but Ringwood (1975) argues in favour of a somewhat lower gradient (20 bars $°C^{-1}$ rather than 27.5 bars $°C^{-1}$) for the LT extrapolation of the plagioclase-out reaction curve for quartz tholeiite B composition (dashed line g in Figure 1.1). Moreover, both thermodynamic reasoning (Jenkins and Newton, 1979) and experimental data (O'Hara *et al* 1971, as shown in Figure 1.1) indicate that the reaction boundary defining the stability of garnet lherzolite relative to spinel lherzolite flattens out to a very low gradient at low temperatures, with a substantial minimum pressure for the eclogite facies assemblage. The LP eclogite facies boundary (d in Figure 1.1) should therefore be viewed as only a crude guideline to the minimum pressures required for the stability of the diagnostic eclogite facies assemblages under essentially 'dry', fluid-absent, conditions.

From the foregoing considerations, it is reasonable to expect to encounter some rock sequences, such as those in East Sudetes, Poland (Kozlowski, 1958; Smulikowski and Bakun-Czubarow, 1973; Pouba *et al*, 1985) and in the Moldanubian zone of southern Czechoslovakia (Dudek and Fediukova, 1974), where certain assemblages which are considered to be characteristic of the granulite facies coexist with other assemblages diagnostic of the eclogite facies in chemically different rocks. Consequently, discussions (such as those by Pouba *et al*, 1985) of whether such transitional paragenetic associations should be assigned to the eclogite or granulite facies are unfruitful unless the eclogite facies-granulite facies boundary is first more rigorously defined in P–T–X space. However, a highly restrictive facies boundary definition would exacerbate rather than ease the problem of seemingly anomalous mineral assemblage associations. In any particular instance, bulk rock and constituent mineral analyses are obviously required in addition to basic petrographic observations before the relationships between spatially associated eclogite and granulite facies assemblages can be reliably interpreted.

It is often assumed that below a depth of about 8 km, pathways to the surface for pore fluids are closed and prograde *metamorphism* of crustal rocks takes place under conditions of near equality of P_S and P_{H_2O}. Thus, most published metamorphic facies P–T petrogenetic grid diagrams (e.g. Turner, 1968, 1981; Miyashiro, 1973; Winkler, 1974) represent the condition where $P \approx P_{H_2O} \approx P_S$, with the facies boundaries referenced to experimentally determined invariant reaction equilibria for water-excess systems. Because of difficulty in equating the P–T conditions for the stability of anhydrous eclogite and granulite facies assemblages, these facies have often been omitted from such facies diagrams.

For water-saturated rock systems, the LP boundary for the stability of eclogite facies assemblages relative to blueschist and amphibolite facies assemblages with hydrate phases is shown in Figure 1.1 as curve *c*, positioned to correspond to the near coincident experimentally determined reaction curves for amphibole elimination in quartz tholeiite *B* composition (Essene *et al* 1970) and for paragonite = jadeite + kyanite + vapour (Holland 1979b). Arguably the minimum pressures required for eclogite stability at $P_{H_2O} = P_S$ in most meta-basaltic compositions may be somewhat lower, and more closely approximated by the calculated stability curve for $Omp(Jd_{50}) + Ky + V$ (after Holland 1979b). What is clear, is that within the potential stability field of amphiboles at $T < c.$ 1050°C, the stability relationships between eclogite facies mineral assemblages and amphibole-bearing blueschist, amphibolite or greenschist facies assemblages are complex and, as convincingly argued by Fry and Fyfe (1969, 1971) and Newton and Fyfe (1976), will be critically dependent upon the chemical potential of H_2O in the rock systems undergoing metamorphism.

The general consistency of metamorphic facies assemblages, in terms of univariant dehydration reaction equilibria experimentally determined under water excess conditions, has been widely accepted to support the assumption that prograde metamorphism, especially of sediments, takes place under conditions where $P_f \approx P_{H_2O}$ ($+P_{CO_2}$, P_{O_2}, P_{CH_4} *etc* in certain rock systems) $\approx P_S$ (e.g. Norris and Henley 1976; Turner 1981). However, as emphasised by Thompson (1983), igneous rocks or previously largely dehydrated meta-sediments in the middle to lower crust may well undergo metamorphism under fluid-absent conditions. Under water deficient conditions, any H_2O present can be expected to be in effect structurally bound within minor amounts of hydroxyl–bearing crystalline phases (such as amphiboles) and insufficient free H_2O molecules will be present to form a discrete H_2O-rich pore fluid (or vapour) phase.

The critical influence of H_2O-activity on the P_S–T stability of clinopyroxene-dominant eclogite facies assemblages relative to amphibole-dominant assemblages of other metamorphic facies is illustrated in Figure 1.2 for both basic (quartz tholeiite) and ultrabasic (pyrolite) compositions. Figures 1.2*a* and *d* demonstrate that for H_2O-saturated conditions, with the partial pressure of water in the pore fluid phase effectively equal to the lithostatic confining pressure on the solid mineral phases (i.e. $P_{H_2O} \approx P_S$), garnet peridotite and eclogite cannot be stabilised at normal crustal depths because of the large stability fields for pargasitic amphibole. On the other hand, for equivalent anhydrous rock compositions (Figures 1.2*c* and *f*) the stability fields for garnet peridotite and eclogite extend to much lower P_S–T conditions.

Experimental studies of phase relationships in rocks for conditions where $P_{H_2O} < P_S$ have been conducted mostly by reducing the water activity (and P_{H_2O}) in the pore fluid phase by dilution with another volatile chemical species (most frequently and appropriately, CO_2). It has been argued, rather controversially, that reduction of a_{H_2O} in the pore fluid phase as a result of CO_2 streaming (i.e. $P_f = P_{H_2O} + P_{CO_2} \approx P_S$) explains the stability of granulite facies relative to amphibolite facies assemblages; but there is little evidence that enhanced CO_2-activity is a significant factor in the stability of eclogite facies assemblages. H_2O-deficient conditions with insufficient water to saturate the rock system and form a separate fluid (or vapour) phase are more appropriate to the deep crust and upper mantle but unfortunately are much more

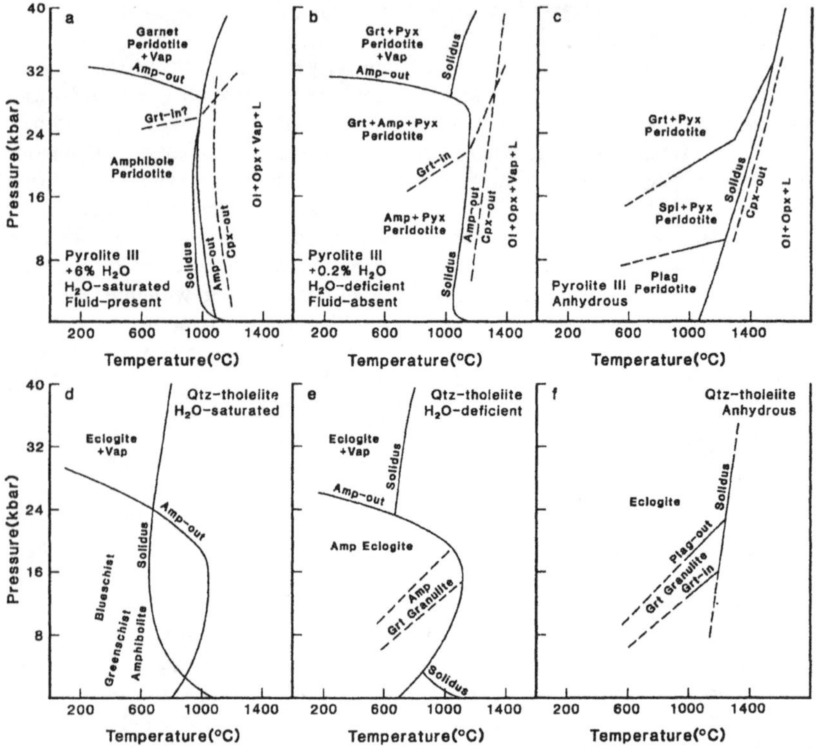

Figure 1.2 P_S–T conditions for the stability in Pyrolite III composition of the eclogite facies garnet peridotite assemblage relative to those for other mineral assemblages under water-saturated (*a*), water-deficient (*b*) and anhydrous (*c*) conditions. Based on data and diagrams presented by Green and Ringwood (1967b) and Green (1973). P_S–T conditions for the stability in quartz tholeiite composition of eclogite relative to garnet granulite, amphibolite, blueschist and greenschist (*d*–*e*). Diagram (*d*) represents water-saturated conditions based largely on the data and diagram presented by Essene *et al* (1970) and diagram (*f*) represents anhydrous conditions based on the experimental work of Green and Ringwood (1967a), but with the plagioclase-out reaction curve extrapolated to lower temperatures with a gradient of 20 bars °C^{-1} as in Ringwood (1975).

difficult to simulate in laboratory experiments. Moreover, it is difficult to assess the influence of variation in water content in terms of thermodynamically significant variables; $X_{H_2O}^{rock}$ is known, but $X_{H_2O}^{fluid}$ cannot be defined if, in effect, there is no fluid phase present.

P_S–T phase relations for a H_2O-deficient peridotite composition are shown in Figure 1.2*b*, based on the experimental data and diagram presented by Green (1973) for the Pyrolite III composition + 0.2% H_2O. These indicate that eclogite facies garnet lherzolite assemblages with minor pargasitic amphibole are expected to be stable at pressures above the garnet-in reaction curve within the amphibole P_S–T stability field. The P_S–T phase relations illustrated in Figure 1.2*e* for an equivalent H_2O-deficient quartz tholeiite composition are unsupported by experimental data but can be predicted by analogy with relationships in the peridotite rock system and from the following considerations. The effect on amphibole stability of lower a_{H_2O} ($X^{rock}_{H_2O}$ and $X^{fluid}_{H_2O}$) is apparently complex (see review by Helz 1982). There are indications that lower a_{H_2O} reduces the thermal stability of amphibole at $P_S < 5$ kbar, but increases stability at $P_S > 10$ kbar (except perhaps at very low a_{H_2O} values). There are some conflicting points among the experimental data of Hill and Boettcher (1970), Allen and Boettcher (1978), Green (1973), and Merrill and Wyllie (1975), but the greater weight of evidence indicates that reduced a_{H_2O} will at most lower the pressure for amphibole breakdown only slightly relative to that in H_2O saturated rocks. It therefore seems unlikely that the amphibole stability field would shrink substantially in rocks with a lower, but still significant, $X^{rock}_{H_2O}$ content.

The change with increased P_S from positive to negative slope of the amphibole breakdown curves in Figures 1.2*d* and *e* reflects the high compressibility and hence reduced molar volume of H_2O at high pressures. As this bend-back occurs at P_S–T conditions close to the plagioclase-out reaction curves for basaltic rock compositions, the transformation of amphibolite (or blueschist)→eclogite is constrained to take place largely in response to increased confining pressures, whereas the amphibolite→ granulite transition is effectively temperature controlled. Together with Figure 1.2*f* these diagrams also demonstrate that the stability of jadeite-bearing clinopyroxene relative to plagioclase provides the fundamental control on the formation of eclogite rather than garnet granulite, whereas amphibole breakdown curves and water activity control the stability of eclogite relative to amphibolite or blueschist.

The phase relations illustrated in Figure 1.2*e* imply that for H_2O-deficient basaltic rock compositions equilibrated at P_S–T conditions within the amphibole stability field, any water present will be incorporated in small amounts of amphibole in equilibrium with the anhydrous phases. This will result in the stability of amphibole-bearing eclogite at pressures above the plagioclase-out reaction curve or amphibole-bearing granulite at lower pressures. It is important to appreciate that the presence of a limited modal content of a hydrate phase, such as amphibole, mica or zoisite in an eclogite, does not necessarily mean that there was H_2O excess in the rocks (in a discrete pore fluid or vapour phase) during metamorphism.

There are some indications that for more alkaline basalt compositions, more sodic amphiboles—notably glaucophane—may be stable to higher pressures than indicated for the pargasitic amphiboles represented in Figure 1.2*d*. For example, thermodynamically calculated phase relations in the NMASH system by Holland (1988) predict that at 400–700 °C, glaucophane remains stable relative to jadeite + talc up to about 31–34 kbars. Hence, it can be expected that the stability of clinopyroxene-dominant eclogite relative to glaucophane-dominant blueschist, although critically dependent upon $X^{rock}_{H_2O}$ values will also be influenced by other chemical parameters such as: $X^{rock}_{Ca}/X^{rock}_{Na}$, $X^{rock}_{Mg+Fe}/X^{rock}_{Al}$ and $X^{rock}_{Al}/X^{rock}_{Si}$ ratios (Chapter 7).

From the foregoing discussion, it is clear that the dehydration reaction equilibria linking hydrous and anhydrous mineral phases will be at least divariant and are represented best as equilibrium surfaces in P_S–P_{H_2O}–T space. In Figure 1.1, the alternative LP eclogite facies boundaries at $T < 1050$°C (curves *c* and *d*, corresponding to water excess and water absent conditions respectively) in effect represent the two extreme locations of the projection of this 3-dimensional surface in P_S–P_{H_2O}–T space onto a P_S–T section. Under the P_S–T conditions between boundaries *c* and *d*, eclogite facies assemblages can be expected to be thermodynamically stable in 'dry' rocks whilst blueschist or amphibolite facies assemblages are stable in 'wet' rocks. With a relatively 'comprehensive' definition of the blueschist facies as including both lower T lawsonite + glaucophane and higher T epidote + glaucophane assemblages (Evans and Brown, 1987) this would then effectively preclude a direct boundary in P_S–P_{H_2O}–T space between the eclogite and greenschist facies.

For rocks with intermediate $X_{H_2O}^{rock}$ values, hybrid, seemingly transitional assemblages containing both clinopyroxene and amphibole will be stable over a broad pressure interval between boundaries *c* and *d*. The precise P–T–X conditions will determine whether amphibole or clinopyroxene is dominant and whether assemblages are most appropriately assigned to the blueschist, amphibolite or eclogite facies. Overlap in the P_S–T conditions for the stability of eclogites relative to blueschists or amphibolites led Coleman *et al* (1965) to propose the abandonment of the eclogite facies. Their suggestion that the eclogite facies might best be restricted to mantle rocks has, however, found little general acceptance. Moreover, more recent studies, especially in the Scandinavian Caledonides (Chapter 8) and in the Alps (Chapter 10), have emphasised the significance of the stabilisation of eclogite facies assemblages in tectonically thickened segments of continental crust (Carswell and Cuthbert, 1986). It makes little sense to abandon the concept of the eclogite facies at a time when there is increasing awareness of the importance of the identification and interpretation of HP eclogite facies assemblages to the understanding of the dynamics of plate collision zones. Furthermore, as demonstrated in several recent studies (e.g. Austrheim, 1987; Mørk, 1985) and discussed in Chapter 5, the penetrative deformation-dependent infiltration of a hydrous fluid is important in overcoming kinetic barriers to the transformation of dry crustal rocks, such as gabbros and granulites, into eclogites. This indicates that the confining pressures necessary for eclogite formation on an appreciable length-scale are likely to be higher than those indicated by the reaction curves in dry rock systems. The increasing recognition of ultra-high pressure mineral polymorphs, coesite and even diamond (Section 13.9) in eclogite facies rocks of undoubted crustal origin also points to exceptionally deep subduction and high lithostatic confining pressures, or at least, in these cases, pressures in excess of those required for the stability of blueschist or amphibolite facies assemblages. The alternative suggestions that the necessary confining pressures for the formation of minerals such as coesite may have been generated in response to a substantial microscale intracrystalline overpressure, or to macroscale, dynamically-stabilised, tectonic overpressure (Smith, 1984, 1985) seem unlikely from physical considerations (see for example Van der Molen and Van Roermund, 1986; Turner, 1968).

The upper pressure boundary to the eclogite facies lies well above the pressure range shown in Figure 1.1. It can be defined by the transformation of eclogite to pyroxene-free garnetite in basaltic rock compositions, reflecting the increased solid solution of clinopyroxene in garnet at ultra-high pressures. Experiments by Irifune *et al* (1986, 1989) at $T = 1200°C$ have demonstrated that the transition to garnetite in olivine tholeiite compositions is only completed at $P = 140$–150 kbar, with again composition control of the precise location in P–T space.

1.5 Subdivision of the eclogite facies P–T regime

The occurrence of eclogites in contrasting geological environments and various mineralogical criteria indicate that eclogites have formed over a wide range of P_S–T conditions. Indeed, taking due note of the stability of eclogite facies assemblages in both deeply subducted oceanic and continental crust rocks, as well as under ambient

Table 1.2 Eclogite classification schemes

Reference	Eskola (1921)	Coleman *et al.* (1965)	Smulikowski (1960, 1964, 1968)	Proposed scheme	
Subdivision criteria	Geological occurrences	Largely mineral compositions	Geological settings and rock and mineral compositions	Equilibration temperature	Likely origin
	Type 1 Xenoliths in kimberlites etc.	*Group A* Garnet > 55 mol.% pyrope Least jadeitic clinopyroxene	*Group I(G)* Griquaites and garnet websterites associated with peridotites	High	Upper mantle
				900°C	
	Type 2 Layers and lenses in alpine-type peridotite bodies			Medium	Tectonically thickened continental crust (but see text for possible further qualifications)
	Type 3 Lenses in migmatitic gneiss complexes	*Group B* Garnet 30–55 mol.% pyrope	*Group III(C)* Common eclogites		
				550°C	
	Type 4 Blocks in blueschist facies terrains	*Group C* Garnet < 30 mol.% pyrope Most jadeitic clinopyroxene	*Group II(O)* Ophiolitic eclogites	Low	Subducted oceanic crust and arc-trench sediments

lower crustal and upper mantle conditions, it would appear that the P_S–T regime encompassed by the eclogite facies (Figure 1.1) exceeds that of all of the other metamorphic facies in total. Some scheme of subdivision is therefore appropriate.

The early classification scheme of Eskola (1921) emphasised the occurrence of eclogites in four contrasted geological settings (Table 1.2), whilst the more recent and more popular three-fold subdivision scheme of Coleman *et al* (1965) into Eclogite Groups A, B and C is based on observed differences in the compositions of the constituent garnets and clinopyroxenes. However, the inferred correlation of each of these Eclogite Groups with a particular geological setting, as further emphasised in the comparable classification scheme of Smulikowski (1960, 1964, 1968), breaks down because of the influence of variable rock compositions. In particular, the most pyropic garnets and least jadeitic clinopyroxenes are encountered in magnesian metaperidotites regardless of whether they occur as xenoliths in kimberlites with mantle derived Group A eclogites or within alpine-type peridotite bodies associated with Group B eclogites.

A simple, three-fold, temperature-based subdivision, which in effect follows the lead of Banno (1970), is preferred here (Table 1.2 and Figure 1.1). This classification is more systematic, but still has the advantage that it broadly correlates with eclogite formation in three fundamentally different geological environments. The appropriate equilibration temperature estimates for the three groups of eclogites can be obtained readily from experimentally-based calibrations of the Fe^{2+}–Mg^{2+} exchange reaction between coexisting garnets and clinopyroxenes (Chapter 4).

The temperatures of $<550°C$ indicated for the stability of the LT eclogite facies assemblages are in line with the likely maximum stability conditions for 'common' natural glaucophane (curve f in Figure 1.1, based largely on Maresch, 1977, and Koons, 1982) in the widely associated blueschists developed in B-type (Benioff) subduction zones. Note, however, that highly magnesian glaucophane, as observed for example in certain eclogites in the Tauern window, may be stable at $T = 550$–$650°C$ and $P > 20$ kbar (Holland, 1979a, 1988). Accordingly, the unusual assemblage Grt + Omp + Mg – Gln + Ky in these rocks is assigned here to the MT rather than the LT eclogite field.

The 900 °C temperature boundary separating the MT and HT eclogites discriminates in the majority of cases between the MT eclogite facies assemblages stabilised in tectonically thickened continental crust in A-type (Ampferer) subduction zones (Hodges *et al* 1982), and xenolithic assemblages (mostly garnet peridotites rather than eclogites *s.s.*) stabilised under ambient conditions in the upper mantle. This temperature-genesis correlation cannot, however, be rigidly or universally applied since with a geotherm of approximately 40 m Wm^{-2} for a stable cratonic region, MT eclogite facies assemblages can be predicted to be stable in the uppermost 80 km of the subcontinental mantle. Moreover, xenolith suite evidence and experimental studies of phase equilibria indicate that, for specific rock compositions, eclogite rather than granulite may be thermodynamically stable under the expected ambient P–T conditions in the lower continental crust beneath such regions (see Chapter 12). Thus, MT eclogite facies assemblages may be stabilised in the continental lower crust and uppermost mantle under anorogenic conditions without subduction, if kinetic factors are favourable.

The minimum pressures required for the stability of the ultra-high pressure eclogite facies assemblages (Chapter 13) containing coesite or diamond can be deduced by reference to the experimentally determined reaction curves for the quartz⇌coesite inversion from Bohlen and Boettcher (1982) and the graphite⇌diamond inversion-curve (Figure 1.1, curve *a*) from Kennedy and Kennedy (1976) or the slightly lower pressure curve (Figure 1.1, curve *b*) from Bundy (1980, after Bundy *et al*, 1961) which is more closely compatible with the thermodynamically calculated inversion conditions of Berman (1979).

1.6 Summary of key points

i) The rock name *eclogite* should be applied only to plagioclase-free rocks of broadly basaltic composition with a modal content of > 70% garnet + jadeitic (omphacite or chloromelanite) clinopyroxene.
ii) Different specific names should be applied to other rocks with contrasting mineralogies (possibly even plagioclase-bearing) stabilised under equivalent (eclogite facies) P–T conditions.
iii) The *eclogite facies* of metamorphism is characterised by high confining pressures, but can be only broadly defined in P_S–P_{H_2O}–T space with reference to the experimentally determined stability conditions for eclogite (*s.s.*) and for garnet peridotite in magnesian ultrabasic rocks.
iv) The stability of eclogite and garnet peridotite relative to compositionally equivalent amphibolite or blueschist facies assemblages is crucially dependent on both P and $X_{H_2O}^{rock}$ values.
v) Small amounts of hydrate phases (e.g. amphibole, mica or zoisite/clinozoisite) may form as primary phases in equilibrium with the dominant anhydrous minerals in eclogite facies mineral assemblages formed under H_2O-deficient conditions (low $X_{H_2O}^{rock}$ values) in the deep crust-uppermost mantle.
vi) As the eclogite facies P–T regime is especially wide, it is appropriate to establish a temperature based subdivision, which broadly correlates with formation in three contrasting geological environments.

Acknowledgements

This chapter has benefited from constructive reviews by Simon Cuthbert, Giles Droop, Bill Griffin, Simon Harley and Manfred Schliestedt.

2 Eclogite facies mineral parageneses

A. MOTTANA, D.A. CARSWELL, C. CHOPIN and
R. OBERHÄNSLI

2.1 Introduction

According to Haüy's original definition (1822), eclogites are rocks distinguished by the paragenesis garnet + clinopyroxene. However, from his very first definition, Haüy also pointed out that they may contain a variety of accessory minerals. Subsequently, these have been used to denominate various eclogite types (quartz eclogite, kyanite eclogite, etc.). Eskola (1920) redefined eclogite as the typical rock type for his 'eclogite facies' but, influenced by later Norwegian studies (1921, 1939), he then came to consider eclogites to be transitional between the HT–HP conditions of the metamorphic realm and those of the upper mantle. Eskola's ambiguity later produced quite a lot of confusion, and occasionally still does (see Chapter 1).

As a result of considerable research interest in the petrogenetic significance of HP rocks, it has now become clear that eclogite (*s.s.*) is just one, albeit the most characteristic, of the many litho-types constituting the eclogite metamorphic facies. Differentiation into three broad groups of eclogites, as initially proposed by Smulikowski (1964, 1965) and Coleman *et al* (1965), has proved useful and popular but such a classification is artificial and restrictive. Although some links are still missing, it now seems that eclogites form over a wide, continuous span of *P–T* conditions, thus necessitating an equally wide and continuous field in the *P–T* grid. Eclogite facies conditions once thought to be typical of only the mantle have in fact been attained even in rocks originally at the earth's surface (*cf* Chopin, 1984, 1985; Schreyer, 1985, 1988). Accordingly, relicts of HP parageneses consistent with those of eclogites (*s.s.*) have been detected in rocks with other than basic or ultrabasic bulk composition. This has been interpreted to mean that eclogite facies conditions were attained regionally over wide areas at a certain stage of metamorphism in subduction zones, but were then mostly obliterated during the ensuing *T* rise often accompanied by infiltration of fluid, at least in rocks susceptible to penetrative deformation (Heinrich, 1982). If this interpretation is correct, eclogite facies parageneses are to be expected in any kind of rock and in any mobile metamorphic belt, provided the rocks have been subducted to sufficient depth and returned rapidly to the surface along a steep dP/dT gradient such that P_{max} metamorphic assemblages were preserved. This seems to have been the case for example in the Dora-Maira massif of the western Alps in Italy (Chopin, 1984) and the Western Gneiss Region of the Caledonides in Norway (Smith, 1985)—where coesite has been reported in a meta-quartzite and in an eclogite, respectively—and the Atbashi range of Uzbekhistan (Sobolev and Shatskii, 1987), where diamond is a widespread accessory phase in biotite-garnet gneisses.

On the basis of these recent developments, it is now necessary to attempt a more complete characterisation of the eclogite facies parageneses stable at different *P–T* conditions for a wide range of rock compositions. As knowledge of the full range of rock compositions is incomplete, attention here is focused on those rock compositions for which most information is available on eclogite facies mineral assemblages. It should be emphasised that the data available for many of these are still incomplete but at least sufficient to allow a more complete petrogenetic framework within which to place future systematic and detailed investigations.

2.2 Ultrabasic rock compositions

The coexistence of forsteritic olivine with pyropic garnet in magnesian meta-peridotites has for a long time been recognised to be diagnostic of eclogite facies *P–T* conditions (e.g. Eskola, 1921; O'Hara, 1967a). Such garnet-bearing peridotites (more specifically and most typically, 4-phase garnet lherzolites) are widely accepted as the principal litho-type in the upper mantle, where they are dominant in the mantle-derived xenoliths suite encountered in kimberlite diatremes (Boyd, 1970; Carswell, 1980;

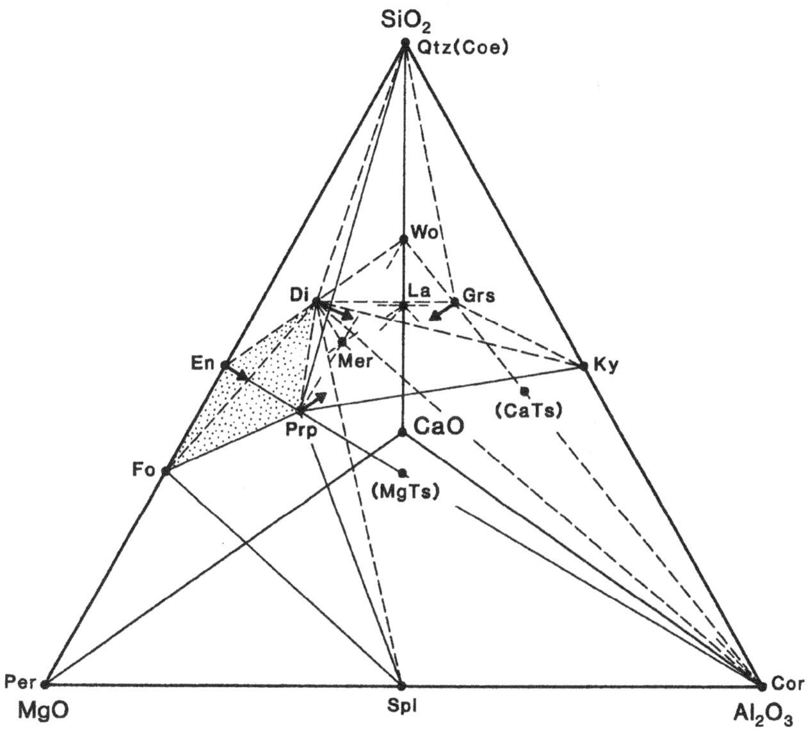

Figure 2.1 Eclogite facies mineral parageneses in the CMAS system. Note the potential solid solutions (indicated by arrows) between pyrope—grossular, enstatite—Mg-Tschermaks and diopside—Ca-Tschermaks. For clarity, the joins to merwinite and larnite in the CaO-rich, SiO_2-depleted portion of the system have not been fully completed.

Carswell and Gibb, 1987b) and occur as tectonically intercalated upper mantle slices in exposed HP gneiss terrains (see Chapter 11). A résumé of experimental data on the actual *P–T* conditions for the stability of such eclogite facies garnet lherzolite assemblages relative to conditions for stability of lower P spinel lherzolite assemblages is provided in Section 3.2.

Most natural 'dry' ultrabasic rock compositions can be closely modelled by the simple 4-component CMAS system. The mineral parageneses which can be expected to be stable in the CMAS system under eclogite facies *P–T* conditions are indicated in Figure 2.1. The sub-tetrahedron for the stability of Fo + En + Prp + Di (*garnet lherzolite*) in MgO-rich ultrabasic compositions is ornamented. In reality, the composition field for the stability of this assemblage extends to more CaO and Al_2O_3 rich compositions, since the pyropic garnet customarily contains about 15 mol% Grs and both En and Di are variably aluminous as a consequence of solid solution (especially at HT) towards MgTs and CaTs, respectively.

Whereas the assemblage Fo + En + Prp + Di in the CMAS system is divariant, the equivalent assemblage Ol + Opx + Grt + Cpx in natural magnesian garnet lherzolites is multivariant, since the latter typically contain appreciable FeO (> 6.0 wt %) and smaller but still significant amounts of Cr_2O_3, MnO, NiO, TiO_2 and Na_2O (mostly 0.05–1.0 wt %). In particular, it is apparent that an increased content of Cr_2O_3 relative to Al_2O_3 favours the stability of Cr-spinel relative to pyropic garnet and consequently spinel stability persists at pressures above those required for garnet stability in many natural magnesian meta-peridotites (see Section 3.2). Such 'transitional', 5-phase Grt + Cr-Spl lherzolite assemblages, stable in Cr-rich peridotite compositions at confining pressures within the eclogite facies *P–T* regime defined in Figure 1.1, are accordingly assigned to this facies. A corollary is that for exceptionally Cr-rich ultrabasic compositions (*cf* Carswell *et al*, 1979; Carswell, 1980), Cr-spinel or Mg-chromite may persist instead of garnet at least within the lower pressure part of the eclogite facies. On the other hand, since increased FeO content reduces the pressure required for garnet stability in peridotite rock compositions (O'Neill, 1981), Grt + Ol assemblages may be stabilised in more ferriferous peridotite compositions (e.g. 100 Mg/(Mg + Fe) < 80) at the lower pressures appropriate for the stability of garnet granulite rather than eclogite in meta-basaltic compositions (Figure 1.1).

Figure 2.2 illustrates the mineral parageneses which are expected to be stable in hydrous magnesian meta-peridotites over the *T*-range of the eclogite facies *P–T* regime. They are shown in the form of pseudo-ternary projections from forsterite and H_2O onto the CAS plane of the mineral assemblages stable in the CMASH system. It is apparent that under H_2O-saturated conditions, the diagnostic 4-phase garnet lherzolite assemblage is only stable above approximately 1000°C. At lower *T*, hydrous phases—pargasitic amphibole and/or chlorite—coexist with forsteritic olivine and pyropic garnet.

From their chemographic analysis of phase relationships in the CMASH system, Obata and Thompson (1981) were uncertain which of Chl or Amp was stable to the highest *T*. However, synthesis experiments in CMASH by Jenkins (1981) indicated that Amp reacts out at lower *T* than Chl. Nonetheless, petrographic observations indicate that in natural magnesian meta-peridotites, Amp is stable to higher *T* than Chl, as depicted in the sequence of assemblage diagrams presented in Figure 2.2. This higher thermal stability of pargasitic Amp relative to magnesian Chl may well reflect the influence of Na_2O as an additional component in natural peridotite compositions

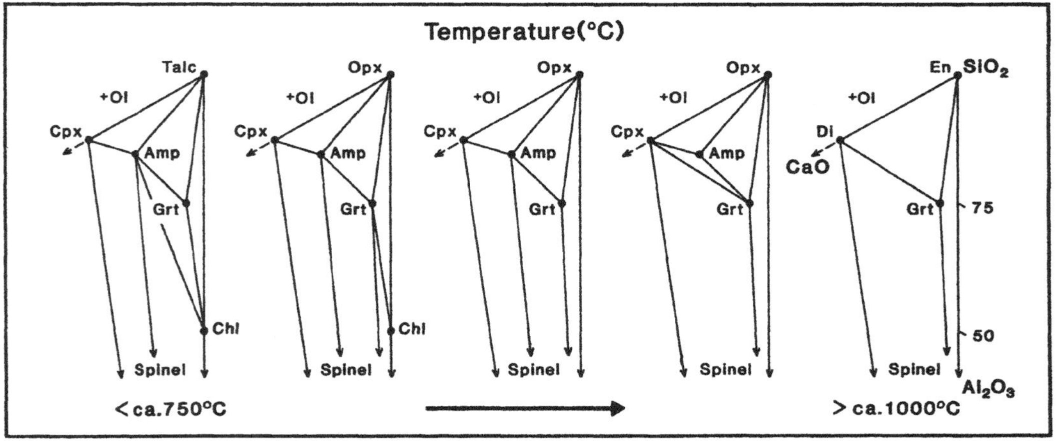

Figure 2.2 Eclogite facies mineral parageneses stable at different temperatures in hydrous magnesian meta-peridotites, presented as pseudo-ternary projections of assemblages from forsterite and H_2O onto the CAS plane in the CMASH system. Orthopyroxene and clinopyroxene compositions have been approximated as enstatite and diopside respectively, ignoring minor Al contents. The garnet composition is plotted as Prp_{85}–Grs_{15}, in line with the observed constrained grossular contents of garnets in natural garnet lherzolite assemblages. The amphibole composition is approximated as $tremolite_{50}$–$tschermakite_{50}$ and the chlorite as clinochlore.

and/or metamorphism under H_2O-deficient rather than H_2O-excess conditions. Moreover, under such chemical conditions a 5-phase Grt + Amp lherzolite assemblage can be expected to be stabilised at $T < c.$ 1000°C. For K_2O and H_2O bearing peridotite compositions, phlogopitic mica rather than pargasitic Amp appears to maintain stability to comparable temperatures as evidenced by certain lower T garnet lherzolite xenoliths (Carswell, 1975; Delaney *et al*, 1980). Note also that in H_2O saturated magnesian peridotites, talc rather than enstatitic Opx is expected to be stable at $T < c.$ 750°C due to the reaction Tc + Fo = En + H_2O.

Figure 2.1 illustrates the eclogite facies mineral assemblages stable in other, less common, ultrabasic rock compositions. The SiO_2 undersaturated, Al_2O_3 and MgO-rich compositions encountered as *alkremite* xenoliths in kimberlites (Ponomarenko, 1975; Nixon *et al*, 1978) lie close to the Prp-Spl join in the MAS plane. In this litho-type the essentially bimineralic assemblage of pyropic garnet + spinel may be accompanied by minor amounts of olivine or corundum.

Progressively more calcic, ultrabasic ($SiO_2 < 45$ wt %) compositions result in the stability of Di + $Prp_{>50}$ − $Grs_{ss.}$ + Ky ± Cor (*kyanite eclogite*) or Di + $Grs_{>50}$ − $Prp_{ss.}$ + Ky ± Cor (*grospydite*), both litho-types again represented in the mantle-derived xenolith suites observed in kimberlites (Sobolev *et al*, 1968). Despite the statement by these authors that there is a complete solid solution between Prp and Grs garnet components under the high P–T conditions for the equilibration of such kyanite eclogites and grospydites, it appears that there remains a significant composition gap between the Prp-rich garnet in kyanite eclogites and the Grs-rich garnet in grospydites. Cpx–Grt–Ky relationships in such xenoliths are complicated by the Alm content in garnet, typically 10–35 mole %, and Jd in clinopyroxene at 30–50 mole %.

Experimental results by Green (1969) on the 3 Di + 2 Ky composition in CMAS at 1200°C indicate that Prp–Grs solid solution is still incomplete at $P = 36$ kbar (with Di + Ky remaining as the stable assemblage) but expected to be complete at somewhat higher P. However, it was further demonstrated that addition of Na_2O component extends the stability field of Jd-bearing Cpx + Ky relative to Grt (Prp_{50} − Grs_{50}) + Qtz/Coe to higher P, thus limiting the Grs and Prp contents of any coexisting garnet. Hence, ultrabasic compositions to the MgO-rich side of the Di–Ky–Cor plane in Figure 2.1 will lie in the Prp–Di–Ky–Cor sub-tetrahedron and, for the natural compositions of these kimberlite xenoliths, will be expected to contain the mineral assemblage omphacitic Cpx + pyropic Grt + Ky ± Cor i.e., kyanite eclogite. By contrast, for compositions to the CaO-rich side of this plane, the equivalent natural assemblage will be omphacitic Cpx + grossularic Grt + Ky ± Cor i.e. grospydite.

The MT kyanite eclogites observed in exposed gneiss terrains commonly contain quartz and are probably formed from both high-Al 'basalt' and argillaceous sedimentary protoliths. The kyanite eclogite and grospydite xenoliths encountered in kimberlites however mostly have ultrabasic compositions within or just to the corundum-bearing side of the Cpx–Grt–Ky plane (Sobolev *et al*, 1968). These probably have formed either from exsolved high T aluminous Cpx-rich mantle cumulates (Lappin, 1978) or from metamorphosed remnants of subducted, rodingitised, oceanic crust (Helmstaedt *et al*, 1979). Rarer, more siliceous, grospydite variants containing the HP polymorph coesite rather than quartz (Smyth and Hatton, 1977; Ponomarenko and Spetsius, 1978) are indicative of $P > c.$ 30 kbar under the prevailing equilibration temperatures.

The diagnostic eclogite facies mineral parageneses in ultrabasic rock compositions are included in Table 1.2. Specific information on mineral assemblages in garnetiferous peridotite occurrences in HP gneiss terrains is included in Chapter 11. Readers are referred to Nixon (1987) for further data and references on xenolith occurrences.

2.3 Basic rock compositions

2.3.1 *General considerations*

Reactions of the type

$$Pl + Cpx \pm Ol \pm Opx \rightleftharpoons Grt + Omp \pm Qtz,$$

by which eclogite forms from compositionally equivalent crustal basic (basaltic) rocks, are multivariant. They take place over a broad range of P–T conditions within the stability fields of a number of other mineral phases. Consequently, these may also occur in the eclogite facies assemblage, provided appropriate chemical conditions exist, at least locally, in the reacting material. The number of additional phases found in eclogites is rather large (about 30) and is larger (over 50) if eclogite facies rocks in general—i.e. rocks whose compositions are out of the strict basaltic range—are included (Smith *et al*, 1982). Their number would be even greater, were eclogitic garnet and clinopyroxene not complex solid solutions. Their variable chemistry effectively guarantees eclogite its characteristic main assemblage over a large range of P and T, and, at the same time, offers the researcher a practical means of determining the conditions under which eclogite formation took place. In addition, their chemical zoning and sequence of inclusions permit the continuous monitoring of P and T conditions during growth, and provide information on the P–T path followed during both prograde and the retrograde stages of eclogite facies metamorphism (e.g. Bryhni and Griffin, 1971, Råheim and Green, 1975).

Mineral assemblages and mineral compositions in eclogites (*s.s.*) depend on the actual P–T–X conditions in the basaltic protoliths. Experiments carried out using simplified anhydrous synthetic basalt compositions produced mostly the simple assemblage Grt + Omp ± Qtz ± Rt (Green and Ringwood, 1967), whereas excess of certain chemical components or addition of water to the reacting system results in the appearance of one or more additional phases (Green, 1967). In addition, minimum P–T conditions for the stabilisation of the eclogite mineral assemblage depend on the bulk chemical composition of the reacting basic material (Ringwood and Green, 1966; Green and Ringwood, 1967, 1972).

Geochemical research over the last twenty years, carried out mainly in relation to the development of the sea-floor spreading concept. It has been shown that basalts, although rather uniform in their bulk composition, do in fact show both definite trends of compositional variation, depending on crystal-liquid fractionation processes (Miyashiro, 1974), and subtle differences in space (geotectonic affinity: Pearce and Cann, 1973; Winchester and Floyd, 1976) and time (O'Nions *et al*, 1978). Their HP metamorphic equivalents—eclogites—must also show similar trends of variation; this may be reflected either in the compositions of the two main phases, or (more frequently) in the mineral assemblage as a result of appearance of additional minerals.

There are, however, other reasons of the variability of the assemblages observed in eclogitic rocks. A major one is selective depletion or addition of elements during metamorphism (*metasomatism*), even to the point of displacing their compositions out of the basaltic range. This is generally not the case during subduction metamorphism (possibly because of the constrained environment and the relative speed with which HP conditions suitable for eclogite formation are attained), but it has been documented where 'oceanic metamorphism' (i.e. rock-water interaction in the oceanic crust) preceded subduction. In such eclogites, changes away from basaltic composition have been explained by a process of elemental migration akin to rodingitisation (Evans *et al*, 1979, 1981), clearly distinguished from metasomatic changes after eclogite facies metamorphism has already taken place (Morten *et al*, 1985).

Further reasons for deviation of the eclogite mineral assemblages away from that found in the simplified basaltic system are related to non-thermodynamic variables: texture of the protolith (topotactic reactions at grain boundaries); availability and local composition of a catalyst (mainly the fluid phase); and pressure distribution (shearing stress enhances reactions while isotropic distribution of P tends to preserve minerals outside of their stability fields). Where any one of these secondary variables has had an effect, it is often difficult to discriminate equilibrium assemblages from those including transient phases. Geothermobarometry based on elemental distribution between garnet and clinopyroxene (the most significant tracer for P–T conditions in eclogites) may give fallacious information; this is particularly so in rocks where only the early stages of eclogitisation occur, such as coarse-grained, originally dry, protoliths that carry corona-type reaction sequences (Morten and Bondi, 1984).

Parageneses and P–T data determined in originally aphanitic or fine-grained rocks (basalts *s.s.*, volcanoclastites) ought to be considered more reliable as a first approximation than those determined in other rocks of identical composition but coarse-grained (e.g. gabbros), because of the more widespread attainment of equilibrium. This should be remembered when determining local conditions of eclogite facies metamorphism, because in an ophiolite sequence (the most likely candidate for subduction and HP metamorphism) the basaltic layer is considerably thinner than the whole of the sheeted dyke complex + isotropic cumulus gabbro, not to mention the ultramafic cumulates (Clague and Straley, 1977; Hopson *et al*, 1981).

Originally 'wet' basic protoliths (amphibolites, glaucophane schists) undergo eclogitisation through a set of reactions different to those of 'dry' protoliths (or of synthetic systems), even though the bulk composition on an anhydrous chemical basis may be exactly the same. However, after the formation of the main eclogite assemblage, via mainly dehydration reactions, equilibration in an eclogite facies prograde P–T regime proceeds essentially through the same reactions in both 'wet' and 'dry' rocks. The same applies to certain rare eclogites thought to be derived from femic metacarbonate rocks i.e. compositionally basic in their immobile element components, but with excess CO_2 (Smulikowski, 1967; Lappin and Smith, 1981). Parageneses in these essentially compare with those of isofacial eclogites derived from basaltic rocks, apart from the occasional presence of minor quantities of carbonate minerals, which are referred to as 'primary' but which actually have been re-equilibrated at higher P–T.

The following account will concentrate first on changes in mineral assemblage resulting from variation in the bulk chemical composition of the protolith (inevitably touching on the subject of topochemical reactions controlled by the protolith texture), secondly on the reactions leading to eclogite formation from 'wet' protoliths, and finally on paragenetic variations arising from prograde reactions under a generalised HP regime within the eclogite facies. A similar stepwise procedure, independently conceived and carried out, was followed recently by Hirajima *et al* (1988); their conclusions and petrogenetic grid largely agree with those proposed here.

The emphasis in this section is on eclogite mineral parageneses stabilised in subducted crustal basic igneous rocks, based largely on the Alpine metamorphic belt. Reference is also made, however, to other terrains where eclogite facies metamorphism has developed regionally along a recognisable dP/dT gradient, and particularly to the Western Gneiss Region of Norway, where eclogite facies metamorphism has been described in particular detail despite of (or indeed, possibly because of) controversies in its interpretation.

2.3.2 *Composition variations in basic rock protoliths*

HP–LT eclogites in the western Alpine belt can be readily divided on the basis of their bulk chemistry (Mottana, 1977) into groups which are believed to reflect different steps of fractional crystallisation at shallow depth of a tholeiitic primary basaltic magma intruded into the oceanic crust of the spreading Piemonte trough (Mottana and Bocchio, 1975). Gabbro dykes injected into both the ultramafics and the massive basalts of the sheeted complex are characterised by high total contents of FeO, TiO_2, P_2O_5, Zr and Y (Mottana and Bocchio, 1975; Cortesogno *et al*, 1977), compared with leucocratic gabbros which occur as masses associated with, but mostly not within, the ultramafics (Messiga *et al*, 1983; Ernst *et al*, 1983). These leucocratic gabbros are enriched in MgO, Al_2O_3, Cr, and Ni and are therefore considered to be early cumulates.

Table 2.1 Average chemical compositions of eclogite types divided according to their recognisable protoliths

Oxide (wt %)	Mg-Al-gabbro	Fe-Ti-gabbro	Rodingitic gabbro	'common' basite	'crustal' basite
SiO_2	51.76	46.78	42.69	51.15	50.40
TiO_2	0.41	4.47	1.22	1.43	1.73
Al_2O_3	17.54	11.30	13.66	16.21	16.35
FeO	5.68	17.02	10.02	10.10	9.75
MnO	0.17	0.26	0.19	0.18	0.17
MgO	10.06	6.72	9.96	7.17	6.86
CaO	11.34	9.18	21.81	10.65	9.98
Na_2O	2.92	3.85	0.22	2.85	4.15
K_2O	0.06	0.10	0.03	0.16	0.40
P_2O_5	0.06	0.32	0.19	0.09	0.21

Notes

All data recalculated to 100 wt % after subtraction of volatiles (mainly H_2O) and reduction of total iron to FeO.

Protoliths:

Mg-Al-gabbros: 18 samples, mainly after sources quoted in Messiga *et al*, 1983.
Fe-Ti-gabbros: 79 samples, mainly after sources quoted in Mottana, 1977; Ernst, 1981; and Morten *et al*, 1985.
Rodingitic gabbros: 36 samples, mainly after sources quoted in Ernst, 1981; Evans *et al*, 1981; and Morten *et al*, 1985.
'Common' basites: 80 samples, the average of 'light' and 'dark' eclogite types included in the basement rocks of the Münchberg Massif, West Germany (Matthes *et al*, 1975).
'Crustal' basites: (i.e. eclogites undoubtedly derived from basalts—mainly pillow-basalts—or bearing clear evidence of derivation from a volcanic or volcanoclastic protolith): 51 samples, mainly from sources quoted in Mottana, 1977 and Messiga *et al*, 1983.

The two chemical types of eclogite initially discovered in the Voltri Massif of the western Alps have worldwide significance. Fe-Ti-rich ('superferrian' according to Mottana and Bocchio, 1975) and Mg–Al-rich eclogites are now known from the eastern Alps, the Caledonides of Norway and Scotland, the Sudetes, and the Sanbagawa belt. Their average chemical compositions, computed mainly on the basis of the Alpine occurrences, are given in Table 2.1. In addition, an intermediate type, possibly derived from an intermediate step of fractionation, has also been recognised: it comprises superferrian eclogites which have TiO_2 contents as low as those of the Mg–Al-gabbros, but total FeO content substantially greater than that of the Fe–Ti gabbros (Griffin and Qvale, 1985; Colombi and Pfeifer, 1986; Nisio *et al*, 1986). Because most TiO_2 concentrates as rutile from the earliest stages of eclogite facies metamorphism, however, this intermediate type will be included in the superferrian group. All such eclogite types are easily distinguished when the gabbroic nature of the protolith can be recognised, although some eclogitised gabbros of intermediate type defy this classification scheme (Mörk, 1986; Morten *et al*, 1987; Miller *et al*, 1988).

Discrimination is more difficult when eclogite bodies, mostly small and interbedded in metasedimentary units, are taken into consideration. These eclogites are probably derived from basic volcanics (tuffs, lavas) or from re-worked volcanoclastics + sediments; indeed, some of them bear structural evidence of derivation from pillow-basalts (Bearth, 1959). Their composition (Table 2.1) is usually transitional between the two extreme types mentioned above, for primary reasons related to the differentiation of the parent magma, or because of contamination by the surrounding sediments, or because of compositional changes undergone during burial and prograde metamorphism. These eclogites are only roughly basaltic in composition and are often anomalously enriched in alumina and/or carbonate material (Messiga *et al*, 1977; Piccardo *et al*, 1979; Mottana *et al*, 1985; Maggetti *et al*, 1987). More unusual eclogite facies parageneses may be found in them. Their paragenetic variations across the eclogite facies *P–T* field are the most difficult to trace because of the variable rock chemistry, and because several researchers describing new and/or critical assemblages have not provided bulk rock composition data.

All eclogite facies metabasites which lack special chemical characteristics, and those which have not been chemically described will be treated together here as 'common' eclogites. This group will include also eclogites derived from previously metamorphosed ('wet'), as well as volatile-rich basic precursors.

2.3.3 *Prograde eclogite facies metamorphism in Fe-rich compositions*

Experiments indicate that the lower the $Mg/(Mg + Fe^{2+})$, (An + Ab)/(Di + Ol + Hy/2 + Mag + Ilm), and Ab/(Ab + An) whole rock ratios are, the lower the *P* (at any given *T*) at which basalts will convert to eclogites (Green and Ringwood, 1967, 1972). The first ratio controls the incoming of garnet, and the latter two the disappearance of plagioclase in eclogite paragenesis. Therefore *superferrian eclogites* will form from Fe–Ti gabbros and basalts at lower P than eclogites derived from Mg–Al gabbros. Their typical assemblage is

$$\text{Alm-rich Grt} + Fe^{3+}\text{-rich Omp (Chl)} + \text{Rt}.$$

Qtz is rarely present as a primary phase, whereas Ap, Py, and Mag (intergrown with Ilm, often inherited from the protolith) are common accessory phases.

Parageneses of superferrian eclogites equilibrated at increasing $P-T$ conditions (as determined via the compositions of their coexisting Cpx and Grt) are given in Figure 2.3 and are summarised in Table 2.2. Additional phases are controlled by subtle changes in both the whole rock chemical composition and the conditions at eclogite formation.

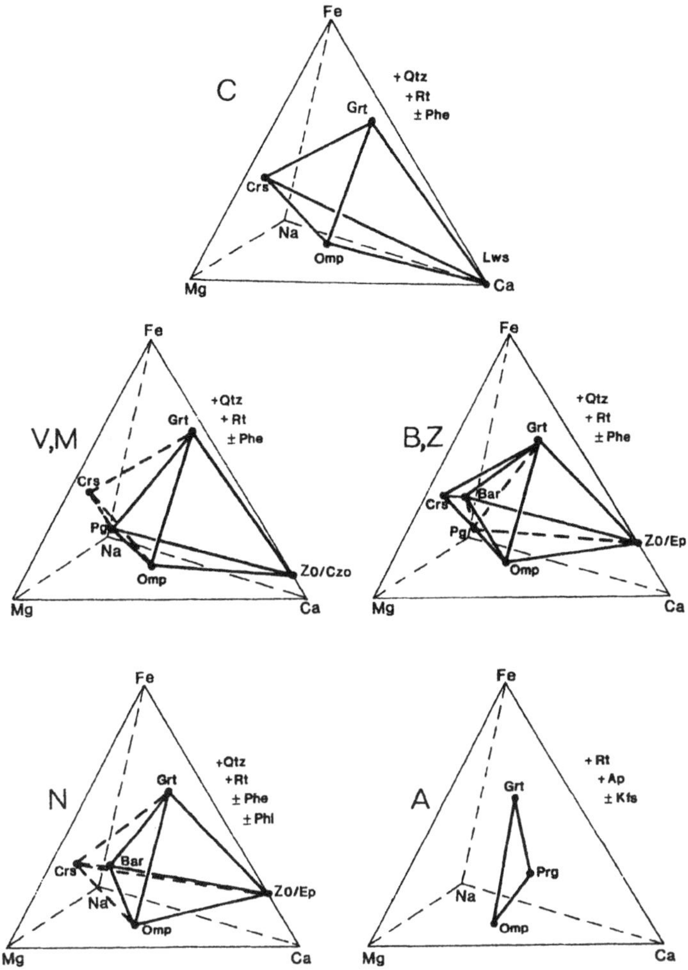

Figure 2.3 Eclogite facies mineral parageneses stable at different pressures and temperatures in Fe-Ti-basaltic compositions as observed in the occurrences of Corsica (*C*), Voltri Group (*V*), Monviso (*M*), Breuil-St. Jacques (*B*), Zermatt (*Z*), all in the western Alps, and Naustdal (*N*) and Almklovdalen (*A*) in the Western Gneiss Region of Norway. Mineral parageneses adjacent to the main assemblage Grt + Omp (+ Qtz + Rt), which extends across the central part of the Ca-Fe-Mg-Na tetrahedron and changes only slightly due to minor changes in composition of the two main phases, include several different additional phases stable on both the Ca and the Na sides, depending on $P-T$ conditions. See text for references and details.

Table 2.2 Summary of diagnostic eclogite facies mineral parageneses in rocks of various basic rock compositions

Locality		Eclogite facies mineral assemblage	Composition type
California		Ep + Bar + Omp + Grt/Gln, Act, Py, Ttn(Arg?)	CR
Corsica		Omp + Grt + Crs + Lws/Qtz, Rt, Phe(Act?, Chl?)	CR
Voltri		Clm + Grt + Zo + Rt/Qtz, Czo, Crs, Pg, Phe, Ap, Ilm, Py	FT
		Omp + Grt + Zo + Qtz/Rt, Pg(Gln?, Tr?)	MA
		Clm + Grt + Zo(Czo-Ep) + Rt/Qtz, Crs, Pg, Phe, Ap, Mt, Py	CR
		Omp(Di) + Grt + Zo(Czo-Ep) + Chl/Ttn, Mt, Act(Cal)	RD
Monviso		Clm + Grt + Rt ± Crs/Pg, Czo, Qtz, Ttn	FT
		Clm + Grt + Rt + Pg ± Crs/Czo, Phe, Ank, Py (Ky?, Tlc?)	CR
		Omp + Grt + Zo + Tlc ± Gln ± Cld/Zo, Ky(Chl?)	MA
Breuil		Omp – Clm + Grt + Zo + Rt/(Gln?, Pg?, Phe?, Ttn?, Cld?)	CR
Zermatt		Omp + Gr + Zo ± Gln ± Hbl ± Pg ± Phe/Cld, Tlc, Qtz, Rt, Ap, Mt, Cal	CR
		Omp + Grt + Ky + Tlc ± Gln ± Cld/Rt, Mt, Hbl, Chl, (Qtz?, Pg?)	MA
Sesia		Grt + Gln + Pg + Phe + Qtz ± Zo/Rt, Czo	CM
Adula	I	Omp + Grt + Qtz ± Gln ± Ab/Czo-Ep, Phe, Hbl, Rt, Py	CM
"	II	Omp + Grt + Qtz + Hbl ± Ky ± Pg/Bar, Zo, Phe, Ep, Rt, Py, Ttn	CM
"	II	Di + Grt + Hbl + Zo + Czo-Ep/Qtz, Rt, Py, Ttn	RD
"	III	Omp + Grt + Ky + Qtz ± Gln ± Hbl/Tlc, Pg, Phe, Rt, Py	CM
"	IV	Omp + Grt + Ky + Qtz + Hbl/Zo, Czo, Rt, Ilm	CM
"	IV	Di + Grt + Hbl + Ep/Rt	RD
Taurn		Omp + Grt + Qtz + Ky ± Gln ± Tlc/Zo, Pg, Phe, Rt, Py, Ap, Dol, Mag	CR
		Omp + Grt + Ky + Cld + Tlc/Pg, Chl, Rt, Py, Ap	MA
Donghai		Grt + Omp + Cor + Ky + Zo/Dol, Rt, Ap(St?, Prg?)	CM
Motalafjella		Grt + Omp + Gln ± Ep ± Pg/Phe, Wnc, Rt, Qtz	CM
Naustdal		Grt + Clm + Ap + Rt ± Gln ± Bar/Qtz, Ilm, Py, Zo-Ep, Phe(Phl?)	FT
Almklovdalen		Omp + Grt/Rt, Ap, Prg(Kfs)	FT
		Grt + Omp + Amp + (Ed-Prg) ± Zo ± Bt/Qtz, Rt, Ap, Py, Cal	MA
Grytting		Omp + Grt + Coe + Dol/Hbl, Rt, Py(± Po), Ilm, Zo(Czo), Zr	CM

Notes
Eclogite compositional types (*cf*. Table 2.1): CM 'common', CR 'crustal', FT ferrotitanian (superferrian); MA magnesian; RD rodingitic. For references, see text.

At the initial stage, in Corsica where Fe-Ti basalts exhibit the full transition from blueschist to eclogite facies (Péquignot *et al*, 1984; Caron and Péquignot, 1986), excess Ca is accommodated in lawsonite, whereas chemically similar, but slightly upgrade rocks in the Voltri Group contain epidote minerals (mostly zoisite, but also Czo). This is probably due to the reaction

$$\text{Lws} + \text{Ab} \rightleftharpoons \text{Pg} + \text{Zo} + \text{Qtz} + H_2O$$

This reaction has been calibrated experimentally and found to occur at 430°C and 10 kbar (Heinrich and Althaus, 1988). Such a T agrees well with previous estimates of the equilibration conditions of the Corsican Lws-eclogite of 420°C (Lardeaux *et al*, 1986), compared with those of the Voltri Group at 435–470°C (Ernst, 1981; Cortesogno *et al*, 1981). Note that these temperatures were computed at the same nominal $P = 10$ kbar, as has become customary in studies of the western Alps, although there is an indication that $P = 15$ kbar would be more realistic. The assemblage detected in the Voltri Group has a comparatively wide field of stability, as it has also been found at Monviso in eclogites equilibrated at 500°C (Lardeaux *et al*, 1986).

Additional phases present in eclogites are white micas—mostly Phe (taking up all available K), Pg (which is present both as a product of the above reaction and as an excess phase taking up all Na not entering Omp at these rather low P–T conditions), Mg-chlorite, and amphiboles. Whilst Mg-chlorite is most likely to have been inherited from the blueschist facies stage, the participation of amphibole in the assemblage is controversial: some authors (Mottana and Bocchio, 1975; Lardeaux *et al*, 1986; Pognante *et al*, 1988) mention glaucophane (actually a crossite) to be present in equilibrium with Grt + Omp + Rt; others consider it to be a late or post-eclogitic phase grown by re-equilibration under conditions declining towards (Godard *et al*, 1981), or within the blueschist facies (Ernst, 1976; Ernst and Dal Piaz, 1978; Messiga, 1987).

Further steps in the prograde P–T evolution of superferrian eclogites are documented from the occurrences at Breuil-St. Jacques, at 525°C (Ernst and Dal Piaz, 1978; Ernst, 1981) and Zermatt, at 550–600°C (Oberhänsli, 1980); on both sides of the Italian–Swiss border in the western Alps; from Naustdal at 630°C (Binns, 1967; Krogh, 1980a); and Almklovdalen in the Western Gneiss Region of Norway at 700–750°C (Griffin and Qvale, 1985). (Note however that T-values in Norway have been calculated for $P > 15$ kbar, as against $P = 10$ kbar in the Alps.) The Grt + Cpx + Rt assemblage changes only cryptically, with Grt becoming more Mg-rich and Cpx moving in composition from the Clm to the Omp field, i.e. losing its Fe^{3+} component (Figure 2.4). Accessory minerals reflect the changes in composition of the main ones, with phase changes dependent on their intake of the new components made available. Zoisite, for instance, tends to acquire Fe^{3+} released from Cpx and changes first to Czo then to Ep. The coexistence of Zo and Czo in many of these rocks can be explained by a compositional gap involving increased Fe^{3+} in the latter phase; this was documented first in 'common' eclogites of higher grade (Enami and Banno, 1980), and later confirmed at lower grade (Yokoyama *et al*, 1982). On the other hand, increasing T and P leads to the breakdown of paragonite, with resultant increased Jd content in the Cpx phase. Consequently, Cpx not only becomes Omp but approaches the ideal $Di_{50}Jd_{50}$ composition (Krogh, 1980a, b). In contrast, phengite appears to remain as the stable K-mica phase to rather high temperatures (*c.* 800°C). Of the above-listed localities, only at Naustdal has biotite been described as the stable mica phase (Krogh, 1980a). However, it is restricted to rock-volumes carrying Na–Ca amphibole (Bar); phengite

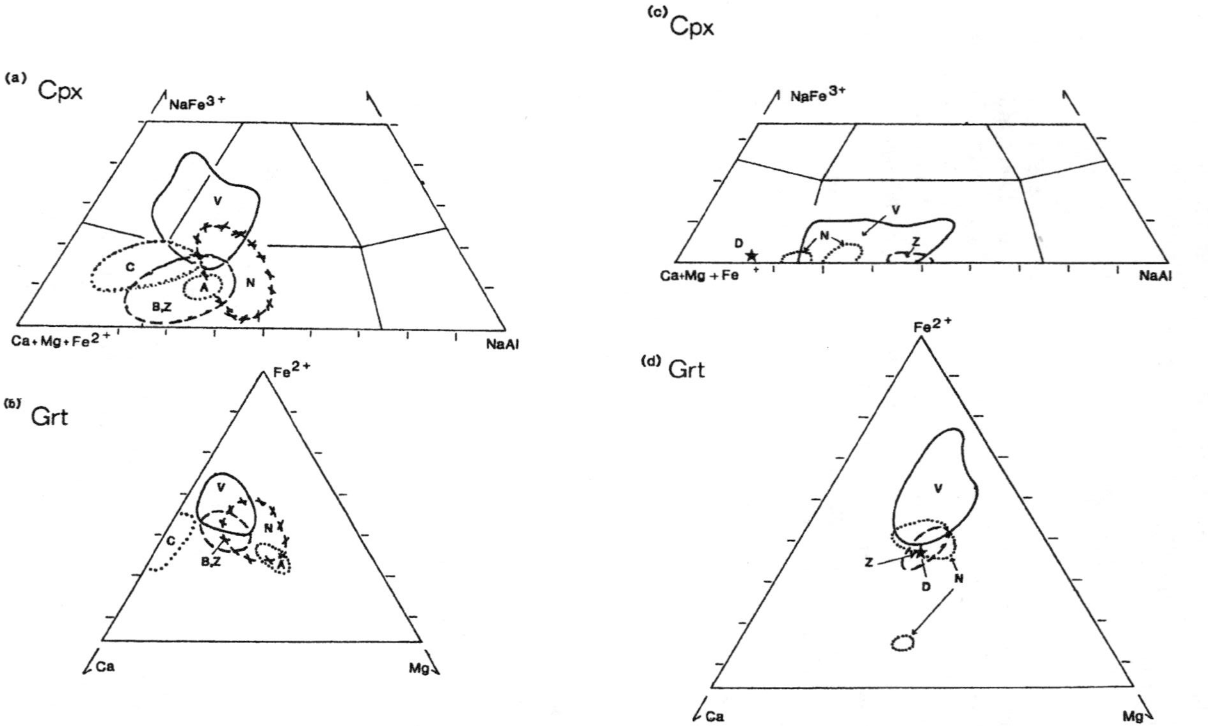

Figure 2.4 Changes in the compositions of clinopyroxenes and garnets with increasing conditions of eclogite facies metamorphism in superferrian eclogites (*a* and *b*, with labels as in Figure 2.3) and Mg-Al rich eclogites (*c* and *d*, with labels as in Figure 2.5). Only data for analyses satisfying Mottana's (1986a) reliability tests have been plotted.

remains the stable mica phase in those carrying crossitic Gln. It is expected that biotite becomes the stable K-mica phase to significantly higher *T*, in accordance with the presence of primary phlogopite in some of the lowest *T* upper mantle ultramafic nodules (Carswell, 1975). However, at Almklovdalen (Griffin and Qvale, 1985) no mica phase is present, and the little K present in the rock is entirely accommodated in K-feldspar inclusions occurring in garnets.

The most conspicuous change in paragenesis shown by superferrian compositions along a prograde eclogite facies *P–T* path is related to the type of amphibole present. At the lowest grade, Gln may be present as a primary amphibole phase, but perhaps more commonly as a secondary replacement phase. Other amphiboles interpreted to be primary at this stage are ferroglaucophane, which was found in a Lws-bearing eclogite (unfortunately stratigraphically out-of-place—it occurs as volcanic ejecta in acid tuffs at Milos; Kornprobst *et al*, 1979), crossite (Stephens and van Roermund, 1984), and magnesiokatophorite (Morten *et al*, 1987). Whilst ferroglaucophane and crossite are probably vicarious of normal Gln in rocks having excess Fe (their alternative presence depending on P_{O_2} conditions), magnesiokatophorite topotactically replaces magmatic Cpx or Amp. Therefore, although described as being in textural equilibrium with Grt and Omp, it may in fact represent a transient phase. With increasing grade (Valtournanche, Val d'Aosta) a Ca-Na-amphibole forms adjacent to Na-amphibole. Although originally interpreted as a product of retrograde metamorphism (Ernst and Dal Piaz, 1978; Ernst, 1981), this Ca-Na-amphibole of barroisitic composition has been re-interpreted as being associated with Gln as a consequence of an equilibrium miscibility gap which closes at about 600°C (Reynard and Ballèvre, 1988). Barroisite and even winchite appear to replace Gln completely under higher *T* eclogite facies conditions, at Naustdal (Binns, 1967; Krogh, 1980a) and Isla Margarita respectively (Maresch *et al*, 1982). A Ca-rich amphibole, Cl-ferropargasite (Griffin and Qvale, 1985), is the only amphibole found at Almklovdalen, in the highest *T* superferrian eclogite described to date.

2.3.4 *Prograde eclogite facies metamorphism in Mg-rich compositions*

Mg-Al gabbros can be expected to undergo eclogitisation along a prograde *P–T* path at higher *P* and *T* than Fe-Ti gabbros. This is shown in the Alps where most Mg-Al gabbros still resemble igneous rocks carrying abundant relicts of Pl $\pm$ Amp + Cpx (partly pseudomorphed e.g. 'smaragditic gabbros'), whilst the enclosed Fe–Ti basic dykes are transformed into eclogites (Compagnoni *et al*, 1984; Reynard and Ballèvre, 1988).

No magnesian gabbros have been described to date in the Corsican sequence, therefore the onset of eclogite facies metamorphism in rocks of this composition is best documented by the Voltri Group occurrences. Their eclogite facies assemblage is

$$\text{Prp–rich Grt} + \text{Di–rich Omp} + \text{Zo} + \text{Qtz}$$

(Messiga *et al*, 1983; Messiga, 1987), with Ap, Zr, Aln as accessory phases (Figure 2.5). Because of their coarse grain size, eclogitised Mg-Al gabbros often display topochemical reactions, which are in part responsible for the preferred growth of additional phases in characteristic textural sites. Among these phases, talc is interpreted as replacing the Opx exsolution lamellae of igneous Cpx (thus contributing to the shiny pseudomorphs known as 'smaragdite'); Pg as incorporating all the Ab component of

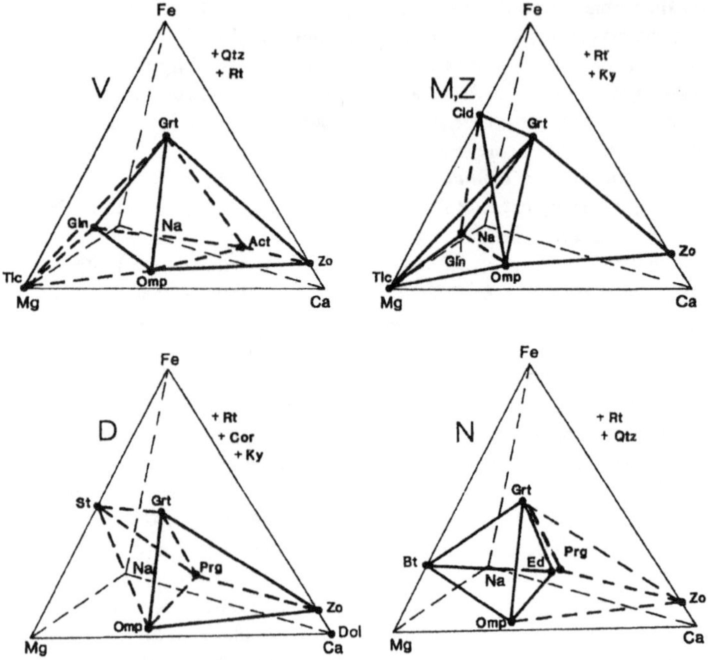

Figure 2.5 Eclogite facies mineral parageneses stable at different pressures and temperatures in Mg-Al-basaltic compositions as observed in the occurrences of the Voltri Group (*V*), Monviso (*M*), and Zermatt (*Z*) in the western Alps, in the Western Gneiss Region of Norway (*N*), and at Donghai in China (*D*). Mineral parageneses adjacent to the main assemblage Grt + Omp (+ Qtz + Rt) include Zo, an almost ubiquitous phase on the Ca-rich side and several possible additional phases on the Mg-Fe side. See text for references and details.

igneous Pl which cannot enter Cpx at comparatively low *P*; and magnesian Chl is believed to substitute for igneous Ol (alternatively it is considered to be a relict of a previous stage of 'oceanic metamorphism': Nisio and Lardeaux, 1987). Primary amphibole is also often considered to be an additional mineral, but most authors insist on interpreting it as a product of a post-eclogitic stage. The first-formed amphibole, in close association with Omp only, is Gln, whilst Tr and Act amphiboles are associated with both Omp and Tlc in textural sites inferred to represent replaced igneous Ol (Messiga *et al*, 1983).

Most minor phases occurring in eclogitised Mg-Al gabbros can be interpreted as transient phases, particularly as their presence is restricted to favourable sites scattered among coarse-grained relicts, and because some of them are associated only with a particular host-mineral. However, they are often present in polyphase assemblages which also contain Grt and Jd-poor Omp (but no Qtz) and can be interpreted as equilibrium parageneses, although only on a local scale. Undoubtedly, they disappear when *P*–*T* conditions increase, but whether this is due to further prograde reaction or simply reflects the attainment of equilibrium on a greater length-scale is often unclear. The latter is certainly the case when grain-coarsening goes hand in hand with the reduction of the number of eclogitic phases, even in closely spaced volumes of the same rock sample.

The prograde HP metamorphism of Mg-Al gabbros is summarised in Figure 2.5 and Table 2.2. Mg-Al gabbros eclogitised under P–T conditions higher than at Voltri can be found at Monviso (Nisio and Lardeaux, 1986; Kienast and Messiga, 1987), at Zermatt (Chinner and Dixon, 1973) and in the Austrian Alps (Miller, 1970, 1977). They are significantly different from the associated superferrian eclogites, not only because of the texture and composition of the main eclogitic minerals, but also because of the type and compositions of their additional phases. Typical parageneses are:

$$\text{Grt} + \text{Omp} + \text{Mg-rich Cld} + \text{Tlc} + \text{Ky}$$
$$\text{Grt} + \text{Gln} + \text{Tlc} + \text{Ky}$$
$$\text{Grt} + \text{Omp} + \text{Ky} + \text{Zo} \pm \text{Cld}$$

The widespread and coarse-grained kyanite is clearly derived from Pg, the breakdown of which releases Na and favours the formation of an omphacite with a higher Jd component, whilst Mg in excess preferentially segregates as talc (or even Mg-Chl). Excess Ca is still accommodated in zoisite. The occurrence of chloritoid is interesting, not only because of its composition—usually Mg-rich (Chinner and Dixon, 1973; Kienast and Pognante, 1988) and occasionally Cr-enriched (Kienast and Messiga, 1987)—but also because its breakdown characterises a further step in the prograde HP path. This is represented by the recent discovery of highly magnesian staurolite associated with corundum in the eclogites from Mengzhong, Donghai district, China, equilibrated at $T = 700$–$750°C$ and $P = 11$–25 kbar (Enami and Zang, 1988). These authors report the assemblage to be

$$\text{Grt} + \text{Cpx} + \text{Cor} + \text{Ky}\ (+\text{Zo} + \text{Dol} + \text{Rt} + \text{Ap})$$

whilst Mg-St and a calcic amphibole (Prg) were interpreted as secondary phases. However, the Mg-St may be interpreted as derived from the Mg-Cld by a prograde reaction of the same type as that described in eclogite facies metapelites (*cf* section 2.4).

Eclogite facies rocks with chemical composition similar to that of Alpine Mg-Al gabbros have also been described from Norway (O'Hara and Mercy, 1963; Medaris, 1980; Carswell, 1981), where they are estimated to have equilibrated at *c.* 700–750°C and 20–22 kbar. They contain the assemblage

$$\text{Grt} + \text{Cpx} + \text{Amp} \pm \text{Zo} \pm \text{Bt}\ (+\text{Qtz} + \text{Rt} + \text{Ap} + \text{Po})$$

where Amp may be either edenitic and in equilibrium with Bi, or pargasitic and in equilibrium with Zo. Grt has a Prp-rich composition and Cpx is Jd-poor. The two main minerals strongly resemble those in the associated garnet websterites, which carry in addition Opx and sometimes O1.

Under *P–T* conditions comparable to those of the Almklovdalen superferrian eclogite occurrence, biotite is the only stable K-mica phase in these rocks of Mg-Al-rich bulk composition, and the amphibole is invariably a calcian variety.

2.3.5 *Prograde eclogite facies metamorphism in 'common' basic compositions*

The mineral assemblages of *'common' eclogites* fall between those of the two extreme basic rock compositions mentioned in the two previous sections. A schematic representation of their paragenetic variation along the eclogite facies prograde *P–T*

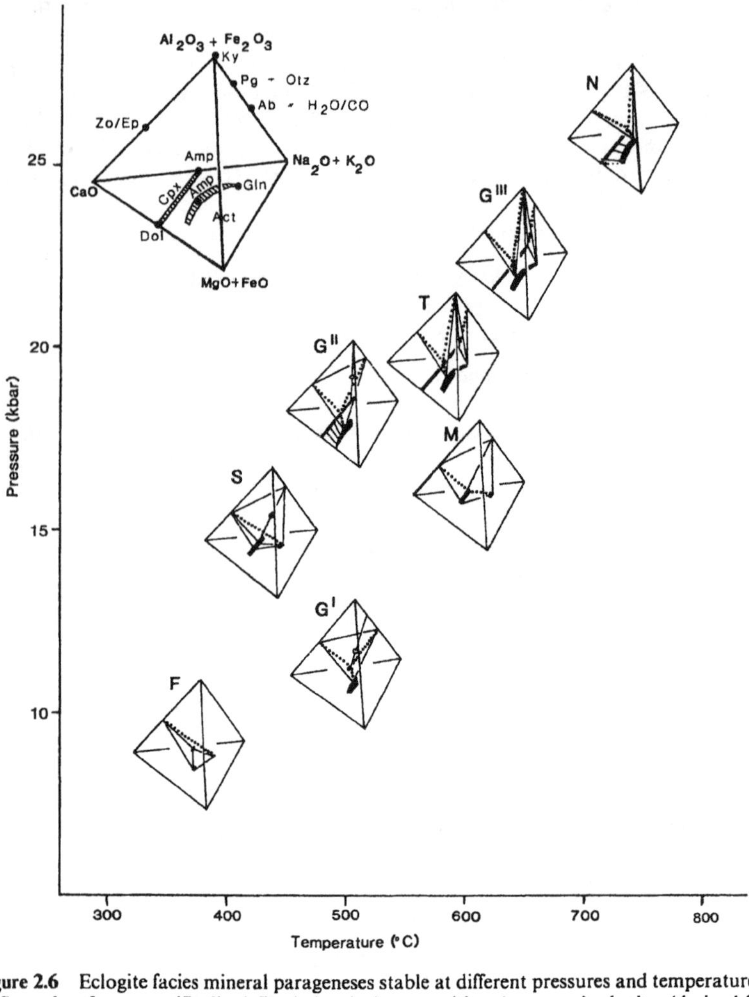

Figure 2.6 Eclogite facies mineral parageneses stable at different pressures and temperatures in mafic rocks of non-specifically defined chemical composition: 'common' eclogites (derived from various types of basic igneous rocks or amphibolites) and 'crustal' metabasites of unequivocal superficial derivation (tuffs, hyaloclastites, volcanics, and perhaps marls, etc.). Observed minerals are represented as perspective plots in the N ($Na_2O + K_2O$) – C (CaO) – M (MgO + FeO) – A ($Al_2O_3 + Fe_2O_3$) tetrahedron projected from quartz and H_2O/CO_2 (modified after Heinrich, 1986). Garnet, rutile and quartz phases are constantly present but are not represented in this type of projection. Typical examples shown are: the *in situ* eclogites of the Franciscan formation (*F*) of California, USA (Maruyama and Liou, 1988); the Sesia-Lanzo Zone (*S*) eclogites of Val Chiusella, Italy (Reinsch, 1979); the Motalafjella (*M*) eclogites of Spitzbergen, Norway (Hirajima *et al*, 1988); the Hohe Tauern (*T*) eclogites from Tyrol, Austria (Holland, 1979); and the three prograde stages of equilibration of the Adula Nappe eclogites of Grisons (*G*′, *G*″ and *G*‴), Switzerland (Heinrich, 1986). Finally (upper right corner, *N*) the extreme assemblage observed at Grytting, Norway (Smith, 1984) is also given. This stage of equilibration corresponds, but for the presence of coesite, to the fourth prograde stage of the Adula eclogite facies metamorphism (Heinrich, 1986). Plotted equilibration conditions are those determined by the authors.

path is given in Figure 2.6 and summarised in Table 2.2. The observed assemblages essentially comprise the following minerals, the amounts and compositions of which depend on *P–T* conditions and the bulk composition of the eclogitised rock:

$$\text{Grt} + \text{Omp} + \text{Rt} \pm \text{Qtz} \ (\pm \text{Py} \pm \text{Po})$$

The effect of bulk composition is less significant than expected for certain components, in that their apparent random variability does not disturb the essential phase relationships of the eclogite facies assemblage compatible with the *P–T* conditions. For example, many 'common' or 'normal' eclogites contain considerable amounts of primary quartz—up to 10 vol.% as in the Sesia Zone, western Alps (Viterbo-Bassani and Blackburn, 1968)—whilst others are completely devoid of it such as some Franciscan exotic blocks (Coleman *et al*, 1965). There are eclogites containing up to 12.5 vol.% primary carbonates and therefore believed to be derived from marls (Smulikowski, 1967). Despite these oddities, there is no question that most common eclogites do indeed possess a basaltic composition, and behave in a consistent manner during the evolution of eclogite facies metamorphism, the odd components becoming segregated as excess phases.

With increasing *P–T* there are changes in garnet composition from Alm-rich to Prp-rich (from roughly $Alm_{60}Prp_{10}$ to $Alm_{20}Prp_{55}$), but the Grs component remains relatively constant at *c.* 30 mol% (Figure 2.7). The cores of many garnets are Sps-rich (up to 8 mol%)—a feature which has been interpreted to indicate their precocious nucleation—whilst the rims may be enriched in either Fe or Mg, depending on the *P–T* course of crystallisation (retrograde or prograde) and/or the types of accompanying reactions (Figure 2.6).

Omphacite changes from $Di_{75}Jd_{25}$ to $Di_{40}Jd_{60}$ with increasing *P–T* conditions of equilibration, but with large deviations from the average trend depending on the bulk Al/Na ratio of the rock (Figure 2.7a). The Acm component of omphacite is normally low, but in some LT eclogites may rise to 10–15 mol%. In contrast to experimental synthetic systems and to eclogitised dry compositions (such as the gabbros and basalts mentioned above), the process of eclogitisation of a 'wet' protolith often involves formation of Omp before Grt, and this remains stable over a fairly large *T*-span before a genuine eclogite facies paragenesis is attained. This has been documented by regional studies on the transition between blueschists and eclogites both in New Caledonia (Black, 1977; Yokoyama *et al*, 1982) and California (Maruyama and Liou, 1988), and that between epidote amphibolites and eclogites at Isla Margarita, Venezuela (Maresch, 1972, 1983). a_{H_2O} must play a crucial role in controlling the stability of both Omp and Grt relative to Amp in eclogites as there is also contradictory evidence that prasinites dehydrate to amphibole eclogites via a zone of garnet amphibolites, such as in the eastern Alps (Abraham *et al*, 1974).

Of the additional minerals which may be stable, some simply depend on oddities in the bulk chemical composition, whereas others are also sensitive indicators of the grade of metamorphism because their solid solution systems are flexible enough to reflect both bulk compositional and *P–T* changes. Amphiboles belong to this second group (Figure 2.7c and d). Those formed early during subduction, at the onset of eclogitisation, are Gln or Crs, depending on the Fe/Al ratio of the rock. With increasing P at low T, they first shift in composition towards pure Gln, then towards nyböite or even Mg-Al-taramite if *T* increases as well but the d*P*/d*T* gradient remains high (Ungaretti *et al*, 1981, 1983). With a lower d*P*/d*T* gradient, Na-Ca-amphiboles such

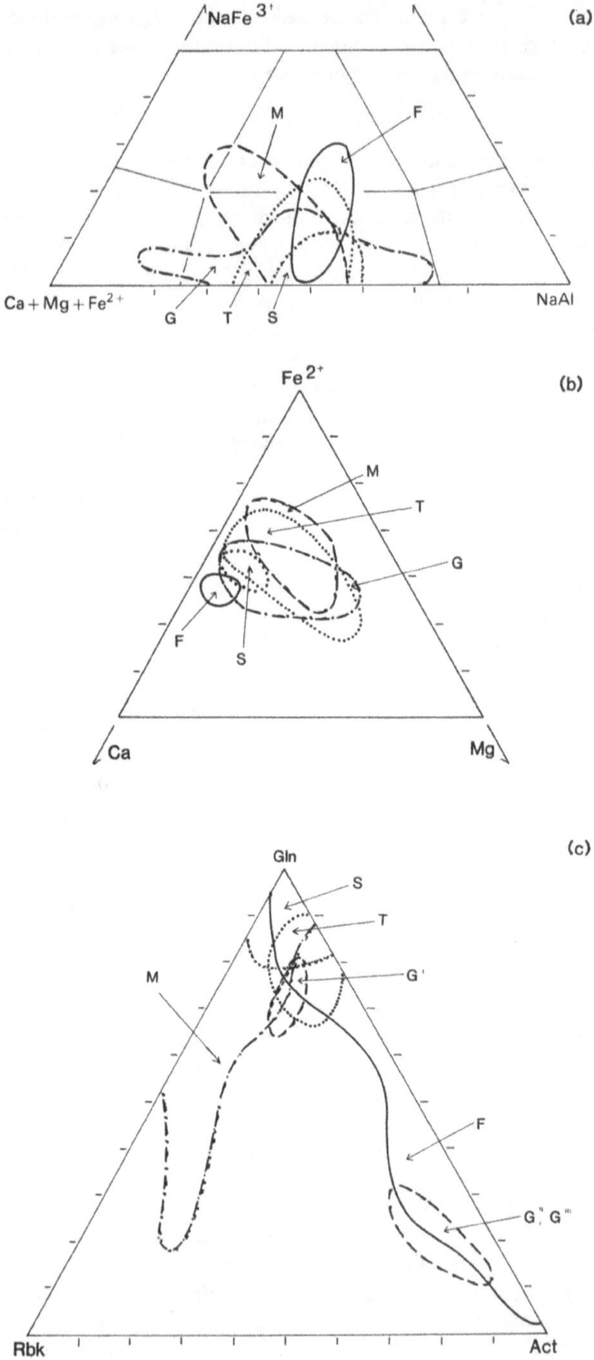
(a)
NaFe 3+
M
F
Ca + Mg + Fe2+
NaAl
G
T
S
(b)
Fe 2+
M
T
G
F
S
Ca
Mg
(c)
Gln
S
T
G'
M
F
G'', G'''
Rbk
Act

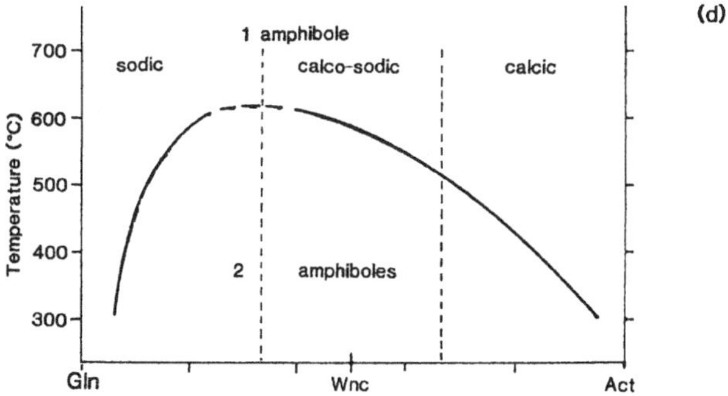

Figure 2.7 Changes in composition of the main mineral phases in 'common' eclogites with increasing conditions of eclogite facies metamorphism. (a) clinopyroxenes; (b) garnets; (c) amphiboles. Data are from the occurrences listed in Figure 2.6 and from other occurrences of the same type. All analyses satisfy Mottana's (1986a) reliability tests. The miscibility gap between sodic and calcid amphiboles in eclogites, showing its gradual closure with increasing T under constantly high P (modified from Reynard and Ballèvre, 1988) is also given (*d*)

as barroisite and winchite may form instead (Ernst, 1979; Maresch *et al*, 1982), followed by brown Ca-amphiboles known as 'carithine' (Mottana *et al*, 1968). These latter are known as 'carithine' and their compositions spread across the fields of Mg-hornblende, Mg-Al-katophorite, edenite, and Fe-Al-pargasite (Bocchio *et al*, 1978), or may even become an unusual complex solid solution (Bocchio *et al*, 1985). Even when primary on textural grounds, amphiboles in eclogites remain very difficult to interpret: they 'display seemingly random compositional variations, and systematic chemical differences as a function of the occurrence remain obscure' (Ernst, 1977).

White mica is restricted to rocks showing Al excess and enriched either in Na or in K. Its type depends essentially on the ratio of K to the available Na remaining after formation of the Cpx stable at a given pressure. The K-rich phase is Phe, often very rich in celadonite component (up to 3.7 Si atoms pfu). Although undoubtedly in equilibrium with the main eclogite phases at times (Krogh and Råheim, 1978), Phe appears to re-equilibrate easily (Lardeaux *et al*, 1983) and is therefore a poor metamorphic indicator. Pg may occur either in association with Phe (Bearth, 1973) or as the only mica phase (Maresch and Abraham, 1981). Its presence depends largely on the bulk composition of the rock, because in itself it is always rather pure.

With prograde eclogitisation, both micas give way to Ky, although this mineral may also form in stable association with them under suitable conditions (Holland, 1979). Whilst the Na previously held in Pg enters Omp (or Gln, in the narrow field of conditions where this mineral is in equilibrium with Omp; Holland, 1988) the K freed by the breakdown of Phe promotes reactions with Grt and/or Ca-Amp. This produces Bt, and may even involve Rt, eventually leading to formation of a Ti-rich, sometimes Na-bearing, Phl (Krogh, 1980a; Enami and Zang, 1988).

At some of the highest temperatures compatible with crustal eclogite facies conditions, such as those claimed for Quinglongshan, China of 820–850°C at P > 19 k bar (Enami *et al*, 1986), kyanite is found in stable association with corundum, as in certain grospydite nodules in kimberlites. However, Cor in crustal eclogites is primary only rarely (Nicollet, 1977)—usually it is a retrograde product from Ky due to pressure release (Smulikowski and Bakun-Czubarow, 1969).

Epidote group minerals initially form as Zo, both in the α and in the β form (Mottana *et al*, 1968); they are later followed by Czo, which can coexist because of its pistacite component. Primary pistacite can be present both in the early stages of eclogitisation (Maruyama and Liou, 1988) and in MT eclogites (Holland, 1979; Heinrich, 1986), but not commonly so.

Other accessory phases of crustal eclogites of note but whose paragenetic relationships are insufficiently known are: high-Al titanite (Smith, 1980a), tourmaline (Smith, 1971), apatite, zircon, moissanite (Golovnya *et al*, 1973), magnesiostaurolite (Enami *et al*, 1986), plus graphite and a variety of sulphides and oxides (e.g. Knauer and Matthes, 1970). These last indicate that eclogite formation and stability can encompass a broad range of oxygen fugacities. Similarly, the presence of carbonates (calcite, dolomite, magnesite) may indicate that some eclogites have been derived from iron-rich dolomitic marls (Smulikowski, 1967); further, they demonstrate the presence of volatiles in excess of those structurally bound in silicates even at the highest *P–T* conditions of the eclogite facies (Lappin and Smith, 1981).

Some gabbro dykes in ultrabasic massifs of the western and central Alps which have become transformed into rodingites (Table 2.1) display eclogite facies mineral parageneses (Table 2.2). Rodingitisation must therefore have preceded subduction, probably when the ultrabasic host-rock was still buried in the ocean floor. In the Voltri Group occurrences (Cortesogno *et al*, 1977; Piccardo *et al*, 1979; see also Morten *et al*, 1985) the paragenesis is dominated by Ca-rich minerals, such as Grs-rich garnet, diopside and Ca-amphibole, but does not contain carbonates. Under the higher grade conditions of the eclogite facies metamorphism in the Cima Lunga Nappe, central Alps (Evans *et al*, 1979, 1981; Heinrich, 1986), epidotes, titanite, vesuvianite and perovskite, as well as carbonates of different types, may become important constituents. The reason is not clear: possibly an increased f_{O_2}, testified by the Adr-rich composition of garnet, plays a significant role. Compositionally similar rocks in the high grade conditions of the Western Gneiss Region of Norway display parageneses similar to those of grospydite nodules in kimberlites (Section 2.1).

Recently reported occurrences of coesite and/or distinctive polycrystalline quartz pseudomorphs after coesite in eclogites—from the Western Gneiss Region, Norway (Smith, 1984) and from the Dora-Maira Massif, Italian western Alps (Chopin, 1984, 1987)—confirm the older, and for a long time forgotten, finding of quartz replacing coesite in an eclogite from south Urals, Ukraine (Chesnokov and Popov, 1965). Similarly, the discovery of diamond in garnetiferous gneisses of the eclogite-bearing Atbashi range, Uzbekhistan, is consistent with its occurrence in an eclogite from the Kokchetav block, Kazakhstan (Rozen *et al*, 1972). These findings are especially important since they demonstrate convergence between the extreme HP conditions attained by regionally distributed eclogite facies, subducted crustal rocks, and the *P–T* equilibration conditions in upper mantle rocks, as evidenced by the coesite-bearing and diamond-bearing eclogite and grospydite nodules entrained in kimberlites (see Chapter 13).

Further details on the mineral parageneses developed in rocks of broadly basic composition under ambient eclogite facies *P–T* conditions in the lower crust and upper mantle are provided in Chapters 12 and 13, together with consideration of their chemistry and origins.

2.4 Pelitic rock compositions

2.4.1 *General considerations and petrogenetic grid*

The mineral evolution of rocks associated with eclogites has been a long-standing problem, because of the question of cofacial development. Late metamorphic events have often obliterated most of the prograde assemblages in pelitic and quartzo-feldspathic compositions, leaving only some relicts in metabasites. This differential overprinting—which reflects a contrasting behaviour of the lithologies in their hydration/dehydration reactions upon decompression (Heinrich, 1982)—may have led to the interpretation of tectonic interdigitation of spatially associated rocks, even when they share a common history. It may also be one of the reasons why pelitic lithologies were thought to be 'uncritical' under eclogite facies conditions and at HP in general. However, in particular geodynamic settings (namely in HP, relatively LT, coherent metamorphic terrains) the counterplay of tectonic uplift and relaxation of the depressed isotherms has favoured the preservation of HP assemblages, thus revealing the cofacial evolution of the different rocks (*cf* Oberhänsli, Section 2.5). Much of our knowledge of the evolution of pelitic compositions under eclogite facies conditions has been gained from such terrains. Since post-metamorphic disruption by nappe tectonics is customary, a continuous prograde path is very seldom recorded, but it can be reconstructed using the different marker features preserved in various tectonic units.

Remarkable features in HP metapelitic rocks are: the relative abundance of chloritoid, often as a Mg-rich variety; the high Si content of the potassic white mica (phengitic substitution); the related scarcity of biotite; and the presence of talc, forming various assemblages with phengite, chloritoid, garnet, kyanite, deerite, etc. These unusual features have prompted a number of experimental studies (Chapter 3). The results obtained are used here to construct a petrogenetic grid for the multisystem K_2O–FeO–MgO–Al_2O_3–SiO_2–H_2O (KFMASH) at HP and intermediate T (Figure 2.8), considering pelitic protoliths as a virtual mixture of Qtz + Ill + Chl (spanning the whole range of Fe/Mg ratios) + variable amounts of Kln. Additional components like Na_2O or CaO will often result in the presence of an additional phase for each of them (such Pg, Gln or Jd, Lws or Zo), but will not alter the paragenetic succession outlined in the following account. Ca may also be accommodated as a Grs component in garnet or combine with Na in omphacite, resulting in less straightforward phase relationships, with many aspects in common with metabasites (*cf* Koons and Thompson, 1985; Schliestedt, 1980). Figure 2.8 gives the main subdivisions of the P–T field. There is a Tlc + Ky + Grt domain at relatively HT; at lower T with increasing pressure, there are Tlc + Phe, Tlc + Phe + Grt, and Tlc + Cld $\pm$ Phe domains at above about 11, 14 and 16 kbar P_{H_2O} respectively. The staurolite field is also of interest, restricted to relatively low P and intermediate T. Natural HP assemblages in metapelitic rocks will now be considered in this petrogenetic framework, starting from the low-grade side.

2.4.2 *From blueschist facies to eclogite facies*

The mineral evolution of metapelites in typical blueschist facies terrains is sensitive to small changes in the intensive parameters as demonstrated by the close succession of assemblages involving kaolinite, chlorite, Fe-Mg-carpholite, the di-trioctahedral

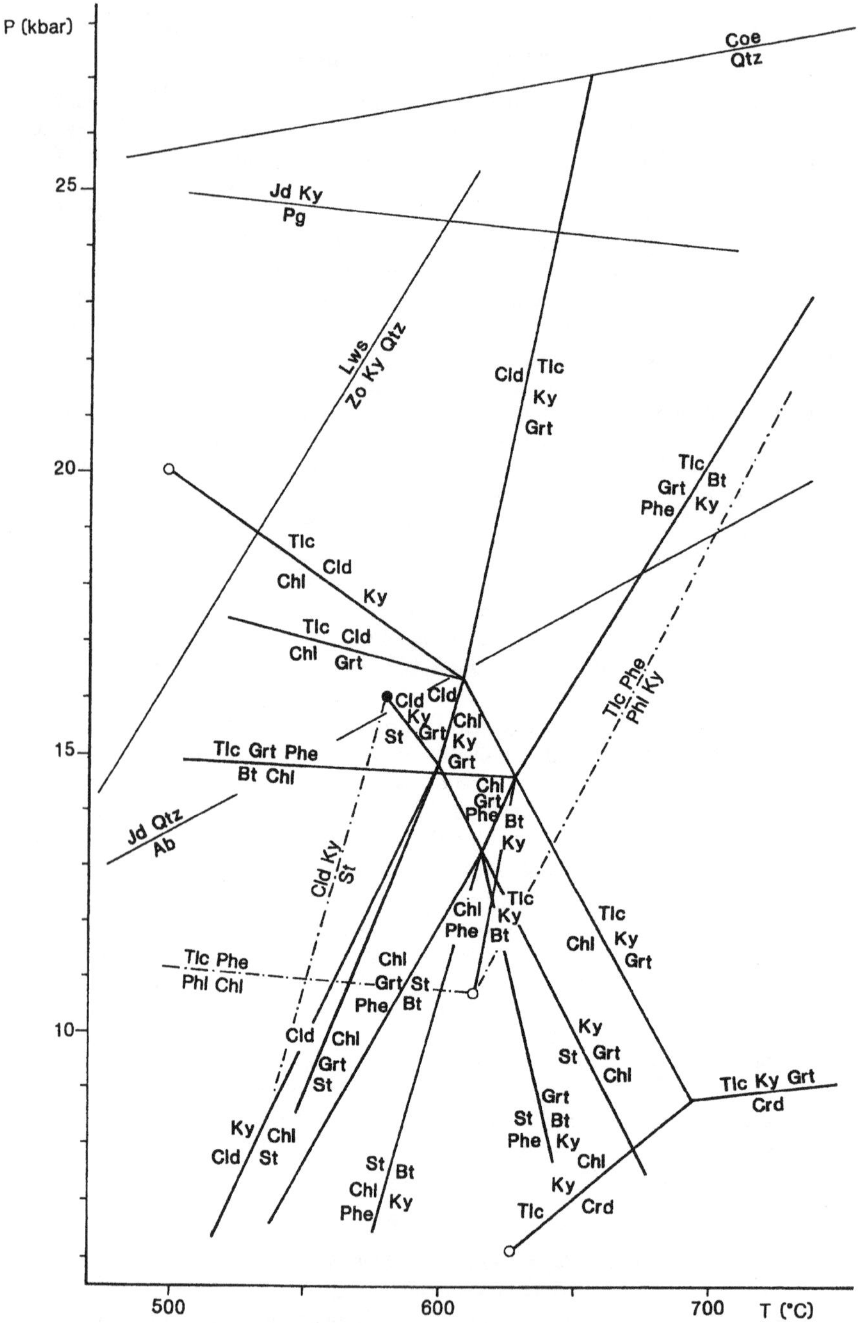
P (kbar)
25
20
15
10
500
600
700
T (°C)
Coe
Qtz
Jd Ky
Pg
Lws
Zo Ky Qtz
Cld
Tlc
Ky
Grt
Tlc
Bt
Grt
Phe
Ky
Tlc
Cld
Chl
Ky
Tlc
Cld
Chl
Grt
Cld
Cld
Ky
St
Grt
Chl
Ky
Grt
Tlc Grt Phe
Bt Chl
Jd Qtz
Ab
Cld Ky
St
Tlc Phe
Phl Ky
Chl
Grt
Phe
Bt
Ky
Tlc
Ky
Bt
Chl
Phe
Tlc
Chl
Ky
Grt
Tlc Phe
Phl Chl
Chl
Grt
St
Phe
Bt
Cld
Chl
Grt
St
Ky
St
Grt
Chl
Tlc Ky Grt
Crd
Grt
St
Bt
Phe
Ky
Chl
Ky
Chl
Cld
St
St
Bt
Chl
Ky
Phe
Ky
Tlc
Crd

chlorite sudoite, pyrophyllite, chloritoid and, at higher grades, Mg-rich chloritoid, garnet, kyanite and talc. Most characteristic at low grade is the progressive breakdown of Chl + Kln or Prl to form the complete series of Fe-Mg-carpholites; the subsequent progressive breakdown of Car to Cld + Qtz + H_2O, then produces the typical coexistence of an increasingly Mg-rich Car with Fe-rich Cld plus either Prl or Chl (+ Qtz + white micas). The most advanced stages of this evolution are represented in Peloponnese—with a Ca- and Mn-bearing garnet appearing beside chloritoid in Fe-rich pelites (Theye, 1988)—and on Samos, where magnesiocarpholite coexists with Mg-rich Cld, Chl and Ky (i.e. the Car-out reaction), and where Grt also occurs in more Fe-rich pelitic compositions (Okrusch, 1981; Okrusch *et al*, 1984). In neither case do eclogites occur in metabasites, but rather in Lws-bearing glaucophanites and Lws-free Grt-Ep-glaucophanites in Peloponnese and Samos respectively.

It would seem that carpholite disappears in the absence of eclogitic assemblages, and therefore Car-bearing metapelites could not belong to the eclogite facies. This conclusion is supported by the low-grade eclogite-bearing terrains such as northern New Caledonia—where Car occurs in the lawsonite zone (Briggs, 1978) but eclogite only in a distinctly higher-grade zone (Brothers, 1985)—and in Oman (Goffé *et al*, 1988). However, in Corsica, where some of the lowest-grade eclogites are known, Lws-bearing eclogite occurs in zone III with Cld and Ca- and Mn-bearing Alm in pelites (Caron and Péquignot, 1986), whereas Car occurs stably in zone I and as relicts and pseudomorphs in zones II and IV. This leaves open the question of a tectonic emplacement of zone III (Goffé, submitted).

Considering this fragmentary evidence it might be stated, somewhat arbitrarily, that the starting topology in LT eclogite facies metapelites does not involve carpholite, but rather chlorite, chloritoid, and incipient garnet and kyanite (*cf* Figure 2.9*a*). Formation conditions estimated, however, for the Corsican Lws-eclogite (420 ± 20°C, at ⩾ 10 kbar) lie just within the stability field of pure magnesiocarpholite (as outlined by Chopin and Schreyer, 1983). Further, they are quite close to those of the Prl ⇌ Ky + Qtz + H_2O reaction at $a_{H_2O} = 1$ so that, in a few instances, pyrophyllite may be the stable Al-silicate in pelitic country-rocks.

2.4.3 *Prograde evolution along a high dP/dT gradient*

An initial situation, with the two divarient assemblages Chl + Cld + Ky and Chl + Cld + Grt (+ Qtz + Phe + H_2O throughout, Figure 2.9*a*) is quite non-specific and may also be encountered in the greenschist and lowermost amphibolite facies. With increased metamorphic grade, this situation persists in what may be called high-grade

Figure 2.8 Petrogenetic grid for the K_2O-FeO-MgO-Al_2O_3-SiO_2-H_2O (KFMASH) multi-system involving (in order of decreasing X_{Mg}) the phases talc (Tlc), chlorite (Chl), biotite (Bt), chloritoid (Cld), staurolite (St), garnet (Grt) and muscovite/phengite (Phe), with excess SiO_2 and $a_{H_2O} = 1$. The construction is based on those of Harte and Hudson (1979) and Grew and Sandiford (1984) for the lower-P part. Implicit is that Tlc + St + Qtz cannot be stable and that Chl is more Mg-rich than in the Tlc-Cld join (*cf.* Vuichard and Ballèvre, 1988; Massonne and Schreyer, 1989). Experimental data are from Rao and Johannes (1979) for the Fe-subsystem; from Massonne *et al* (1981), Massonne (1989), Massonne and Schreyer (1989), and Chopin and Schreyer (1983, as modified by Chopin and Monié, 1984) for the Mg-subsystem. Open and filled circles denote invarient points in the limiting Mg and Fe systems, respectively; dash-dotted lines are reaction curves in these subsystems. For comparison, stability limits of lawsonite (Lws) after Chatterjee *et al.* (1984), of albite (Ab) and paragonite (Pg) after Holland (1979b) are also given.

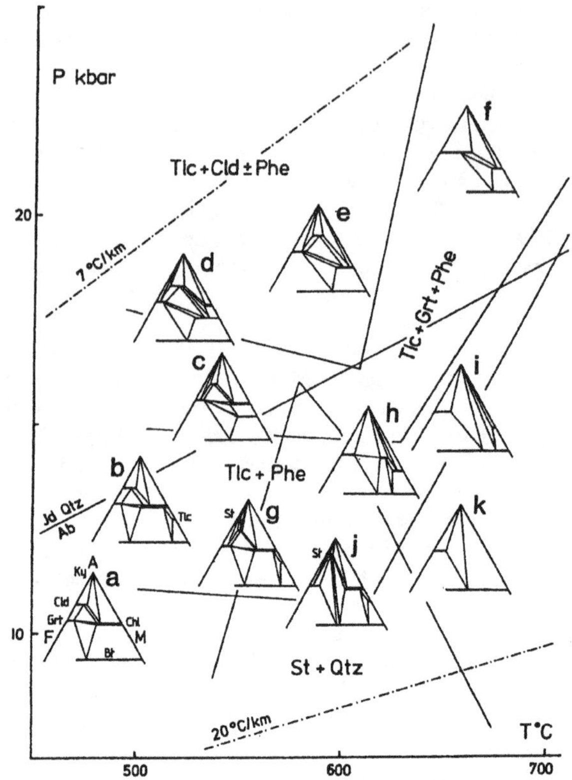

Figure 2.9 Phase relationships in the KFMASH system, projected through the 'muscovite', SiO_2, and H_2O components as a function of *P* and *T*, based on natural assemblages and Figure 2.8. The correlation between both data sets is discussed in the general text. Geothermal gradients assume a mean density of 2.7 g cm^{-3}; the 10°C/km gradient is nearly coincident with the Ab⇌Jd + Qtz reaction curve.

blueschists, i.e. terrains with no Lws, but with Gln-eclogites and Grt-Ep-glaucophanites as metabasites. The only indicators of HP are the relatively high Si contents of Phe (Massonne and Schreyer, 1987), the presence of Rt rather than Ilm besides Alm-rich Grt (Bohlen *et al*, 1983), and the presence of deerite in the most Fe-rich and somewhat oxidised pelites (Lattard and Schreyer, 1981). In our simplified system, the only unusual feature is the relatively high Mg-content of chloritoid, typically between 25 and 40 mol% (Kiénast and Triboulet, 1972; Ohta, 1979; Okrusch *et al*, 1984; Forbes *et al*, 1984; Ghent *et al*, 1987). This content is controlled by the divariant reaction

$$\text{Cld} + \text{Qtz} \rightleftharpoons \text{Chl} + \text{Ky} + H_2O$$

for which Cld lies on the LT, HP side (see three-phase field in Figure 2.9*a* to *d* and *g*). Increase in *T* will result in a decrease in the Mg-content of Cld; increase in *P* will result in increased Mg-content. In the type of dP/dT gradient envisaged in this section, the latter effect is dominant. The shift of the second divariant assemblage Chl + Cld + Grt is controlled by the reaction

$$\text{Chl} + \text{Cld} + \text{Qtz} \rightleftharpoons \text{Grt} + H_2O$$

which favours the increase in pyrope component in garnet upon increase in *P* and *T*, as documented in New Caledonia (Ghent *et al*, 1987).

With increasing pressure, the next step will be the appearance of the Tlc + Phe assemblage (Figure 2.9*b*; see limiting reaction in Figure 2.8). This will first be restricted to highly magnesian and relatively Al-poor compositions as, for example, schists linked to highly oxidised Fe-Mn deposits (Abraham and Schreyer, 1976; Chopin, 1981); the divariant field Chl + Tlc + Bt will then move towards more Fe-rich compositions, allowing the coexistence of a more Fe-rich Tlc with Phe, until the Chl + Bt pair gives way to Tlc + Grt (+ Phe; Figure 2.9*b–c* and see univariant reaction in Figure 2.8). At these stages, the assemblages Cld + Grt + Chl is still stable—as in northern New Caledonia at 500–550°C and 12–13 kbar (Ghent *et al*, 1987) or in the Sesia Zone of the western Alps at 550°C and 15–18 kbar (Vuichard and Ballèvre, 1988)—and the Tlc + Phe ± Grt assemblage is still restricted to relatively Al-poor compositions. The pair Alm + Ky may become stable through incipient breakdown of Fe-rich Cld, an essentially T-dependent reaction expected to occur above about 580°C and 16 kbar in the Ca- and Mn-free system. This has been achieved in the Sesia Zone (≡ Figure 2.9*c*) but not quite in New Caledonia (≡ Figure 2.9*b*).

The important next step, known from the Penninic western Alps (Chopin, 1981) and the Tien Shan region (Korikovskiy *et al*, 1983), is reached with the appearance of the Tlc + Cld assemblage through breakdown of the Chl-Grt tie-line (Figure 2.9*c* *cf. d*). De may still be present, coexisting with Tlc and Grt (Chopin, 1985). Tlc + Ky is not yet stable: the range of stable chlorite compositions is progressively restricted from the Fe-rich side by an essentially *P*-dependent reaction (Chopin and Schreyer, 1983) but extends to the pure Mg-side (Figure 2.9*a* to *d*). Chloritoid composition is further restricted on the Fe-rich side by the same Grt + Ky forming reaction as identified above. Grt with Ky is known in the Gran Paradiso massif (Dal Piaz and Lombardo, 1986; Figure 2.9*d*), but is likely to be Ca- and/or Mn-rich because of the low temperature. Chloritoid composition extends progressively through a *P*-dependent reaction towards the Mg-side: from 39 mol% of the Mg-end-member in Tien Shan with Tlc + Grt, to reach about 65 mol% in Gran Paradiso in the Chl + Ky + Cld assemblage (Pognante, 1984). There, the most Mg-rich Grt coexisting with Tlc and Cld contains 20 mol% Prp.

A further step is reached with the incipient breakdown of Mg-Chl to Tlc + Ky, which then restricts the range of chlorite compositions on the Mg-rich side as well, a process favoured by both *P* and *T* increase (Massonne *et al*, 1981; Massonne, 1989); this stage is presently being documented in an eclogite-bearing, coesite-free unit of the Dora-Maira Massif, western Alps (Chopin, *present research*). In the Monte Rosa Massif, Chl has started to form the characteristic assemblage Tlc + Ky + Mg-Cld (560°C, 18 kbar for $a_{H_2O} = 1$; Chopin and Monié, 1984). In the associated metabasic eclogites, it is typical for kyanite to be absent but for paragonite to be common. In contrast, kyanite eclogite is characteristic in the Hohe Tauern, where Tc + Ky + Mg-Cld occurs without Chl and represents a slightly higher grade (600°C, 19 kbar; Figure 2.9*e*) than Monte Rosa (Miller, 1977; Holland, 1979; Figure 2.10*e*). There, as well as in other kyanite eclogite-bearing terrains (Ecuador: Feininger, 1980).

Alm + Ky + Cld is the typical assemblage of Fe-rich pelites (Spear and Franz, 1986), but St is not present unless stabilised by considerable Zn content (Leupolt and Franz, 1986). The low-variance assemblage St + Ky + Cld + Grt reported by Brown and Forbes (1986; 600 ± 25°C, 15 ± 2 kbar) in metapelites associated with Pg-bearing, Ky-free eclogites corresponds to somewhat lower *P* and d*P*/d*T* gradient than shown in Figure 2.9.

The completion of this evolution is the disappearance of chloritoid (Mg-rich) through the two *T*-dependent reactions to form first the assemblage Tlc + Ky + Grt, which then moves towards the Mg-rich side upon continuous *P* or *T* increase. As a result, Ky + Alm–Prp–garnet (+ Phe) becomes the typical metapelitic assemblage for a broad range of Fe/Mg ratios. Possible examples might be the Münchberger Gneissmasse, Bohemian Massif, or the Champtoceaux nappe, Armorican Massif (650–700°C, 18–20 kbar; Ballèvre *et al*, 1987), even though sufficiently magnesian compositions to express Tlc + Grt + Ky were not found in these regions.

In a tectonic unit of the southern Dora-Maira Massif, the conditions reached were such that coesite became the stable silica polymorph (*c.* 30 kbar, 750°C; Chopin, 1984, 1987). The Tlc + Ky + Grt divariant field has moved completely to the Mg-side (garnet contains 96 mol% Prp) with continued *T* and *P* increase so that Ky + Grt (+ Phe + Coe ± Jd) is the assemblage in every common pelitic composition, whatever its Fe/Mg ratio. This implies incompatibility of chloritoid or staurolite with quartz or coesite. In Al-poor compositions, most of the Fe/Mg range is now covered by the undiagnostic Bt + Grt pair, through the move of the Tlc + Grt + Bt field toward the Mg-side. This is essentially a *T*-dependent process (Chopin, 1984), which makes the occurrence of the Tlc + Phe pair rare, on bulk-compositional grounds. Biotite + Prp–Grs–garnet also forms the highest grade 'pelitic' assemblage so far reported, in which metamorphic diamond is present (northern Kazakhstan: Sobolev and Shatskii, 1987).

2.4.4 *Extension to the whole eclogite facies field*

Considering this evolution, the following features are characteristic under high d*P*/d*T* gradients:

i) the Mg-content of chloritoid in otherwise unconspicuous assemblages at relatively low grade;
ii) at higher grade, the appearance of the unusual Tlc + Phe ± Grt ± Cld assemblages;
iii) at still higher grades, the rather inconspicuous Grt + Ky + Phe *without Bt*, or Grt + Bt + Phe assemblages in which, unless coesite or diamond are found, only the possible high Prp-content of garnet would reveal unusual HP conditions.

The grid in Figure 2.8 has been shown to be applicable to natural assemblages in pelitic compositions for the highest d*P*/d*T* gradients (*b–f* in Figure 2.9). Its use may also be extended to somewhat lower gradients (*g–i* and *j–k* in Figure 2.9) to characterise pelite mineral parageneses for the entire eclogite facies *P–T* field. The following features should be noted:

i) In Fe-rich rocks, staurolite or St + Bt indicate both *T* and d*P*/d*T* gradients (Figures 2.8, 2.9).
ii) At greater than 11 kbar, Tlc + Phe should be a common feature of eclogite

facies metapelites. This pair may coexist with Alm–Prp–garnet only above 15 kbar in this lithology, but might coexist with Ca–bearing garnet in metabasic eclogites at lower *P*, as in Norway (Bryhni and Griffin, 1971) and Togo (Ménot and Seddoh, 1985). In all cases, Tlc + Phe remains a HP, relatively LT assemblage (Massonne and Schreyer, 1985, 1989).

iii) The 'whiteschist' assemblage Tlc + Ky (Schreyer, 1977) has a considerable stability field extending over almost 300°C from about 6 to at least 40 kbar (Schreyer, 1988); low- and high-*P* whiteschists should therefore be distinguished (Chopin and Schreyer, 1983). The additional presence of phengite, or of chloritoid in particular, points to HP–LT conditions. The assemblage Tlc + Ky + Grt (Råheim and Green, 1974c; Udovkina *et al*, 1978; Chopin, 1984) also has a considerable stability field, from about 9 kbar to within the coesite field (Figure 2.8). Only in the additional presence of phengite, i.e. in more truly pelitic compositions, is it constrained to HP and a relatively narrow *T* interval (Figures 2.8, 2.9), in which the Prp content of garnet is a measure of pressure.

iv) The Bt + Ky assemblage has a broad stability range extending to relatively LP and HT (Figure 2.8). It is bounded towards higher *P* and lower *T* by the reaction

$$\text{Tlc} + \text{Grt} + \text{Phe} = \text{Ky} + \text{Bt} + H_2O,$$

which seems to be a very important thermal marker. Indeed, in eclogitic metapelites, in which the fundamental assemblage is Grt + Ky + Phe, the additional presence of biotite coexisting with kyanite (for the same bulk composition) will be the only readily detectable feature that will allow crude *T* distinctions. It may also be quite useful for constraining uplift paths. Furthermore, the divariant assemblage Bt + Phe + Grt + Ky offers interesting thermobarometric possibilities in the large Bt + Ky field: thermometry through Fe-Mg partitioning among the phases, and barometry related to the Prp content of garnet and the extent of phengitic substitution in white mica.

v) Towards higher *T* and/or lower *P*, white mica would disappear and bring in phases like Kfs and then possibly melt, sapphirine or orthopyroxene; thus, the distinction between eclogite and HP granulite facies assemblages is less clear in metapelites than in metabasites. In this context, pseudomorphs of Bt + Kfs ± Ky ± Pl after Phe are often the only clues that HP conditions once prevailed in metapelitic or quartzo-feldspathic country rocks enclosing eclogite bodies (Heinrich, 1982; Griffin, 1987).

vi) Quartz-free, yet pelite-related rock systems may be useful additional indicators under relatively high *T* conditions. They may be either highly aluminous protoliths or very Chl-rich rocks in which the small amount of quartz initially present was consumed during prograde metamorphism. In particular the considerable Mg incorporation in staurolite may help in identifying the transition from eclogite facies ($X_{Mg} = 0.95$ in St inclusions in Dora Maira pyrope megacrysts, 0.74 in a Grt + Cor rock of eastern China; Enami and Zang, 1988) to granulite facies rocks ($X_{Mg} \approx 0.5$; Schreyer *et al*, 1984). In such Qtz- or Coe-free systems, the presence of the newly discovered silicate ellenbergerite (Chopin *et al*, 1985; Chopin, 1986) will reveal unusually low *T* conditions at mantle depths, such as $<770°C$ at $P>27$ k bar (Chopin *et al*, 1988 and unpubl.; Schreyer, 1988a).

2.5 Metagranitoids and other acid-intermediate rock compositions

Many eclogite bodies occur within acid country rocks. For a long time, these bodies have been considered to be tectonically emplaced fragments whose metamorphic grade is unrelated to that of the country rocks. However, more and more evidence has been found that granitic country rocks in the Sesia-Lanzo Zone (western Alps) have undergone HP metamorphism. Further, basement gneisses of NW Scotland, the Schwarzwald in Germany, the Norwegian Western Gneiss Region, and some Alaskan gneisses, as well as acid rocks of New Caledonia, Japan, and the USSR (Sobolev *et al*, 1986) have been reported to reveal relict mineral assemblages (Table 2.3) and textures of a HP metamorphic regime.

In deformed intermediate acid gneisses at Glenelg, Scotland (Sanders, 1982) and in the Schwarzwald (Wimmenauer and Stenger, 1987), the coexistence of Grt, Ky, 3T–Phe and sometimes Omp is evidence of a HP event. Often relictic corona reaction textures can be observed. Norwegian mangerites from the Western Gneiss Region containing Grt ± Cpx + Pl + Kfs + Qtz ± Ky reveal *P–T* conditions identical to those of the enclosed eclogites (Krogh, 1977). Anatectic backveins of gneiss in the eclogites again show the HP assemblages (Griffin and Carswell, 1985) Grt + Ky + Pl + Qtz ± Kfs ± Omp (Griffin and Carswell, 1985) or Grt + Omp + Phe ± Qtz (Krogh, 1980 a,b; Cuthbert *et al*, 1983).

In the Seward peninsula, Alaska, granitic orthogneisses underwent blueschist facies metamorphism. The stable paragenesis reported is Qtz + Ab + Mc + Bt + Phe.

In rocks of tonalitic composition, Alm-rich garnet occurs (Evans and Patrick, 1987). High *P* is indicated by the Si content of the phengite mica. In acid metavolcanic rocks of New Caledonia, a HP overprint is recorded by the assemblage

$$\mathrm{Qtz + Jd_{87-98} + Kfs + Phe \pm Alm\text{-}rich\,Grt \pm Fe\text{–}Gln/Crs \pm Aln \pm Ttn + Ab + Ep}$$

in the high grade epidote zone, and by the assemblage

$$\mathrm{Qtz + Omp\ or\ Agt\text{–}Jd\ and/or\ Clm + Or + Phe \pm Fe\text{–}Gln/Crs \pm Stp + Ab + Ep}$$

in the omphacite zone (Yokoyama *et al*, 1986). In the Kanto Mts. of the Sanbagawa metamorphic belt (Hirajima, 1983) as well as in the Kumuikotan metamorphic rocks of Hokkaido (Takayama, 1986), plagiogranites associated with ophiolites contain the HP assemblage Jd + Qtz together with Ab, Agt and Gln.

The ACM projection space (Figure 2.10*a*) shows that different acid to intermediate rock compositions plot within the stability fields of HP phases such as Ky, Phe, Grt, Omp and Jd. In such rocks, the HP breakdown of Pl produces Jd + Zo + Qtz. Mafic phases react under HP to form Omp, Grt and/or Gln, and in Al-rich compositions Ky may form.

The Sesia-Lanzo Zone of the western Alps—an exceptionally well-preserved HP metamorphic area—exhibits granitic bodies which even after their complex tectono-metamorphic history, still preserve primary textures and contact phenomena with respect to their country rocks (Callegari *et al*, 1976). Among these intrusive bodies, the Monte Mucrone granodiorite is particularly notable for its well preserved primary features, transformations from static to dynamic HP recrystallisation and later

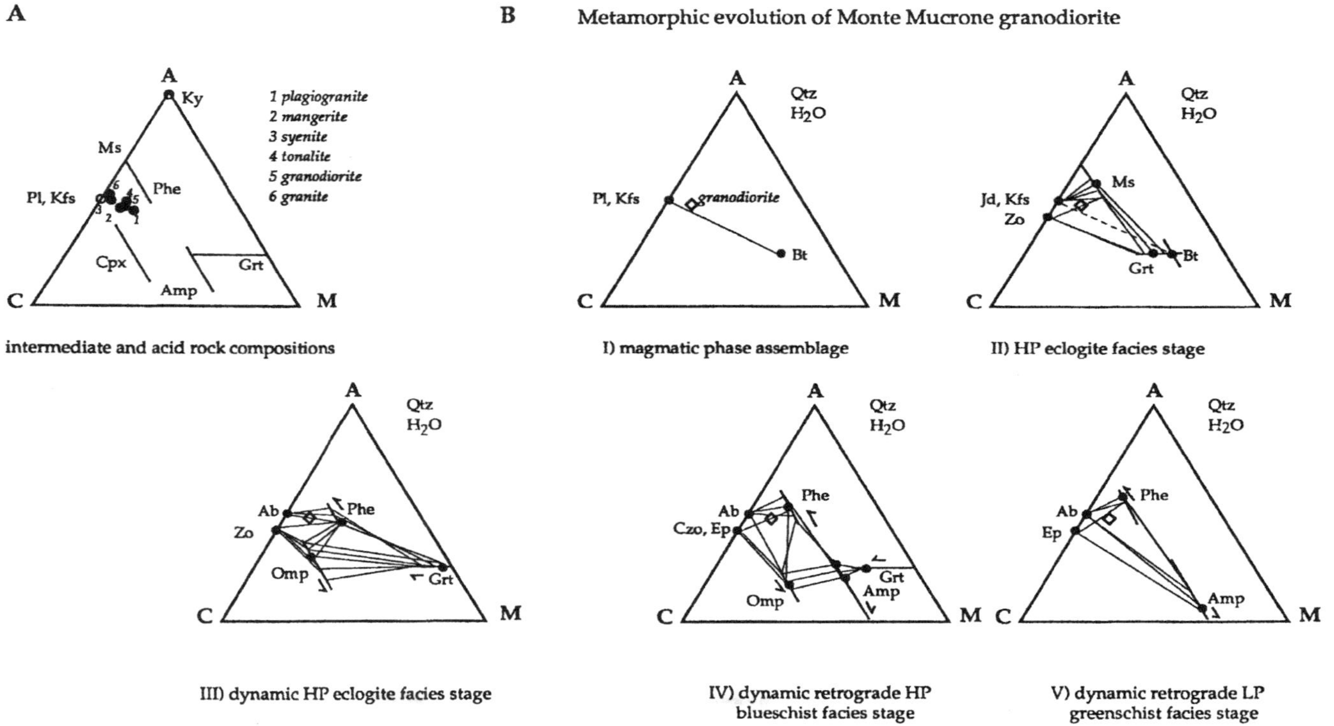

Figure 2.10 ACM diagrams (after Thompson, 1981, Figure 28) illustrating the HP phases of acid and intermediate igneous rock compositions in condensed CMASH composition space, with projection from Qtz and H_2O and condensation along Na Si Ca_{-1} Al_{-1}, Fe^{2+} Mg_{-1} and Fe^{3+} Al_{-1} exchange vectors. Different average rock compositions are plotted along with the compositions of the most significant HP phases, and their range of possible variations (*A*). The evolution of the Monte Mucrone granodiorite (*B*), from its magmatic assemblage (I) to its three HP metamorphic stages: (II) static HP recrystallisation with formation of Jd; (III) dynamic HP recrystallisation with Omp; and (IV) dynamic HP recrystallisation with Gln. The final diagram (V) shows the assemblages derived from the final greenschist facies overprint. Note that other granitoid rock compositions would allow for a similar HP evolution in mineral parageneses (*A*).

Table 2.3 Compilation of HP mineral assemblages in different granitoid compositions. Within one rock type and locality, HP metamorphic conditions decrease downwards for numbered assemblages I–IV, as indicated by the Jd contents of Omp and Si^{4+} contents of Phe.

Rock type	Region	HP assemblages	X_{Jd}^{Omp}	Si_{Phe}^{4+}	P/T range (k bar/°C)
Plagiogranite	Kanto Mts. Sanbagawabelt, Shikoku	I) Jd + Qtz + Gln	0.90		
		II) Ab + Agt + Wnc-Act	0.19		
	Kamuikotan, Hokkaido, Japan	Jd + Qtz + Gln + Lws ± Ms ± Ttn	0.90		9–11/120–300
Tonalite	Kristiansund, Norway	Pl + Qtz + Grt + Cpx + Fsp + Bt + Ep + Rt ± Phe		6.46	18.5 ± 3/750 ± 50
	Førde, Sunnfjord, Norway	Bt + Ksp + Mag + Qtz + Fsp + Grt + Pl + Cpx + Mag + Qtz			15.0 ± 2.5/630 ± 35
		Grt + Acm + Pl + Mag			12.5 ± 2.5/540 ± 35
		Grt + Phe + Bt + Qtz			
	Seward Peninsula, Alaska	Qtz + Ab + Phe + Bt ± Grt		7.2–6.4	
Mangerite	Flatraket, Stadlandet, Maløy, Norway	Grt ± Cpx + Pl + Kfs + Qtz ± Ky			
Intermediate gneiss	Glenelg, Scotland	Qtz + Pl + Grt + Omp + Ky + Rt ± Kfs ± Bt	0.45		14.0 ± 2/700 ± 50

Trondhjemite	Black Forest	Omp sympl., Grt coronas			
Granodiorite	Mte. Mucrone Sesia-Lanzo-Zone W Italian Alps	I) Qtz + Jd + Zo	0.93		14/500
		II) Omp + Qtz + Grt + Phe + Czo	0.68	6.8	11/600
		III) Ab + Omp/Clm + Gln + Czo + Phe + Qtz	0.56	6.6	
		IV) Ab + Clm/Agt + Gln/Hbl + Czo + Phe + Qtz		6.5	
	Seward Peninsula, Alaska	Qtz + Ab + Kfs + Phe + Bt ± Grt		7.2–6.4	
Granite	Seward Peninsula, Alaska	Qtz + Ab + Kfs + Phe + Bt ± Grt		7.2–6.4	
Syenite	Tafjord, Norway	Sympl. Cpx, Grt coronas			>9/700–800
Anatectic back veins in eclogites	Sunnfjord, Norway	Grt + Omp/Di-Pl sympl. + Rt + Qtz			
		Omp + Qtz + Rt + Ap			
	Midøy, Norway	Omp + Phe + Qtz			
		Grt + Ky + Pl + Qtz ± Kfs ± Omp			
Acid volcanics	New Caledonia	I) Jd + Qtz + Kfs + Phe + Ab + Ep ± Alm ± Fe-Gln/Crs ± Aln ± Ttn	0.87–0.98		9.5/390
		II) Omp/Acm-Jd + Qtz + Ab + Ep	0.30–0.40		11/430
		and/or Clm + Kfs + Phe + Ab + Ep ± Fe-Gln/Crs ± Stp			10.5/550
	Sinfos, Greece	Jd + Qtz + Gln + Pg + Grt + Ep ± Cld + Phe + Rt	0.85–0.67	6.8–6.4	

dynamic greenschist facies overprint (Compagnoni and Maffeo, 1973; Oberhänsli *et al*, 1982, 1985; Rubie, 1984). The sequence of mineral assemblages stable in the tectonometamorphic evolution of this original granodiorite body is shown in a series of composition-paragenesis diagrams (Figure 2.10*b*). The various evolutionary stages correspond to the different rock textures (cf. Figures 5.1, 5.6 and 5.8) which are juxtaposed not only on an outcrop scale but even on a length-scale of a few decimetres. The least-transformed rocks have well-preserved relict granitic textures. With the exception of Pl, they contain all the relict minerals of the protolith: Bt, Kfs, Qtz, Ap, Zrn and opaques. Pl is pseudomorphed by aggregates of Jd + Qtz + Zo. Biotites are rimmed by coronae of strongly zoned Grt, and the reaction zone towards the pseudomorphs after Pl contains the sequence

$$\text{Bt–white mica I (Ms)–Grt } (\text{Alm}_{70-60}\text{Grs}_{14-30}\text{Prp}_{16-10}\text{–}$$
$$\text{white mica II (Phe)} + \text{Qtz–Jd} + \text{Qtz} + \text{Zo}.$$

Other structurally preserved metagranitoids have completely recrystallised to the HP assemblage

$$\text{Qtz} + \text{Omp} + \text{Grt} + \text{Phe} + \text{Zo} + \text{Czo} + \text{Ttn} + \text{Rt}.$$

These rocks have evolved (see Figure 5.8) directly or gradually into gneisses in which Omp is replaced by aggregates consisting of Ab + Clm + Glm + Czo + Phe. Glaucophane is often rimmed by blue-green Hbl and occasionally by Act. With increasing deformation the rocks become schistose and have similar mineral associations but no textural relicts. In these rocks, Clm can be replaced by Agt. Further deformation leads to the formation of mylonites with white mica (Pg, Phe, and Ms), Ab, Qtz and Ep. These transformations are thought to reflect retrogression under lower confining pressures, as well as increasing deformation and subsequent influx of metamorphic fluids. In the Monte Mucrone area, the early Eo-Alpine (110 Ma) static HP recrystallisation took place at minimum $P \approx 14$ kbar and $T \approx 500$°C. It was followed by a second dynamometamorphic HP stage, which probably involved increased fluid pressure but lower total minimum P, 11 kbar, and higher T, up to 600°C. Final uplift and cooling are recorded by further deformation and retrograde recrystallisation, to blueschist and ultimately greenschist facies assemblages (Figure 2.10b).

2.6 Carbonate rock compositions

Eclogite-bearing terrains almost invariably include carbonate rocks of different types; yet the eclogite facies paragenesis of these is poorly known. It is significant that calcite, their essential constituent under present-day conditions, probably was not the stable $CaCO_3$ phase at the time of eclogite facies metamorphism (see Figure 1.1). However, the rapid inversion of aragonite to calcite at $T >$ c. 400°C (Brown *et al*, 1962) indicates that aragonite preservation is likely only in the very lower–T eclogite facies assemblages. This has not been documented to date, but it accords with the extensive partial preservation of aragonite in carbonate veins and lenses associated with Californian blueschists (Coleman and Lee, 1963). Although primary aragonite has generally not survived, there is some evidence from the peculiar microtextures now displayed by calcite and from its anomalously high Sr content, that aragonite was the stable eclogite facies $CaCO_3$ phase (Brown *et al*, 1978).

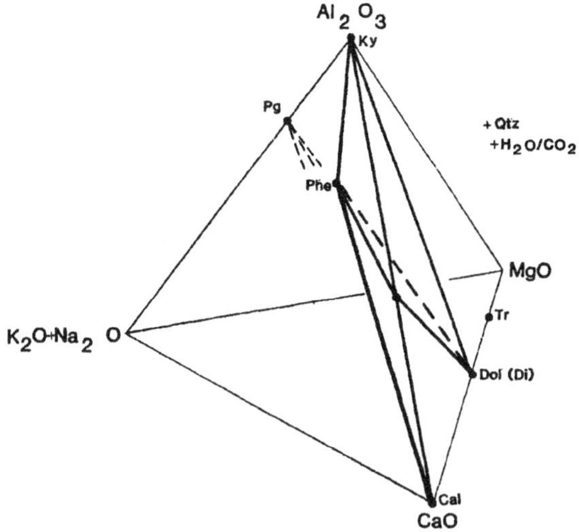

Figure 2.11 Phase relations of primary mineral assemblages in kyanite-zoisite marbles of the Eclogite Zone, Tauern Window, Austria, plotted in the system Al_2O_3-CaO-MgO-($Na_2O + K_2O$). Projection is from quartz at a fixed H_2O/CO_2 (redrawn from Spear and Franz, 1986).

A few parageneses in carbonate rocks under eclogite facies conditions have been studied where HP subduction metamorphism has developed under essentially static conditions and where the eclogite-bearing terrain has behaved as a coherent unit. This is the case in the Triassic-Jurassic stratigraphic succession of the Tauern Window, Austria (Miller, 1977; Franz and Spear, 1983; Spear and Franz, 1986; Dachs, 1986), where a carbonate sequence with different types and amounts of impurities underwent uniform conditions of eclogite facies metamorphism ($T \approx 600°C$, $P \approx 20$ kbar). The most significant well-preserved parageneses occur in siliceous dolomites, which have been transformed into marbles carrying the assemblages (Figure 2.11):

$$\text{Cal + Dol + Zo + Di + Tr + Qtz},$$

$$\text{Dol + Zo + Ky + Qtz} \pm \text{Phe},$$

$$\text{Cal + Dol + Zo + Qtz + Phe}.$$

The assemblage Cal + Qtz + Mg–Chl is also thought to have been present, although this could not be undoubtedly proven (Spear and Franz, 1986). The first assemblage given above is significant in that it is invariant under the high P and T conditions and constrains X_{CO_2} at < 0.02. Other impure marbles contain in addition a variety of silicate phases such as Omp (Jd_{30-35} Acm_{5-10}), Ep, Al-rich titanite (Franz and Spear, 1985) and Pg, as well as non-silicates such as Ank, Mgs, and Py (Table 2.4). The most aluminous layers may also contain pseudomorphs after lawsonite, in the form of prismatic inclusions now consisting of margarite + chlorite within the kyanite and/or zoisite porphyroblasts (Spear and Franz, 1986).

The assemblages found in the Tauern impure dolomitic marbles compare well with those of similar rocks in the Adula nappe prograde sequence (Dol + Zo + Tr + Jd-

Table 2.4 Type of protolith and (diagnostic) eclogite facies mineral parageneses in the carbonate sequence of the Eclogite Zone, Hohe Tauern, Austria

Protolith (age)	Eclogite facies mineral paragenesis
Siliceous limestone (Jurassic)	Qtz + Dol ± Cal/Phe, Czo–Ep (Ttn, Py)
Impure limestone (Jurassic)	Cal + Omp/Ep, Qtz, Hbl (Ttn, Py)
Marl (Jurassic)	Cal + Qtz + Phe ± Dol/Czo, Zo, Grt, Act ± Phl (Ttn, Ap, Rt, Tur, ± Mag)
	Ank + Qtz + Phe + Grt/Zo, Chl, Bi, Hbl (Ttn, Rt, Tur, Py)
Siliceous dolostone (Triassic)	Cal + Dol + Qtz/Phe, Pg, Chl, Phl (Ttn, Ap, Py, Tur)
	Cal + Dol + Tr ± Phe ± Qtz/Chl, Zo, Phl (Py, Tur)
	Cal + Dol + Tr + Tlc + Qtz/Zo (Py)
	Cal + Tr/Zo, Tlc, Phl (Ttn, Py, Fl)
	Cal + Dol + Qtz + Tr + Di/Tlc, Zo (Py, Ttn)
	Dol + Qtz + Ky + Zo ± Phe/Chl

Notes
Minerals are listed as: major phases/minor (accessory) phases.
Sources: Miller (1977), Franz and Spear (1983, 1985), Spear and Franz (1986)

poor Omp; Heinrich, 1986), as well as in the eclogite-bearing terrains of Sifnos (Matthews and Schliestedt, 1984) and the western Alps. Here, however, the most common assemblage is Cal ± Dol + Ank with scattered Omp (Jd_{50-55} Acm_{1-2}), garnet (Grs with Sps-rich cores), Phe, Gln, Zo, Cld, Ttn, Py and at times abundant graphite (Compagnoni, 1977; Pognante *et al*, 1987; Castelli, 1987). Furthermore, the most magnesian compositions contain Mg-Chl and Tr, and pseudomorphs after Lws have also been found in pelitic layers. The assemblage Cal + Dol + Grt + Phe ± Omp ± Tr is present also in the polymetamorphic dolostones of the southern Dora-Maira massif, associated with the well-known pyrope- and coesite-bearing quartzite (Chopin, 1984). It appears therefore, that either such an assemblage has a very broad field of stability, or the appropriate eclogite facies typomorphic assemblages were obliterated by retrogression during uplift.

Another bulk composition which is well represented in the Hohe Tauern sequence is carbonate-bearing quartzite to pelite. Through metamorphism, this has been transformed into Cal ± Dol + Qtz + Phe (with accessory Zo and/or Czo, Grt, Phl, Tr and rectangular pseudomorphs after Lws, now consisting of a felt of mostly Czo + Qtz + Pg + Phe + opaque minerals ± Cld) and acquires the conspicuous sheen typical of *schistes lustrés* (calcescisti, Kalkglimmerschiefer).

'Calcescisti' form the overwhelming majority of the Piemonte rocks subducted during the Europe-to-Africa collision. However, most studies carried out on them to date are hardly adequate for present purposes, because they mostly failed to distinguish parageneses formed during the eclogite facies stage from those of the later blueschist facies stage. Most calcescisti now contain Cal + Qtz + Phe ± Pg ± Alm-rich garnet ± Czo, and occasional pseudomorphs after Lws (Chatterjee, 1971; Pognante *et al*, 1987). Layers enriched in aluminium mostly exhibit the assemblage Qtz + Cal (± Ank) + Phe + Zo + Mg–Chl ± Pg ± Gln and, in exceptional cases, also Grt and Omp (Chiesa *et al*, 1972; Compagnoni, 1977). However, it has been proved that micas are almost invariaby recrystallised. Only first-generation phengite is typically eclogitic, characteristically Si-rich (up to 3.48 atoms pfu), and is the 3T polymorph (Liborio and Mottana, 1975; see also Lardeaux *et al*, 1985; Stöckert, 1985). Paragonite, although a primary phase in the enclosed eclogite pods, may form later in calcescisti. Garnet shows Alm-rich rims on Sps-rich cores, and even calcite (perhaps originally aragonite) is recrystallised with exsolution of dolomite and/or ankerite (Liborio and Mottana, 1973). Therefore little confidence can be placed in phase assemblage evaluations which would indicate $X_{CO_2} < 0.02$. The preservation of Grt + Omp ± Gln in the mafic layers interbedded in calcescisti indicates that the entire sequence underwent eclogite facies metamorphism, but the evidence given by the calcescisti themselves does not enable compilation of a full P–T–t loop for rocks of carbonate composition. Even calcescisti which contain Omp and Grt, and therefore are suitable for exchange reaction geothermobarometry, are usually recrystallised, as are those which contain two carbonates and are suitable for geothermometry using the carbonate solvus.

A peculiar carbonate-bearing rock consisting of ophiolite debris in a carbonatic matrix ('oficalce') occurs in association with eclogites and calcescisti within the Voltri Massif (Cortesogno *et al*, 1981). The carbonate-bearing assemblage consists of Cal + Dol ± Di ± Grt (Grs_{40-50} Adr_{35-45}) ± Tr (or, rarely, pargasitic Hbl) ± Mg–Chl; Mag is the ubiquitous opaque phase but Hem may also be present. The rare Ep is unusually Cr-rich. Phase analysis indicates X_{CO_2} to be in the range 0.01–0.03, as elsewhere in carbonate-bearing rocks.

In summary, the state of knowledge on eclogite facies parageneses in carbonate rock compositions is far from complete. For some of them (impure dolostones) there is the firm evidence given by the Tauern occurrence, from which it may be possible to deduce a prograde evolutionary *P–T* path (see Spear and Franz, 1986). In other cases, and particularly in mixed carbonate-pelite sequences such as the calcescisti, retrograde reactions are commonplace, possibly driven by fluids according to the model proposed by Heinrich (1982).

2.7 Metamorphosed Fe–Mn Rocks

Siliceous metalliferous muds enriched in either Fe or Mn (plus base metals and Ba) are minor but significant constituents of layer 1 of the oceanic crust. As such, they can undergo subduction and attain conditions of the eclogite facies. Unfortunately, although these rocks are conspicuous, they are rarely good indicators of the eclogite facies, since Fe- and Mn-rich minerals appear to be prone to reaction under the variable conditions of subduction metamorphism, particularly when changes occur in oxygen fugacity via infiltration of fluids. Preservation of eclogite facies assemblages is restricted to a few favourable occurrences, where a combination of rapid uplift and local armouring (for instance by being surrounded by ductile greenstones) protected Fe-Mn rocks against the pervasive effects of later metamorphic events and enabled them to escape re-equilibration at lower *T* and *P*.

Chemical reasons separate Mn from Fe at the discharge sites of their carrier solutions (Bonatti, 1980). Separation is then maintained both in their depositional environments, and later in metamorphism, giving rise to distinct occurrences of meta-ironstones and manganiferous metacherts. Even where the separation mechanism did not occur on a regional scale, distinct Fe-rich and Mn-rich layers can be traced in the same chert sequence. However, metamorphism may tend to counteract the sedimentary separation and to obliterate compositional layering via infiltration of metasomatising fluids. Abrupt changes in Fe and Mn contents, as well as in oxidation conditions, can always be detected across such sequences. Consequently, metamorphosed Fe-Mn rocks give information only on local oxidation conditions or on sulphur fugacity (particularly in meta-ironstones, which are normally enriched in sulphides), but this information can rarely be extrapolated to a regional scale.

Where oxygen fugacities were originally high, both rock compositions acquire rather uninteresting mineral assemblages at the very onset of metamorphism, under zeolite facies conditions: Qtz + Hem ($\pm$ Py $\pm$ Po) and Qtz + braunite. These assemblages ('oxidised', in the sense of Chopin, 1978, and Reinecke, 1983) would normally be maintained up to the eclogite facies, unless silicate-forming reactions take place primed essentially by the catalytic effects of circulating fluids. Where conditions were originally 'reduced', or well-balanced ('neutral', Mottana, 1986b), a prograde sequence of minerals develops which, although not yet entirely understood and therefore unsuitable to precise determination of P–T conditions, can at least be used to bracket them, as well as to fix P_{H_2O}, X_{CO_2}, f_{O_2}, and f_{S_2}.

Eclogite facies assemblages in *meta-ironstones* have been described from near Zermatt (Bearth and Schwander, 1981), and at Chuc-Servette, Créton-Les Vorpilles and Champ de Praz (Dal Piaz and Omenetto, 1978) in the western Alps; at Sifnos (Schliestedt, 1980; Evans, 1986) and Andros (Reinecke, 1983) in the Cycladic belt; and at Laytonville (Muir Wood, 1982) in the Franciscan belt (Figure 2.12). Observed minerals include as main phases aegirine-augite, Alm-or Adr-rich garnet, often with a substantial calderite component, deerite, epidote, crossite, riebeckite, cummingtonite, Na-actinolite, and ubiquitous quartz. Opaque minerals include, besides Hem, Mag, Py, and Rt, a large number of oxides and/or sulphides, often Cu-, Co-, Ni-, or V-rich, mostly intergrown and in small quantities. An interpretation of their actual significance has been attempted (Itaya *et al*, 1985), but is still practically unknown.

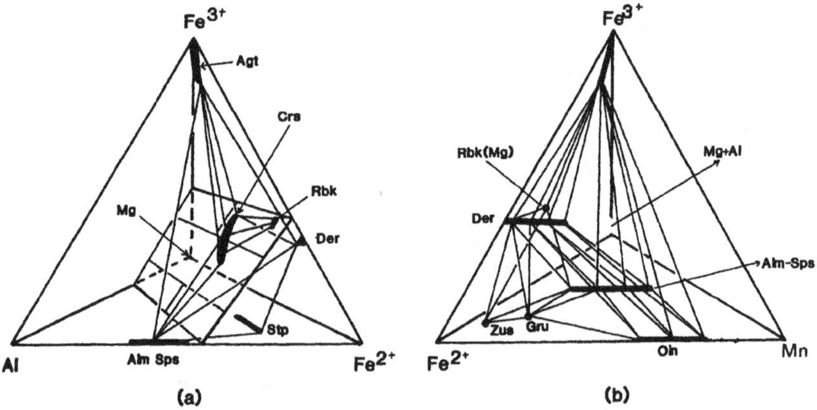

Figure 2.12 Essential mineral compatibilities in a meta-ironstone involved in subduction metamorphism, at the middle stage of its evolution when maximum P and T are reached: the relationships shown are for the Fe^{2+}-Fe^{3+}-Mg-Al (*a*) and Fe^{2+}-Fe^{3+}-(Mg + Al)-Mn (*b*) quaternary systems. The example is from the Laytonville Quarry, Franciscan belt, California, USA (redrawn and modified from Muir Wood, 1982).

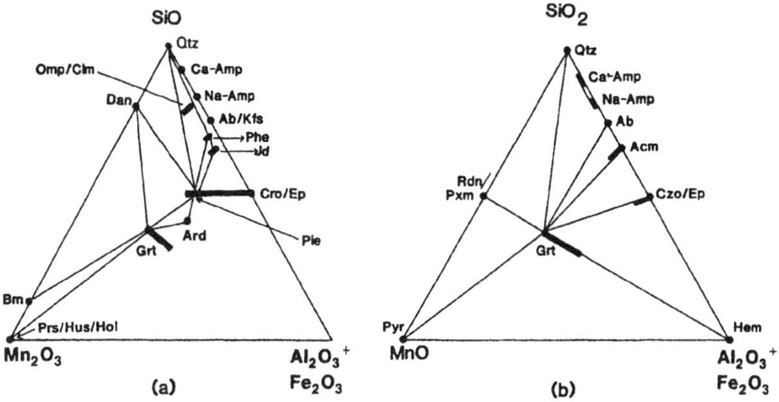

Figure 2.13 Essential mineral compatibilities in eclogite facies metamorphosed manganiferous cherts: under 'oxidising' conditions (*a*—pseudoternary system Mn_2O_3-($Al_2O_3+Fe_2O_3$)-SiO_2, and under 'neutral' and 'reducing' conditions (*b*—pseudoternary system MnO-($Al_2O_3+Fe_2O_3$)-SiO_2.

At the present state of knowledge, meta-ironstones do not give an indication of a prograde evolution along a HP–HT path. This does not mean that they do not exhibit it: it only means that studied occurrences are too scattered and phase relations not explicit enough to permit such a reconstruction at the present time. Attempts based on the very unusual simultaneous presence of four amphiboles, such as that of Evans (1986) at Sifnos, are promising, but, as he explicitly states, are still 'highly conjectural'.

Mineral assemblages in HP-metamorphosed manganiferous cherts (Figure 2.13) have been reviewed recently (Mottana, 1986b), on the basis of previous detailed studies carried out jointly with W.L. Griffin on the well-known locality St. Marcel in the western Alps and of studies by Chopin (1978), Dal Piaz *et al* (1979), Martin-Vernizzi (1982) and Reinecke (1983). This review, and new evidence by Perseil (1985), Mottana and Griffin (1986), and Martin and Kienast (1987), indicate at least three eclogite facies typomorphic parageneses depending mostly on the local redox conditions, but also on composition (availability of Ca, Na, Mg in the chert layer). These three types are described as:

(a) *oxidised assemblages.* These contain minerals where both Mn and Fe are predominantly in the trivalent state (Figure 2.13*a*): braunite, hausmannite, piemontite, ardennite, manganoan deerite, phengite ('alurgite'), riebeckite, as well as Cr- ('kosmochlor') or Mn^{3+}- ('violan') or Fe^{3+}-rich ('chloromelanite') clinopyroxenes intermediate between Di and Jd, or close to the latter end-member;
(b) *reduced assemblages.* These contain minerals where Mn is divalent and Fe mostly divalent and bound to carbonates (Figure 2.13*b*). They either have carbonates (kutnahorite, rhodochrosite) as main phases—associated with Sps-rich garnet, tephroite, friedelite, pyrophanite, tirodite, ilmenite, pyroxmangite (usually intergrown with rhodonite) and Ilm or Mag—or they are peraluminous and consist mostly of chlorites, sonolite, alleghanyite, jacobsite and galaxite. Reduced assemblages are characteristically Qtz-free;
(c) *neutral assemblages.* These contain minerals such as rhodonite, aegirine

clinopyroxene, dannemorite, winchite, calderite- or Sps-rich garnet, as well as minor quantities of other minerals of either the oxidised or the reduced assemblages. Thus they contain Mn and Fe in both states of oxidation and reflect the intrinsic f_{O_2} of the layer under the prevailing P–T conditions.

In these Mn-rocks, haematite plays a role as an f_{O_2} indicator. In highly oxidised assemblages, it could be expected to trend in composition towards Mn_2O_3 (bixbyite). However, in these natural assemblages (Mottana, 1986b; Reinicke, 1983) with rather lower f_{O_2} it appears to contain significant Mn^{2+} together with Ti, thus exhibiting solid solution towards $MnTiO_3$ (pyrophanite). In more reduced assemblages, with notable Ti enrichment, compositions trend towards ilmenite. Indeed for particularly Fe and Ti-rich bulk compositions, discrete ilmenite is present and this haematite composition redox indicator is no longer valid.

At the eclogite facies peak, under static conditions, an inflow of water-rich fluids has occurred almost everywhere in Fe–Mn rocks and has complicated the phase relations further. Interaction of fluids with the walls of the fractures deposited vein minerals such as deerite, zussmanite, and crossite in the meta-ironstones; and Mn-omphacite ('violan'), piemontite, àrdennite, Mn-titanite ('greenovite'), and néltnerite, in the manganiferous cherts. Often, the same fractures remained open during the subsequent retrograde stages of metamorphic evolution. This caused not only retrogression of the vein minerals, but also widespread alteration of the components of the groundmass equilibrated under eclogite facies conditions, thus producing symplectic intergrowths in clinopyroxenes, kelyphitic rims around garnets and widespread formation of clinozoisite.

3 Experimental studies on the stability of eclogite facies mineral parageneses

S.L. HARLEY and D.A. CARSWELL

3.1 Introduction

Numerous experimental studies concerned with the delineation of the $P-T$ conditions of eclogite facies mineral assemblages in various model and natural rock compositions have been conducted since the important reviews by Newton and Fyfe (1976) and Ringwood (1975) summarised the state of knowledge over a decade ago. In more recent times experimental studies of natural rock compositions have been neglected in favour of experiments on simple-system reactions and equilibria that involve a limited number of solid-solutions, which reflects the strong emphasis now placed on the extraction of thermodynamic data and development of realistic activity–composition models from experiments. Another major focus of recent experimentation at high pressures has been the elucidation of phase relationships in a number of simple systems (e.g. NASH, MASH, KMASH) which are more or less directly relevant to unusual but important rock compositions containing diagnostic assemblages.

This review and critique of recent experimental petrology relevant to eclogite facies metamorphism strongly reflects the emphasis on the determination of equilibria in simple systems. As a consequence, experimental constraints on eclogite facies assemblages in various important rock types (ultrabasics, basics, intermediate and acid rocks, pelites, and impure quartzites) are here discussed from a systems approach (e.g. NASH, CASH), with the relevance of such systems to the natural rock cases also noted.

3.2 P–T stability conditions for the eclogite facies garnet lherzolite assemblage in magnesian metaperidotites

Direct experimental and thermodynamically calculated data that most reliably indicate the $P-T$ conditions for stability of garnet lherzolite relative to spinel lherzolite are summarised in Figure 3.1.

In magnesian metaperidotites the up-pressure transition from a spinel-bearing to garnet-bearing assemblage can be modelled in the MAS system by the univariant reaction:

$$\underset{\text{En}}{2Mg_2Si_2O_6} + \underset{\text{Spl}}{MgAl_2O_4} \rightleftharpoons \underset{\text{Fo}}{Mg_2SiO_4} + \underset{\text{Prp}}{Mg_3Al_2Si_3O_{12}} \qquad \text{(R3.1)}$$

This equation oversimplifies the transition reaction because the enstatite is variably

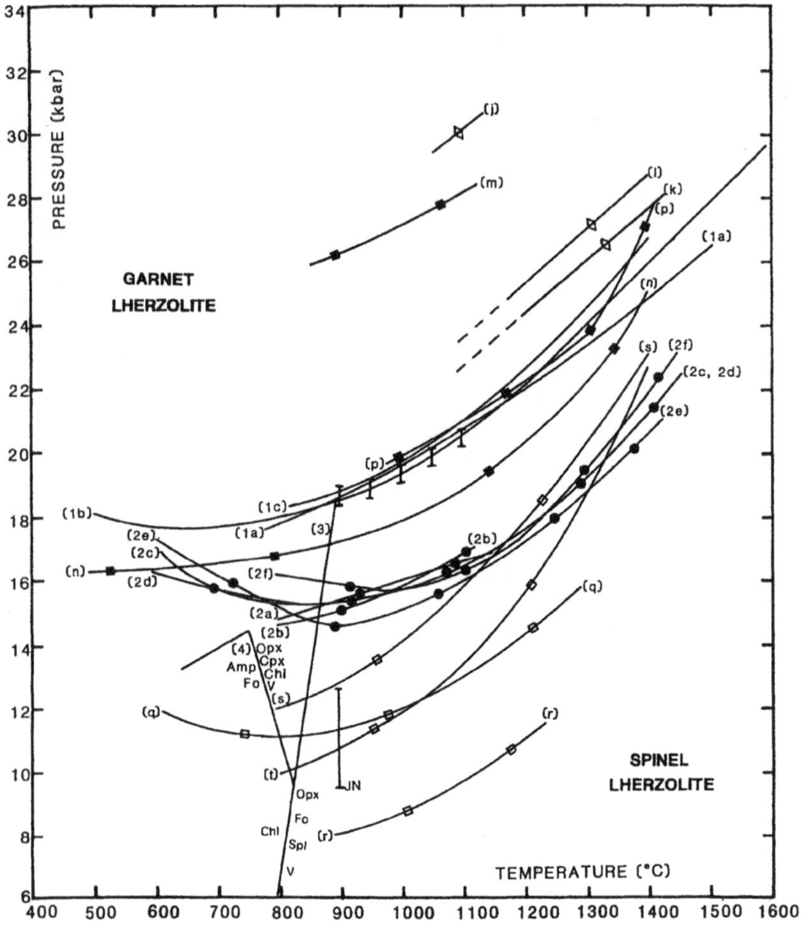

Figure 3.1 Experimental and calculated data relevant to the stability of garnet lherzolite relative to spinel lherzolite.
Lines (*1*) MAS system reaction R3.1. (1a) from Perkins *et al* (1981) (1b) after Gasparik (1984a); (1c) calculated result of Saxena and Eriksson (1983); I-beams are data of Danckwerth and Newton (1978); Lines (*2*), CMAS system reaction R3.2 (curves with circle ornament). (2a) after Jenkins and Newton (1979); (2b) after O'Neill (1981) (with −10% friction correction); (2c) after Gasparik (1984a); (2d) after Gasparik (1987); (2e) after Wood and Holloway (1984); (2f) after Saxena and Eriksson (1983). (*j*) in the SMACCR system (◇) with Cr* = 12 (Cr* of Spl = 48) (Webb and Wood, 1986). (*k*) in the SMACCR system (◇) with Cr* = 9.5, typical of many lherzolites (Nickel, 1986). (*l*) in the SMACCR system with Cr* = 20 (Nickel, 1986), (*m*) in the SMACCR system with Fe present (Mg* = 90) and with Cr* = 12 (Cr* of Spl = 48), based on O'Neill (1981). (*n*) in the natural system (■) with Mg* = 90 and Cr* = 7–10 (O'Hara *et al*, 1971. (*p*) in synthetic pyrolite III model mantle composition with Mg* = 89 and Cr* = 7.5 (Green and Ringwood, 1967b). (*q*), (*r*) in model CFMAS system (□) with Mg* (Opx) = 80 and 60 respectively (Gasparik, 1987). (*s*) in a model 'pyrolite' as calculated by Saxena and Eriksson (1983). Note this composition has a very low Cr*. (*t*) in a model lherzolite as calculated by Saxena and Eriksson (1983). I-beam JN shows the appearance of garnet in a model composition with Mg* = 84 and Cr* = 1 (Jenkins and Newton, 1979) Line (*3*): reaction R3.3 as determined by Jenkins (1981). Line (*4*): reaction R3.4 (Jenkins, 1981).

aluminous depending on P–T conditions. Reliable direct experimental data on R3.1 in the MAS system over a wide temperature range (900–1450°C) have been provided by Danckwerth and Newton (1978), Perkins *et al* (1981) and Gasparik and Newton (1984) (I-beams and curve la, Figure 3.1). Gasparik (1984a) has calculated a transition curve (curve lb) which is consistent with the data of Perkins *et al* (1981) but which shows greater curvature at lower (< 900°C) and higher (> 1400°C) temperatures. The calculated equilibrium curve given by Saxena and Eriksson (1983) shows less curvature at lower temperature but is reasonably consistent with available experimental data (curve lc).

Natural peridotite assemblages frequently contain an additional Ca bearing phase (diopsidic clinopyroxene) and are thus more closely modelled by experimental data for the CMAS system. The spinel lherzolite to garnet lherzolite reaction is complex (where both pyroxenes have P–T dependent Al contents and $CaMg_{-1}$ exchange occurs between pyroxenes) but may be written in simplified form as:

$$\underset{\text{En}}{Mg_2Si_2O_6} + \underset{\text{Di}}{CaMgSi_2O_6} + \underset{\text{Spl}}{MgAl_2O_4}$$

$$\rightleftharpoons \underbrace{\underset{\text{Prp}}{\tfrac{2}{3}Mg_3Al_2Si_3O_{12}} \cdot \underset{\text{Grs}}{\tfrac{1}{3}Ca_3Al_2Si_3O_{12}}}_{\text{in garnet ss.}} + \underset{\text{Fo}}{Mg_2SiO_4} \qquad \text{(R3.2)}$$

The direct experimental determinations of R3.2 in CMAS by Jenkins and Newton (1979), and O'Neill (1981) are closely compatible (curves 2a and 2b, respectively, in Figure 3.1), so long as a − 10% friction correction is applied to the quoted pressures of O'Neill (1981). The experimental determination in CMAS by O'Hara *et al* (1971) is probably satisfactory at highest temperatures but less certain at $T < 1300$°C, where a very marked curvature of the univariant reaction curve was indicated.

Thermodynamic evaluations of the low temperature extrapolation of the CMAS univariant reaction by Gasparik (1984a, 1987), and Wood and Holloway (1984), and Saxena and Eriksson (1983) (curves 2c, 2d, 2e, and 2f respectively, Figure 3.1) have all indicated extensive curvature leading to a negative dP/dT slope at $T < 900$–1000°C. The analysis of Saxena and Eriksson (1983) generates a negative dP/dT slope at near 1000°C, inconsistent with the direct experimental observations. Curvature in reaction R3.2 results from the changes in S and V associated with P–T dependent variations in the Al-content of pyroxenes and from ΔCp effects which are significant in such solid–solid reactions. The CMAS data show that the addition of Ca leads to a reduction in the pressure required for garnet stability relative to that necessary in the MAS system, by 3–4 kbar in the temperature range 800–1400°C and perhaps only 2 kbar at $T < 800$°C.

The experimental investigation by O'Neill (1981) has demonstrated that in the Cr-bearing SMACCR (CMAS + Cr_2O_3) system Cr is heavily partitioned into spinel relative to garnet, with the consequence that a divariant assemblage of garnet + spinel + orthopyroxene + clinopyroxene + olivine is stable between the stability fields of spinel lherzolite and garnet lherzolite. Moreover, the pressure interval over which the 5-phase assemblage is stable can be expected to widen with increasing 100 Cr/(Cr + Al) ratio (Cr*) in the rock (MacGregor, 1970; O'Neill, 1981; Webb and Wood, 1986). From his experimental data O'Neill (1981) derived the following linear empirical expression for calculation, in terms of the Cr content of the coexisting spinel ($X_{Cr}^{Spl} = Cr/Cr + Al$), of the pressure P required for garnet stability in a Cr-bearing lherzolite assemblage relative to that indicated P^o for garnet stability in the CMAS system

$$P_{\text{kbar}} = P^{o}_{\text{kbar}} + 27.9(X^{\text{Spl}}_{\text{Cr}}) \quad \text{(E3.1)}$$

within the range of P–T–X conditions ($T = 900$–$1100°C$, $P < 40\,\text{kbar}$, $X^{\text{Spl}}_{\text{Cr}} < 0.7$) covered by the experiments.

In a more detailed analysis of the dependence of spinel–garnet stability relationships in peridotite compositions on Cr content, Webb and Wood (1986) calculated the compositions of coexisting spinels and garnets expected to be stable with orthopyroxene + clinopyroxene + olivine in the system SMACCR at T of 1100°C and high pressures. Their results indicate that the 5-phase assemblage is stable over a wide pressure interval. However, these calculations ignore the fact that Cr is also significantly partitioned into the clinopyroxene. Given the experimental evidence of Green and Ringwood (1967b) that the spinel lherzolite to garnet lherzolite reaction in their model mantle pyrolite III composition—with Cr* = 7.5—occurred over only a small pressure interval (around 2 kbar), Webb and Wood (1986) concluded that the presence of a substantial $NaCrSi_2O_6$ component in natural clinopyroxene must buffer the Cr content of spinel. This result may also be important in the simple SMACCR system for clinopyroxene-rich compositions if the clinopyroxenes dissolve appreciable Cr as $CaCrAlSiO_6$ or similar substitutions. The incoming of garnet would be delayed in bulk compositions with high Ca/(Ca + Mg) relative to those with low Ca/(Ca + Mg), even at constant Cr* (bulk), because of the presence of these Cr solid-solutions. Such an effect and a relatively narrow pressure interval (2–3 kbar at 1200–1400°C) for the 5-phase assemblage has been confirmed in recent experimental data on the SMACCR system (Nickel, 1986) for a composition with Cr* = 9.5 (line k, Figure 3.1).

The SMACCR system experimental data of Nickel (1986) demonstrate that, contrary to the conclusions of O'Neill (1981) and of Webb and Wood (1986), the influence of Cr on the pressure required for garnet stability in lherzoliite assemblages is far from linear. His data indicate that the presence of minor amounts of Cr shifts the garnet-in boundary to appreciably higher pressures, but that at Cr* > 10 the pressure increase becomes quite small. This behaviour has been modelled in terms of regular and reciprocal solution effects in garnets and spinels which suggest strong non-ideality at low Cr*. However, the role of Cr in clinopyroxene is again important here, and any dependence of clinopyroxene–spinel Cr partitioning on bulk Cr* will strongly influence the incoming of garnet. The assumption by Webb and Wood (1986) that Cr* (spinel) = 4Cr* (bulk) is indicated to be invalid by the data of Nickel (1986).

The garnet-in reaction curves determined at 1200–1400°C by Nickel (1986) for SMACCR system compositions with Cr* = 9.5 and 20 are reproduced in Figure 3.1 as curves k and l, respectively. These lie at appreciably lower pressures than the corresponding reaction curve m calculated from equation E3.1, according to O'Neill (1981), for a typical upper mantle peridotite composition with Cr* = 12. This is despite the fact that the latter calculated curve involves a 2 kbar pressure reduction (cf. O'Neill, 1981) to take account of the presence of Fe^{2+} in natural upper mantle lherzolite with Mg*(= 100 Mg/(Mg + Fe^{2+})) ≈ 90 (compare with line j). The curves derived by Nickel (1986), however, are consistent with the natural system (pyrolite III) experiments of Green and Ringwood (1967b) when a similar correction for Fe is taken into account.

Experimental determination of garnet–spinel stability relationships in Fe-bearing peridotite compositions is hampered by difficulties in precisely controlling the oxygen fugacity in long-term experiments. Whilst increased Fe^{2+} content is predicted to decrease progressively the pressure required for garnet stability (O'Neill, 1981) the presence of Fe^{3+} will favour spinel stability (MacGregor, 1970) and thus have an opposing effect. The net result is likely to be a reduction in the pressure required for garnet stability of only 1–2 kbar relative to curve k of Nickel (1986) for the SMACCR system composition with Mg* = 100 and Cr* = 9.5. This is in reasonably close agreement with the independent experimental determinations of O'Hara *et al* (1971) for natural peridotite mineral mixtures (curve n) with Mg* = 90 and Cr* = 7–10 and of Green and Ringwood (1967b) for their synthetic, multi-component, pyrolitê III composition (curve p) with Mg* = 89 and Cr* = 7.5. Therefore together these three sets of experimental data constrain fairly closely the pressure required for garnet stability in typical natural magnesian peridotite compositions. The Mg* = 90, Cr*(bulk) = 12 curve calculated from E3.1 of O'Neill (1981) lies at pressures which are anomalously high and inconsistent with the natural rock experiments.

The experimental determination of the spinel-lherzolite to garnet-lherzolite transition by Jenkins and Newton (1979) (I beam JN, Figure 3.1) for a natural garnet–olivine mixture occurs at significantly lower pressures than obtained in other natural rock studies primarily because of the Fe-rich (Mg* = 84) and Cr-poor (Cr* = 1) characteristics of the bulk composition (i.e. essentially CFMAS). Recent independently calculated CFMAS models bearing on reaction R3.2 (Gasparik, 1987; Saxena and Eriksson, 1983), shown as curves q, r, s and t in Figure 3.1, are in close agreement with the data of Jenkins and Newton (1979) at appropriate bulk Mg*. The calculated curves of Saxena and Eriksson (1983) for 'pyrolite' should not be directly compared with the results of Green and Ringwood (1967b) because the pyrolite III composition studied by the latter workers has a much higher Cr* than the pyrolite as modelled by Saxena and Eriksson (1983). The model of Gasparik (1987) does not pretend to emulate any particular natural rock composition, but rather yields CFMAS curves which are consistent with the MAS, CMAS, and natural rock data.

Many experimental studies have been directed at the melting relations of hydrous peridotites and lherzolites under pressure conditions appropriate to the eclogite facies. Of interest here are the subsolidus assemblages developed at $P > 12$–15 kbar. In natural peridotites the most important result (e.g. Green, 1973) is that amphibole (pargasitic) is a major additional phase under subsolidus conditions up to 23–30 kbar or more. In simpler systems such as CMASH the situation has been shown by Jenkins (1981) to be slightly more complex because of the contraction of the amphibole stability field relative to the stability of Mg-chlorite. The stability of a chlorite that approximates Mg-clinochlore is governed by the MASH equilibrium:

$$\text{Chl} \rightleftharpoons \text{Opx} + \text{Fo} + \text{Spl} + H_2O \qquad \text{(R3.3)}$$

which has been reversed by Jenkins (1981) and previous workers in the P range 10–16 kbar (line 3, Figure 3.1). Equilibrium R3.3 intersects R3.1 in an invariant point inferred to occur at about 900°C at 18.5 kbar. Therefore below 900°C the spinel lherzolite model assemblage will be replaced by chlorite lherzolite for water saturated conditions. The breakdown of tremolitic amphibole in CMASH ultramafic analogues is given by:

$$\text{Amp} + \text{Fo} \rightleftharpoons \text{Opx} + \text{Cpx} + \text{Chl} + H_2O \qquad \text{(R3.4)}$$

(line 4, Figure 3.1), which is inferred on the basis of two reversals, near 10 and 14 kbar (Jenkins, 1981), to have a negative dP/dT slope. Reaction R3.4 occurs at lower temperatures (up to 120°C) than R3.3.

3.3 P–T stability conditions for eclogite facies assemblages in natural basalt and related rock types

The experimental study of the transformation of gabbro to eclogite presented by Green and Ringwood (1967a) is an important benchmark in petrology. It provides significant insights into the nature of the transition, the effects of variations in bulk rock chemistry on the incoming of garnet and the disappearance of plagioclase, and the relationships between eclogites and high-pressure granulites. Apart from defining broad fields of eclogite stability in a range of basaltic rock-types, this study identified and emphasised the importance of a spectrum of continuous reactions that involve pyroxenes, plagioclase, olivine, spinel and garnet which have since been extensively studied themselves in simpler systems. Ringwood (1975) has summarised the results of Green and Ringwood (1967a), Ito and Kennedy (1971), and other workers on the position and characteristics of eclogite-forming reactions in natural basalts and related rocks. Although further experiments have been performed since Ringwood's (1975) synthesis, particularly on more evolved and acid–intermediate rock-types (e.g. Wyllie, 1979; Huang and Wyllie, 1973) where kyanite and K-Na alkali feldspar may be important phases at high pressures, the major conclusions derived from the earlier work under anhydrous conditions remain the same. As discussed by Newton and Fyfe (1976) and in Chapter 1, high-pressure assemblages obtained in dry experiments will be replaced by amphibole-bearing assemblages at lower temperatures ($<$ 1000°C–800°C) in hydrous systems.

Rocks of broadly basaltic chemistry attain eclogite facies (garnet + pyroxene $\pm$ quartz) on the disappearance of plagioclase (plagioclase-out curves and points A2–K2, Figure 3.2) from the high-pressure granulite assemblage garnet + pyroxenes + plagioclase. The latter assemblage can be regarded as transitional between pyroxene granulite and eclogite, and is bounded at lower P by the garnet-in curves (A1–K1) of Figure 3.2. The incoming of garnet results from the simplified reaction

$$\underset{\text{En}}{2Mg_2Si_2O_6} + \underset{\text{An}}{CaAl_2Si_2O_8} \rightleftharpoons \underset{\text{Prp}}{Mg_3Al_2Si_3O_{12}} + \underset{\text{Di}}{CaMgSi_2O_6} + \underset{\text{Qtz}}{SiO_2} \qquad \text{(R3.5)}$$

together with continuous reactions involving tschermaks substitutions ($Al^{[4]}\,Al^{[6]}\,Si_{-1}\,Mg_{-1}$) in pyroxenes, in silica-oversaturated compositions. In silica-deficient rock types the simplified reaction:

$$\underset{\text{An}}{CaAl_2Si_2O_8} + \underset{\text{En}}{2Mg_2Si_2O_6} + \underset{\text{Spl}}{MgAl_2O_4} \rightleftharpoons \underset{\text{Grt}}{2(\tfrac{1}{6}Ca_3Al_2Si_3O_{12}\cdot\tfrac{5}{6}Mg_3Al_2Si_3O_{12})} \qquad \text{(R3.6)}$$

causes the incoming of garnet at lower pressures than R3.5, given that other chemical parameters (e.g. Mg*) are constant. Within the transitional garnet-granulite field, therefore, garnet increases in modal abundance whereas plagioclase decreases and becomes more sodic until the albite component begins to react with pyroxenes to generate omphacite. The outgoing of plagioclase is effectively controlled by reactions such as:

$$\underset{\text{Ab}}{NaAlSi_3O_8} + \underset{\text{Di}}{xCaMgSi_2O_6} \rightleftharpoons \underset{\text{Omp}}{xCaMgSi_2O_6\cdot NaAlSi_2O_6} + \underset{\text{Qtz}}{SiO_2} \qquad \text{(R3.7)}$$

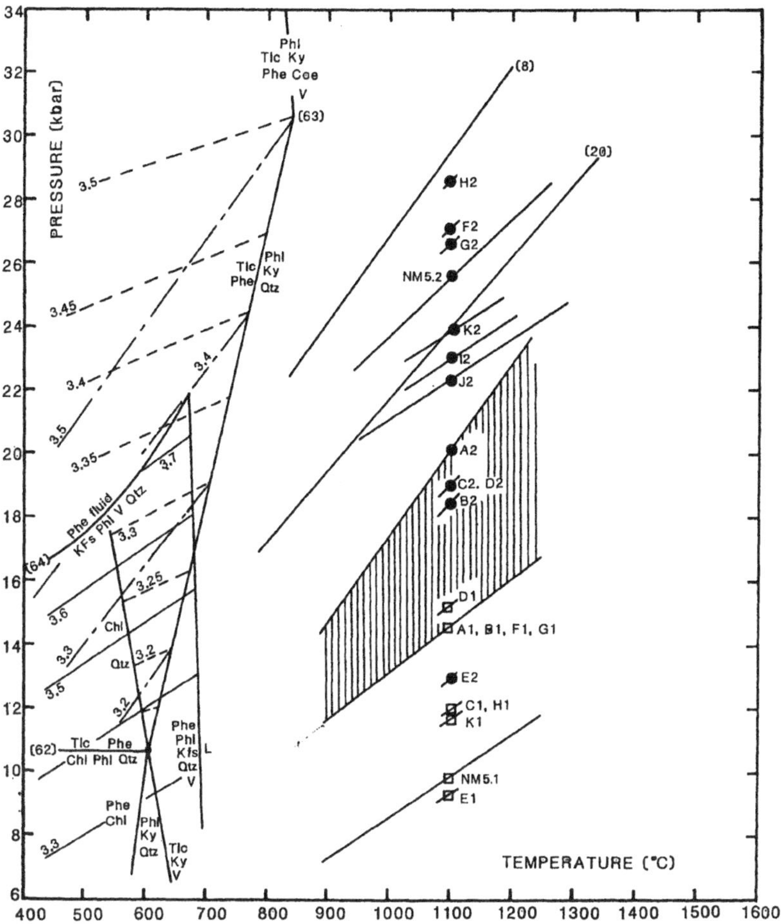

Figure 3.2 Granulite to eclogite transition (vertical ruled lines) determined in natural basaltic and intermediate rock compositions (higher-*T* curves and points at 1100°C), and important reactions involving phengite in KMASH (lower-*T* lines and isopleths).
Natural rock data: A, quartz tholeiite; B, alkali-poor quartz tholeiite; C, alkali olivine basalt; D, oxidised alkali olivine; basalt; E, alkali-poor olivine tholeiite (Green and Ringwood, 1967a); F, rhyodacite (Green and Lambert, 1965); G, andesite (Ringwood, 1975); H, gabbroic anorthosite; I, grospydite; J, kyanite eclogite; K, high alumina basalt (Green, 1967); NM5, high alumina basalt (Ito and Kennedy, 1971).

Curves and points 1 denote garnet-in positions, curves and points 2 denote plagioclase-out positions. (8) NAS system reaction R3.8 (Holland, 1980). (20) CAS system reaction R3.20 (Gasparik, 1984).

KMASH equilibria after Massonne and Schreyer (1985, 1987, 1989), Massonne *et al* (1984) and Schreyer *et al* (1987). Reactions numbered as in the text. Isopleths of Si-content in phengite are labelled with Si cations per formula unit. Lines: —— assemblage Phe + Kfs + Phl + Qtz; --- assemblage Phe + Tlc + Ky + Qtz; —-— assemblage Phe + Phl + Tlc + Qtz.

The pyroxenes contain significant tschermaks component; however, the *Jd*/*Ts* ratio will be high.

Pressures required for the incoming of garnet in a spectrum of basalts are in the range 9–15 kbars at 1100°C (Ringwood, 1975). Lower pressures are required for undersaturated rocks and compositions with lower Mg*, whereas higher pressures are necessary in oxidised basalt (Green and Ringwood, 1967a; Figure 3.2). The outgoing of plagioclase and hence transformation into eclogite is strongly dependent on whole-rock composition and varies from only 12 kbar for alkali-poor olivine tholeiites (point E2, Figure 3.2) to 23–25 kbar for high-alumina basalts and similar rock types (Green, 1967; Ito and Kennedy, 1971). This variation is principally a function of silica-saturation, Ca/(Ca + Na) ratio and total normative plagioclase component in the rock type. Typical quartz tholeiites and alkali olivine basalt compositions are transformed to eclogites at 1100°C and pressures of 18.5–20 kbar, conditions where more aluminous basaltic rock types would still retain plagioclase despite being within the eclogite facies.

All of the basaltic rock types summarised by Ringwood (1975), including the high-alumina basalts (point H2 and line NM5, Figure 3.2), attain eclogite mineralogy at pressures less than the simple NAS system reaction:

$$\underset{\text{Ab}}{NaAlSi_3O_8} \rightleftharpoons \underset{\text{Jd}}{NaAlSi_2O_6} + \underset{\text{Qtz}}{SiO_2} \qquad \text{(R3.8)}$$

(line 8, Figures 3.2 and 3.3), indicating that the constituent clinopyroxenes are omphacites in all cases. The dP/dT slopes of the plagioclase-out boundaries—27.5 bars/°C or 24 bars/°C for Qtz-tholeiite (Green and Ringwood, 1967a) and 20 bars/°C for high-alumina basalt (Ito and Kennedy, 1971)—are in good agreement with the dP/dT slopes of reactions which define the stability of plagioclase in the NAS and CAS simple systems.

Ringwood (1975) also reviewed examples of the transition from garnet-granulite to eclogite assemblages in acid to intermediate rock types. Highly aluminous compositions such as gabbroic anorthosite and diorite retain plagioclase to very high pressures (26–28 kbar at 1100°C) at or near the position of R3.8 (Figure 3.2), and well within the eclogite facies. The assemblage quartz-garnet-omphacite-alkali feldspar in such rock types therefore denotes HP eclogite facies conditions, and is also diagnostic of eclogite facies conditions in granitic rock types (Green and Lambert, 1965) where it may appear at 21 kbar and 1100°C.

3.4 P–T stability conditions for eclogite facies assemblages in basaltic and related rock types based on simple system studies

Although the natural rock experimental data outlined by Ringwood (1975) and in the previous section are useful in defining broad regions of stability for anhydrous assemblages, they do not provide an adequate basis for understanding the roles of hydrous phases in eclogite metamorphism or for the determination and extraction of thermodynamic data on mineral phases. Simple system experimental studies have been emphasised in the past decade in order to develop a comprehensive data set of thermodynamic properties through which other useful equilibria can be calculated or extrapolated. The bulk of the remainder of this chapter reviews these simple system studies and their relevance to natural rock suites or groups, beginning with systems relevant to broadly basaltic rock types.

3.4.1 *NAS and NASH systems*

Important equilibria in NAS and NASH are summarised in Figure 3.3. The critical reactions:

$$\underset{\text{Ab}}{NaAlSi_3O_8} \rightleftharpoons \underset{\text{Jd}}{NaAlSi_2O_6} + \underset{\text{Qtz}}{SiO_2} \quad \text{(R3.8)}$$

$$\underset{\text{Ab}}{NaAlSi_3O_8} + \underset{\text{Ne}}{NaAlSiO_4} \rightleftharpoons \underset{\text{Jd}}{2NaAlSi_2O_6} \quad \text{(R3.9)}$$

provide limits on the stability of jadeite in quartz-present and absent cases respectively.

Reaction R3.8 has been investigated by numerous workers (see review by Newton and Fyfe, 1976), and indeed has become a standard reaction for piston-cylinder (PC) and pressure cell calibration. Holland (1980) reversed R3.8 over the T range 600–1200°C in a 1.9 cm diameter PC apparatus using salt cells. His expression for the reaction involving high albite:

$$P(\text{kbar}) = 0.35 + 0.265\,T(°\text{C}) \quad \text{(E3.2)}$$

is applicable to most eclogites as it has been shown that high albite does not undergo significant Al–Si ordering down to 600°C. Good agreement is also found between the experimentally determined dP/dT (26.5 bars/°C) and that calculated from calorimetric measurements (27.6 bar/°C). Of other recent determinations of R3.8, only the study by Mirwald and Massonne (1980) yields data inconsistent with Holland (1980). These authors' reversal of R3.8 at 31 to 31.7 kbar and 1090°C yields a greater dP/dT (31 bar/°C) for the reaction and lies 2 kbar higher than the curve derived by Holland (1980). The discrepancy may in part result from differences in starting materials.

Experiments by Windom and Unger (1988), carried out using a mixed salt + soft glass pressure medium, yield pressures for R3.8 slightly below those of Holland (1980) (line 8c, Figure 3.3) when the author's ~2 kbar rigidity (friction) correction is applied. As the dP/dT slopes of both experimental data sets are very similar, it is suggested that the pressure correction factor used by Windom and Unger (1988) is marginally too large. Precise determination of several equilibria, including R3.8, in salt cells indicates that a -9% ($\pm 1.5\%$) P correction is applicable to more rigid pressure cells such as Talc-BN, at $T = 800$–1200°C.

The extent of offset of reaction R3.8 in more complex systems depends critically on a–X relations in sodic pyroxene and plagioclase. Holland (1980) indicated that jadeite-diopside a–X relations at 600°C could be described by a two-site regular solution model with $a_{Jd} = X_{Jd}^2 \cdot \gamma_{Jd}$, where γ_{Jd} is evaluated through regular solution theory with W_H equal to 5800 cal. This result is compatible with solution calorimetric results at 697°C (Wood *et al*, 1980), which were instead modelled with the use of a reciprocal solution model that accounts for the interdependence of pyroxene M1 and M2 site occupancies. The balance between the enthalpy of mixing and the ideal two-site configurational entropy terms furthermore lead to an energy of mixing at 1000–1300°C which is similar to that of an ideal one-site solid solution with $a_{Jd} \simeq X_{Jd}$. As discussed by Wood *et al* (1980), the assumption of pseudo-ideality in C2/c sodic Jd-Di pyroxenes will not lead to serious errors in P calculations for higher T eclogites, and will even be a reasonable approximation for ordered omphacites at 600°C.

At higher T conditions relevant to some HT ecologites, clinopyroxenes may contain appreciable CaTs ($CaAl_2SiO_6$) solid solution. Gasparik (1985b) has investigated

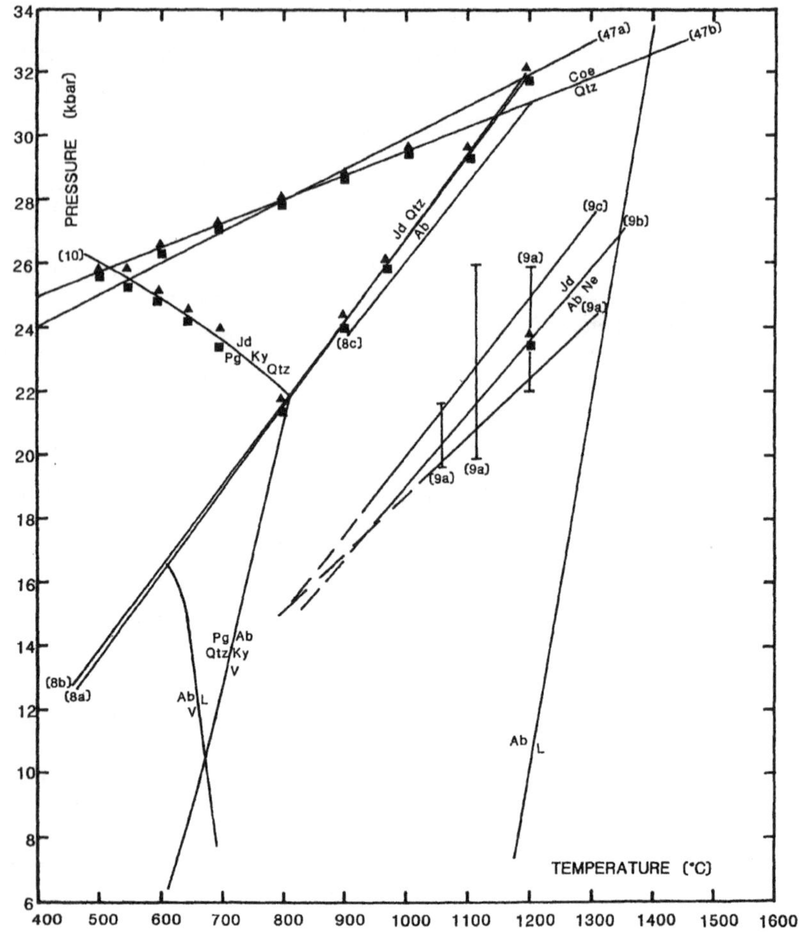

Figure 3.3 Experimentally determined equilibria in NASH system. Selected run results: ■ low-*P* assemblages; ▲ high-*P* assemblages. (8) reaction R3.8: (8a) after Holland (1980); (8b) from Gasparik (1985b); (8c) from Windom and Unger (1988). (9) reaction R3.9: (9a) from Robertson *et al* (1957); (9b) from Gasparik (1985b); (9c) from Hlabse and Kleppa (1968). (10) R3.10 and (11) R3.11 from Holland (1979b). Melting reactions for Ab based on Bohlen *et al* (1982). (47) R3.47: (47a) from Mirwald and Massonne (1980); (47b) from Bohlen and Boettcher (1982).

the offset of the CAS equilibrium

$$\underset{\text{Grs}}{Ca_3Al_2Si_3O_{12}} + \underset{\text{Cor}}{2Al_2O_3} \rightleftharpoons \underset{\text{CaTs}}{3CaAl_2SiO_6} \qquad \text{(R3.24)}$$

with the addition of Na at 1000–1400°C, using oxide mixes seeded with minerals. Modelling of the pyroxene compositional data using a Redlich–Kister polynominal method indicates only small deviations from ideality on the CaTs–Jd join at 1000°C because of a balance between excess enthalpy and excess entropy (order–disorder) effects.

The lower-P stability of jadeite for quartz-absent assemblages in NAS is defined by the equilibrium R3.9, studied experimentally at 600–1200°C in a gas apparatus by Robertson *et al* (1957) (line 9a, Figure 3.3), and more recently reversed in a PC apparatus at 1200°C (Gasparik, 1985b) (line 9b, Figure 3.3). The latter result agrees within experimental uncertainties with the early study, in which the reaction was not tied down accurately through closely-spaced reversals. Solution calorimetry data (Hlabse and Kleppa, 1968) corrected for the low to high albite transition and coupled with the reversal of Gasparik (1985b), yields a position for R3.9 (line 9c, Figure 3.3) which has a dP/dT slope (23.2 bars/°C) compatible with the available thermochemical data.

Paragonite, $NaAl_3Si_3O_{10}(OH)_2$, in association with quartz occupies a large P–T field at temperatures below 800°C and pressures up to 26 kbar. The stability of paragonite + quartz is defined by reactions

$$\underset{Pg}{NaAl_3Si_3O_{10}(OH)_2} \rightleftharpoons \underset{Jd}{NaAlSi_2O_6} + \underset{Ky}{Al_2SiO_5} + \underset{V}{H_2O} \qquad (R3.10)$$

$$\underset{Pg}{NaAl_3Si_3O_{10}(OH)_2} + \underset{Qtz}{SiO_2} \rightleftharpoons \underset{Ab}{NaAlSi_3O_8} + \underset{Ky}{Al_2SiO_5} + \underset{V}{H_2O} \qquad (R3.11)$$

which intersect in a (metastable) invariant point on reaction R3.8, at 22 kbar and 800°C (Figure 3.3). Under water-saturated conditions this invariant point is metastable with respect to melting, because the solidi for Ab and Jd + Qtz in the presence of pure water occur at 600–700°C (Bohlen *et al*, 1982) (Figure 3.3).

Holland (1979b) reversed R3.10 with mineral mixes, at temperatures of 550–700°C (line 10, Figure 3.3), and obtained a negative dP/dT slope. Under water-saturated conditions the assemblage Jd + Ky is restricted to very high P (> 22 kbar). The assemblage Pg + Omp + Ky is itself a useful barometer or a_{H_2O} indicator as paragonite is usually nearly pure and the a–X relations of omphacite are reasonably well defined at 600°C or more.

3.4.2 *NMASH and related systems*

3.4.2.1 *Stability and phase relations of glaucophane, $Na_2Mg_3Al_2Si_8(OH)_2$.*

Early experimental studies on glaucophane stability are reviewed by Newton and Fyfe (1976) and Maresch (1977). The initial interpretation (Ernst, 1963a) of two glaucophane polytypes (GI at lower T, GII at higher T) related by Al order–disorder effects has not been supported by later experiments. Maresch (1977) correctly reinterpreted GI as a ternary amphibole displaced from glaucophane by the incorporation of Mg-richterite and eckermannite components, and new data indicate that GII is also displaced from glaucophane composition.

Koons (1982) synthesised glaucophanitic amphiboles at $T < 800$°C and P conditions of 18–35 kbar (curves labelled ‘K’, Figure 3.4). His observation of Na-mica phases in water-excess experiments led him to conclude that glaucophanitic amphibole is only stable at lowered a_{H_2O}. Product amphiboles were displaced from glaucophane through substitutions along $NaAlSi_{-1}$ (edenite) and $MgMgNa_{-1}Al_{-1}$ (anthophyllite) vectors. Such ternary amphiboles were found to coexist with quartz at all conditions, and additionally with albite + talc, jadeite, and jadeite + talc with progressive increase in P

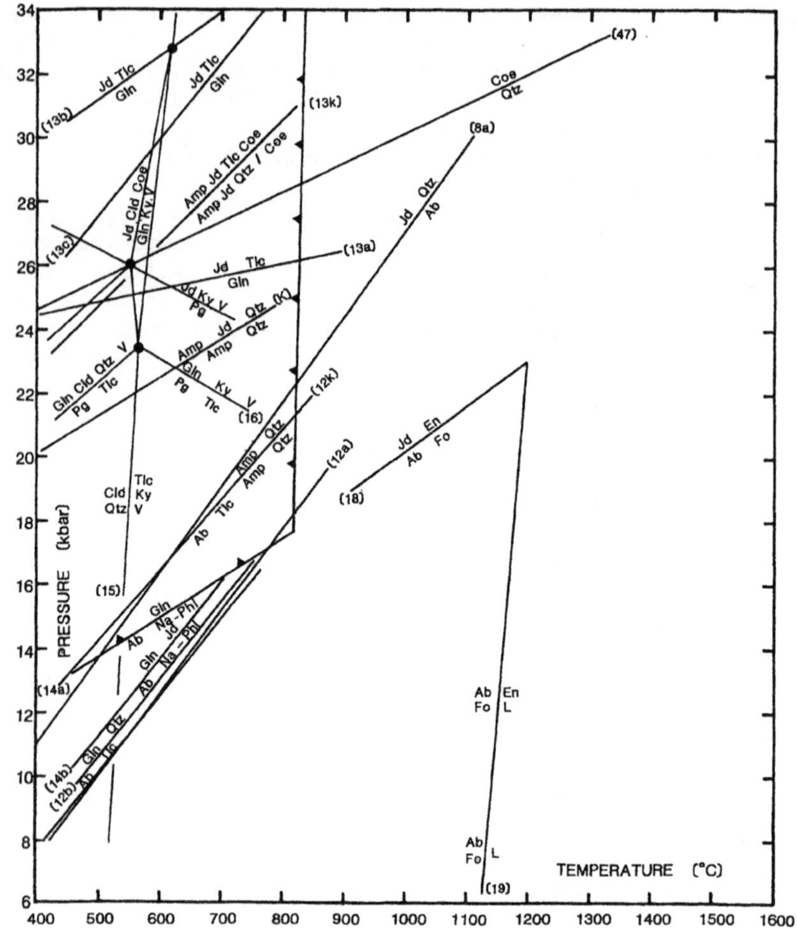

Figure 3.4 Equilibria in NMASH system. Line numbers correspond to reactions as defined in text. (8a) as for Figure 3.3. (12) reaction R3.12: (12a) Boettcher and Wyllie (1969); (12b) Holland (1988); (12k) Koons (1982). (13) reaction R3.13: (13a) Boettcher and Wyllie (1969); (13b) Holland (1988); (13c) Carman and Gilbert (1983); (13k) Koons (1982). (14) reaction R3.14: (14a) Carman and Gilbert (1983); (14b) Holland (1988). (15) and (16) calculated reactions R3.15 and R3.16 from Holland (1988). (18) and (19) NMAS reactions R3.18 and R3.19 from Windom and Unger (1988). Glaucophane stability field according to Carman and Gilbert (1983) is bounded by the reactions with toothed ornament.

(Figure 3.4). As a result, the main reactions controlling the lower- and higher-*P* stability limits of glaucophane:

$$\underset{\text{Gln}}{Na_2Mg_3Al_2Si_8O_{22}(OH)_2} + \underset{\text{Qtz}}{2SiO_2} \rightleftharpoons \underset{\text{Ab}}{2NaAlSi_3O_8} + \underset{\text{Tlc}}{Mg_3Si_4O_{10}(OH)_2} \quad \text{(R3.12)}$$

$$\underset{\text{Gln}}{Na_2Mg_3Al_2Si_8O_{22}(OH)_2} \rightleftharpoons \underset{\text{Jd}}{2NaAlSi_2O_6} + \underset{\text{Tlc}}{Mg_3Si_4O_{10}(OH)_2} \quad \text{(R3.13)}$$

are at least divariant in NMASH.

Carman and Gilbert (1983) defined a P–T stability field for glaucophane which is broadly similar to those of Koons (1982) and Maresch (1977) but differs in several important details. The on-composition synthesis of glaucophane was claimed to occur in a field below 825°C and bounded by the high-P reaction R3.13 at 26–35 kbar for T in the range 450–800°C, and by the lower-P reaction:

$$\mathrm{Gln} = \mathrm{Ab} + \text{Na-mica}_{(ss)} \tag{R3.14}$$

at 10–18 kbar for a similar T range. The divariant relations documented by Koons (1982) were not observed, and R3.14 was found to have a shallower dP/dT than the Ab + Tlc and Na-mica forming reactions given by Koons (1982) and other workers (Figure 3.4).

In a study of the thermodynamic properties of glaucophane based on V, Cp and S data obtained by use of a natural near-end member sample, Holland (1988) has applied the Carman and Gilbert (1983) data on reaction R3.13 to calculate a ΔH_f^0 of glaucophane. Univariant equilibria between glaucophane and other phases in NMASH calculated by Holland (1988) using this data and the dataset of Holland and Powell (1985) are presented in Figure 3.4 (curves R3.15, R3.16 and related lines). These calculated reactions indicate that Gln + Qtz is stable at approximately 3 kbar less than Jd + Qtz; glaucophane is stable with excess water present; and the assemblage Gln + Ky is restricted to $P > 20$–24 kbar in NMASH. This grid, however, does not take into account experimental data in MASH (Section 3.6.2) which indicate that Mg-Cld + Qtz is metastable with respect to carpholite (Car), and hence that reactions such as

$$\underset{\mathrm{Cld}}{3\mathrm{MgAl_2SiO_5(OH)_2}} + \underset{\mathrm{Qtz}}{4\mathrm{SiO_2}} \rightleftharpoons \underset{\mathrm{Tlc}}{\mathrm{Mg_3Si_4O_{10}(OH)_2}} + \underset{\mathrm{Ky}}{3\mathrm{Al_2SiO_5}} + \underset{\mathrm{V}}{\mathrm{H_2O}} \tag{R3.15}$$

(Holland, 1988) are metastable. The lower-P limit on the stability of Gln + Ky:

$$\underset{\mathrm{Gln}}{\mathrm{Na_2Mg_3Al_2Si_8O_{22}(OH)_2}} + \underset{\mathrm{Ky}}{2\mathrm{Al_2SiO_5}} + \underset{\mathrm{V}}{\mathrm{H_2O}} \rightleftharpoons \underset{\mathrm{Pg}}{2\mathrm{NaAl_3Si_3O_{10}(OH)_2}} + \underset{\mathrm{Tlc}}{\mathrm{Mg_3Si_4O_{10}(OH)_2}} \tag{R3.16}$$

will not be affected by these considerations, but this and all the calculated reactions that involve glaucophane may be shifted in P–T position if the glaucophanes of Carman and Gilbert (1983) are offset from the ideal Gln composition.

Welch (1987), in a thorough study of fluor-glaucophane stability, has defined continuous changes in amphibole composition as a function of P and assemblage. Apart from the absence of talc in the NMAS + fluorine system, the sequence of assemblages with increased pressure 20–35 kbar are analogous to those of Koons (1982) [i.e. Amp + Na-mica + Qtz, Amp + Qtz, Amp + Jd + Qtz] but offset to higher P and T. Product glaucophanitic amphiboles in this system are essentially ternary and offset from Gln through substitutions along $\mathrm{MgMgSi_{-1}}$ to Mg-kataphorite (K) and along $\mathrm{NaAlSi_{-1}}$ towards nyböite (N). Amphiboles synthesised in all previous NMASH studies, including Carman and Gilbert (1983), are deduced to be such ternary GNK amphiboles (Welch, 1987, 1989). The $\mathrm{MgMgSi_{-1}}$ substitution dominates when 'glaucophane' coexists with Jd + Qtz:

$$5\,\mathrm{Gln} \rightleftharpoons 3\mathrm{K} + 4\mathrm{Jd} + 11\mathrm{Qtz} + 2\mathrm{V} \tag{R3.17a}$$

whereas amphiboles coexisting only with quartz lie well within the GNK composition space because of the reaction:

$$9\mathrm{Gln} \rightleftharpoons 4\mathrm{N} + 3\mathrm{K} + 23\mathrm{Qtz} + 2\mathrm{V} \qquad \text{(R3.17b)}$$

(Welch, 1989). As can be seen from R3.17a and R3.17b, lowering a_{H_2O} favours the formation of GNK amphiboles rather than true glaucophane which would therefore be most stable at water saturated conditions, in contradiction to the interpretation of Koons (1982). It is also clear that the high-*P* 'glaucophanes' reported by Carman and Gilbert (1983) and indirectly used by Holland (1988) in his calculations of NMASH equilibria, are glaucophane–Mg-katophorite solid solutions which may not be suitable for the extraction of precise thermodynamic data.

3.4.2.2 *Higher temperature reactions in NMAS.* At temperatures above 800–850°C glaucophanitic amphiboles and equivalent assemblages break down to Jd + En in addition to other phases (L, V, Qtz) (Carman and Gilbert, 1983). The stability of the simple-system eclogite assemblage Jd + En (ss) has been determined in the *P–T* range 16–26 kbar and 800–1200°C by Windom and Unger (1988). The *P–T* position of the equilibrium:

$$\underset{\mathrm{Jd}}{\mathrm{NaAlSi_2O_6}} + \underset{\mathrm{En}}{\mathrm{Mg_2Si_2O_6}} \rightleftharpoons \underset{\mathrm{Ab}}{\mathrm{NaAlSi_3O_8}} + \underset{\mathrm{Fo}}{\mathrm{Mg_2SiO_4}} \qquad \text{(R3.18)}$$

(Figure 3.5) is nearly coincident with the NAS reaction R3.9, mainly because of solid solution along the Jd–En join. The status of the possible assemblage nepheline (Ne) + enstatite could not be unambiguously established, but Windom and Unger (1988) favoured the interpretation that Ne + En and the NMAS invariant point between Jd, En, Ab, Ne and Fo are both metastable with respect to melting through the reaction:

$$\mathrm{Ab} + \mathrm{Fo} = \mathrm{En} + \mathrm{L} \qquad \text{(R3.19)}$$

which intersects R3.18 at 23 kbar and 1200°C (Figure 3.4, lines 18 and 19).

3.4.3 *CAS and NCAS systems*

Early studies of the reaction:

$$\underset{\mathrm{An}}{3\mathrm{CaAl_2Si_2O_8}} \rightleftharpoons \underset{\mathrm{Grs}}{\mathrm{Ca_3Al_2Si_3O_{12}}} + \underset{\mathrm{Ky}}{2\mathrm{Al_2SiO_5}} + \underset{\mathrm{Qtz}}{\mathrm{SiO_2}} \qquad \text{(R3.20)}$$

summarised by Newton and Fyfe (1976), have been largely superseded by the experiments of Goldsmith (1980), Gasparik (1984b), and Koziol and Newton (1988). These independent studies are broadly consistent in the *T* range 500–1100°C, and are also consistent with calculations by Chatterjee *et al* (1984), which indicate similar d*P*/d*T* gradients of 22–24 bars/°C.

Goldsmith (1980) studied R3.20 at 1100–1400°C in various PC apparatuses with both salt (lower-*T*) and talc/soft glass (higher-*T*) pressure mediums. Reversals that used mineral mixes yielded the following fit for R3.20:

$$P(\mathrm{kbar}) = -2.1 + 0.0232\,T(^\circ\mathrm{C}) \qquad \text{(E3.3)}$$

(Figure 3.5, line 20a), which indicated the presence of some Al–Si disorder even in pure anorthite. Gasparik (1984b) obtained an essentially similar result in experiments over

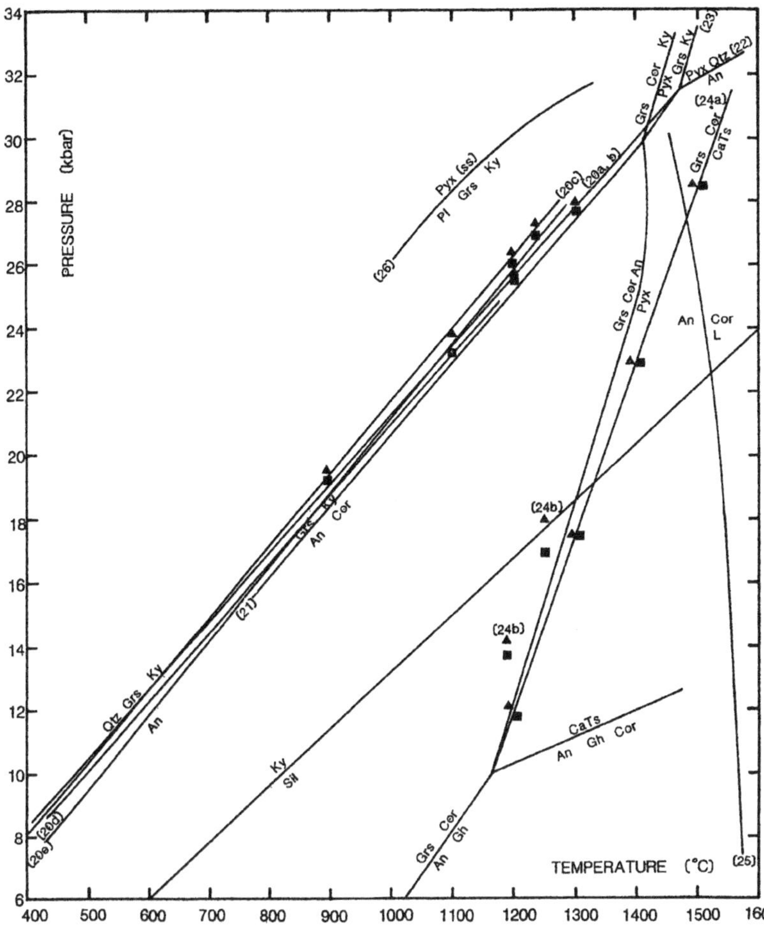

Figure 3.5 Equilibria in the CAS and NCAS systems. Line numbers correspond to reactions defined in the text. *LP* (■) and *HP* (▲) reaction products for selected equilibria indicated. (20) reaction R3.20: (20a) Goldsmith (1980); (20b) Gasparik (1984b); (20c) Koziol and Newton (1988); (20d) calculated by Chatterjee *et al* (1984); (20e) Wood and Holloway (1984). (21), (22) and (23) are reactions from Gasparik (1984b). (24) reaction R3.24: (24a) Gasparik (1984b); (24b) Hays (1966). (25) anorthite melting from Goldsmith (1981). (26) reaction R3.26 in the NCAS system from Gasparik (1985); Pyx in this case is rich in CaTs and CaEs components. Ky = Sil reaction from Chatterjee *et al* (1984).

the same *T* interval and with the use of a PbO flux to improve reaction rates at lower temperatures (Figure 3.5, line 20b):

$$P(\text{kbar}) = -0.7 + 0.022\,T(°\text{C}) \qquad \text{(E3.4)}$$

The latest calibration of R3.20, based on eight reversals between 900 and 1250°C (Koziol and Newton, 1988), significantly improves the uncertainty in location of the reaction and yields the following relation:

$$P(\text{kbar}) = -1.093 + 0.0228\,T(^\circ\text{C}) \qquad \text{(E3.5)}$$

(Figure 3.5, line 20c), which predicts anorthite stability to about 0.5–0.7 kbar greater pressure at $T > 900^\circ$C than other recent work, but rather similar stability limits at lower T. Anorthite breakdown to Grs + Ky + Qtz at 1100°C in all CAS system studies (23.2–24 kbar) is in remarkable agreement with the incoming of Ky and loss of plagioclase (and an increase in Grs content of garnet) in 'natural rock' experiments on high-alumina basalt, kyanite-eclogite, and grospydite compositions (Green, 1967).

At high P–T conditions generally inaccessible to eclogite metamorphism (> 1400°C and 30 kbar), anorthite breakdown by way of

$$\text{An} \rightleftharpoons (\text{CaTs–CaEs})\text{Pyx} + \text{Qtz} \qquad \text{(R3.22)}$$

and the breakdown of Grs + Ky:

$$\text{Grs} + \text{Ky} \rightleftharpoons (\text{CaTs–CaEs})\text{Pyx} + \text{Qtz} \qquad \text{(R3.23)}$$

leads to the stability of CaTs-rich aluminous clinopyroxene (Wood, 1978; Gasparik, 1984b) which contains a significant Ca-Eskola (CaEs) component, $Ca_{0.5}AlSi_2O_6$. The maximum-T stability limits of Grs + Cor, important in grospydites, are defined by the reaction

$$\underset{\text{Grs}}{Ca_3Al_2Si_3O_{12}} + \underset{\text{Cor}}{2Al_2O_3} \rightleftharpoons \underset{\text{CaTs}}{3CaAl_2SiO_6} \qquad \text{(R3.24)}$$

(Figure 3.5, line 24). Reversals of R3.24 at 1170–1500°C by use of oxide mixes and synthetic phases (Gasparik, 1984b) yield the relation:

$$P(\text{kbar}) = -53.9 + 0.055\,T(^\circ\text{C}) \qquad \text{(E3.6)}$$

which is strongly at variance by 2–5 kbar with the results of Hays (1966). The latter experiments may need some systematic calibrational pressure correction, and appear to require too high an entropy of disorder in CaTs (Wood and Holloway, 1984), whereas the curve of Gasparik (1984b) implies significant but not complete Al–Si disorder in CaTs.

Addition of Na to CAS stabilises CaTs–Jd solid solutions and hence shifts R3.24 to lower T. Gasparik (1985) obtained compositional data for CaTs–Jd that coexisted with Grs, Cor, and PbO melt in the P–T field 20–28 kbar and 1100–1400°C. Only small departures from ideality on the CaTs–Jd join are apparent if these data are modelled using the a–X formulation $a_{\text{CaTs}} = X_{\text{CaTs}} \cdot \gamma_{\text{CaTs}}$. A model that involves complete disorder on the M2 and T pyroxene sites also fits the data, but implies more non-ideality at lower temperatures of 600°C.

Complementary experiments (Gasparik, 1985) on coexisting plagioclase and CaTs–Jd pyroxene at 25 kbar have enabled the calculation of plagioclase (Ab–An) mixing properties at high-T by use of the equilibrium

$$\underset{\text{Grs}}{3Ca_3Al_2Si_3O_{12}} + \underset{\text{Ky}}{6Al_2SiO_5} \rightleftharpoons \underset{\text{CaTs}}{7CaAl_2SiO_6} + \underset{\text{CaEs}}{4Ca_{0.5}AlSi_2O_6} \qquad \text{(R3.26)}$$

P–X sections (Gasparik, 1985) illustrate that albite is the most stable feldspar with respect to pressure, and that CaEs is not significant at temperatures relevant to most eclogite facies rocks (< 1000°C). The high-P stability of albite is only marginally greater than that of Ab_{70} at 1000–1200°C, whereas Ab_{40} (An_{60}) feldspar stability is only extended by about 1 kbar with respect to pure anorthite at $T < 1000$°C.

3.4.4 *CASH and NCASH systems*

Early studies on the reaction that defines the maximum stability of lawsonite:

$$\underset{\text{Lws}}{4CaAl_2Si_2O_7(OH)_2 \cdot H_2O} \rightleftharpoons \underset{\text{Zo}}{2Ca_2Al_3Si_3O_{12}(OH)} + \underset{\text{Ky}}{Al_2SiO_5} + \underset{\text{Qtz}}{SiO_2} + \underset{\text{V}}{7H_2O} \quad \text{(R3.27)}$$

have been summarised by Newton and Fyfe (1976). PC experiments (Newton and Kennedy, 1963) yielded equilibration temperatures for R.27 which were 20–30°C higher than experiments in gas apparatus (Nitsch, 1968) (lines 27a and 27b, Figure 3.6). Chatterjee *et al* (1984) have reversed R3.27 at 10–20 kbar using synthetic lawsonite and zoisite. These results (Figure 3.6, line 27c) indicate equilibrium at temperatures 30–40°C below the Newton and Kennedy (1963) curve, and define the approximate relation:

$$P(\text{kbar}) = -23.37 + 0.0784\,T(°\text{C}) \quad \text{(E3.7)}$$

Lawsonite is therefore restricted to low-T eclogites (< 550°C) even at pressures of 20 kbar.

The stability of margarite within the eclogite facies is controlled by the reaction:

$$\underset{\text{Mrg}}{4CaAl_4Si_2O_{10}(OH)_2} \rightleftharpoons \underset{\text{Zo}}{2Ca_2Al_3Si_3O_{12}(OH)} + \underset{\text{Ky}}{2Al_2SiO_5} + \underset{\text{Cor}}{3Al_2O_5} + \underset{\text{V}}{3H_2O} \quad \text{(R3.28)}$$

(Figure 3.6, line 28), investigated by Chatterjee *et al* (1984) in four PC experiments at 10–13 kbar. Although the direct experimental data allow considerable latitude in the position of R3.28, requirements of internal consistency with other CASH equilibria (Halbach and Chatterjee, 1984) fix its position well. These experiments and calculations demonstrate a larger (by about 4 kbar) stability field for margarite at high-P than predicted by some earlier CASH datasets (Halbach and Chatterjee, 1984).

The stability of margarite in the presence of quartz:

$$\underset{\text{Mrg}}{4CaAl_4Si_2O_{10}(OH)_2} + \underset{\text{Qtz}}{3SiO_2} \rightleftharpoons \underset{\text{Ky}}{5Al_2SiO_5} + \underset{\text{Zo}}{2Ca_2Al_3Si_3O_{12}(OH)} + \underset{\text{V}}{H_2O} \quad \text{(R3.29)}$$

has been the subject of some controversy because calculations based on other CASH equilibria indicate shallower dP/dT than obtained in early experimental studies (Jenkins, 1984). New reversals of R3.29 in the P–T range 7.5–12 kbar and 500–600°C using mineral mixes (Jenkins, 1984) allow dP/dT slopes in the range -2.2 to -17 kbars/°C (vertical ruled area in Figure 3.6) which are consistent with other CASH equilibria. Jenkins (1984) found that optical or X-ray diffraction (XRD) techniques were poor at resolving the direction of reaction, and instead used a much more sensitive IR spectroscopy method for product identification. The equilibrium R3.29 is a potential geobarometer in NCASH where paragonite solid-solution stabilises margarite to higher pressures within the realm of the eclogite facies.

The stability of anorthite in relation to higher-P CASH assemblages involving zoisite, grossular and kyanite is of greater importance to eclogite metamorphism and

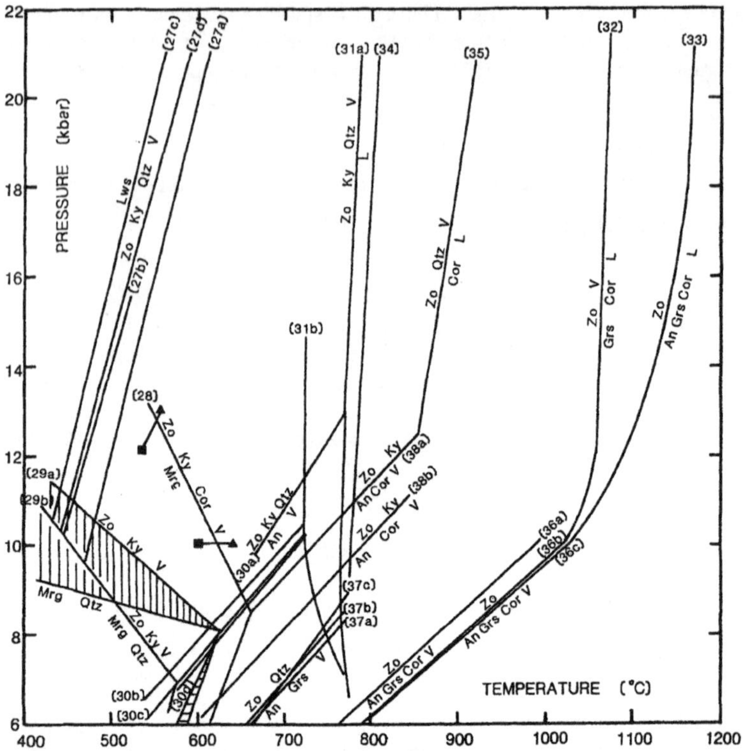

Figure 3.6 Experimentally determined and calculated equilibria in CASH. (27) reaction R3.27 based on (27a) Newton and Kennedy (1963); (27b) Nitsch (1968); (27c) Chatterjee *et al* (1984); (27d) earlier CASH data sets. (28), (29b), (30c), (36a), (37a), (38a) reactions R3.28, R2.29, R3.30, R3.36, R3.37 and R3.38 respectively, from Chatterjee *et al* (1984). (29a) reaction R3.29 after Jenkins (1984), with uncertainties. (30) reaction R3.30: (30a) Boettcher (1970); (30b) Johannes (1980); (30d) Goldsmith (1981, 1982). (31) reaction R3.31: (31b) Goldsmith (1981); (31a) Boettcher (1970). (32), (33), (34) and (35) reactions R3.32, R3.33, R3.34 and R3.35 respectively, after Boettcher (1970). (36) reaction R3.36; (36b) Boettcher (1970); (36c) Newton (1966). (37) reaction R3.37: (37c) Boettcher (1970); (37b) Newton (1966). (38) reaction R3.38: (38b) Boettcher (1970).

has been the subject of many experimental studies since the analysis of Boettcher (1970) and the generally low-P (<10 kbar) studies by Newton and Kennedy (1963), Newton (1966) and others. Some calculated CASH phase relations (Chatterjee *et al*, 1984; Halbach and Chatterjee, 1984) are presented in Figure 3.6 along with recent experimental calibrations for comparison with the work of Boettcher (1970). Although there is generally good agreement between the updated calculated equilibria and the early experiments in terms of dP/dT slopes and for pressures less than 14 kbar, the stability of Zo + Ky is now inferred to occur to 50°C lower T than indicated by Boettcher (1970) and Newton (1966).

The reaction:

$$\underset{\text{Zo}}{2Ca_2Al_3Si_3O_{12}(OH)} + \underset{\text{Ky}}{Al_2SiO_5} + \underset{\text{Qtz}}{SiO_2} \rightleftharpoons \underset{\text{An}}{4CaAl_2Si_2O_8} + \underset{\text{V}}{H_2O} \qquad \text{(R3.30)}$$

will only occur in H_2O-rich fluid systems as zoisite is unstable relative to anorthite-calcite (and scapolite) at moderate to high CO_2^* ($CO_2^* = CO_2/(H_2O + CO_2)$) (Storre *et al*, 1982). Early experiments on R3.30, summarised by Newton and Fyfe (1976) and represented by line 30a (Figure 3.6) have been superseded by more recent studies (Goldsmith, 1981, 1982; Johannes, 1980, 1984; Chatterjee *et al*, 1984) which yield closely consistent data. Goldsmith (1981) reversed R3.30 over the T range 550–700°C in a salt pressure medium PC apparatus, using synthetic zoisite and anorthite. The best-fit relation for R3.30 based on these experiments:

$$P = -4.59 + 0.0204\,T(°C) \qquad \text{(E3.8)}$$

is within error of the results of Chatterjee *et al* (1984) and Johannes (1984), obtained at 400–720°C also with the use of synthetic zoisite and anorthite. All these results vary considerably with the anomalously steep reaction curve for R3.30 obtained by Best and Graham (1978), and limit anorthite stability to about 1 kbar lower pressures than inferred by Boettcher (1970). Furthermore, melting in the CASH system at 10–12 kbar through the eutectic reaction

$$\text{Zo} + \text{Ky} + \text{Qtz} + \text{V} \rightleftharpoons \text{L} \qquad \text{(R3.31)}$$

(Figure 3.6, lines 31a, 31b) occurs at 720°C (Goldsmith, 1981) rather than at 770–800°C as previously obtained by Boettcher (1970). Zoisite may persist to much higher temperatures dictated by the zoisite melting curves

$$\text{Zo} + \text{V} \rightleftharpoons \text{Grs} + \text{Cor} + \text{L} \qquad \text{(R3.32)}$$

$$\text{Zo} \rightleftharpoons \text{An} + \text{Grs} + \text{Cor} + \text{L} \qquad \text{(R3.33)}$$

which occur at temperatures well above 1000°C (Boettcher, 1970) (Figure 3.6, lines 32, 33). These reactions, and the melting reactions

$$\text{Zo} + \text{Qtz} + \text{V} \rightleftharpoons \text{Grs} + \text{L} \qquad \text{(R3.34)}$$

$$\text{Zo} + \text{Ky} + \text{V} \rightleftharpoons \text{Cor} + \text{L} \qquad \text{(R3.35)}$$

have not been the subject of recent experimental investigations, and are positioned in Figure 3.6 based on the work of Boettcher (1970), but with the appropriate invariant points shifted because of relocations of the related subsolidus equilibria necessitated by recent experiments.

Chatterjee *et al* (1984) have re-examined the reactions

$$\underset{\text{Zo}}{6Ca_2Al_3Si_3O_{12}(OH)} \rightleftharpoons \underset{\text{An}}{6CaAl_2Si_2O_8} + \underset{\text{Grs}}{2Ca_3Al_2Si_3O_{12}} + \underset{\text{Cor}}{Al_2O_3} + \underset{\text{V}}{3H_2O} \qquad \text{(R3.36)}$$

$$\underset{\text{Zo}}{4Ca_2Al_3Si_3O_{12}(OH)} + \underset{\text{Qtz}}{SiO_2} \rightleftharpoons \underset{\text{An}}{5CaAl_2Si_2O_8} + \underset{\text{Grs}}{Ca_3Al_2Si_2O_{12}} + \underset{\text{V}}{2H_2O} \qquad \text{(R3.37)}$$

by use of pure synthetic phases in the pressure range 2–8 kbar in a gas apparatus, at temperatures of 480–890°C. These new results are compatible with the previous experiments of Boettcher (1970) and Newton (1966) but require a shallower dP/dT slope for R3.37 than inferred by Boettcher (1970). Requirements of internal consistency in the CASH thermodynamic data for Zo, Ky, Grs, Qtz, An, Cor and Mrg (Halbach and Chatterjee, 1984) yield a calculated position for the reaction:

$$2Ca_2Al_3Si_3O_{12}(OH) + 2Al_2SiO_5 \rightleftharpoons 4CaAl_2Si_2O_8 + Al_2O_3 + H_2O \quad (R3.38)$$

Zo Ky An Cor V

(Figure 3.6, line 38a) which is at a temperature 50°C lower (or at 1 kbar higher P) than the position obtained by Boettcher (1970).

The equilibium R3.30 has been investigated in the NCASH system by Goldsmith (1982) in the T-range 400–800°C and at 8–11 kbar. Experiments in which the compositions of plagioclase coexisting with Zo + Ky + Qtz were reversed using synthetic and natural plagioclases in seeded mixes indicate that the isobaric offset of R3.30 is negligible in NCASH for plagioclases more calcic than about An_{40}. This remarkable result implies that, at 8–10 kbar and 615–715°C and under water-saturated conditions, the microscopic activity of anorthite in plagioclase is approximately equal to unity ($a_{An} = 1$) for plagioclases more calcic than An_{40}. Goldsmith (1982) has interpreted the results in terms of molecular ordering of feldspar into An-like and Ab-like domains and has further suggested that these domains control the behaviour of plagioclases in the region An_{40} to An_{100}, where no single plagioclase is therefore stable. Reaction 3.30 is progressively displaced towards higher pressures with decreasing X_{An} for plagioclases more sodic than An_{40}.

3.5 Experiments in CMAS and CMSH relevant to mineral assemblages in eclogite facies metabasites, grospydites, and intermediate rocks

3.5.1 *Silica deficient CMAS system*

The assemblage clinopyroxene + garnet + corundum, stable over a wide range of P–T conditions in the CMAS system and controlled by the CAS end-member reaction R3.24 (Figure 3.7), has been studied by Gasparik (1984c) at 1300°C and 20–32 kbar. These experiments allow modelling of both Prp-Grs solid-solution properties over a wide compositional range (Grs_{100} to Grs_{20} Prp_{80}), and CaTs-Di activity–composition relationships. Two stability fields of Grt + Cpx + Cor are established below 33 kbar at 1300°C because of the intervening appearance of a Cpx + Cor (CaTs–Di + Cor) field (see also Figure 2.1). The crest of the solvus that separates the two Grt + Cpx + Cor divariant loops occurs at Grs_{64} and the reaction

$$Cpx_{ss} \rightleftharpoons Grt_{ss} \quad (R3.39)$$

occurs at 18 kbar and 900°C, 25 kbar and 1100°C, and at 32 kbar and 1300°C (Figure 3.7, line 39). The Grs_{ss} + Cpx + Cor CMAS assemblage is therefore stable down to 800°C at pressures relevant to eclogite facies metamorphism, and occurs in a P–T field with positive dP/dT and bounded by reaction R3.24 at high T. The Mg-richer assemblage Prp_{ss} + Cpx + Cor is stable over a similar temperature range but is precluded at low pressures by the reaction:

$$Cpx + Grt + Cor \rightleftharpoons Spl + An \quad (R3.40)$$

reversed using crystalline starting materials at 1300°C and 17.9 kbar (Gasparik, 1984c) and calculated to have a positive dP/dT.

CaTs-Di solid solution modelling (Gasparik, 1984c) indicates that a_{CaTs} is less than X_{CaTs} at 1300°C, in contrast to the conclusions of Wood (1979) and Wood and Holloway (1984) that CaTs activity approximated its mole fraction in CaTs–Di solid solutions because of a balance between reciprocal solution effects and disordering on the two pyroxene sub-lattices. It is important to emphasise that the modelling by Gasparik (1984c) relies on fitted data for equilibria other than R3.24, and hence must be treated with caution. Properties of garnet solid solutions as deduced by Gasparik (1984c) vary from those determined by Haselton and Newton (1980) in that the excess parameters show an opposite sense of compositional dependence, although activities are always greater than mole fractions. The *a*-*X* relations proposed by Gasparik (1984c) are not applicable to Mg-rich garnets with $X_{Grs} < 20$. Unmixing of Grs-Prp solid solutions about a solvus crest at $X_{Grs} = 70$ could be inferred (Gasparik, 1984c) to occur at temperatures below 850–900°C (Figure 3.7). This critical temperature is higher than that inferred by Hensen *et al.* (1975) (630 ± 90°C) but is in accord with the recalculated CMAS model of Cressey (1981).

3.5.2 *Silica saturated CMAS and CMSH systems*

Important equilibria in CMAS at high temperatures involving the phases Cpx, Opx, Grt, Qtz, An and Ky or Sil have been studied by Hensen (1976), Hansen (1981), and Perkins (1983). These experimental studies have identified an invariant point between the six phases which occurs at about 14.5 kbar and temperatures of 850–950°C (Figure 3.7, point IP). Thermodynamic calculations (Wood and Holloway, 1984; Aranovich and Podlesskii, 1989) which incorporate calorimetric and *Cp* data suggest a somewhat lower *P* for this point, 13–14 kbar, at similar temperatures. The range of reasonable *P–T* conditions for IP is indicated by the stippled area on Figure 3.7. The simplified reaction:

$$\underset{\text{Di}}{CaMgSi_2O_6} + \underset{\text{En}}{Mg_2Si_2O_6} + \underset{\text{Ky}}{Al_2SiO_5} \rightleftharpoons \underset{\text{Grt}}{CaMg_2Al_2Si_3O_{12}} + \underset{\text{Qtz}}{SiO_2} \qquad \text{(R3.41)}$$

which defines the lower-*T* boundary of Ky-eclogite in CMAS, has not been experimentally determined and hence is not precisely located in terms of *T* (Figure 3.7, lines 41a and 41b), although the negative d*P*/d*T* slopes calculated by various workers are in good agreement. As with all the univariant equilibria discussed in this section, variation in phase compositions along the equilibria results in curvature.

The maximum stability of plagioclase (An) in a typical rock-analogue composition within CMAS is defined by the simplified reaction:

$$\underset{\text{Di}}{2CaMgSi_2O_6} + \underset{\text{Ky/Sil}}{2Al_2SiO_5} \rightleftharpoons \underset{\text{An}}{CaAl_2Si_2O_8} + \underset{\text{Grt}}{CaMg_2Al_2Si_3O_{12}} + \underset{\text{Qtz}}{SiO_2} \qquad \text{(R3.42)}$$

which has been studied experimentally in the *T* range 1000–1300°C (Hensen, 1976; Perkins, 1983). A shallower d*P*/d*T* is indicated for R3.42 by the reversed data of Perkins (1983) than inferred from Hensen (1976), mainly as a result of differences in the 1000°C reversal pressures (15.2 kbar compared with 17.2 kbar). Part of the discrepancy (Figure 3.7, lines 42a, 42b) may result from differences in the pressure cell characteristics of salt + soft glass compared with talc.

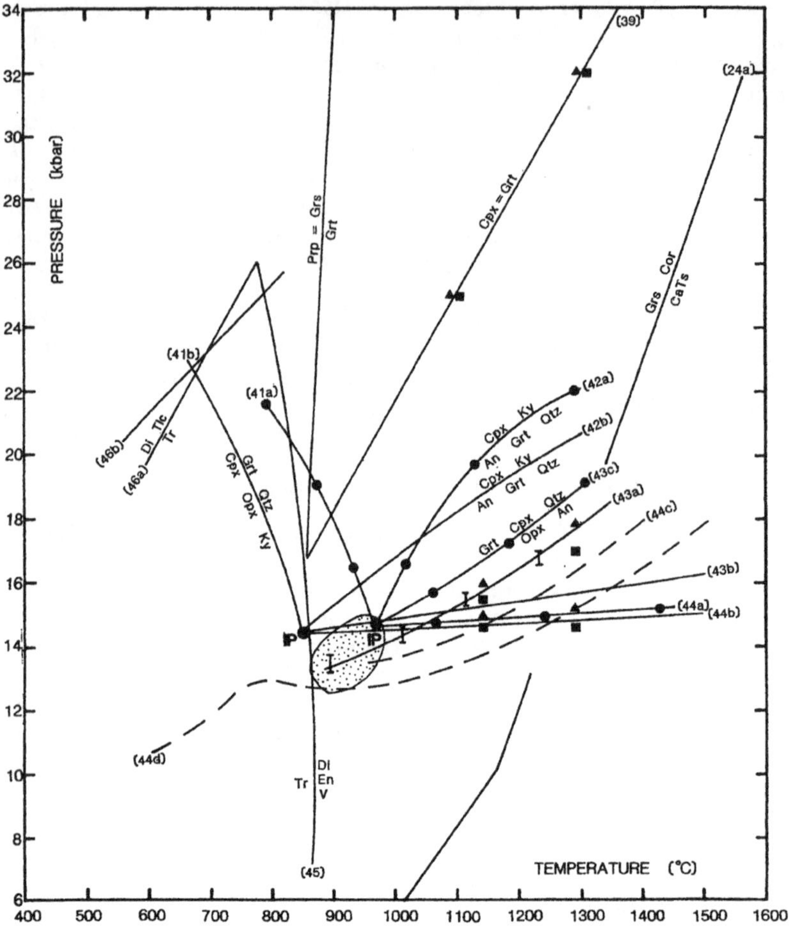

Figure 3.7 Experimentally determined and calculated equilibria in CMAS and CMSH relevant to eclogite facies mineral assemblages in metabasites, grospydites, and intermediate rocks. Lower-*P* (■) and higher-*P* (▲) run products indicated for selected reactions. Line numbers correspond to reactions as defined in the text.

(39), (24a) and the reaction Grt = Prp = Grs are from Gasparik (1984c). (41a), (42a), (43a) and (44a) reactions R3.41, R3.42, R3.43 and R3.44 respectively, based on Perkins (1983) [lines with ●]. (43a) also fits the data of Hansen (1981) [I-beams] and calculations by Wood and Holloway (1984) if ca. 12 mol% CaEs occurs in the Cpx. (41b), (42b), (43b) and (44b) reactions as above based on data of Hensen (1976). (43c), (44c) calculated reaction curves after Wood and Holloway (1984). Compare with the calculated curve (44d) for R3.44 by Aranovich and Podlesskii (1989). (45) CMSH reaction R3.45 from Welch (1987), (46) CMSH reaction R3.46: (46a) Welch (1987); (46b) Gilbert and Troll (1974).

IP is the CMAS invariant point. Stippled region represents the possible *P–T* field in which IP could occur.

The simplified garnet-granulite model reaction:

$$\underset{\text{Prp}}{Mg_3Al_2Si_3O_{12}} + \underset{\text{Di}}{CaMgSi_2O_6} + \underset{\text{Qtz}}{SiO_2} \rightleftharpoons \underset{\text{En}}{MgSiO_3} + \underset{\text{An}}{CaAl_2Si_2O_8} \qquad \text{(R3.43)}$$

has been reversed using synthetic crystalline starting materials at temperatures of 900–1300°C by Perkins (1983) and Hansen (1981) (Figure 3.7, line 43a). These two studies yield consistent results for the P–T location of R3.43, but are not compatible with the seeded-glass (synthetic Grt + glass) experiments of Hensen (1976) (Figure 3.7, line 43b) which indicated a shallower dP/dT for R3.43. Modelling by Wood and Holloway (1984) yields results consistent with the recent experiments (Perkins, 1983; Hansen, 1981) provided that clinopyroxene is a ternary Di-CaTs-CaEs pyroxene (Figure 3.7, line 43c). Estimates of the garnet composition on R3.43 vary from Grs_{11} (Perkins, 1983) to Grs_{20} (Hensen, 1981), despite the consistency in the P–T location of the reaction for these studies.

The CMAS model granulite reaction:

$$\underset{\text{Grt}}{CaMg_2Al_2Si_3O_{12}} + \underset{\text{Qtz}}{SiO_2} \rightleftharpoons \underset{\text{En}}{2MgSiO_3} + \underset{\text{An}}{CaAl_2Si_2O_8} + \underset{\text{Sil/Ky}}{Al_2SiO_5} \qquad \text{(R3.44)}$$

is displaced to lower pressures by about 4 kbars compared with the MAS system plagioclase-absent analogue Prp + Qtz = En + Sil. Experiments in talc–glass (Hensen, 1976) and salt–glass (Perkins, 1983) pressure cells and using both seeded glass and synthetic mineral mix starting materials yield essentially identical results (Figure 3.7, lines 44a, 44b). These experimental curves, which occur at about 14.5 kbar and have very shallow dP/dT slopes, are at variance with calculated equilibrium boundaries for R3.44 (Wood and Holloway, 1984; Aranovich and Podlesskii, 1989) (Figure 3.7, lines 44c, 44d). The latter show pronounced curvature as a result of changes in the alumina content of orthopyroxene and X_{Grs} with temperature, and allow Grt + Qtz stability at lower pressures (12 kbar near 900°C) than hitherto indicated by experimental data.

Also presented in Figure 3.7 are some CMSH experimental and calculated equilibria, relevant to the stability of tremolitic amphibole, which lie at temperatures lower than the CMAS invariant point IP. Experiments at low P by Welch (1987), combined with calculations using best-fit thermochemical data for Tr, Di, and H_2O, yield the position of the reaction:

$$\underset{\text{Tr}}{Ca_2Mg_5Si_8O_{22}(OH)_2} \rightleftharpoons \underset{\text{Di}}{2CaMgSi_2O_6} + \underset{\text{En}}{3MgSiO_3} + \underset{\text{V}}{2H_2O} \qquad \text{(R3.45)}$$

as given in Figure 3.7, which is applicable to natural tremolites with some $MgCa_{-1}$ substitution. The data extracted by Welch (1987) yields a calculated position for the reaction defining the high-pressure stability of tremolite:

$$\underset{\text{Tr}}{Ca_2Mg_5Si_8O_{22}(OH)_2} \rightleftharpoons \underset{\text{Di}}{2CaMgSi_2O_6} + \underset{\text{Tlc}}{Mg_3Si_4O_{10}(OH)_2} \qquad \text{(R3.46)}$$

which is in reasonable agreement with the experimental results of Gilbert and Troll (1974) (Figure 3.7, lines 46a and 46b). The assemblage Di + Tlc would be expected to occur in pyroxenites metamorphosed in the highest pressure eclogite facies under water-saturated conditions.

3.6 Experiments in simple systems relevant to eclogite facies metamorphism of quartzites, Mg-quartzites, pelites and other rock types

3.6.1 *Quartz to coesite transition*

Numerous studies of the reaction:

$$\underset{\text{quartz } (\alpha \text{ or } \beta)}{\text{Qtz}} \rightleftharpoons \underset{\text{coesite}}{\text{Coe}} \qquad \text{(R3.47)}$$

have been performed in piston cylinder apparatus. Two recent experimental studies (Mirwald and Massonne, 1980; Bohlen and Boettcher, 1982) supersede previous work and yield broadly consistent and precise P–T data when differences in starting materials are taken into account. Mirwald and Massonne (1980) studied R3.47 in the T range 600–1600°C, using an in situ 'differential pressure analysis' technique that involved the precise measurement of the piston displacement required to maintain pressure on crossing of the reaction, given by the relation (for low quartz):

$$P(\text{kbar}) = 20.73 + 0.00986\,T\,(^\circ\text{C}) \qquad \text{(E3.9)}$$

(Figure 3.3, line 47a, and Figure 3.4). These workers also reported a shift in the equilibrium position of R3.47 in both water-present and -absent conditions when $NaAlO_2$ was added. This remarkable result has since been dismissed by Bohlen and Boettcher (1982), who found no effect of either H_2O or $NaAlO_2$ (or jadeite) on the reaction. The reversals of Bohlen and Boettcher (1982), obtained exclusively in salt cells over the T range 400–1000°C, can be fitted by the relation:

$$P(\text{kbar}) = 21.6 + 0.0079\,T(^\circ\text{C}) \qquad \text{(E3.10)}$$

(Figure 3.3, line 47b) and suggest that the higher-T data of Mirwald and Massonne (1980) are too high (by < 1 kbar). All the extant experimental data indicate that R3.47 has a positive $\mathrm{d}P/\mathrm{d}T$ (8–10 bars/°C) at all temperatures down to 400°C, contrary to thermochemical data (Holm *et al*, 1967) which suggest a negative $\mathrm{d}P/\mathrm{d}T$ at the lowest temperatures. Existing low-T thermochemical data on coesite are therefore probably seriously in error.

3.6.2 *MASH system*

Major advances in understanding of blueschist–eclogite facies metamorphism in the past decade have been made possible by experimental studies focused on the MASH system. These studies were prompted by the recognition of the widespread occurrence of whiteschist assemblages such as talc + kyanite and talc + phengite, the discovery of important index minerals in magnesian pelites at high pressures (carpholite, Mg-chloritoid), and the finding of the Dora Maira pyrope-coesite-kyanite quartzites and their associated unusual phases. The experimental results discussed here are principally those of Schreyer and co-workers, considered in more detail by Schreyer in recent publications (Schreyer, 1985, 1988a). Both experimentally determined and calculated or inferred equilibria that involve the MASH phases Car, Cld, Chl, Ky, Tlc, Yo, St, Cor, Prp, En, and Qtz/Coe are presented (Figure 3.8) along with the P–T fields that define the stabilities of critical parageneses.

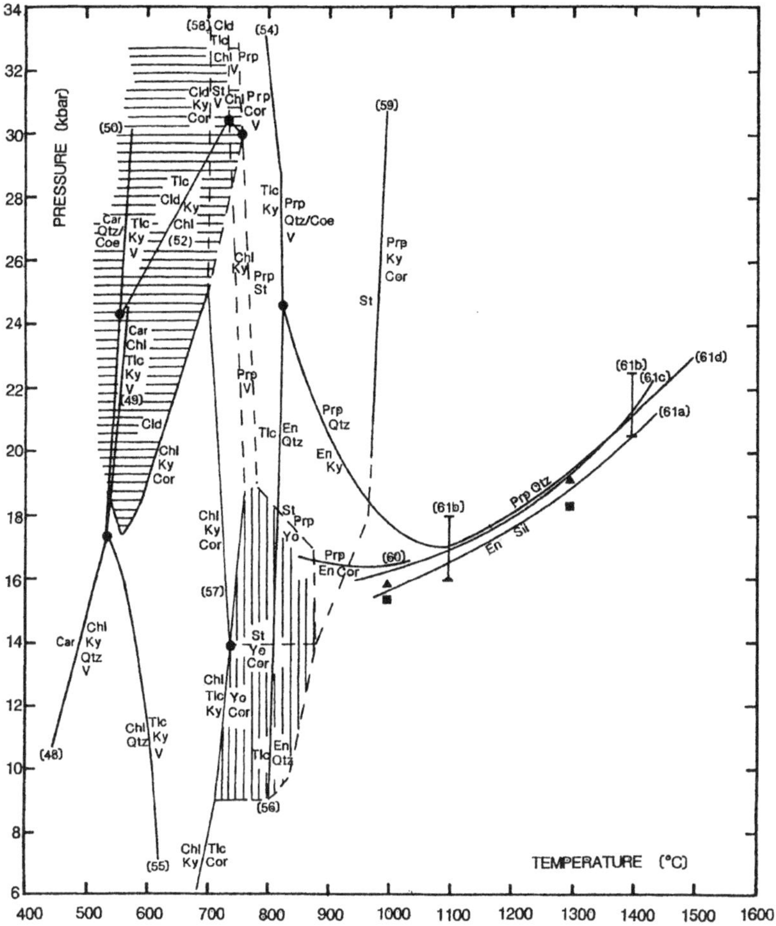

Figure 3.8 Experimentally determined and inferred or calculated equilibria in the MASH system relevant to the formation of eclogite facies assemblages in magnesian metaquartzites and metapelites. Reactions represented by lines (48), (49), (50), (51), (52), (54), (55), (56), (57), (58) and (59) are from Schreyer (1988) and earlier papers (Schreyer and Seifert, 1969; Massonne *et al*, 1981; Chopin and Schreyer, 1983). (60) MAS reaction R3.60 from Gasparik and Newton (1984). (61) MAS reaction R3.61: (61a) Perkins (1983); (61b) Hensen and Essene (1971) (I-beams); calculated by (61c) Wood and Holloway (1984), (61d) Aranovich and Podlesskii (1989). Horizontal ruled field = Mg-Chloritoid; vertical ruled field = Yoderite.

Phase relationships involving Mg-carpholite (Car), $MgAl_2Si_2O_6(OH)_4$, and Mg chloritoid (Cld), $MgAl_2SiO_5(OH)_2$ (Chopin and Schreyer, 1983; Schreyer, 1988a) indicate that Mg-carpholite is stable under water-saturated conditions at 7–35 kbar and temperatures up to 500–580°C. At pressures greater than at least 35 kbar, Car becomes unstable with respect to the new phase MgMgAl-pumpellyite (Schreyer, 1988a). Carpholite can coexist with quartz over most of its stability field. Chopin and

Schreyer (1983) inferred the existence of a [Cld] invariant point that involves the phases Car, Tlc, Chl, Ky, Qtz and V at 18 kbar and 540°C based on reversed experiments on the equilibria

$$\underset{\text{Car}}{5MgAl_2Si_2O_6(OH)_4} \rightleftharpoons \underset{\text{Chl}}{Mg_5Al_2Si_3O_{10}(OH)_8} + \underset{\text{Ky}}{4Al_2SiO_5} + \underset{\text{Qtz}}{3SiO_2} + \underset{\text{V}}{6H_2O} \quad \text{(R3.48)}$$

$$\underset{\text{Car}}{14MgAl_2Si_2O_6(OH)_4} \rightleftharpoons \underset{\text{Chl}}{Mg_5Al_2Si_3O_{10}(OH)_8} + \underset{\text{Tlc}}{3Mg_3Si_4O_{10}(OH)_2} + \underset{\text{Ky}}{13Al_2SiO_5} + \underset{\text{V}}{21H_2O} \quad \text{(R3.49)}$$

$$\underset{\text{Car}}{3MgAl_2Si_2O_6(OH)_4} + \underset{\text{Qtz/Coe}}{SiO_2} \rightleftharpoons \underset{\text{Tlc}}{Mg_3Si_4O_{10}(OH)_2} + \underset{\text{Ky}}{3Al_2SiO_5} + \underset{\text{V}}{5H_2O} \quad \text{(R3.50)}$$

using gel and seeded gel starting materials, in the P–T range 12–24 kbar and 500–600°C. Reactions R3.48–R3.50 occur at temperatures similar to those of the calculated reaction R3.15 (Holland, 1988) and hence render that reaction and the association Cld + Qtz metastable.

Existing indirect experimental evidence (Chopin and Schreyer, 1983) indicates Mg-chloritoid to be stable only at pressures greater than 18 kbar (Figure 3.8) and at temperatures of 400–760°C. As reactions bounding the stability field of Cld involve silica undersaturated phases (e.g. diaspore, corundum), pure Mg-chloritoid is inferred to be stable only in quartz-absent compositions. Of the reactions relevant to Mg-chloritoid presented in the partial grid of Figure 3.8, only

$$\underset{\text{Cld}}{5MgAl_2SiO_5(OH)_2} \rightleftharpoons \underset{\text{Chl}}{Mg_5Al_2Si_3O_{10}(OH)_8} + \underset{\text{Ky}}{2Al_2SiO_5} + \underset{\text{Cor}}{2Al_2O_3} + \underset{\text{V}}{H_2O} \quad \text{(R3.51)}$$

$$\underset{\text{Tlc}}{2Mg_3Si_4O_{10}(OH)_2} + \underset{\text{Cld}}{14MgAl_2SiO_5(OH)_2} \rightleftharpoons \underset{\text{Chl}}{4Mg_5Al_2Si_3O_{10}(OH)_8} + \underset{\text{Ky}}{10Al_2SiO_5} \quad \text{(R3.52)}$$

are constrained by experimental data (Chopin and Schreyer, 1983), the details of which are currently unpublished.

The assemblage Chl + Ky has a large P–T stability field within the eclogite facies, bounded at lower-T by the incoming of Mg-carpholite (R3.48) and at higher-T (about 700°C) by reactions forming Tlc + Cor, Yo + Cor, and particularly Mg-staurolite at higher pressures. The upper-pressure limits on Chl + Ky defined by reaction R3.52, are in the region of about 24 kbar at 700°C (Figure 3.8, line 52) based on extrapolation of the results of Chopin and Schreyer (1983).

The classic whiteschist assemblage Tlc + Ky has a very large stability field under water-saturated conditions, from 6 kbar to probably 40 kbar, where the reactions:

$$\text{Tlc} + \text{Ky} \rightleftharpoons \text{MgMgAl} - \text{pumpellyite} + \text{Qtz} \quad \text{(R3.53)}$$

$$\underset{\text{Tlc}}{Mg_3Si_4O_{10}(OH)_2} + \underset{\text{Ky}}{Al_2SiO_5} \rightleftharpoons \underset{\text{Prp}}{Mg_3Al_2Si_3O_{12}} + \underset{\text{Qtz/Coe}}{SiO_2} + \underset{\text{V}}{H_2O} \quad \text{(R3.54)}$$

are postulated to meet at an invariant point at about 800°C (Schreyer, 1988a). The lower-T stability limits of Tlc + Ky are defined by the reaction R3.50 at high pressures, and by the reaction:

$$\underset{\text{Chl}}{3Mg_5Al_2Si_3O_{10}(OH)_8} + \underset{\text{Qtz}}{14SiO_2} \rightleftharpoons \underset{\text{Tlc}}{5Mg_3Si_4O_{10}(OH)_2} + \underset{\text{Ky}}{3Al_2SiO_5} + \underset{\text{V}}{7H_2O} \quad \text{(R3.55)}$$

experimentally located by Massonne *et al* (1981) at pressures between 6 and 20 kbar (Figure 3.8, line 55). The high-T limit of Tlc + Ky is defined by reaction R3.54 at pressures greater than 24–26 kbar, and by the MSH reaction:

$$\underset{\text{Tlc}}{Mg_3Si_4O_{10}(OH)_2} \rightleftharpoons \underset{\text{En}}{3MgSiO_3} + \underset{\text{Qtz}}{SiO_2} + \underset{\text{V}}{H_2O} \quad \text{(R3.56)}$$

defined in many studies including those of Schreyer and Seifert (1969).

Mg-staurolite has been synthesised at temperatures of 740–1000°C and at pressures greater than 14–18 kbar from a variety of starting materials (Schreyer and Seifert, 1969). Its lower-T stability limit is defined by

$$\underset{\text{Chl}}{4Mg_5Al_2Si_3O_{10}(OH)_8} + \underset{\text{Ky}}{28Al_2SiO_5} + \underset{\text{Cor}}{13Al_2O_3} \rightleftharpoons \underset{\text{St}}{5Mg_4Al_{18}Si_8O_{46}(OH)_2} + \underset{\text{V}}{11H_2O} \quad \text{(R3.57)}$$

(Figure 3.8, line 57) at pressures up to about 25 kbar, and by the inferred reaction:

$$\underset{\text{Cld}}{4MgAl_2SiO_5(OH)_2} + \underset{\text{Ky}}{4Al_2SiO_5} + \underset{\text{Cor}}{Al_2O_3} = \underset{\text{St}}{Mg_4Al_{18}Si_8O_{46}(OH)_2} + \underset{\text{V}}{3H_2O} \quad \text{(R3.58)}$$

at higher P. Schreyer and Seifert (1969) in unreversed synthesis experiments using talc-BN pressure cells and gel or oxide mix starting materials, determined the upper thermal stability of Mg-staurolite at pressures greater than 20 kbar as:

$$\underset{\text{St}}{3Mg_4Al_{18}Si_8O_{46}(OH)_2} \rightleftharpoons \underset{\text{Prp}}{4Mg_3Al_2Si_3O_{12}} + \underset{\text{Ky}}{12Al_2SiO_5} + \underset{\text{Cor}}{11Al_2O_3} + \underset{\text{V}}{3H_2O} \quad \text{(R3.59)}$$

(Figure 3.8, line 59). Phase relations at lower pressures are poorly known because of inadequate experimental evidence on the stability of yoderite, $Mg_2Al_6O_2[SiO]_4(OH)_2$, and the broad interpretation presented by Schreyer (1988a) is followed in Figure 3.8. As both Mg-staurolite and yoderite are stable only in silica-deficient bulk compositions, eclogite facies metamorphism of Mg-pelites and aluminous pelites at pressures of 18 kbar or more will yield two distinct sequences of index minerals at temperatures of 500–900°C:

(a) in quartz-saturated compositions the sequence Car, Chl + Ky, Tlc + Ky, (En + Ky), Prp with increasing T
(b) in quartz-deficient compositions the sequence Car, Cld, Chl + Ky + Cor, St, Prp with increasing T.

Many serious gaps still exist in the MASH grid. The reaction relations between Cld, St and Prp at high pressures are as yet not experimentally defined; and the many reactions possible at 800–900°C and 10–20 kbar have not been clarified (Schreyer,

1988a). Furthermore, there are very few published data on the compositions of the $Al_2Si_{-1}Mg_{-1}$ solid solutions (Chl, Tlc, En), which may vary as functions of *P*, *T* and assemblage and hence cause curvature in reaction lines. Such data are required if the basic experimental relations are to be used in thermochemical calculations.

Experimental data pertaining to pyrope and Prp + Qtz in MAS and MASH are also summarised in Figure 3.8. Reaction R3.54 has been located at 800°C by Chopin (1986). The higher temperature anhydrous breakdown of pyrope:

$$\underset{\text{Prp}}{Mg_3Al_2Si_3O_{12}} \rightleftharpoons \underset{\text{En}}{3MgSiO_3} + \underset{\text{Cor}}{Al_2O_3} \quad \text{(R3.60)}$$

has been reversed by Gasparik and Newton (1984) at 850°C and 16.2 kbar using a salt pressure medium (Figure 3.8, line 60), and is computed to have a shallow positive dP/dT at all temperatures greater than 850°C.

The reaction:

$$\underset{\text{Prp}}{Mg_3Al_2Si_3O_{12}} + \underset{\text{Qtz}}{SiO_2} \rightleftharpoons \underset{\text{En}}{3MgSiO_3} + \underset{\text{Sil/Ky}}{Al_2SiO_5} \quad \text{(R3.61)}$$

has been experimentally investigated over the *T*-range 1000–1300°C in salt + soft glass (Perkins, 1983) and at 1100 and 1400°C in talc (Hensen and Essene, 1971) pressure media (Figure 3.8, lines 61a, 61b). These reversed experiments, when corrected appropriately for pressure-cell rigidity, are within error at 1000°C but diverge by about 1 kbar at higher temperatures, perhaps because of the formation of metastable aluminous enstatites in experiments where seeded oxide mixes were used. Thermodynamic modelling of R3.61 by Wood and Holloway (1984) (Figure 3.8, line 61c) yields a close fit to the experiments of Perkins (1983), but suggest that R3.61 occurs at pressures about 0.5 kbar higher than deduced by Perkins (1983). *P–T* curvature in R3.61 resulting from ΔCp effects and *T*-dependent changes in the alumina contents of enstatite, suggested by the calculations of Wood and Holloway (1984), is very pronounced in more recent calculated fits (Aranovich and Podlesskii, 1989; Holland and Powell, 1985) which indicate a negative dP/dT slope at temperatures below 1000–1100°C. The calculated curve of Aranovich and Podlesskii (1989) (Figure 3.8, line 61d) yields a *P–T* position at 800–900°C consistent with experiments by Massonne (1983), but is not compatible with the 1000°C reversal by Perkins (1983) which occurs at 15.5 kbar. The latter experimental result, however, is itself not consistent with the position of R3.60. More experimental data are required with the *P–T* region of maximum curvature.

3.6.3 *KMASH system*

Incorporation of K_2O as an extra component modifies MASH equilibria relevant to quartz-bearing rock types in the eclogite facies through stabilisation of the phases phlogopite, K-feldspar, and phengitic muscovite. Studies in KMASH have focused on the stability of the diagnostic whiteschist assemblage talc + phengite, and on the extent of $MgSiAl_{-1}Al_{-1}$ substitution in phengites as a function of *P*, *T* and assemblage. The latter aspect and its implications for geobarometry are considered in more detail in Chapter 4.

3.6.3.1 *Talc + phengite stability and phase relations.* Experimental constraints on the stability of talc + phengite have been reported by Massonne and Schreyer (1985, 1989) and summarised by Schreyer (1988). Experiments over the P–T range 5–40 kbar and 500–800°C in salt pressure cells indicate the existence of an invariant point (Figure 3.2, IP) between the phases Chl, Ky, Tlc, Phl, Qtz and V at 11 kbar and 610°C, where the Si content of phengite is calculated as 3.23 pfu from X-ray powder diffraction data. The assemblage talc + phengite is restricted to pressures greater than this invariant point by the low-T reaction:

$$\text{Chl} + \text{Phl} + \text{Qtz} \rightleftharpoons \text{Tlc} + \text{Phe} \qquad \text{(R3.62)}$$

(Figure 3.2, line 62), which is considered to be water-conserving. The high-T stability limit of talc + phengite is controlled by the reaction

$$\text{Tlc} + \text{Phe} \rightleftharpoons \text{Phl} + \text{Ky} + \text{Qtz} + \text{V} \qquad \text{(R3.63)}$$

(Figure 3.2, line 63). Attempted reversals of R3.62 using synthetic mineral mixes proved inconclusive in determining its dP/dT slope because of kinetic and metastability problems, and the P-brackets on this reaction at $T < 600$°C are relatively large. However, four tight reversals of R3.63 at 10–25 kbar define a dP/dT slope of + 85 bars/°C for this reaction (Massonne and Schreyer, 1989).

3.6.3.2 *Phengite stability and compositional variations.* Experiments in the P–T range 2–30 kbar and 350–800°C (Massonne and Schreyer, 1987) utilising a variety of starting materials (gels, gel mixes, oxides and synthetic minerals) and high-pressure apparatus have allowed both the definition of the stability field of the KMASH assemblage phengite + phlogopite + K-feldspar + quartz under water-saturated conditions and determination of the compositional variations in phengites as a function of P and T in this assemblage.

An important upper-pressure constraint on the stability of the assemblage Phe + Phl + Ksp + Qtz under water-saturated conditions is the appearance of a K–Mg–Si-bearing fluid (KMS), of inferred approximate composition $K_2O.3MgO.7SiO_2.(nH_2O)$, at pressures above 16–22 kbar in the T-range 350–700°C (Schreyer *et al*, 1987):

$$\text{Phe} + \text{KMS(fluid)} \rightleftharpoons \text{Kfs} + \text{Phl} + \text{Qtz} + \text{V} \qquad \text{(R3.64)}$$

(Figure 3.2, line 64). This reaction opens up important possibilities for K-metasomatism and magmatism in subduction zone environments (Schreyer *et al*, 1987) and for the buffering of fluid compositions in eclogite facies metagranitoids at pressures greater than those of R3.64. Under water-saturated conditions the Phe + Phl + Kfs + Qtz assemblage is limited to temperatures below 700°C by melting reactions (Figure 3.2).

Data on phengite compositions at 22 P–T locations, obtained through XRD measurements, indicate that phengites in the limiting assemblage (+ Phl + Kfs + Qtz) are not binary dioctahedral micas. Si-contents of phengite in this assemblage increase with increasing P (Massonne and Schreyer, 1987). Synthesis data on phengite compositions at 24 P–T locations (Massonne and Schreyer, 1989) also indicate that the Si-contents of phengites increase with increasing P in the assemblages Phe + Tlc + Ky + Qtz and Phe + Tlc + Phl + Qtz, but also decrease significantly with increasing T in the latter assemblage. Isopleths of constant Si-content for phengites in all three assemblages (Massonne and Schreyer, 1987, 1989), depicted in Figure 3.2, show that phengite composition provides a good geobarometer when the mica coexists with Phl + Kfs + Qtz or Tlc + Ky + Qtz. It should also be noted, however, that the KMASH reactions controlling the Si-contents of phengite in the various assemblages are not water-conserving:

$$\text{Phe} \rightleftharpoons \text{Phl} + \text{Kfs} + \text{V} \qquad \text{(R3.65)}$$

$$\text{Ms} + \text{Tlc} \rightleftharpoons \text{Phe} + \text{Ky} + \text{V} \qquad \text{(R3.66)}$$

$$\text{Ms} + \text{Phl} \rightleftharpoons \text{Phe} + \text{Tlc} + \text{V}, \qquad \text{(R3.67)}$$

As a consequence, the positions of the Si-isopleths will vary as a function of a_{H_2O}. Thus, decreased a_{H_2O} will displace Si-isopleths to higher-P in the case of R3.65, and to lower-P for reactions R3.66 and R3.67.

3.7 Concluding remarks

Although it is clear that substantial progress has been made in the experimental study of eclogite facies mineral assemblages, particularly in simple systems, there are still important gaps in our knowledge and avenues to follow for further experimentation. More experiments are required even in some simple systems (KNASH, MASH) to tie up and complete existing data, and there is much scope for studies in more complex systems where the influence of additional components needs to be evaluated directly. In particular, high-pressure experiments in Fe-bearing systems are required in order to allow the links between the MASH and KMASH studies currently in vogue and lower-P FMASH and FASH petrogenetic grids to be firmly established.

4 Mineral barometry and thermometry

D.A. CARSWELL and S.L. HARLEY

4.1 Background

The compositions of key minerals, such as clinopyroxene and garnet, stable in eclogite facies rocks show considerable variation dependent upon the $P-T-X$ conditions for formation (cf. Chapter 1 and Table 1.1). For particular litho-types differing mineral compositions must essentially reflect varied $P-T$ conditions within the wide overall $P-T$ regime of the eclogite facies (Figure 1.1). Accordingly, considerations of element distribution between coexisting mineral phases are important in the deduction of $P-T$ conditions of formation. In this Chapter the most firmly established mineral thermometers and barometers applicable to eclogite facies mineral assemblages are documented and discussed.

Mineral reactions which record the straightforward exchange of elements with similar ionic sizes, between crystal lattice sites with the same co-ordination arrangement in different minerals, generally involve relatively large entropy changes but only very small volume changes. This is especially the case when the exchanging elements are isovalent—such as with Mg_1Fe_{-1} substitution. The univariant equilibria (with low dT/dP) that express such mineral exchange reactions form the basis of useful thermometers, as they are relatively insensitive to pressure changes.

By contrast more complex mineral reactions (such as coupled $Al_2Mg_{-1}Si_{-1}$ substitution) which record exchange of elements with different valencies between sites with different co-ordination arrangements involve much larger volume changes. Such mineral exchange reaction equilibria (with higher dT/dP) are more sensitive as barometers, although they remain significantly temperature dependent. In addition simple solid–solid mineral reactions, such as those that link the respective polymorphs of SiO_2, Al_2SiO_5 and $CaCO_3$, involve large changes in volume relative to entropy and are thus also valuable indicators of pressures for formation.

The mineral reaction equilibria considered here are designated as either thermometers or barometers but it should be appreciated that they are all in effect both P and T dependent, although to varying degrees. Hence, determination of absolute $P-T$ values requires the establishment of intersections in $P-T$ space of at least a pair of univariant reaction equilibria curves with reasonably divergent dT/dP gradients. In practice derived $P-T$ estimates for a particular mineral assemblage will be encompassed within a $P-T$ polygon or ellipse, the size and shape of which determines the error brackets which should be attached to the $P-T$ values (Hodges and McKenna, 1987; Powell and Holland, 1988).

Mineral reaction equilibria that involve pyroxenes (both clino- and orthopyroxenes) with or without garnet are especially important for the interpretation of eclogite facies mineral assemblages. It is therefore relevant to summarise here the fundamental features of the crystal chemistry of these minerals.

The stoichiometric chemical formula of garnets is customarily expressed as $X_3^{[8]} Y_2^{[6]} Z_3^{[4]} O_{12}$, with eight formula units per unit cell (Deer *et al*, 1982; Meagher, 1982). In the common garnets, divalent cations (Ca^{2+}, Mg^{2+}, Fe^{2+}, Mn^{2+}) are sited in X, trivalent cations (Al^{3+}, Fe^{3+}, Cr^{3+}) in Y, and Si^{4+} in Z. Complications arise in titaniferous garnets with Ti^{4+} generally considered to be sited in Y rather than Z and coupled either to Fe^{2+} in Y or Na^+ in X. Contents of Al^{3+} or Fe^{3+} in Z sites of garnets are minor, and data on the P–T dependence of the intercrystalline partitioning of these elements between Y and Z sites are currently lacking. However, there are some data (e.g. Bishop *et al*, 1978) which indicate that the coupled substitutions of Na^+ in X with Ti^{4+} in Y or P^{5+} in Z are favoured by high pressures.

The stoichiometric formula unit for pyroxenes may be expressed as $(X^{[6-8]} Y^{[6]})_2 Z_2^{[4]} O_6$. Table 4.1 summarises the conventionally accepted nomenclature and occupancies of the common cations in these various lattice sites of pyroxenes (Deer *et al*, 1978; Cameron and Papike, 1980; Ganguly and Ghose, 1979). Whilst cation size considerations restrict certain elements to particular pyroxene sites, others are partitioned between different sites according to P–T conditions of formation. Consideration of the latter relationships in pyroxenes can therefore provide important thermobarometric information.

P–T calibrations of reaction equilibria appropriate to the interpretation of eclogite facies mineral assemblages have been obtained in a number of different ways. A limited amount of experimental data exists for the direct P–T calibration of particular reaction equilibria in natural rock systems (e.g. for Fe^{2+}–Mg^{2+} exchange between garnet and clinopyroxene (Råheim and Green, 1974a; Ellis and Green, 1979). Such data can be applied directly to comparable rocks but problems are to be expected with application outside of the P–T–X range of the experiments. In order that mineral equilibrium can be experimentally achieved in a matter of hours, such experiments have been conducted mainly at high temperatures. Hence uncertainties arise from extrapolations to the lower equilibration temperatures concerned in many natural mineral assemblages.

For eclogite facies mineral reaction equilibria considerably more experimental data

Table 4.1 Site occupancies of common cations in the various lattice sites of pyroxenes

Site nomenclature	$X^{[6-8]}$	$Y^{[6]}$	$Z^{[4]}$	
Cation nomenclature	M2	M1	T(A)	T(B)
Site restricted occupancies	Ca^{2+} Na^+ Li^+	Cr^{3+} Fe^{3+} Ti^{4+}	Si^{4+}	Si^{4+}
Site occupancies dependent on P–T conditions	Mg^{2+} Fe^{2+} Mn^{2+}	Mg^{2+} Fe^{2+} (Mn^{2+})* Al^{3+}		Al^{3+}

*Some analyses of pyroxene crystal chemistry (e.g. Wood and Banno, 1973) assume that Mn^{2+} is restricted to only the larger M2 sites.

are available for simpler, synthetic, compositions—especially in the MAS and CMAS systems. These data are of limited direct application to natural rocks. However, extrapolation of such data to more chemically complex natural mineral reaction equilibria can be achieved through thermodynamic modelling and incorporation of empirically-determined or estimated thermodynamic data (molar volumes V, entropies S, heat capacities Cp, enthalpies of formation ΔH, volume thermal expansion α and compressibility β coefficients) in mineral end-member compositions. Where for a particular mineral assemblage a full thermodynamic data set exists, which is internally consistent with available empirically-determined mineral equilibria, P–T conditions of equilibration can be calculated from consideration of a number of reactions that link the observed mineral phases (e.g. Powell, 1985a; Powell and Holland, 1988). Such calculations are based on the fundamental thermodynamic equilibrium relation that:

$$\Delta G_R = \Delta H^\circ - T\Delta S^\circ + (P-1)\Delta V^\circ + RT\ln K = 0 \qquad \text{(E4.1)}$$

for a balanced chemical reaction between hypothetical end-number compositions (superscript $^\circ$) of the mineral phases concerned, and assuming that ΔV° is independent of P.

As well as calculation errors which often result from uncertainties in the enthalpy of reaction ΔH data, further problems with the equilibrium thermodynamics approach to calculation of P–T conditions arise from differences of opinion over which mineral activity–composition relationships and types of solution models are appropriate for the element exchange reactions concerned. For example, garnets of varied compositions are important phases in most eclogite facies mineral assemblages. As garnet-bearing reaction equilibria typically involve large volume changes they are important pressure indicators in the eclogite facies. However, a serious obstacle to their use arises from complex solution properties of different cations in the garnet structure and consequent uncertainty over values of cation mixing parameters W to be used to calculate end-member activities.

At the current state of the art therefore it seems inadvisable to place total reliance on thermodynamically-calculated P–T estimates. However, close correspondence between such values and those derived from direct experimental calibration of mineral reaction equilibria (even if extrapolated) strengthens confidence in the reliability of deduced P–T conditions for formation. An ultimate test is how reasonable the P–T values are with respect to a petrogenetic grid which defines the relative locations in P–T space of univariant reaction equilibria (such as Figure 1.1) and to general geological constraints (Essene, 1982; Powell, 1985a).

4.2 Garnet–clinopyroxene Al barometer

The Al_2O_3 contents of clinopyroxenes in equilibrium with garnets in eclogite facies mineral assemblages are markedly P–T dependent. In the CMAS system the garnet–clinopyroxene equilibrium which controls the Al_2O_3 content of the clinopyroxene can be represented by the reaction:

$$\underset{\text{Di}}{3CaMgSi_2O_6} + \underset{\text{CaTs}}{3CaAl_2SiO_6} \rightleftharpoons \underset{\text{Grs}}{2Ca_3Al_2Si_3O_{12}} + \underset{\text{Prp}}{Mg_3Al_2Si_3O_{12}} \qquad \text{(R4.1)}$$

Experimental data and thermodynamic evaluation of the CMAS system have permitted calculation of isopleths (equal composition lines) for the Al content of clinopyroxenes in equilibrium with garnet (O'Hara, 1967b; Herzberg, 1978; Gasparik, 1984; Wood and Holloway, 1984). The relatively steep dT/dP gradients of these isopleths enable them to be used for pressure evaluation if an independent temperature estimate is available. However, Nickel *et al* (1985) have shown that Al solubility in clinopyroxene is complex, even in the CMAS system, and that the Al isopleths are not related by linear increments with pressure.

Further difficulties arise because a substantial proportion of the Al in eclogite facies clinopyroxenes is invariably present as jadeite rather than as substitution in a Ca Tschermaks (CaTs) molecule. This problem is especially acute in the clinopyroxenes present in garnet lherzolite assemblages. As these contain only very small amounts of CaTs molecule, determination of the quantity of this component present requires extremely precise and accurate analyses of Al and Si contents. In practice the customary presence of significant amounts of Cr in such clinopyroxenes makes it impossible to determine unequivocally how much Al is present in CaTs relative to that in Jd. The non-stoichiometry of some high P–T clinopyroxenes (Wood and Henderson, 1978; Smyth, 1980; Gasparik and Lindsley, 1980), as expressed by solid solution towards a Ca–Eskola pyroxene molecule ($\square\, Ca_{0.5}AlSi_2O_6$) with X site vacancies, can be a further complication.

Thus, difficulties over the evaluation of the quantity and activity of CaTs in clinopyroxenes seriously impair extrapolation and application of formulations of this particular barometer, based on CMAS system data, to more chemically complex natural garnet–clinopyroxene equilibria.

4.3 Garnet–orthopyroxene Al barometer

The strong pressure dependence of the Al_2O_3 content of orthopyroxenes in equilibrium with garnet was first demonstrated by Boyd and England (1984) and MacGregor and Ringwood (1964). This equilibrium now forms the basis of the most tried and tested mineral reaction barometer applicable to eclogite facies assemblages.

The advent of a large volume of experimental data on garnet–orthopyroxene equilibria in both synthetic and natural rock systems has resulted in extensive application of this barometer, especially to the garnet lherzolite assemblages stable in magnesian meta-peridotites under eclogite facies conditions. In particular this barometer has been widely used, in combination with either the two-pyroxene solvus thermometer or with thermometers based on Fe^{2+}–Mg^{2+} exchange between mineral pairs, to establish upper mantle palaeogeotherms from the P–T arrays calculated for suites of mantle-derived garnet lherzolite xenoliths in kimberlites (e.g. Boyd, 1973; Carswell and Gibb, 1987a, b; Finnerty and Boyd, 1984, 1987). These and other related publications demonstrate that controversy over the shape and interpretation of such palaeogeotherms has arisen from application of the different formulations of the garnet-orthopyroxene Al barometer which are currently available.

In the MAS system the reaction that controls the Al content of orthopyroxene in equilibrium with garnet can be expressed by the following transfer reaction involving Mg-Tschermaks molecule component:

$$\underset{\text{En}}{Mg_2Si_2O_6} + \underset{\text{MgTs}}{MgAl_2SiO_6} \rightleftharpoons \underset{\text{Prp}}{Mg_3Al_2Si_3O_{12}} \qquad \text{(R4.2)}$$

X-ray data and crystal-chemistry analysis have, however, led Ganguly and Ghose

(1978) to propose the following alternative transfer reaction, which involves pyrope component in both orthopyroxene and garnet:

$$\underset{\text{in orthopyroxene}}{Mg_3Al_2Si_3O_{12}} \rightleftharpoons \underset{\text{in garnet}}{Mg_3Al_2Si_3O_{12}} \qquad \text{(R4.3)}$$

The argument behind this alternative representation is that $Al^{[4]}$ is restricted to T(B) sites (Table 4.1) and further to only half of these sites, in order to avoid unstable tetrahedral Al–O–Al linkages (Takeda, 1972). The imposed limit of $Al^{[4]}$ to one quarter of the pyroxene tetrahedral sites leads to the stoichiometric end-member $Mg_3Al_2Si_3O_{12}$ for aluminous orthopyroxenes. The choice of end-members to describe the reaction controlling Al solubility in orthopyroxene in equilibrium with garnet is nonetheless arbitrary, and reaction R4.2 is generally preferred. However, it has been argued by Ganguly and Ghose (1978), Lane and Ganguly (1980) and Bertrand *et al* (1987) that thermodynamic analysis of aluminous orthopyroxene solid solution in the MAS system is simplified if it is expressed in terms of $Mg_4Si_4O_{12}$ and $Mg_3Al_2Si_3O_{12}$ end-member components. Insufficient site distribution data currently exist to justify rejection of the customary view (as expressed in Table 4.1) that $Al^{[6]}$ is effectively restricted to M1 sites in natural orthopyroxenes, in favour of the suggestion by Ganguly and Ghose (1979) that octahedral Al cations have an ordered distribution between M1 and M2 sites in a near 7:3 ratio.

Direct experimental and/or thermodynamic analysis of garnet–orthopyroxene equilibria for SiO_2 undersaturated compositions in the MAS system has been provided by Boyd and England (1964), MacGregor (1974), Wood and Banno (1973), Dankwerth and Newton (1978), Lane and Ganguly (1980), Perkins *et al* (1981), and Kawasaki and Matsui (1983). A review of MAS system data by Bertrand *et al* (1987) indicates that the Al_2O_3 isopleths for enstatite in equilibrium with pyrope determined by Perkins *et al* (1981) are superior to those of MacGregor (1974). Perkins *et al* (1981) indicated slight curvatures to these Al_2O_3 isopleths for enstatite in P–T space. However, their data are effectively compatible with a linear fit to the isopleths as represented by the following (rearranged) barometric expression derived by Finnerty and Boyd (1987):

$$P_{\text{kbar}} = 0.007918 \quad [-T(\text{K})(\ln K - 0.74) - 1976] \qquad \text{(E4.2)}$$

where K is the wt% of Al_2O_3 in enstatite/100.

Data for the SiO_2 undersaturated portion of the 4-component CMAS system simulate more closely mineral reaction equilibria in natural 4-phase garnet lherzolite assemblages, such as observed in kimberlite xenoliths. Direct experimental data on garnet–orthopyroxene equilibria in this system have been provided by Akella (1978), Howells and O'Hara (1978), Perkins and Newton (1980), Yamada and Takahashi (1984) and Nickel *et al* (1985). There are doubts over experimental procedures and satisfactory equilibration with some of the earlier data but the data contained in the latter three publications are more satisfactory and mostly compatible, as indicated and discussed in further papers by Nickel and Green (1985) and Brey *et al* (1986).

Al_2O_3 isopleths for orthopyroxene in equilibrium with garnet + clinopyroxene + olivine in the CMAS system are shown in Figure 4.1. These have been calculated from the expression:

$$P_{\text{kbar}} = 0.01058 \quad [-T(\text{K})(\ln K - 0.94) - 3473] \qquad \text{(E4.3)}$$

where K is wt% of Al_2O_3 in orthopyroxene/100, derived from the linear least-squares

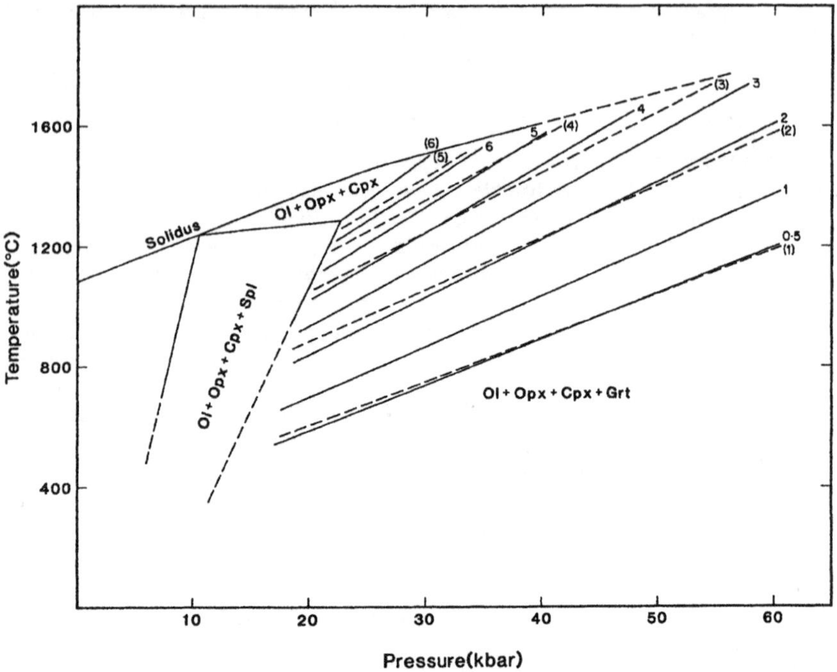

Figure 4.1 *P–T* plot of isopleths (———) for wt% Al_2O_3 in orthopyroxene in equilibrium with garnet + clinopyroxene + olivine for ultrabasic compositions in CMAS system, calculated from linear expression derived by Finnerty and Boyd (1987) from data of Perkins and Newton (1980). Shown for comparison are various pyrolite (peridotite) assemblage stability fields and wt% Al_2O_3 isopleths (– – –) for orthopyroxene in equilibrium with garnet from Green and Ringwood (1967b) for their pyrolite III composition.

fit by Finnerty and Boyd (1987) of the CMAS data from Perkins and Newton (1980). It is clear that the addition of Ca substantially shifts the *P–T* locations of the Al_2O_3 in orthopyroxene isopleths relative to those for the MAS system.

Natural garnet lherzolite assemblages are mostly highly magnesian. For example whole rock 100 Fe/(Fe + Mg) ratios in xenolith samples are mostly in the range 7–11 (Carswell, 1980). Experimental data on the influence of Fe on Al solubility in orthopyroxene in equilibrium with garnet + olivine currently exists for only a limited range of *P–T–X* conditions in the FMAS and CFMAS systems (Kawasaki and Matsui, 1983; Wood, 1974). Harley (1984b) has provided additional data for more Fe and Si rich compositions in these systems, more relevant to mineral equilibria in SiO_2 saturated or oversaturated granulites. These various data suffice to indicate that the addition of Fe reduces the solubility of Al in orthopyroxene in equilibrium with garnet but that the effect is non-linear and most marked in Fe-rich compositions (see Figure 3 in Bertrand *et al*, 1986).

The paper by Wood and Banno (1973) pioneered the use of thermodynamic modelling of garnet–orthopyroxene equilibria in order to correct *P–T* calibrations, derived from experimental data for simple chemical systems, for the influence of the additional chemical components encountered in natural rocks. Their thermodynamic analysis was based on the garnet–orthopyroxene Al equilibrium as expressed in the MgTs transfer reaction indicated in R4.2. It led to the publication by Wood (1974) of a formulation for the garnet–orthopyroxene Al barometer purported to yield equilibration pressures for most natural garnet–orthopyroxene assemblages accurate to within 2–3 kbar. This barometer has since been applied widely to natural assemblages but evaluation tests for garnet lherzolite assemblages (Carswell and Gibb, 1987a) have indicated that the optimistic accuracy estimate by Wood (1974) cannot be justified.

Subsequent alternative formulations of the garnet–orthopyroxene Al barometer, designed to be applicable to natural multi-component rock systems, have been provided by Harley and Green (1982), Harley (1984b), Nickel and Green (1985) and Bertrand *et al* (1986). Additional formulations proposed and tested by Finnerty and Boyd (1984, 1987) have been derived directly from the experimental data for simple synthetic systems by MacGregor (1974), Akella (1978), Lane and Ganguly (1980), Perkins and Newton (1980), Perkins *et al* (1981), Yamada and Takahashi (1984). In their uncorrected forms the latter cannot reasonably be expected to yield accurate P–T values for more chemically complex natural garnet–orthopyroxene assemblages, unless by chance the influences of additional components on the solubility of Al in orthopyroxene cancel each other out. However, the limited experimental data (see Figure 4.1) of Green and Ringwood (1967b) on Al solubility in orthopyroxene in equilibrium with garnet + olivine in synthetic pyrolite compositions, which closely model natural garnet lherzolite rock compositions, suggest that such fortuitous effects do not exist to the extent that an entirely MAS or CMAS data based barometer formulation is likely to be satisfactory.

At first discussed by Wood (1974) a further problem with all applications of this barometer to natural garnet–orthopyroxene equilibria arises from uncertainty over the calculation of the key composition parameter, namely the amount of Al in analyses of natural orthopyroxenes that is involved in Mg-Tschermaks molecule type substitution. In most formulations this is expressed by the term $X^{M1}_{Al(Ts)}$—the mole fraction of Al cations in the M1 sites. However, in the formulations derived by Finnerty and Boyd (1984, 1987) from synthetic, simple system, data (e.g. MacGregor, 1974) it has been more crudely assessed as simply the total weight fraction of Al_2O_3 present in the orthopyroxene analysis. For orthopyroxene–garnet equilibria in the MAS and CMAS systems, $X^{M1}_{Al(Ts)}$ in orthopyroxene can be calculated accurately as half the total Al cation content and Finnerty and Boyd's alternative formulation is acceptable. However, both calculation procedures are unsatisfactory for natural orthopyroxenes with significant contents of Cr, Fe^{3+}, or Ti in M1 sites and/or Na^+ in M2 sites.

Nickel and Green (1985) have proposed that for such orthopyroxenes the amount of Al in M1 sites specifically involved in Tschermaks molecule substitution should be calculated by the following algorithm.

$$X^{M1}_{Al(Ts)} = [Al^{Total} - Cr - 2Ti + Na]/2 \qquad (E4.4)$$

The rationale behind this algorithm is that it corrects for the amount of Al involved in $Cr_1Al^{[4]}_1Mg_{-1}Si_{-1}$ and $Ti_1Al^{[4]}_2Mg_{-1}Si_{-2}$ substitutions (as expressed in $MgCrAlSiO_6$ and $MgTiAl_2O_6$ molecules), after first taking into account the amount of Cr which can be coupled with Na in kosmochlor ($NaCrSi_2O_6$). Nickel and Green (1985) made the assumption that the content of Fe^{3+} cations in orthopyroxenes is negligible. If not, then

E4.4 should be expanded to:

$$X^{M1}_{Al(Ts)} = [Al^{Total} - Cr - Fe^{3+} - 2Ti + Na]/2 \tag{E4.5}$$

to take account of $Na_1Fe^{3+}_1Mg_{-2}$ or $Fe^{3+}_1Al^{[4]}_1Mg_{-1}Si_{-1}$ substitutions (as expressed in $NaFe^{3+}Si_2O_6$ and $MgFe^{3+}AlSiO_6$ molecules).

Carswell and Gibb (1987b) have demonstrated that high *P–T* orthopyroxenes in certain garnet lherzolite xenoliths from the Thaba Putsoa and Mothae kimberlites of northern Lesotho contain significant amounts of jadeite component ($NaAlSi_2O_6$). In these cases there are excess Na cations present over the number which can be coupled with Cr, Fe^{3+} and Ti and algorithms E4.4 and E4.5 break down with the excess Na adding to Al. Accordingly, for sodic orthopyroxenes Carswell and Gibb (1987b) used the alternative algorithm:

$$X^{M1}_{Al(Ts)} = [Al^{Total} + Cr + Fe^{3+} + 2Ti - Na]/2 \tag{E4.6}$$

However, further consideration of the crystal chemistry of these sodic orthopyroxenes has led now to the conclusion that the most valid algorithm for the evaluation of the number of Al cations involved in Tschermaks molecule substitution in orthopyroxenes with $Na > Cr + Fe^{3+} + Ti$ is:

$$X^{M1}_{Al(Ts)} = [Al^{Total} + Cr + Fe^{3+} - Na]/2 \tag{E4.7}$$

This algorithm gives appropriate priority in sodic orthopyroxenes to $NaCrSi_2O_6$ over $MgCrAlSiO_6$, to $NaFe^{3+}Si_2O_6$ over $MgFe^{3+}AlSiO_6$ and to $NaTiAlSiO_6$ over $MgTiAl_2O_6$.

Carswell and Gibb (1987a, b) have reported results of evaluation tests on various proposed formulations of the garnet–orthopyroxene Al barometer, based on their ability to reproduce the experimentally recorded equilibration conditions of natural or synthesised multi-component garnet lherzolite assemblages. The formulation provided by Nickel and Green (1985) was found to give the most accurate results, perhaps predictably since it most closely models the compositions of natural garnet lherzolite assemblages. It had been demonstrated earlier by MacGregor (1970) and O'Neill (1981) that the Cr content of natural garnet lherzolite has a profound influence on the mineral equilibria but the Nickel and Green (1985) barometer formulation was the first to incorporate appropriate experimental data on a Cr-bearing system (SMACCR). These data were combined with CMAS results and the FMAS and CFMAS system expressions developed by Harley and Green (1982) to derive the following geobarometric equation (with *T* in K), which includes both thermodynamic and empirical terms expressing the activity–composition relations of the phases:

$$\begin{aligned} P_{kbar} = {} & \{1/-[183.3 + 178.98 X^{M1}_{Al}(1 - X^{M1}_{Al})]\} \\ & \cdot\{-RT\ln[[(1 - X^{Grt}_{Ca})^3(X^{Grt}_{Al})^2]/[X^{M1}_{Mg.Fe}(X^{M2}_{Mg.Fe})^2 X^{M1}_{Al}]] \\ & - 9000(X^{Grt}_{Ca})^2 - 3400[2(X^{Grt}_{Cr})^2 - X^{M1}_{Mg}X^{M1}_{Cr}] \\ & - X^{Grt}_{Ca}X^{Grt}_{Cr}(90853 - 52.1\,T) \\ & - 7590 X^{Grt}_{Fe}X^{Grt}_{Ca} + 5157 X^{M1}_{Mg}X^{M1}_{Fe} + 6047 - 3.23\,T\} \end{aligned} \tag{E4.8}$$

This formulation is currently considered to yield the most accurate and geologically meaningful equilibration pressure values for garnet lherzolite assemblages in magnesian meta-peridotites. Determination of absolute *P–T* values does depend,

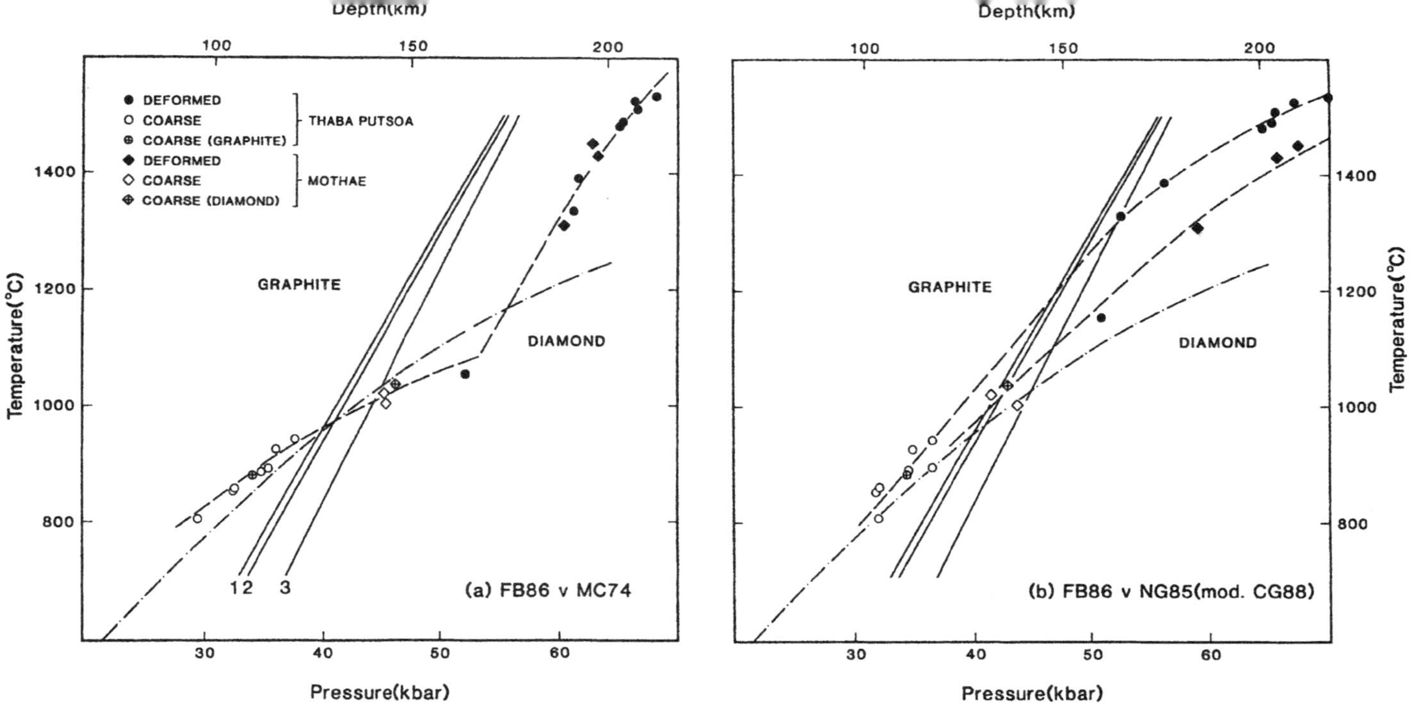

Figure 4.2 Comparative *P–T* equilibration estimates for garnet lherzolite xenoliths from Thaba Putsoa and Mothae kimberlites in northern Lesotho, calculated by two-pyroxene solvus thermometer of Finnerty and Boyd (1987; labelled FB86) with alternative formulations of garnet-orthopyroxene Al barometer: (a) based on experimental data of MacGregor (1974) as formulated and favoured by Finnerty and Boyd (1984, 1987); (b) based on experimental data of Nickel and Green (1985) with modified evaluation of $X^{M1}_{Al(Ts)}$ as outlined in text.

Note the contrasting forms of deduced palaeogeotherms (------) relative to the predicted shield geotherm (------) corresponding to 40 mWm^{-2} surface heat flow of Pollack and Chapman (1977) and diamond–graphite transition curves: (1) from Berman (1979); (2) from Bundy *et al* (1981) and Bundy (1980); (3) from Kennedy and Kennedy (1976).

however, on the availability of a reliable independent temperature estimate. This can be obtained through combined use with any of the superior thermometer calibrations for either the two pyroxene solvus or for Fe^{2+}–Mg^{2+} exchange reactions as outlined later.

Figure 4.2 demonstrates the marked difference between P–T values calculated for the garnet lherzolite xenolith suites from the Thaba Putsoa and Mothae kimberlites in northern Lesotho by the use of E4.8 (together with appropriate evaluation of X_{Al}^{M1} as specifically $X_{Al(Ts)}^{M1}$ from either E4.5 or E4.7, with Fe^{3+} assumed as zero) and those obtained by Finnerty and Boyd (1987) who used their preferred thermometer (two pyroxene solvus)–barometer (after MacGregor, 1974) pairing. These contrasting P–T estimates have led to considerable controversy over the form (inflected or non-inflected) and interpretation (thermally perturbed or not) of the derived upper mantle palaeogeotherms—as comparison of Finnerty and Boyd (1987) with Carswell and Gibb (1987b) will show.

Experimental evidence in the SMACCR system (Nickel, 1986) that for appropriate compositions the assemblage Grt + Opx + Cpx (orthopyroxene eclogite or garnet websterite) is stabilised at lower pressure than Grt + Opx + Cpx + Ol (garnet lherzolite) cautions whether the Nickel and Green (1985) barometer formulation will yield similarly accurate results for more siliceous olivine-free assemblages. In particular, for more Fe- and Si-rich rocks, such as quartz-bearing orthopyroxene eclogites, the barometer formulations of Harley and Green (1982), and Harley (1984b), based on experimental data for more equivalent compositions in the FMAS and CFMAS systems, may well be superior especially at lower pressures. Moreover, there are some indications (Harley, 1984c) that the empirical formulation of Harley and Green (1982) is more satisfactory for Fe-rich compositions than the Harley (1984b) formulation, despite the fact that the latter involves a rigorous thermodynamic treatment of garnet–orthopyroxene equilibria from the same experimental data base. This paradox arises because of the large uncertainties which propagate in fitting complex, multi-parameter, solid solution models to even fairly extensive experimental data bases.

Further complications with applications of all current formulations of the garnet–orthopyroxene Al barometer may arise in particularly Cr-rich rocks. Various experimental data (O'Neill, 1981; Nickel, 1986; Webb and Wood, 1986) show that increased 100 Cr/(Cr + Al) rock ratios extend the stability of spinel to higher pressures and results in progressive expansion of the stability field of an intermediate 5-phase garnet + spinel lherzolite assmblage. It is known that in garnet–spinel equilibria Cr is strongly partitioned into the spinel but nevertheless its presence is likely to have a detectable influence on the Al solubility in orthopyroxene coexistent with garnet. Indeed there is limited evidence (Carswell *et al*, 1984; Carswell, 1986) that for garnet + spinel lherzolite assemblages the Nickel and Green (1985) barometer formulation is unsatisfactory and that more realistic pressure estimates can be calculated by following the procedure outlined in O'Neill (1981).

4.4 Jadeite + quartz barometer

The reaction

$$\underset{\text{albite}}{NaAlSi_3O_8} \rightleftharpoons \underset{\text{jadeite}}{NaAlSi_2O_6} + \underset{\text{quartz}}{SiO_2} \qquad (R4.4)$$

provides the basis of an important barometer applicable to the jadeite-bearing clinopyroxenes in many eclogite facies assemblages, especially the litho-type eclogite (s.s.). The assemblage jadeite + quartz has an appreciably lower molar volume than albite and accordingly is favoured at high pressures. Tight pressure brackets (± 0.5 kbar) for this reaction at $T = 600 - 1200$°C were experimentally established by Holland (1980) and provide the best data on the location in P–T space of the univariant curve for this reaction. These data (reproduced in Figure 4.3) correspond closely with those for the earlier experimental reversal of this reaction at 500–600°C by Newton and Smith (1967) but lie 1–2 kbar higher at corresponding temperatures (1050–1350°C) than the experimental results of Kushiro (1969). Extrapolation to lower temperatures is complicated by the change from the high albite to low albite structural state but available thermodynamic data suggest that any deflection in a linear extrapolation of the univariant reaction curve will be small (Newton and Smith, 1967; Holland, 1980).

More crucial to application of these experimental data for barometry is evaluation of the effect of diluent in diopside, hedenbergite or acmite components with respect to jadeite. Clinopyroxenes in low-T eclogites associated with blueschist terranes commonly have only about 20 mol% of non-jadeite substituents (Di + Hd + Acm) and the experimental data of Newton and Smith (1967), Kushiro (1969) and Holland (1983) all indicate that such amounts are likely to reduce the pressure required for stability by only about 1–2 kbar. However, the clinopyroxenes in most medium–high-T eclogites are omphacites with $X_{Jd} \approx 50$ or less.

Activity–composition relationships along the Jd–Acm join appear to be straightforward with substitution of Fe^{3+} for Al^{3+} in one structural site (M1) and therefore can be described satisfactorily by a simple ideal mixing model. However, relationships along the Jd–Di join are more complex because with $NaAlCa_{-1}Mg_{-1}$ substitution, two different pairs of cations are mixed in two different structural sites (M2 and M1, Table 4.1). Moreover, there is evidence that omphacite (with compositions close to the ideal $Jd_{50}Di_{50}$) comprises a structurally distinct end-member within the Jd–Di join (Rossi, 1982; Holland, 1983) and that at $T < 600$°C there are appreciable miscibility gaps between both Jd–Omp and Omp–Di. In addition it is apparent that omphacites undergo a structural change from a disordered C2/c space group symmetry at $T > 850$°C to an ordered P2/n symmetry at lower temperatures. Wood *et al* (1980) have shown from solution calorimetry that Jd–Di mixing in high temperature (C2/c) pyroxenes is significantly non-ideal. Therefore when the pressure required for stability of eclogite facies clinopyroxenes of mixed Jd–Di composition is calculated from thermodynamic parameters, structural symmetry changes and non-ideality must be taken into account.

The pressure P required for the stability of a natural jadeitic clinopyroxene + albitic plagioclase + quartz assemblage, relative to that required for pure Jd + Ab + Qtz in the simple end-member reaction (R4.4), can be calculated from the thermodynamic equilibrium relationship:

$$\Delta G = O \approx (P - P^{\circ})\,\text{bar} \quad \Delta V^{\circ} + RT(\text{K}) \ln K \qquad \text{(E4.9)}$$

where ΔV° and P° are the volume change and equilibrium pressure, respectively, for the simple end-member reaction and K is the equilibrium constant. $\Delta V^{\circ} = -1.734\,\text{kJ}\,\text{kbar}^{-1}$ for the reaction with high albite and $-1.698\,\text{kJ}\,\text{kbar}^{-1}$ with low albite (Powell and Holland, 1985).

The equilibrium pressure for the end-member reaction (E4.9) has been experimentally determined by Holland (1980) as:

$$P^{\circ} = 0.35 + 0.0265\,T(^{\circ}\mathrm{C}) \pm 0.50\,\mathrm{kbar} \quad \text{(E4.10)}$$

For natural jadeitic clinopyroxene + albitic plagioclase + quartz equilibria

$$K(\text{equilibrium constant}) = (a_{\mathrm{Jd}}^{\mathrm{Cpx}} \cdot a_{\mathrm{SiO_2}}^{\mathrm{Qtz}})/a_{\mathrm{Ab}}^{\mathrm{Pl}}$$

Rearranging E4.9 and noting that for quartz $a_{\mathrm{SiO_2}}^{\mathrm{Qtz}} = X_{\mathrm{SiO_2}}^{\mathrm{Qtz}} = 1$ results in

$$(P - P^{\circ})\,\mathrm{bar} = -[RT(\mathrm{K})/\Delta V^{\circ}] \cdot \ln(a_{\mathrm{Jd}}^{\mathrm{Cpx}}/a_{\mathrm{Ab}}^{\mathrm{Pl}}) \quad \text{(E4.11)}$$

Experimental activity–composition data on the Jd–Di join were considered by Holland (1983) to indicate that for C2/c structured clinopyroxenes a regular disordered two-site solution model with the magnitude of the non-ideality expressed in terms of a large interaction energy ($W = 24 \pm 2\,\mathrm{kJ\,mol^{-1}}$) is most realistic. Accordingly plotted in Figure 4.3 are $X_{\mathrm{Jd}}^{\mathrm{Cpx}}$ isopleths for jadeitic clinopyroxene + albite ($a_{\mathrm{Ab}}^{\mathrm{Pl}} = X_{\mathrm{Ab}}^{\mathrm{Pl}} = 1$) + quartz equilibria calculated from equations E4.10 and E4.11 (effectively as in Holland, 1983) on the basis of his favoured activity–composition model with $a_{\mathrm{Jd}}^{\mathrm{Cpx}} = (X_{\mathrm{Jd}}^{\mathrm{Cpx}})^2 \cdot \gamma_{\mathrm{Jd}}$ and the activity coefficient $\gamma_{\mathrm{Jd}} = \exp[(W/RT)(1 - X_{\mathrm{Jd}}^{\mathrm{Cpx}})^2]$. It should be noted that these calculated isopleths diverge considerably at low $X_{\mathrm{Jd}}^{\mathrm{Cpx}}$ and low T values from the earlier isopleths determined empirically at high temperatures (1050–1350°C) by Kushiro (1969). Certainly it is apparent that equilibrium pressure estimates for

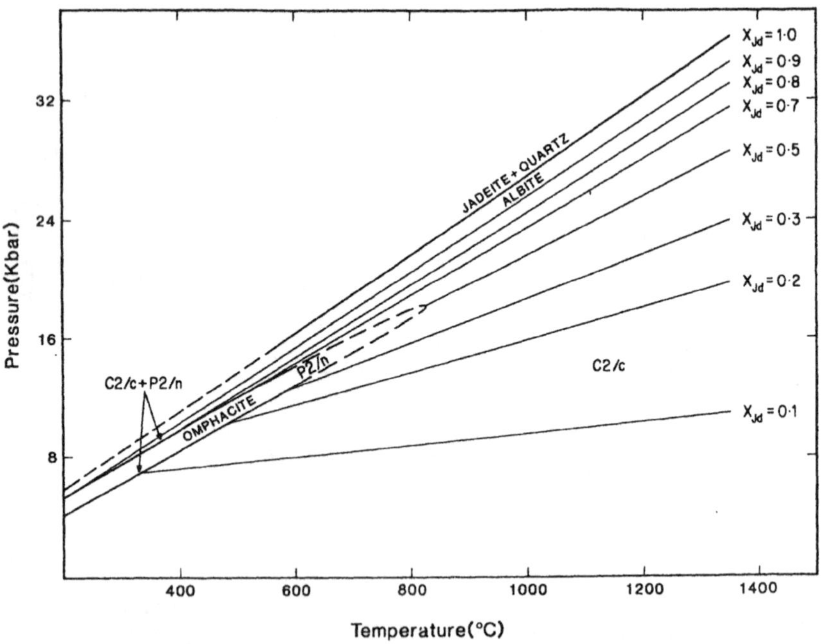

Figure 4.3 Univariant reaction curve for albite = jadeite + quartz (after Holland, 1980) and isopleths of jadeite content X_{Jd} in disordered (C2/c) and ordered (P2/n) clinopyroxenes coexisting with albite and quartz, calculated as outlined in text (after Holland, 1983).

eclogite facies clinopyroxenes with low X_{Jd}^{Cpx} values are the most critically dependent upon the mixing model employed in the calculation of activity values. They are therefore subject to the greatest uncertainty. Independent evidence, for example of $P-T$ conditions for eclogite formation in the Western Gneiss region of Norway (Chapter 8) suggests that with respect to natural eclogite facies clinopyroxenes the $X_{Jd}^{Cpx} = 0.3$ isopleth at $T = 700$–$800°C$ in Figure 4.3 may be 2–4 kbars too low.

For structurally ordered P2/n omphacites close to the ideal $Jd_{50}Di_{50}$ composition reasonable pressure estimates can probably be obtained on the assumption that $a_{Jd}^{Cpx} = X_{Jd}^{Cpx}$ (Holland, 1983). For assemblages in which the coexisting albitic plagioclase has significant anorthite component diluent (i.e. $X_{Ab}^{Plag} \ll 1$) the activity of albite component in the plagioclase can be calculated from the following expression derived from solution calorimetry by Perkins and Newton (1981):

$$a_{Ab} = X_{Ab}^2(2 - X_{Ab})\exp[(X_{An}^2/RT)(6746 - 9392X_{Ab})]$$

In reality, few eclogite facies assemblages—and certainly few eclogites (*s.s.*)—contain albitic plagioclase in clear textural equilibrium with a jadeitic clinopyroxene and quartz. In the large majority of cases sodic plagioclase in eclogites is of secondary origin and is related to breakdown of the primary jadeitic clinopyroxene during retrogressive decompression. Hence, in most eclogites the jadeite content of the primary clinopyroxene can be used only to calculate a minimum pressure for the equilibration of the eclogite facies assemblage.

4.5 Phengite barometer

Previous discussion in this chapter of pyroxene crystal chemistry has demonstrated that Tschermaks component substitution $Al^{[6]}Al^{[4]}M^{2+}_{-1}Si_{-1}$ is strongly $P-T$ dependent: X_{Ts} increases with T but decreases with P. Also as a general principle higher confining pressures can be expected to favour aluminium occupancy in octahedrally co-ordinated rather than in tetrahedral co-ordinated mineral lattice sites. Such crystal chemical considerations are also the basis of the phengite barometer, which is especially important for the evaluation of equilibration pressures in relatively low-T–high-P metapelites.

Phengite is essentially an intermediate composition dioctahedral mica within the muscovite–celadonite solid solution series as follows:

muscovite	$K\,Al_2^{[6]}[Si_3Al]^{[4]}O_{10}(OH)_2$
phengite	$K[Al_{1.5}(MgFe^{2+})_{0.5}]^{[6]}[Si_{3.5}Al_{0.5}]^{[4]}O_{10}(OH)_2$
celadonite	$K[(MgFe^{2+})_{1.0}(AlFe^{3+})_{1.0}]^{[6]}[Si_4]^{[4]}O_{10}(OH)_2$

The phengite barometer as originally proposed and formulated by Velde (1967) considers the Si^{4+} content per formula unit (>3.00 but <4.00) as a measure of the extent of inverse Tschermaks molecule substitution $(MgFe^{2+})^{[6]}Si^{[4]}Al^{[6]}_{-1}Al^{[4]}_{-1}$ in muscovitic micas towards the ideal celadonite composition. Petrographic observations, especially of blueschist facies metapelites (e.g. Ernst, 1963b; Velde, 1967; Moore and Liou, 1979), support crystal chemical prediction—from the higher $Al^{[6]}/Al^{[4]}$ ratio in phengite relative to muscovite—that the celadonite content of such dioctahedral micas is positively correlated with pressure.

Velde (1967, figure 1) provided the first experimentally-based calibration (with data from Velde, 1965) on the P–T dependence of the Si^{4+} content in dioctahedral potassic white micas. However, this calibration has been superseded by that of Massonne and Schreyer (1987), based on more detailed synthesis experiments up to much higher pressures (24 kbar) in the water excess KMASH system. Their calibration, which demonstrates a near linear increase in the Si^{4+} content of phengitic micas with P and a moderate decrease with T is reproduced in Figure 3.2.

Massonne and Schreyer (1987) have demonstrated that phengitic micas in the KMASH system are not simply binary solid solutions along the muscovite–celadonite join. They are more complex ternary solid solutions between muscovite, celadonite and phlogopite—$K\,Mg_3^{[6]}[Si_3Al]^{[4]}O_{10}(OH)_2$. Instead of having an ideal dioctahedral mica composition, the octahedral cation content is generally > 2.00 per formula unit, with values up to about 2.10 recorded. They caution that only the ternary phengite composition that coexists in a limiting assemblage with K feldspar + phlogopitic mica + quartz can be used to evaluate equilibration P–T conditions from these Si^{4+} isopleths (Figure 3.2). They also noted that the composition of the coexisting trioctahedral mica in the KMASH system differs significantly from ideal phlogopite and has $Si^{4+} > 3.00$ and octahedral site occupancies as low as 2.8 per formula unit at high pressures. Accordingly, it is expected that inverse Tschermaks substitution is markedly pressure dependent in both the dioctahedral and trioctahedral mica solid solutions.

The reaction that controls the celadonite content of an assumed true dioctahedral potassic mica in equilibrium with the limiting assemblage of K feldspar + trioctahedral phlogopite + quartz can be expressed in the KMASH system as follows:

$$\begin{aligned}
&\underset{\text{phengite}}{6K[Al_{1.5}Mg_{0.5}]^{[6]}[Si_{3.5}Al_{0.5}]^{[4]}O_{10}(OH)_2} \\
&\rightleftharpoons \underset{\text{muscovite}}{3KAl_2^{[6]}[Si_3Al]^{[4]}O_{10}(OH)_2} + \underset{\text{K feldspar}}{2KAlSi_3O_8} \\
&\quad + \underset{\text{phlogopite}}{KMg_3^{[6]}[Si_3Al]^{[4]}O_{10}(OH)_2} + \underset{\text{quartz}}{3SiO_2} + \underset{\text{water}}{2H_2O} \qquad \text{(R4.5)}
\end{aligned}$$

In natural high pressure metapelites this reaction can be generalised as:

$$\text{phengite}_1 \rightleftharpoons \text{phengite}_2 + \text{K feldspar} + \text{phlogopitic mica} + \text{quartz} + \text{water vapour} \qquad \text{(R4.6)}$$

with phengite of higher Si^{4+} content, favoured at higher pressures over the higher molar volume assemblage, on the right hand side of this reaction equation.

It is clear from these reaction equations that the Si^{4+} content of phengitic micas will depend upon a_{H_2O} as well as P_s–T conditions. A reduction in a_{H_2O} will shift the Si^{4+} isopleths for phengites in this limiting assemblage to higher P compared with the water saturated system ($a_{H_2O} = 1$). For example, thermodynamic calculation by Massonne and Schreyer (1987) predicts a 3 kbar pressure increase for the $Si^{4+} = 3.30$ isopleth with $a_{H_2O} = 0.6$. On the other hand, the substitutions $Fe^{2+}Mg_{-1}$, $Fe^{3+}Al_{-1}^{[6]}$ and NaK_{-1} in natural phengites are all expected to reduce the Si^{4+} content at any pressure relative to that indicated in Figure 3.2 for pure MgAl phengite in the limiting assemblage with Kfs + Phl + Qtz.

Phengites that coexist with alternative HP assemblages, Tlc + Ky + Qtz or Phl +

Ky + Qtz, are also expected to have Si^{4+} contents which increase with pressure. Stability fields and predicted Si^{4+} isopleths for phengites in these kyanite-bearing assemblages in the KMASH system are also shown in Figure 3.2 (see also discussion in section 3.6.3). These provisional Si^{4+} isopleths are consistent with the occurrence of phengite ($Si^{4+} = 3.55$) in a kyanite + talc + nearly pure pyrope + coesite assemblage from the Dora Maira Massif (Chapter 10) for which pressures of 30–35 kbar are indicated for stability (Chopin, 1984).

The HP stability of phengite with Kfs + Phl + Qtz is limited under H_2O-saturated conditions by the appearance of a K and Mg-rich siliceous fluid (R3.64 in Figure 3.2). So the stability in metapelites of phengite with this assemblage is constrained to P_s–T conditions which can be expected to stabilise blueschists rather eclogites in any associated H_2O-saturated metabasaltic rocks (cf. Figure 1.1). Accordingly such phengite bearing metapelitic assemblages are perhaps more appropriately assigned to the blueschist rather than eclogite facies. On the other hand, the Phe + Ky + Tl + Prp + Coe assemblage from Dora Maira has apparently equilibrated at confining pressures sufficiently high for eclogite stability even in water-saturated metabasaltic rocks. Phengites occur in eclogitic mica schists (e.g. Compagnoni, 1977; Lombardo *et al*, 1977; Koons and Thompson, 1985) and in HP jadeite-bearing metagranitoids (e.g. Oberhänsli *et al*, 1985) in field association with true meta-basaltic eclogites in the Sesia-Lanzo Zone, Western Alps (Chapter 10). Stability of phengite is reported in certain K_2O-bearing aluminous low–medium-*T* eclogites (e.g. Råheim and Green, 1975; Krogh and Råheim, 1978; Ernst and Dal Piaz, 1978; Krogh, 1980). Such findings are all also indicative of phengite stability within the eclogite facies *P–T* regime.

4.6 Other barometers

Minimum equilibration pressure at any particular *T* can be estimated for eclogite facies assemblages that contain any of the HP mineral polymorphs: aragonite, kyanite, coesite or diamond—with reference to the appropriate experimentally determined inversion curves in Figure 1.1.

At medium–high temperatures the assemblage grossular-rich garnet + kyanite + anorthite-rich plagioclase + quartz may remain stable in metamorphosed marly sediments up to pressures in excess of those required to stabilise meta-basaltic eclogites (Figure 1.1). The relatively low d*P*/d*T* slope of the univariant curve for the reaction:

$$\underset{\text{in plagioclase}}{\underset{\text{An}}{3CaAl_2Si_2O_8}} \rightleftharpoons \underset{\text{in garnet}}{\underset{\text{Grs}}{Ca_3Al_2Si_3O_{12}}} + \underset{\text{Ky}}{2Al_2SiO_5} + \underset{\text{Qtz}}{SiO_2} \quad \text{(R4.7)}$$

means that it constitutes a valuable barometer for such assemblages. There have been various experimental studies of this end-member reaction and several different formulations of this barometer proposed, of which that by Newton and Haselton (1981), based on the experimental data of Goldsmith (1980), has probably been most frequently applied. However, the more recent detailed experimented study of this reaction at $T = 900–1250°C$ and $P = 19–28$ kbar by Koziol and Newton (1988) has indicated that the univariant curve for the end-member reaction R4.7 is best described by the equation:

$$P^{\circ}(\text{bar}) = 22.80\,T(°\text{C}) - 1093 \quad \text{(E4.12)}$$

From this and the approximated thermodynamic calculation of the free energy change for the reaction as outlined by Newton and Haselton (1981), the following barometric expression for this reaction can be derived:

$$P(\text{bar}) = \{-141.59[T(\text{K}) - 273] + 6788 - RT(\text{K})\ln[a_{\text{Grs}}^{\text{Grt}}/(a_{\text{An}}^{\text{Pl}})^3]\}/\Delta\bar{V}_{\text{R}} \quad \text{(E4.13)}$$

Procedures for the calculation of the activity values and $\Delta\bar{V}_R$ (the partial molal volume change of the reaction at 1 bar) are outlined in Newton and Haselton (1981) but overall uncertainties in resultant P estimates are largely attributable to these parameters. In the absence of plagioclase, at least minimum pressures can be estimated from this barometer for eclogite facies assemblages that contain grossular-bearing garnet + kyanite + quartz.

Medium–high T eclogites may contain the assemblage omphacite (about Jd_{50}) + kyanite + paragonite in relatively aluminous meta-basaltic lithologies (e.g. Tauern window, Eastern Alps—Miller 1977, 1986; Holland, 1979a). The univariant curve for the reaction:

$$\underset{\text{Pg}}{NaAl_3Si_3O_{10}(OH)_2} = \underset{\text{Jd}}{NaAlSi_2O_6} + \underset{\text{Ky}}{Al_2SiO_5} + \underset{\text{V}}{H_2O} \quad \text{(R4.8)}$$

has been determined by Holland (1979b) from hydrothermal reversal experiments at $T = 550$–$740°C$ and is reproduced in Figure 1.1. The shallow negative slope ($-10\,\text{bar}°\text{C}^{-1}$) of this reaction curve means that it constitutes a valuable barometric indicator for kyanite eclogites formed under H_2O-saturated ($a_{H_2O} = 1$) conditions. Holland (1979a) has calculated, on the basis of an assumed one-site ideal solution model for Jd–Di mixtures, that omphacite ($Jd_{50}Di_{50}$) will coexist with kyanite and paragonite at pressures about 5 kbar lower than will pure jadeite. Reduced water activity ($a_{H_2O} < 1$) can be expected to lower further the pressures required for the stability of natural kyanite eclogites (*cf* Barnicoat and Fry, 1986, figure 9).

The higher Na content in garnets from HP diamond-bearing eclogite xenoliths and in garnet inclusions of eclogitic affinity within diamonds, relative to the observed contents in garnets lower P crustal eclogites (Sobolev and Laurent'ev, 1971; Meyer and Boyd, 1972; Reid *et al*, 1976), has prompted suggestions that Na substitution into garnet is significantly pressure dependent. Experimental calibration of the P–T dependence of Na partitioning between garnet and clinopyroxene is not yet available but may prove useful for evaluation of equilibration pressures of eclogites (*s.s.*), where the absence of coexisting orthopyroxene precludes the use of the garnet–orthopyroxene Al barometer. However, other composition factors will probably have to be taken into account in eclogite assemblages as indications are that the entry of Na into garnet reflects the coupled substitution of $NaTiCa_{-1}Al_{-1}$ (Reid *et al*, 1976; Bishop *et al*, 1978) or perhaps Na P $Ca_{-1}Si_{-1}$ (Thompson, 1975) rather than $NaSiCa_{-1}Al_{-1}$ as proposed earlier by Sobolev and Laurent'ev (1971). Moreover, the review of garnet compositions in diamond eclogites presented in Chapter 13 indicates that the values of both X_{Na}^{Grt} and $K_{Na} = X_{Na}^{Grt}/X_{Na}^{Cpx}$ are primarily controlled by bulk rock composition rather than by the P–T conditions of formation.

4.7 Two-pyroxene solvus thermometer

Mutual solubility between coexisting Ca-rich pyroxene (Cpx) and Ca-poor pyroxene (mostly Opx, but pigeonite at high T with $P <$ about 15 kbar) has long been recognised to increase with T and hence provide the basis of a thermometer (e.g. Davis and Boyd, 1966; Boyd, 1973) widely applicable to mineral assemblages in metabasic and meta-ultrabasic compositions. The shape and location of the two-pyroxene solvus in the simple CMS system is controlled by two coupled transfer reactions:

$$\underset{\text{En in Opx}}{Mg_2Si_2O_6} \rightleftharpoons \underset{\text{En in Cpx}}{Mg_2Si_2O_6} \qquad \text{(R4.9)}$$

$$\underset{\text{Di in Opx}}{CaMgSi_2O_6} \rightleftharpoons \underset{\text{Di in Cpx}}{CaMgSi_2O_6} \qquad \text{(R4.10)}$$

These reactions imply the transfer of En and Di components between the coexisting Ca-poor and Ca-rich pyroxene phases but can be combined into a single reaction that expresses the Ca–Mg exchange in M2 sites between the two pyroxenes:

$$\underset{\text{Di in Opx}}{CaMgSi_2O_6} + \underset{\text{En in Cpx}}{Mg_2Si_2O_6} \rightleftharpoons \underset{\text{En in Opx}}{Mg_2Si_2O_6} + \underset{\text{Di in Cpx}}{CaMgSi_2O_8} \qquad \text{(R4.11)}$$

i.e. $CaMg_{-1}(Cpx) = Ca_{-1}Mg(Opx)$

As the Di limb of the solvus is more temperature dependent than the En limb, formulations of this thermometer which consider only a single limb of the solvus are preferably based on the Cpx composition—for example those derived by Finnerty and Boyd (1984, 1987). Where the compositions of both coexisting pyroxenes are considered, the transfer reaction R4.9 is preferred to the complementary reaction R4.10 because any imprecision in the determination of the low content of Ca in orthopyroxene at $T <$ about 1000°C will introduce a large error into the calculation of the equilibrium constant for reaction R4.10.

The equilibrium constant for the preferred transfer reaction R4.9 is calculated as:

$$K = a_{En}^{Cpx}/a_{En}^{Opx} = X_{En}^{Cpx}\cdot\gamma_{En}^{Cpx}/X_{En}^{Opx}\cdot\gamma_{En}^{Opx} = X_{En}^{Cpx}/X_{En}^{Opx}$$

if a crude assumption of ideal mixing of Di and En components is made (Wood and Banno, 1973). On the reasonable assumption that all Ca is ordered into M2 sites then pyroxenes in the CMS system have:

$$X_{En}^{Pyx} = X_{Mg}^{M1}\cdot X_{Mg}^{M2} = X_{Mg}^{M2} = (1 - X_{Ca}^{M2}) \qquad \text{(E4.14)}$$

However, for complex multi-component pyroxenes the mole fraction of enstatite present in pyroxenes is probably most appropriately calculated as (after Wood and Banno, 1973):

$$\begin{aligned} X_{En}^{Pyx} &= X_{Mg}^{M1}\cdot X_{Mg}^{M2} \\ &= \frac{Mg^{2+}}{(Mg^{2+}+Fe^{2+})}[1.00 - (Al^{3+[6]} + Cr^{3+} + Ti^{4+} + Fe^{3+})] \\ &\cdot \frac{Mg^{2+}}{(Mg^{2+}+Fe^{2+})}[1.00 - (Ca^{2+} + Na^{+} + Mn^{2+})] \qquad \text{(E4.15)} \end{aligned}$$

The two-pyroxene solvus thermometer is especially valuable for the assessment of equilibration temperatures in certain of the highest T eclogite facies assemblages—such as in garnet lherzolite xenoliths with HT-deformed textures documented by Nixon and Boyd (1973). However, at $T <$ about 900°C the temperature sensitivity of X_{En}^{Pyx} is substantially reduced and thus this thermometer is less reliable for temperature evaluation of MT eclogite facies assemblages and inapplicable to LT assemblages.

Early experimental data (Atlas, 1952; Boyd and Schairer, 1964; Davis and Boyd, 1966) provided some insight into the temperature dependence of the two pyroxene solvus in the CMS system. More recent reversal experiments by Mori and Green (1975), Lindsley and Dixon (1976), Perkins and Newton (1980), Brey and Huth (1984) have provided a superior calibration of the two pyroxene solvus over a wide range of *P–T* conditions in the CMS system. These have indicated that the pressure influence on the solvus is minimal at low *T* but becomes significant at *T* > about 1100°C where the extent of mutual solubility decreases with increased *P* (Figure 4.4). So the frequently used formulation of Wells (1977), representing a best-fit to the then available CMS experimental data which had an assumed ideal mixing model and no *P* correction term, cannot be expected to yield reliable *T* estimates for the highest *P–T* eclogite facies assemblages. The fact that it appears significantly to underestimate the equilibration temperatures of the HT garnet lherzolite xenolith assemblages, compared with other 'superior' themometer formulations, confirms this (Carswell and Gibb, 1987b, table 1). Importantly, however, the Wells (1977) formulation does incorporate an empirically derived correction term to take account of increased mutual solubility of pyroxenes that results from the presence of Fe as an additional component (Lindsley and Munoz, 1969; Mori, 1978; Lindsley, 1983). However, although the Wells (1977) formulation takes no note of observed non-ideality or *P*-dependence in pyroxene miscibility, the averaging method of its calibration appears to yield acceptable temperatures under conditions where *T* < 1100°C and *P* < 30 kbar.

None of the attempts to establish a thermodynamic solution model which can predict the observed phase equilibria on the join Di–En over a wide range of *P–T* conditions has been entirely successful (for example Holland *et al*, 1979; Lindsley *et al*,

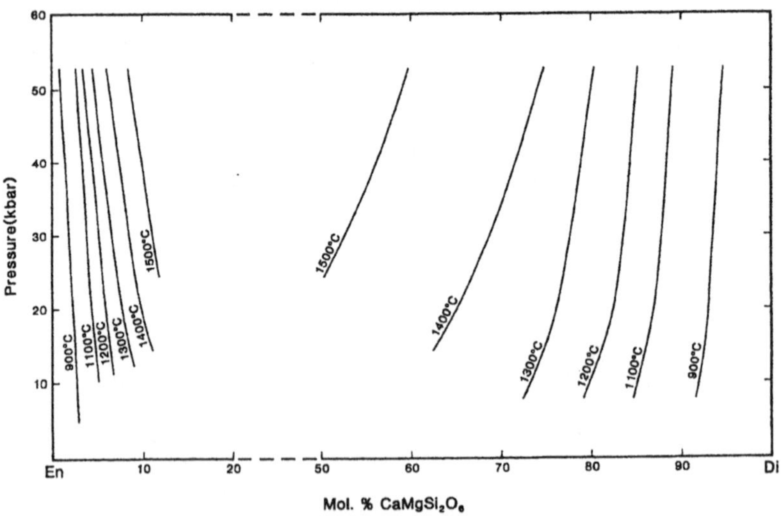

Figure 4.4 *P–T* dependence of the two pyroxene solvus in the enstatite–diopside system based on reversed experimental data from Lindsley and Dixon (1976), Brey and Huth (1984), Mori and Green (1975), Nickel and Brey (1984).

1981; Nickel and Brey, 1984). Particular problems arise because of the complexities of polymorphism with the Ca-poor pyroxene phase, especially under low-P–high-T conditions where a monoclinic Fe-free pigeonite structure is stable instead of one of the more customary orthorhombic structures (orthoenstatite or protoenstatite). Even the most recent, highly sophisticated, thermodynamic calibration of the two-pyroxene solvus by Carlson and Lindsley (1988), although satisfactory from 800–1100°C at all pressures, fails to replicate adequately the experimentally determined pyroxene phase relations at higher temperatures. Moreover, as none of the available thermodynamically modelled formulations includes the additional terms necessary for extrapolation from CMS system data to the more complex multi-component compositions of natural pyroxenes, they cannot reasonably be expected to yield meaningful equilibration T estimates for most natural pyroxene pairs.

Tests on the reproducibility of experimentally recorded equilibration conditions for natural multi-component garnet lherzolite assemblages (Bertrand and Mercier, 1985; Carswell and Gibb, 1987a, b) indicate that the two-pyroxene solvus thermometer formulation provided by Bertrand and Mercier (1985) generates more reliable T estimates for such assemblages than other then available formulations. The Bertrand and Mercier (1985) formulation is based on 'superior' experimentally reversed data sets for the CMS system with extrapolation to natural pyroxene compositions through incorporation of an empirical correction for Fe content and a theoretical correction for Na. Their Na correction is equivalent to normalising the Ca content of the pyroxene to a Na-free system and assumes no significant Na–Ca or Na–Mg interactions in M2 sites. Like the earlier formulation of Wells (1977) that of Bertrand and Mercier (1985) does not include any correction term to take account of Al influence on the two pyroxene solvus as indicated, for example, by the experimental data on pyroxene compositions in the CMAS system, within both the spinel (Fujii, 1977) and garnet (Perkins and Newton, 1980) stability fields.

Nickel *et al* (1985) have shown that in the CMAS system pyroxene behaviour, especially of clinopyroxene, is complex. The different observed styles of Al incorporation into the pyroxene structure for low and high Al concentrations strongly influences the X_{En} values of coexisting pyroxenes. There was noted a P–T dependent crossover (Nickel *et al*, 1985, figures 2 and 9), from reduced pyroxene miscibility with increased Al content at low temperatures at any particular P to increased pyroxene miscibility with increased Al content at higher temperatures. The crossover occurs near 1100°C at 30 kbar and 1300°C at 50 kbar. Bertrand and Mercier (1985) concluded that it was unnecessary to include in their thermometer formulation any correction for the influence of any Al present, which may in part reflect the fact that most of the experimentally equilibrated natural garnet lherzolite assemblages which they used to test their thermometer equilibrated close to P–T conditions with no discernible Al effect on X_{En}. Moreover, eclogite facies garnet lherzolite assemblages have pyroxenes with relatively low Al contents and, especially in clinopyroxenes, much of the Al is incorporated in Jd rather than CaTs components. Nonetheless the evidence from the larger test data set employed by Carswell and Gibb (1987a, b) appears to substantiate the prediction from the CMAS system data of Nickel *et al* (1985) that the Bertrand and Mercier (1985) formulation temperatures for garnet lherzolite assemblages, equilibrated at greater depths than those corresponding to P–T conditions of about 50 kbar and 1300°C, may be slight overestimates, whereas temperatures for assemblages equilibrated at shallow depths can be expected to be underestimated.

Brey *et al* (1989) have determined the following expression for the two-pyroxene solvus thermometer, based on reversed experimental data for CMS and natural peridotite systems at 900–1400°C and 2–60 kbar:

$$T(\mathrm{K}) = [23664 + (24.9 + 126.3\, X_{\mathrm{Fe}}^{\mathrm{Cpx}}) P_{\mathrm{kbar}}]/[13.38 + (\ln K)^2 + 11.59\, X_{\mathrm{Fe}}^{\mathrm{Opx}}] \qquad \text{(E4.16)}$$

with

$$K = [1 - \{X_{\mathrm{Ca}}^{\mathrm{Cpx}}/(1 - X_{\mathrm{Na}}^{\mathrm{Cpx}})\}]/[1 - \{X_{\mathrm{Ca}}^{\mathrm{Opx}}/(1 - X_{\mathrm{Na}}^{\mathrm{Opx}})\}]$$

and

$$X_{\mathrm{Fe}}^{\mathrm{Pyx}} = \mathrm{Fe}/(\mathrm{Fe} + \mathrm{Mg})$$

Interestingly, Brey *et al* (1989) report that substitutions of Cr and Al have counter-balancing effects on the transfer reaction R4.9 and that, for natural pyroxenes in 4-phase peridotites, deviations from the pyroxene miscibility relations in the CMS system can be attributed essentially to Fe.

No clear preference between this calibration (E4.16) and that of Bertrand and Mercier (1985) can be established from currently available test data sets (Carswell and Gibb, 1987a and b). It yields higher T estimates for the lower T xenolith assemblages and closely similar, although again almost invariably higher, estimates for the higher T xenolith assemblages. The Bertrand and Mercier (1985) and Brey *et al* (1989) two pyroxene solvus thermometer formulations both appear to yield reasonably acceptable T estimates for natural garnet lherzolite assemblages, but the considerations outlined caution against their application to olivine-free garnet pyroxenite assemblages which have distinctly more aluminous pyroxenes, especially those with high Al/Cr ratios.

4.8 Garnet–clinopyroxene Fe^{2+}–Mg exchange thermometer

Fe^{2+}–Mg partitioning between coexisting garnet and clinopyroxene is strongly temperature dependent and constitutes an important mineralogical thermometer from which a T based classification of eclogites (*s.s*) and other eclogite facies litho-types (e.g. garnet lherzolites) can be established (Banno, 1970; Mysen and Heier, 1972; and Chapter 1). The P–T dependence of the $K_{\mathrm{D}_{\mathrm{Fe}^{2+}-\mathrm{Mg}}}^{\mathrm{Grt-Cpx}}$ distribution coefficient (commonly designated simply as K_D) was first experimentally calibrated by Råheim and Green (1974a) for a series of basaltic rock compositions crystallised to eclogite at 20–40 kbar and 600–1400°C. Further experimental data and re-evaluation have resulted in a number of revised calibrations (most notably by Ellis and Green, 1979; Powell, 1985b; Krogh, 1988) which indicate that the dependence of K_D on P is much less than deduced by Råheim and Green (1974a).

The exchange of Fe^{2+} and Mg between garnet and clinopyroxene can be expressed by the reaction:

$$\underset{\mathrm{Prp}}{\tfrac{1}{3}\mathrm{Mg_3Al_2Si_3O_{12}}} + \underset{\mathrm{Hd}}{\mathrm{CaFeSi_2O_6}} \rightleftharpoons \underset{\mathrm{Alm}}{\tfrac{1}{3}\mathrm{Fe_3Al_2Si_3O_{12}}} + \underset{\mathrm{Di}}{\mathrm{CaMgSi_2O_6}} \qquad \text{(R4.12)}$$

The equilibrium constant K for this reaction can be defined as:

$$K = \frac{(a_{\mathrm{Alm}}^{\mathrm{Grt}})^{1/3} \cdot (a_{\mathrm{Di}}^{\mathrm{Cpx}})}{(a_{\mathrm{Prp}}^{\mathrm{Grt}})^{1/3} \cdot (a_{\mathrm{Hd}}^{\mathrm{Cpx}})} = \frac{(X_{\mathrm{Fe}^{2+}}/X_{\mathrm{Mg}})^{\mathrm{Grt}}}{(X_{\mathrm{Fe}^{2+}}/X_{\mathrm{Mg}})^{\mathrm{Cpx}}} = K_\mathrm{D} \qquad \text{(E4.17)}$$

if these minerals are assumed to be ideal solid solutions.

From the standard thermodynamic eqilibrium relation E4.1.

$$\Delta H^{\circ} - T\Delta S^{\circ} + (P-1)\Delta V^{\circ} = -RT \ln K_{D}$$

Therefore, at constant pressure:

$$\ln K_{D} = -(\Delta H^{\circ}/RT) + (\Delta S^{\circ}/R)$$

Thus $\ln K_D = A/T(\mathrm{K}) + B$ (where A and B are constants) and a linear relationship between $\ln K_D$ and $1/T(\mathrm{K})$ is to be expected.

From re-evaluation of the experimental data of Råheim and Green (1974a, b) Mori and Green (1978), and Ellis and Green (1979) for Fe^{2+}–Mg exchange between garnet and clinopyroxene in various basaltic, peridotitic and CFMAS system compositions, (normalised to $X_{Ca}^{Grt} = 0.2$ and $P = 30$ kbar), Krogh (1988) determined the following linear relationship (Figure 4.5):

$$\ln K_{D} = 3278/T(\mathrm{K}) - 1.393 \qquad \text{(E4.18)}$$

The magnitude of the P effect on K_D is related to the volume change for reaction R4.12 with $d \ln K_D/dP = \Delta V^{\circ}/RT$ (Banno, 1970). Molar volume data indicate that $\Delta V^{\circ}/R = -10$ (Powell, 1985b). Thus modification of E4.18 through addition of a pressure correction term gives:

$$\ln K_{D} + 1.393 = 3278/T(\mathrm{K}) + 10(P_{\mathrm{kbar}} - 30)/T(\mathrm{K}) \qquad \text{(E4.19)}$$

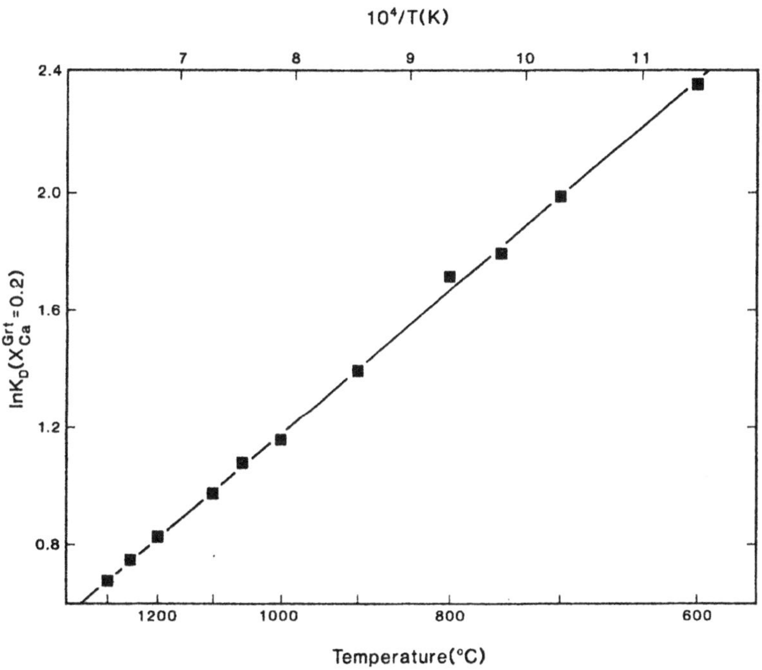

Figure 4.5 Plot of $\ln K_D[K_D = (X_{Fe^{2+}}/X_{Mg})^{Grt}/(X_{Fe^{2+}}/X_{Mg})^{Cpx}$ versus $1/T$ (K) for reaction R4.12 based on experimental data normalised to $X_{Ca}^{Grt} = 0.20$ and $P = 30$ kbar. Diagram reproduced with permission from Krogh (1988, figure 2).

The experimental study of Ellis and Green (1979) demonstrated that ln K_D varies considerably with X_{Ca}^{Grt} [Ca/(Ca + Mg + Fe) in garnet]. This composition dependence reflects the non-ideality of Ca–Mg substitutions in garnet and clinopyroxene. On the other hand, K_D was found to be essentially independent of the Mg/(Mg + Fe) ratios in garnet and clinopyroxene (Ellis and Green, 1979; Krogh, 1988).

The empirical relation expressing K_D as a function of T, P and X_{Ca}^{Grt} derived by Ellis and Green (1979) involves a rectilinear relationship between ln K_D and X_{Ca}^{Grt}. Powell (1985b) presented a slightly modified K_D thermometer equation, formulated on a similar basis to that of Ellis and Green (1979) but with two influential data outliers excluded. Tests conducted by Carswell and Gibb (1987a) have indicated that differences between temperatures calculated from the respective K_D thermometer equations of Ellis and Green (1979) and Powell (1985b) are small (probably within uncertainties derived from other sources such as Fe^{2+}/Fe^{3+} ratios in minerals) but that the Powell (1985b) calibration yields slightly lower and seemingly superior equilibration temperature estimates at $T < 1200°C$. The alternative calibrations of Råheim and Green (1974a), Mori and Green (1978), Ganguly (1979) and Saxena (1979) have all been found to yield less satisfactory results.

Krogh (1988) has demonstrated that at least in the composition range $X_{Ca}^{Grt} = 0.10$–0.50 there is a curvilinear rather than rectilinear relationship between ln K_D and X_{Ca}^{Grt}. Combining his deduced quadratic regression equations for the variation in ln K_D with X_{Ca}^{Grt} at various temperatures with equation E4.19, Krogh (1988) derived the following revised expression for the K_D thermometer:

$$T(\mathrm{K}) = [-6173(X_{Ca}^{Grt})^2 + 6731\,X_{Ca}^{Grt} + 1879 + 10\,P(\mathrm{kbar})]/(\ln K_D + 1.393) \quad \text{(E4.20)}$$

Tests indicate that at $T < 1000°C$ this revised thermometer expression yields lower temperatures than the equations of Ellis and Green (1979) and Powell (1985b), the differences increasing for lower T and lower X_{Ca}^{Grt} values. Further experimental data are required to establish more fully the relationships between ln K and X_{Ca}^{Grt} at low X_{Ca}^{Grt} values (especially at < 0.10), but E4.20 (after Krogh, 1988) is judged to be the best K_D thermometer calibration currently available for LT–MT eclogites. There are some indications, however, that the alternative expression of Powell (1985b) may yield superior T estimates for HT eclogites (Chapter 13). Additional uncertainty over application of all currently available variants of the $K_{Fe^{2+}-Mg}^{Grt-Cpx}$ thermometer arises from experimental data by Pattison and Newton (1987) which suggest that K_D may be dependent on Mg/(Mg + Fe) in garnet (and whole rock), contrary to the indications from the earlier experimental data of Råheim and Green (1974), Ellis and Green (1979).

For omphacitic clinopyroxenes in meta-basaltic eclogites there appears to be no obvious influence on K_D of the jadeite content of the clinopyroxene (Koons, 1984; Krogh, 1988). However, for more jadeitic clinopyroxenes ($X_{Jd}^{Cpx} > 0.7$) in quartzo-feldspathic eclogite facies assemblages (e.g. metagranitoids) K_D appears to vary inversely with jadeite content, probably due to non-ideality in $Fe^{2+}MgNa_{-1}Al_{-1}$ substitution resulting in preferential ordering of Fe^{2+} into M2 sites in the clinopyroxenes (Koons, 1984). Thus the thermometer equation E4.20 should not be applied to garnet–clinopyroxene pairs in such lithologies without incorporation of an additional composition correction factor to take account of the influence of such high X_{Jd}^{Cpx} values on K_D.

A serious problem with application of the K_D thermometer, and indeed for all thermometers based on Fe^{2+}–Mg^{2+} exchange between coexisting mineral pairs, arises from the lack of knowledge of $Fe^{2+}/(Fe^{2+} + Fe^{3+})$ ratios in minerals when analysed by electron microprobe techniques. Wet chemical analyses or Mössbauer spectroscopy on mineral separates have often indicated significant Fe^{3+} contents in eclogite facies mineral assemblages with lower $Fe^{2+}/(Fe^{2+} + Fe^{3+})$ ratios in clinopyroxenes than in coexisting garnets (e.g. Medaris, 1980; Sanders *et al*, 1984). Consequently the frequently employed assumption that all Fe is Fe^{2+} in both minerals will generally result in the calculation of a *minimum* K_D value, which will in turn yield a *maximum* equilibration T estimate for that assemblage. An alternative approach is to calculate Fe^{3+} contents from electron microprobe analyses of the minerals either by sequential allocation of cations to end member molecules (Essene and Fyfe, 1967; Mysen and Griffin, 1973; Cawthorn and Collerson, 1974) or by an assumption of overall stoichiometry and charge balance (Neumann, 1976; Droop, 1987). The problem with all such Fe^{3+} calculation procedures is that they are highly sensitive to any analytical errors (especially in SiO_2 content), a fact which effectively prohibits their use for low Fe clinopyroxenes, as in garnet lherzolite assemblages in magnesian metaperidotites. Further difficulties arise with the former types of calculation procedures because selection of different end-member calculation sequences can either maximise (Essene and Fyfe, 1967; Mysen and Griffin, 1973) or minimise (Cawthorn and Collerson, 1974) the determined content of Acm and Ts molecules relative to Jd. On balance the more generally applicable charge balance calculation schemes of Neumann (1976) or Droop (1987), which both yield similar results, are preferable but are dependent on perfect mineral stoichiometry which may not always be the case (e.g. Wood and Henderson, 1978).

4.9 Garnet–orthopyroxene Fe^{2+}–Mg exchange geothermometer

The exchange of Fe^{2+} and Mg between coexisting garnet and orthopyroxene

$$\underset{\text{Prp}}{\tfrac{1}{3}Mg_3Al_2Si_3O_{12}} + \underset{\text{Fs}}{FeSiO_3} \rightleftharpoons \underset{\text{Alm}}{\tfrac{1}{3}Fe_3Al_2Si_3O_{12}} + \underset{\text{En}}{MgSiO_3} \qquad \text{(R4.13)}$$

forms the basis of a useful geothermometer applicable to MT–HT eclogites and cofacial peridotites. This exchange equilibrium has been calibrated as a geothermometer in independent experimental studies (Harley, 1984a; Lee and Ganguly, 1988) and also through calculations from thermodynamic data (Sen and Bhattyacharya, 1984).

Harley (1984a) experimentally determined the dependence of $K_{D\,Fe^{2+}-Mg}^{Grt-Opx}$ (K_D^{GO}) on T in both the FMAS and CFMAS systems, in the P–T range 5–30 kbar and 800–1200°C. This extensive data set, which comprises 88 experimental data points at 36 P–T conditions, provides an excellent basis for construction of the thermometer. Whilst seeded-glass mixes were used in most runs, some experiments utilised mineral-mixes and reversibility of the Fe–Mg partitioning was demonstrated at selected P–T conditions. Preferential addition of Fe from iron capsules into orthopyroxene during the course of some experiments caused a lowering of K_D^{GO} (by 0.1–0.2) compared with similar experiments carried out in graphite capsules, and hence led to some scatter in the K_D^{GO} data even when the arbitrary adjustment is made. Using mainly the K_D^{GO} data obtained from experiments performed in graphite capsules, Harley (1984a) derived the following geothermometric expression:

$$T(\mathrm{K}) = \frac{1882(\pm 307) + 11.5P(\mathrm{kbar}) + 705(X_{\mathrm{Ca}}^{\mathrm{Grt}} + X_{\mathrm{Mn}}^{\mathrm{Grt}})}{\ln K_{\mathrm{D}}^{\mathrm{GO}} + 0.986(\pm 0.02)} \qquad \text{(E4.21)}$$

where $K_{\mathrm{D}}^{\mathrm{GO}} = (\mathrm{Fe/Mg})^{\mathrm{Grt}}/(\mathrm{Fe/Mg})^{\mathrm{Opx}}$

$X_{\mathrm{Ca}}^{\mathrm{Grt}} = (\mathrm{Ca/Ca + Mg + Fe + Mn})$ in garnet

$X_{\mathrm{Mn}}^{\mathrm{Grt}} = (\mathrm{Mn/Ca + Mg + Fe + Mn})$ in garnet

This result may be compared with that directly obtained by Lee and Ganguly (1988) who presented $K_{\mathrm{D}}^{\mathrm{GO}}$ data at 10 P–T conditions in the range 20–45 kbar and 975–1400°C. These reversed $K_{\mathrm{D}}^{\mathrm{GO}}$ data, obtained using mineral mixes and with the aid of a $\mathrm{PbO{-}PbF_2}$ flux, complement those of Harley (1984a) and moreover are consistent with the earlier data within the limits of experimental uncertainty. The calculated best-fit to the data of Lee and Ganguly (1988):

$$T(\mathrm{K}) = \frac{1981(\pm 166) + 11.9\,P(\mathrm{kbar}) + 1510(X_{\mathrm{Ca}}^{\mathrm{Grt}} + X_{\mathrm{Mn}}^{\mathrm{Grt}})}{\ln K_{\mathrm{D}}^{\mathrm{GO}} + 0.97(\pm 0.015)} \qquad \text{(E4.22)}$$

yields temperatures 60–90°C higher than those calculated using the relation after Harley (1984a). Both equations are reproduced in Figure 4.6 for the FMAS case and normalised to a pressure of 25 kbar. It is clear that both equations fit the FMAS data, including that of Kawasaki and Matsui (1983) obtained at higher pressures, when experimental uncertainties are taken into account. Treatment of all the data sets together, with extra weighting given to reversed $K_{\mathrm{D}}^{\mathrm{GO}}$ data from each study, yields the following thermometric expression generalised from FMAS into more complex systems by adding an appropriate Ca + Mn correction factor as in equations E4.21

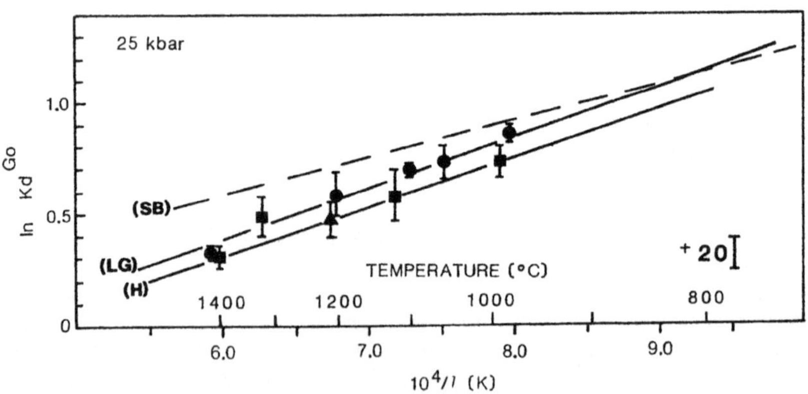

Figure 4.6 Plot of $\ln K_{\mathrm{D}}^{\mathrm{GO}}$ versus $1/T$ K for the garnet–orthopyroxene exchange reaction at 25 kbar in FMAS system. Experimental data with uncertainties are from Harley (1984a, ■) Lee and Ganguly (1988, ●), and Kawasaki and Matsui (1983, ▲). Lines (———) LG and H refer to the geothermometer relationships of Lee and Ganguly (1988), and Harley (1984a) respectively. Line (– – –) SB is the thermometer based on Sen and Bhattacharya (1984) simplified.

and E4.22:

$$T(\mathrm{K})=\frac{1920(\pm 210)+11.9\,P(\mathrm{kbar})+[\Delta W/R](X_{\mathrm{Ca}}^{\mathrm{Grt}}+X_{\mathrm{Mn}}^{\mathrm{Grt}})}{\ln K_{\mathrm{D}}^{\mathrm{GO}}+0.975(\pm 0.02)} \qquad \text{(E4.23)}$$

where ΔW is a net interaction parameter related to non-ideal mixing in garnet.

Whilst there is no obvious compositional dependence of $K_{\mathrm{D}}^{\mathrm{GO}}$ on Fe/Mg (Harley, 1984a; Lee and Ganguly, 1988), $K_{\mathrm{D}}^{\mathrm{GO}}$ increases with increasing Ca in the garnet (Harley, 1984a). The only direct experimental data on the effect of $X_{\mathrm{Ca}}^{\mathrm{Grt}}$ on $K_{\mathrm{D}}^{\mathrm{GO}}$ is that of Harley (1984a) who found that the increase in $K_{\mathrm{D}}^{\mathrm{GO}}$ is not as great as would be expected by comparison with garnet–clinopyroxene data (Ellis and Green, 1979) and typical garnet solid solution models. Thus Harley (1984a) derived a value of 1400 ± 500 cal/mol site for the net interaction term $W_{\mathrm{CaMg}}^{\mathrm{Grt}} - W_{\mathrm{CaFe}}^{\mathrm{Grt}}$ whereas Lee and Ganguly (1988) used a value of 3000 cal/mol site obtained from independent sources. The magnitude of ΔW to be chosen in E4.22 probably lies nearer the lower value, based on application to natural rock data (Carswell, unpubl.; Brey *et al*, 1989). However, ΔW probably varies with $X_{\mathrm{Mg}}^{\mathrm{Grt}}$ and to a limited extent with temperature.

The experimentally based geothermometer relations have ΔH values within error (-3740 ± 610 cal/mol and -3916 ± 282 cal/mol) and identical ΔS values (about -1.96 cal/mol K). The enthalpy is in excellent agreement with some independent calorimetric results (Lee and Ganguly, 1988), which suggests that the presence of Al in orthopyroxene has little effect on the Fe–Mg partitioning. These consistent experimental calibrations of the garnet–orthopyroxene thermometer are at variance with the calculated thermometer of Sen and Bhattyacharya (1984) which has a much stronger T dependence than the experiments indicate is real. The Sen and Bhattyacharya (1984) geothermometer will yield similar temperatures to those obtained from E4.22 at $T < 900$°C, but will diverge to unrealistically high temperatures at $T > 900$°C. The geothermometer expression E4.22, with $\Delta W/R$ evaluated near 700, is recommended here for application to eclogites and garnet peridotites.

4.10 Garnet–olivine Fe^{2+}–Mg exchange thermometer

Fe^{2+}–Mg partitioning between garnet and olivine can be represented by the reaction:

$$\underset{\text{Prp}}{2Mg_3Al_2Si_3O_{12}} + \underset{\text{Fa}}{3Fe_2SiO_4} \rightleftharpoons \underset{\text{Alm}}{2Fe_3Al_2Si_3O_{12}} + \underset{\text{Fo}}{3Mg_2SiO_4} \qquad \text{(R4.14)}$$

The apparent equilibrium distribution coefficient (assuming ideal solid solutions)

$$K_{\mathrm{D}_{Fe^{2+}-Mg}}^{\mathrm{Grt-Ol}} = (X_{Fe^{2+}}/X_{Mg})^{\mathrm{Grt}}/(X_{Fe^{2+}}/X_{Mg})^{\mathrm{Ol}}$$

is less sensitive to temperature than $K_{\mathrm{D}_{Fe^{2+}-Mg}}^{\mathrm{Grt-Cpx}}$ (Mori and Green, 1978) but has the advantage that Fe^{3+}/Fe^{2+} ratios in olivines are expected to be lower and therefore less problematic than in clinopyroxenes.

Experimental calibration of this thermometer at $P = 30$ kbar and $T = 900$–1400°C for FMAS and CFMAS system compositions by O'Neill and Wood (1979) has demonstrated that $K_{\mathrm{D}_{Fe^{2+}-Mg}}^{\mathrm{Grt-Ol}}$ is strongly dependent both on the Fe/Mg ratio of the minerals and on $X_{\mathrm{Ca}}^{\mathrm{Grt}}$. Expectations are that this thermometer is most accurate for relatively magnesian compositions (such as many natural eclogite facies garnet peridotite assemblages) at $T < 1300$°C. The pressure influence on $K_{\mathrm{D}_{Fe^{2+}-Mg}}^{\mathrm{Grt-Ol}}$ has been estimated to raise calculated temperatures by 3–6°C kbar^{-1}. Derivation of equilibration temperatures from the thermometric expression of O'Neill and Wood (1979) requires iterative calculation because of a complex volume term (DV). It is also necessary to incorporate corrections detailed by O'Neill and Wood (1980) to the equations (26) and (24) of O'Neill and Wood (1979).

An alternative formulation of the $K_{D_{Fe^{2+}-Mg}}^{Grt-Ol}$ thermometer has been derived by Kawasaki (1979) from thermodynamic analysis of experimental data for the FMAS system. On the basis of only a limited number of tests on experimentally equilibrated garnet lherzolite assemblages (Carswell and Gibb, 1987a) it was tentatively concluded that the calibration of O'Neill and Wood (1979, 1980) yields slightly more accurate equilibration temperatures than that of Kawasaki (1979).

4.11 Garnet–phengite Fe^{2+}–Mg exchange thermometer

Phengitic white mica is an important HP mineral phase (Section 4.6) and consequently Fe^{2+}–Mg partitioning between phengite and garnet is of potential value for the evaluation of equilibration temperatures in certain eclogite facies assemblages. Krogh and Råheim (1978) provided a provisional calibration of this thermometer, based largely on experimental data by Råheim and Green (1974a), for a limited range of basaltic compositions with minor K_2O contents. Whilst their calibration was judged to yield acceptable temperatures for phengite-bearing eclogites with low Fe_2O_3 contents, application to radically different rock compositions such as garnet-phengite-quartzs schists (eclogite facies meta-pelites) is questionable.

The garnet phengite Fe^{2+}–Mg exchange reaction may be expressed as:

$$\underset{\text{Prp}}{\tfrac{1}{6}Mg_3Al_2Si_3O_{12}} + \underset{\text{Fe–Phe}}{KFe_{0.5}Al_2Si_{3.5}O_{10}(OH)_2}$$

$$\rightleftharpoons \underset{\text{Alm}}{\tfrac{1}{6}Fe_3Al_2Si_3O_{12}} + \underset{\text{Mg–Phe}}{KMg_{0.5}Al_2Si_{3.5}O_{10}(OH)_2} \qquad \text{(R4.15)}$$

with the equilibrium constant:

$$K = \frac{(a_{Mg-Phe}^{Phe})\cdot(a_{Alm}^{Grt})^{1/6}}{(a_{Fe-Phe}^{Phe})\cdot(a_{Prp}^{Grt})^{1/6}} = [(X_{Fe^{2+}}/X_{Mg})^{Grt}/(X_{Fe^{2+}}/X_{Mg})^{Phe}]^{1/2} = [K_{D_{Fe^{2+}-Mg}}^{Grt-Phe}]^{1/2}$$

if ideal solutions are assumed.

In reality the garnet and phengite solid solutions display considerable non-ideality and $K_{D_{Fe^{2+}-Mg}}^{Grt-Phe}$ is dependent upon T, P and X. Further experimental studies by Green and Hellman (1982) for a range of phengite mica + quartz compositions have demonstrated marked dependence of $K_{D_{Fe^{2+}-Mg}}^{Grt-Phe}$ on the CaO content and MgO/(MgO + FeO) ratio of the bulk composition. Accordingly, Green and Hellman (1982) have thermodynamically derived separate calibrations of this thermometer for low CaO pelitic rock compositions, with alternatively low and high MgO/(MgO + FeO) ratios, and for basaltic rock compositions.

$K_{D_{Fe^{2+}-Mg}}^{Grt-Phe}$ is also appreciably dependent upon pressure but the dP/dT slopes of the $K_{D_{Fe^{2+}-Mg}}^{Grt-Phe}$ and $K_{D_{Fe^{2+}-Mg}}^{Grt-Cpx}$ isopleths are sufficiently disparate to enable P–T values to be established for phengite bearing eclogites within geologically acceptable precision limits (Krogh and Råheim, 1978). However, more serious errors may arise from uncertainties over the Fe^{3+} contents of natural phengitic micas. This may not be too serious in certain basaltic rock compositions with low Fe_2O_3 content but it effectively precludes application of this thermometer to many pelitic rock compositions, since these have high Fe_2O_3/FeO ratios.

If natural phengites contain Fe^{3+} then an assumption that all Fe in analysed phengites is present as Fe^{2+} will mean that calculated temperatures at any pressure will only be *maximum* values. However, if natural phengites have high Fe^{3+}/Fe^{2+} ratios relative to coexisting garnet and clinopyroxene (and all Fe is assumed to be Fe^{2+} in all three minerals) then combined use of $K_{D_{Fe^{2+}-Mg}}^{Grt-Phe}$ $K_{D_{Fe^{2+}-Mg}}^{Grt-Cpx}$ isopleths may well result in P–T estimates which are too low (Krogh and Råheim, 1978, figure 3). Unfortunately, because of the complexities of possible cation site vacancies and variable H_2O contents, it is not feasible to calculate Fe^{3+} contents in phengites on an assumption of an ideal formula and charge balance with any acceptable degree of certainty. Such problems signal extreme caution in the application of the garnet–phengite Fe^{2+}–Mg exchange thermometer to assemblages analysed only by electron microprobe, even within the limited rock composition ranges for which this thermometer has so far been calibrated.

4.12 Trace element exchange thermometers

As yet the possibilities for such thermometers have not been fully explored and very few have been adequately calibrated. A serious, but surmountable, drawback is that superior analytical routines to most current procedures are required, in order to determine the trace element contents with sufficient precision to yield accurate equilibration temperature estimates.

Mysen (1976) presented an experimental calibration for Al^{VI}–Cr partitioning between coexisting pyroxenes of potential value for T evaluation of garnet lherzolite and garnet websterite (orthopyroxene eclogite) assemblages. Experience with application of this thermometer to natural garnet lherzolite assemblages is that it yields erratic, frequently inaccurate, results (Carswell and Gibb, 1980a and b). Some of these may be due to analytical imprecision in the determination of Cr contents in the pyroxenes but uncertainty over the $Al^{[6]}/(Al^{[6]}+Al^{[4]})$ ratios in the pyroxenes is also a serious problem. An assumption that $Al^{[6]}=Al^{Total}/2$ is invalid for most eclogite facies pyroxenes, especially in the clinopyroxenes. Alternative calculation of $Al^{[4]}$ as 2.000-Si^{4+} and $Al^{[6]}=Al^{Total}-Al^{[4]}$ is extremely sensitive to even small errors in the SiO_2 analysis. In practice such problems severely limit the accuracy of this thermometer.

Carswell (1974) noted that the distribution coefficient

$$K_{D_{Mn-Mg}}^{Grt-Cpx}=(X_{Mn}/X_{Mg})^{Grt}/(X_{Mn}/X_{Mg})^{Cpx}$$

increases significantly with temperature in garnet–clinopyroxene pairs from garnet lherzolites. Delaney *et al* (1979) indirectly calibrated a modified formulation of this thermometer over the range $T=900$–$1300°C$ against temperatures determined from the two pyroxene solvus (Wells, 1977) and $K_{D_{Fe^{2+}-Mg}}^{Grt-Ol}$ (O'Neill and Wood, 1979) thermometers. Extrapolation to $T<900°C$ is uncertain and exacerbated by the low content of Mn in clinopyroxenes at such temperatures. Further problems with the formulation and application of this thermometer stem from non-ideal Ca–Mg substitutions in garnet and clinopyroxenes, uncertainty over the site occupancy of Mn in pyroxenes (Table 4.1) and the fact that the Mn present in minerals may not all be divalent.

Ni–Mg partitioning between olivine and orthopyroxene may provide a more satisfactory trace element thermometer, applicable to many eclogite facies assemblages in ultrabasic rocks. The recent study by Podvin (1988) has shown that the distribution coefficient $K_{D_{Ni-Mg}}^{Ol-Opx} = (X_{Ni}/X_{Mg})^{Ol}/(X_{Ni}/X_{Mg})^{Opx}$ is strongly temperature dependent with $\ln K_{D_{Ni-Mg}}^{Ol-Opx} = [3801/T(K) - 1.815]$. The small volume change in the exchange reaction involved indicates little dependence of $K_{D_{Ni-Mg}}^{Ol-Opx}$ on pressure. No significant dependence on Fe/Mg value of the minerals was discerned but it remains to be shown whether the contents of other minor and trace elements (such as Ca, Cr, Al, Ti and Mn) in olivines and orthopyroxenes affect the activity–composition relations. Certainly this thermometer merits testing on natural assemblages in which the Ni contents of olivines and orthopyroxenes have been determined with suitable precision.

5 Role of kinetics in the formation and preservation of eclogites

D.C. RUBIE

5.1 Introduction

Numerous studies have shown that an evaluation of the *P–T–t* evolution of metamorphic rocks provides important information and constraints on tectonic evolution (e.g. see volume edited by Daly *et al*, 1988). The evaluation of *P–T–t* paths is generally dependent on the application of geothermometry and geobarometry to metamorphic mineral assemblages (see Chapter 4), with radiometric dating techniques providing time data. The most highly developed methods for quantifying *P–T* paths are based on the assumption that equilibrium was maintained between the various mineral growth sites throughout metamorphism. Thus, zoning patterns in minerals, such as garnet, are considered to develop primarily in response to changing *P–T* conditions and are therefore regarded as preserving a decipherable record of such changes (Spear and Selverstone, 1983; Spear *et al*, 1984). One of the purposes of this chapter is to demonstrate that this type of approach cannot always be applied to eclogites because the rates of mineral reactions, both during eclogite formation and retrogression, can be slow compared with the rates of change in *P* and *T*. Examples will be reviewed to demonstrate that eclogitisation can be dominated by irreversible processes and disequilibrium. In principle, evidence of such processes, combined with experimental kinetic data, can also be used to constrain *P–T–t* paths and rates of tectonic processes (e.g. Carlson and Rosenfeld, 1981; Gillet and Goffé, 1988).

The kinetics of eclogitisation are also of interest in a broader geodynamical context. The large density increase (about 15%) associated with the transformation of gabbro to eclogite must have a significant effect on the dynamics of subduction. Ahrens and Schubert (1975a), for example, have estimated that the downward body force on a descending slab due to eclogitisation of oceanic crust is comparable to the body forces due to both the elevation of the olivine–spinel phase boundary and the low temperature of the slab relative to the surrounding mantle. The exact depth of the gabbro to eclogite transformation and the size of the depth interval over which it occurs will be important factors in models of the subduction dynamics and will also affect rheology (Rubie, 1983), stress distribution and possibly seismicity in subduction zones. The depth of transformation depends not only on the equilibrium phase relations (e.g. Green and Ringwood, 1967a) but may also be greatly affected by reaction kinetics.

Although Ahrens and Schubert (1975a, 1975b) have treated the subject theoretically, there have been few, if any, experimental studies of the kinetics of mineral reactions which commonly occur during eclogite formation. For this reason, a quantitative evaluation of the rates of eclogitisation is not yet possible. The main objective of this chapter is to assess qualitatively the role of kinetics in controlling mineral assemblage evolution. The factors which control the rates and timing of the relevant mineral reactions during the formation and retrogression of eclogites will be discussed, with particular reference to LT and MT eclogites from subduction zone and continental collision zone environments.

5.2 Kinetics and reaction pathways during eclogite formation

Rocks which have completely equilibrated in the eclogite facies can provide no information about kinetics. Such data are obtained from rocks which have only partially equilibrated and which therefore contain evidence of disequilibrium. Coarse-grained metagabbros and metagranitoids, which have been eclogitised without being strongly deformed, have proved the most suitable lithologies for this purpose (e.g. Mørk, 1985a; Pognante, 1985; Koons *et al*, 1987). Features indicative of disequilibrium include relicts of protolith minerals, the development of domainal or mosaic equilibrium, reaction coronas around relict minerals or textural domains (Figure 5.1), thermodynamically incompatible phases in different domains, newly formed metastable minerals, and an absence of Fe–Mg partition equilibrium between different phases.

When eclogites have been strongly deformed, as is often the case, it is generally difficult to determine if deformation and eclogitisation occurred simultaneously or if the deformation postdated an earlier disequilibrium stage of mineral reaction. In the former case, diffusion and therefore reaction rates are likely to be enhanced by deformation so that equilibrium may have been more readily achieved. In the second case, however, relicts of the disequilibrium stage may survive deformation but are likely to be difficult to recognise (Koons *et al*, 1987).

5.2.1 *Domainal equilibrium*

In the absence of deformation, coarse-grained rocks such as gabbros and granitoids frequently preserve their precursor igneous textures during HP metamorphism (Figure 5.1). Eclogitisation initially involves localised reactions which occur either within textural/mineralogical domains defined by the original (precursor) mineral grains or at the boundaries between such domains. For example, apart from the addition of H_2O, original plagioclase grains can react to form Jd + Qtz + Zo + Ky under isochemical or near-isochemical conditions, i.e. with very limited chemical exchange with other precursor minerals. Similarly, localised reactions occur within domains defined by original olivine, augite and biotite crystals (Figures 5.1–5.3) although these reactions are seldom isochemical.

If characteristic diffusion distances, over the appropriate time scale (e.g. 10 Ma), are small compared with the grain size of the original rock, the reaction domains defined by the original mineralogy will not be in equilibrium with each other and will therefore contain thermodynamically incompatible phases. For example, olivine gabbros from Rocciavre (Western Alps) contain sodic pyroxenes with three different compositions—jadeite, omphacite and relict igneous augite (Figures 5.2a, 5.3a). Although such phases may be thermodynamically stable with respect to the chemistry of their local domain, they will often be metastable with respect to the bulk rock composition (Section 5.6). In contrast, local equilibrium will be reached comparatively easily within an individual domain because necessary diffusion distances will be short, especially for isochemical

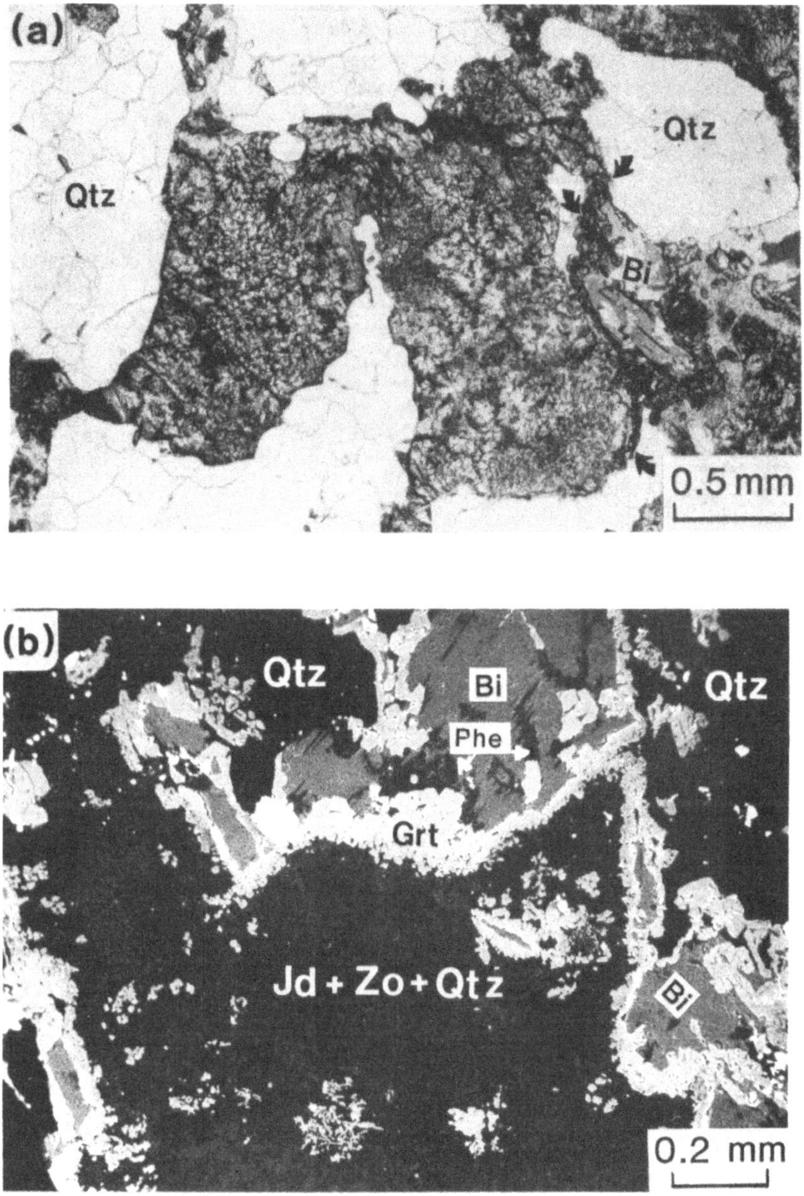

Figure 5.1 Micrographs of eclogitised quartz diorite showing relict igneous textures and reaction coronas. Plagioclase and biotite sites define domains of local equilibrium (Monte Mucrone, Sesia Zone). (a) Optical micrograph showing two pseudomorphs after plagioclase, which consist of fine-grained Jd + Zo + Qtz. Quartz and biotite of igneous origin are present and a narrow corona (arrowed—also Figure 5.1b) has developed between biotite and pseudomorphed plagioclase. (b) Backscattered electron image showing the different domains and coronas (consisting mainly of Grt ± Phe) between biotite and plagioclase sites.

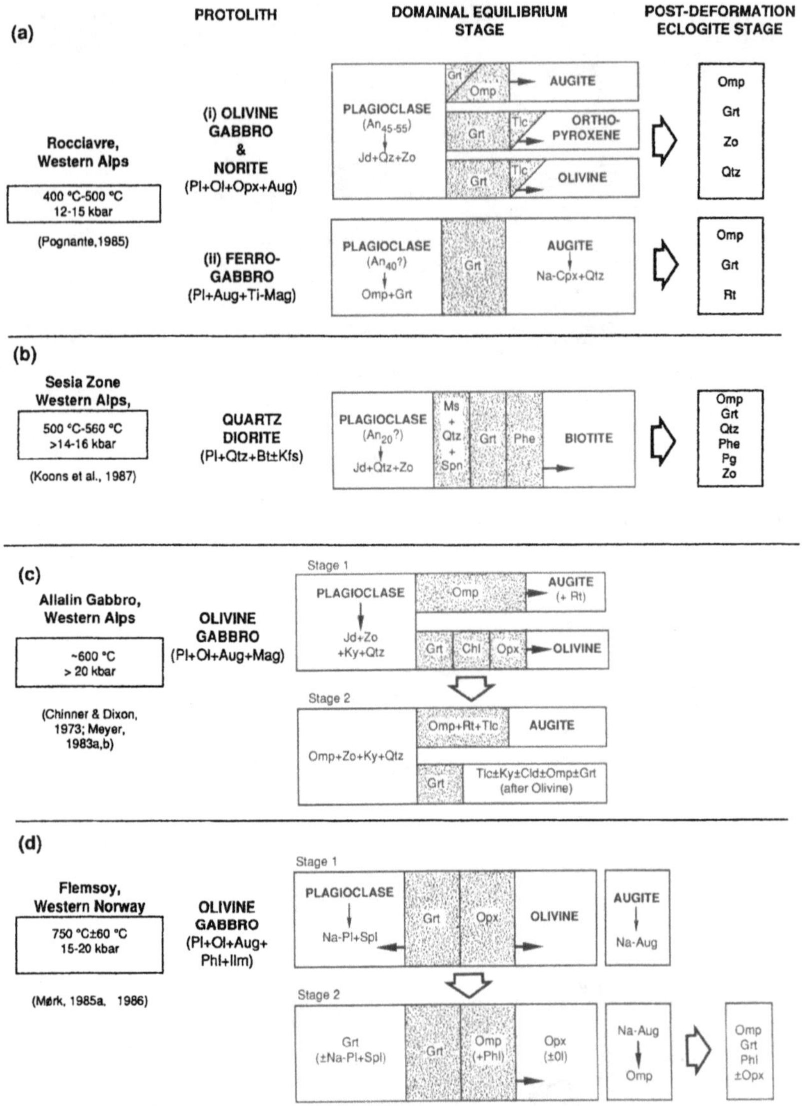

Figure 5.2 Examples of reaction pathways during eclogitisation, from the Western Alps and Norway. Precursor igneous minerals, grains of which form domains of local equilibrium, are shown by bold upper case lettering. Coronas are indicated by shaded areas and arrows show the direction of movement of layer interfaces during the replacement of precursor minerals by corona growth. Layers which are discontinuous (see Figure 5.3a, for example) are bounded by diagonal lines.

reactions (see Section 5.3, however). Therefore, the reaction behaviour under discussion, which results in disequilibrium on a thin section scale, is frequently referred to as 'domainal' or 'mosaic' equilibrium (Blackburn, 1968; Vernon, 1976).

Localized reactions at interfaces between compositionally-distinct domains invariably produce reaction coronas (Figures 5.1–5.3). The fact that these coronas often have a layered structure indicates that their growth rate is diffusion-controlled (Joesten, 1977; Fisher, 1978). With slow diffusion rates, it is to be expected that there will be significant chemical potential gradients between domains of different composition (e.g. plagioclase and olivine) during metamorphism, and this is confirmed by corona geometry and the nature of zoning patterns in some corona minerals (Section 5.5).

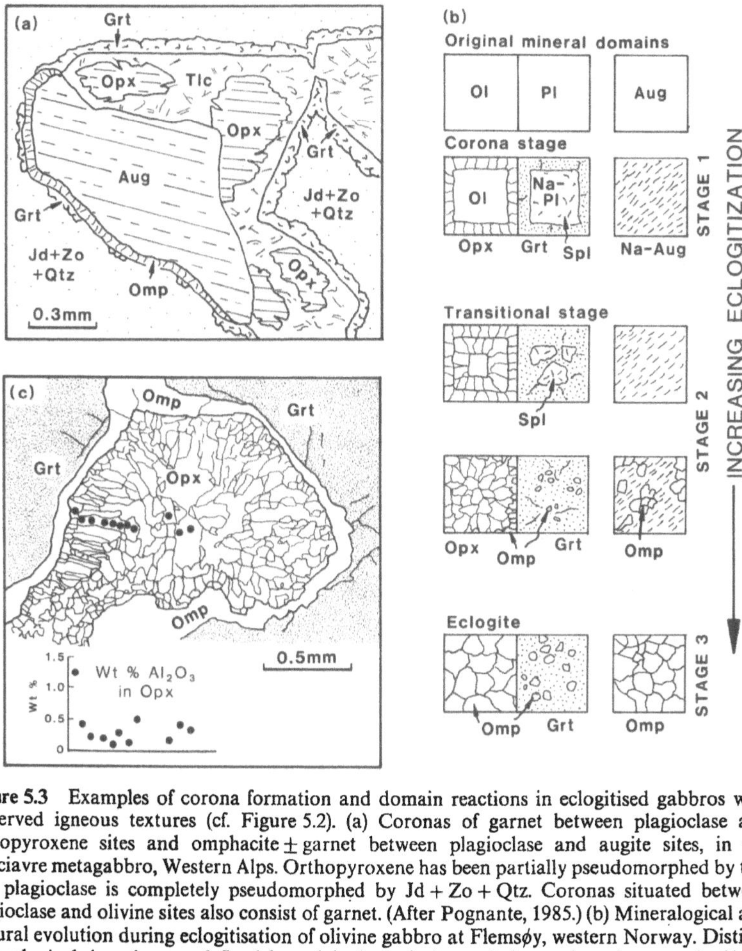

Figure 5.3 Examples of corona formation and domain reactions in eclogitised gabbros with preserved igneous textures (cf. Figure 5.2). (a) Coronas of garnet between plagioclase and orthopyroxene sites and omphacite ± garnet between plagioclase and augite sites, in the Rocciavre metagabbro, Western Alps. Orthopyroxene has been partially pseudomorphed by talc and plagioclase is completely pseudomorphed by Jd + Zo + Qtz. Coronas situated between plagioclase and olivine sites also consist of garnet. (After Pognante, 1985.) (b) Mineralogical and textural evolution during eclogitisation of olivine gabbro at Flemsøy, western Norway. Distinct mineralogical domains are defined by original plagioclase, olivine and augite crystals. (After Mørk, 1985a). (c) Zoned orthopyroxene aggregate which has pseudomorphed olivine in the Flemsøy metagabbro. There is a marked increase in the Al_2O_3 content of orthopyroxene towards the original plagioclase sites, which now consist of garnet. (After Mørk, 1985a).

5.2.2 *Reaction pathways*

Some examples of the results of typical domain and corona reactions in eclogitic rocks from the Western Alps and Norway are summarised in Figure 5.2. Where known, the final equilibrium (usually post-deformation) eclogitic mineral assemblage is also shown. This emphasises that the corona or domainal equilibrium stage represents an intermediate disequilibrium state along a reaction pathway between the protolith (e.g. gabbro) and fully equilibrated eclogite. The rate of the overall process of eclogitisation is controlled by the rate of each localised reaction in the sequence of individual reactions which constitute the reaction pathway. Thus, an understanding of reaction pathways is of paramount importance in studies of rates of metamorphic reactions (see also Rubie and Brearley, 1987).

The reaction pathways shown in Figure 5.2 show considerable variability. This results primarily from the variability of reactions that occur in the localised domains during the initial domainal equilibrium stage. For example, at 14–20 kbar, plagioclase breaks down by quite different reactions depending on temperature (Figure 5.2b, c and d) and bulk rock chemistry (Figure 5.2a, (i) and (ii)). The same factors affect the type of domain in which sodic pyroxene first nucleates, which in turn controls the chemistry of this phase. Although the different reaction pathways can be explained partly by phase equilibrium relations applied on a localised scale, it will be shown in the following sections that the role of kinetics is of major importance.

5.3 Plagioclase-breakdown reactions

Plagioclase reacted to Jd + Zo + Qtz $\pm$ Ky in localised domains during HP metamorphism of some lithologies in the Western Alps (Figures 5.1–5.3). Calculated phase relations for the breakdown of plagioclase in a closed system as a function of P at 600°C, assuming equilibrium, are shown in Figure 5.4. In the presence of H_2O, calcic plagioclase breaks down initially with increasing pressure by the continuous reaction:

$$\mathrm{Pl + H_2O \rightarrow Zo + Ky + Qtz + (NaSiCa_{-1}Al_{-1})_{Pl}} \qquad \text{(R5.1)}$$

(Figure 5.4a; Goldsmith, 1982). This reaction proceeds to the right with increasing pressure so that the plagioclase composition becomes progressively more sodic. At a certain pressure (about 17 kbar at 600°C), plagioclase disappears from the assemblage by the discontinuous reaction:

$$\mathrm{Pl + H_2O \rightarrow Jd + Zo + Ky + Qtz.} \qquad \text{(R5.2)}$$

In the absence of H_2O, the equilibrium breakdown of calcic plagioclase (in a closed system) occurs initially with increasing pressure by the continuous reaction:

$$\mathrm{Pl \rightarrow Grs + Ky + Qtz + (NaSiCa_{-1}Al_{-1})_{Pl}} \qquad \text{(R5.3)}$$

Figure 5.4 Equilibrium phase relations for (a) plagioclase + H_2O, and (b) plagioclase, as a function of pressure at 600°C (after Wayte *et al*, 1989). Equilibrium boundaries were calculated using the self-consistent data set of Holland and Powell (1985) and the plagioclase activity model of Newton *et al* (1980). Because this activity model takes no account of the complex exsolution and order–disorder behaviour shown by plagioclase, these phase diagrams are only semi-quantitative. Numbered reaction curves correspond to reactions listed in the text. Phase relations are also shown as Al_2O_3-$NaAlO_2$-CaO projections from Qtz + H_2O (a) and Qtz (b). Arrows show the migration directions of three-phase triangles on these projections with increasing pressure as the result of continuous reactions.

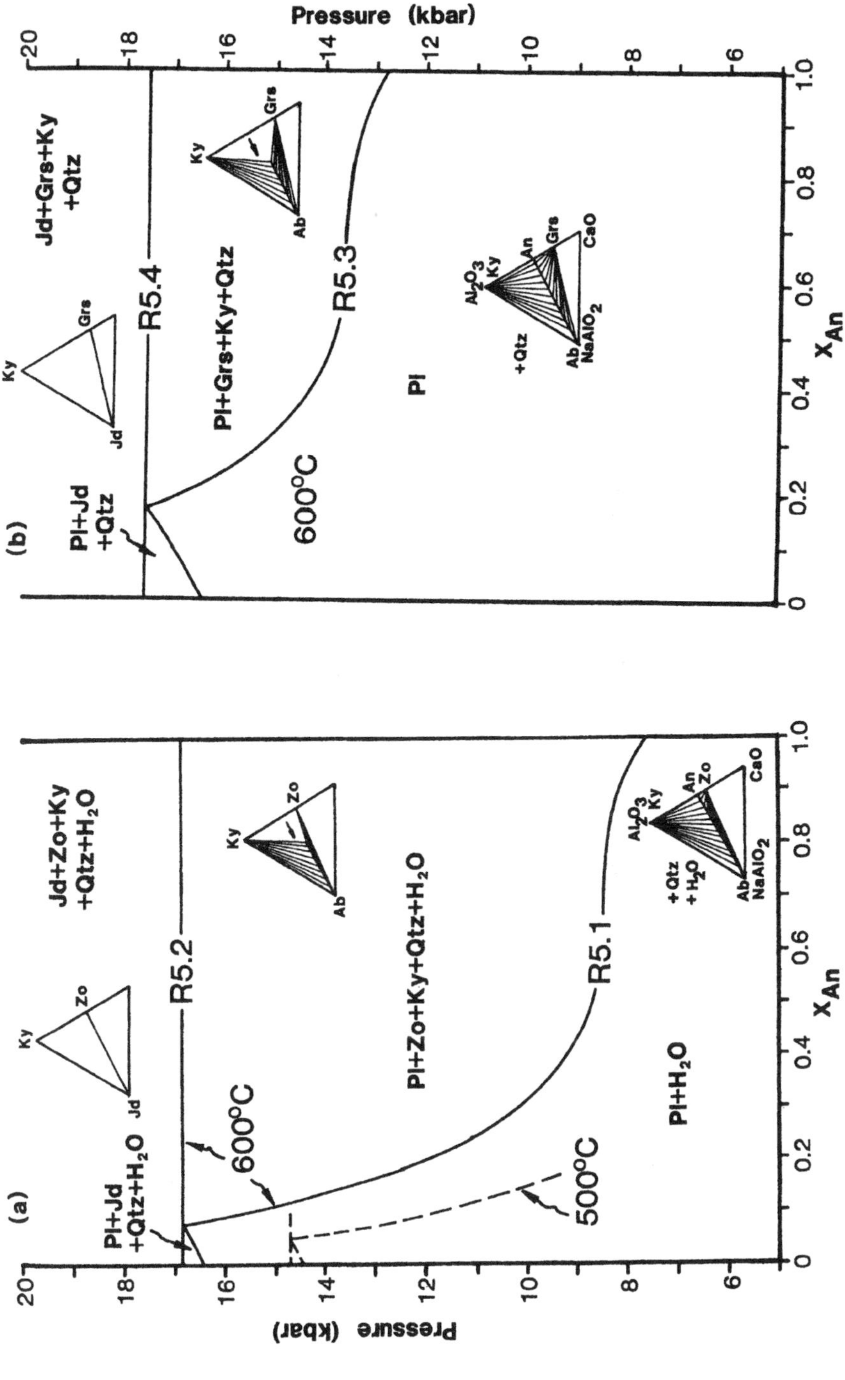
(a)
(b)
Pressure (kbar)
X_{An}
Jd+Zo+Ky
+Qtz+H_2O
Pl+Jd
+Qtz+H_2O
Pl+Zo+Ky+Qtz+H_2O
Pl+H_2O
R5.2
R5.1
600°C
500°C
Jd+Grs+Ky
+Qtz
Pl+Jd
+Qtz
Pl+Grs+Ky+Qtz
Pl
R5.4
R5.3
600°C
Ky
Zo
Grs
Jd
Ab
An
Al_2O_3
CaO
$NaAlO_2$
+Qtz
+H_2O

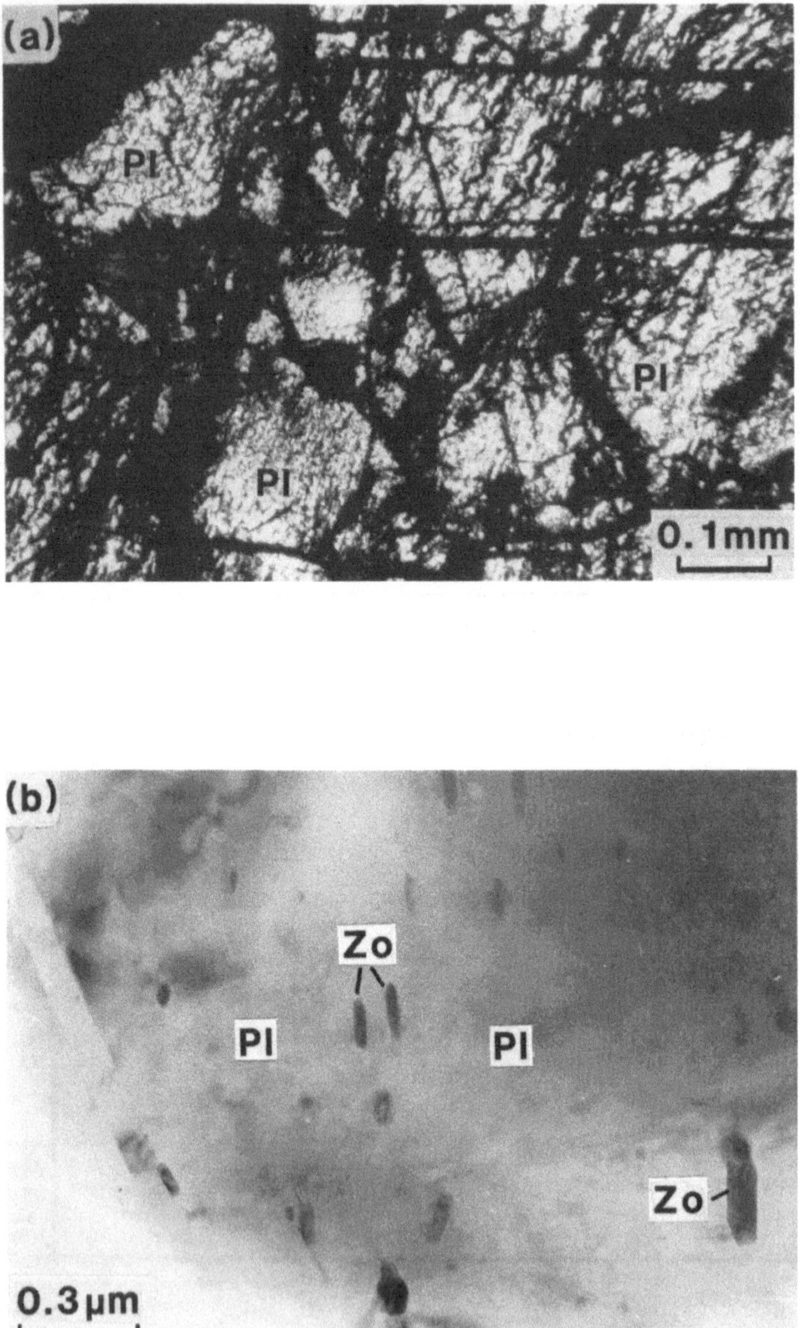

(**Figure 5.5** *Continued*)

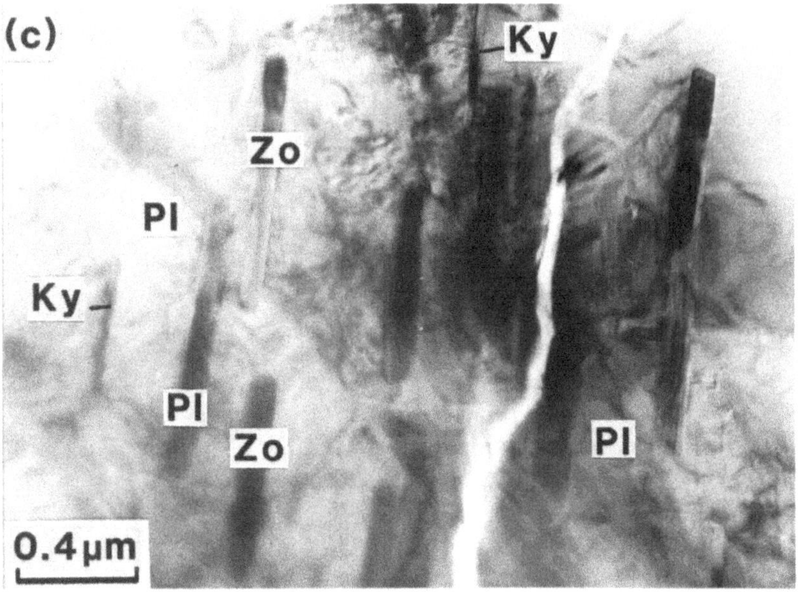

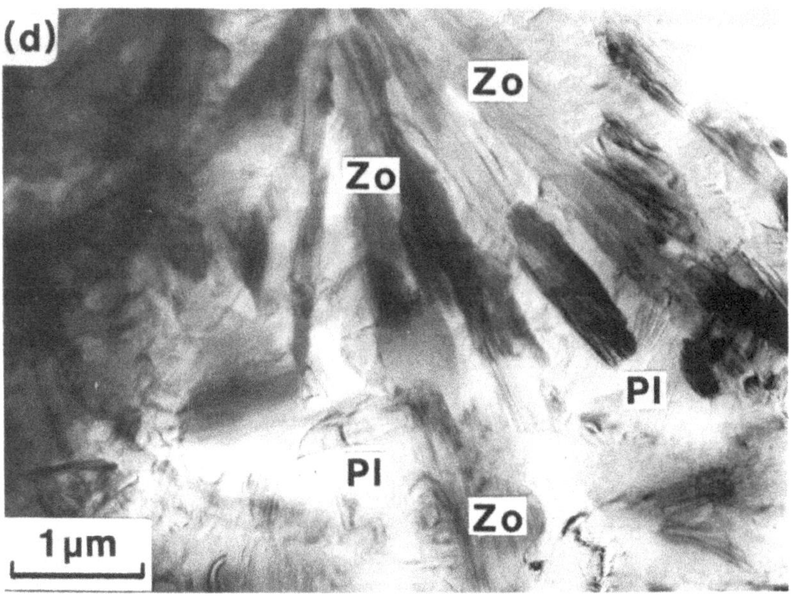

(**Figure 5.5** *Continued*)

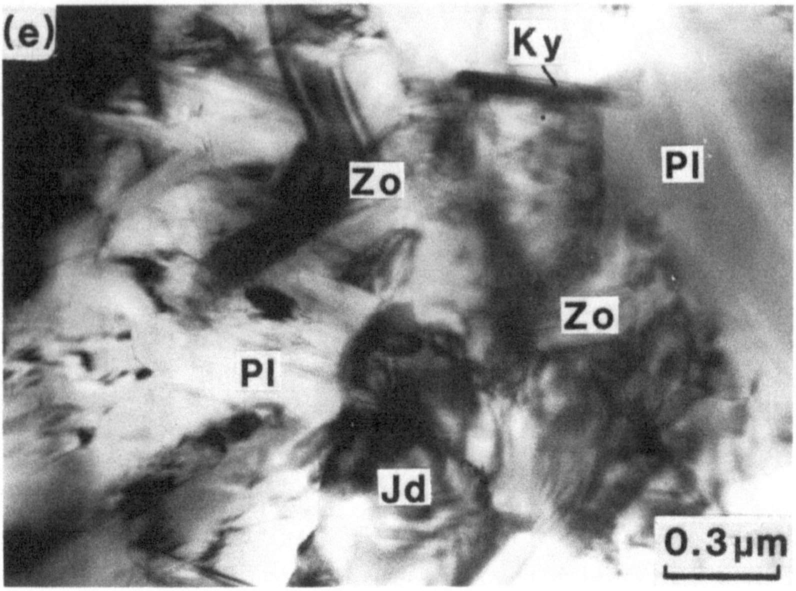
(e)
Ky
Zo
Pl
Zo
Pl
Jd
0.3µm

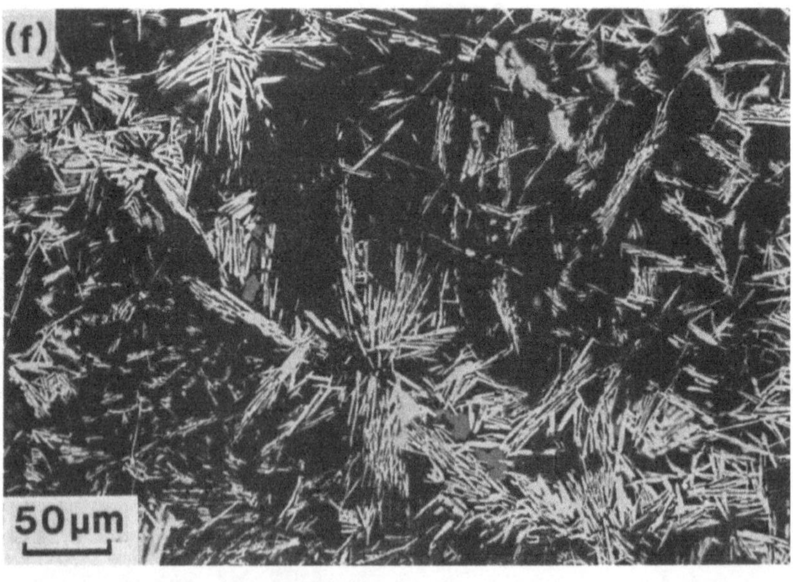
(f)
50µm

and finally by the discontinuous reaction (Figure 5.4b):

$$Pl \rightarrow Jd + Grs + Ky + Qtz \qquad (R5.4)$$

The results of a study of partially-reacted plagioclase from the Allalin gabbro (Western Alps) suggest that plagioclase is more likely to break down under disequilibrium conditions during HP metamorphism (Wayte *et al*, 1989). Plagioclase, which originally had a composition An_{60-65} is often completely pseudomorphed by Jd + Zo + Ky + Qtz in the Allalin gabbro (Figure 5.2c) as a result of metamorphism at about 600°C and $>$20 kbar (Meyer 1983a, b). In rare cases where reaction is incomplete, reaction products are concentrated adjacent to cracks and grain margins in plagioclase (Figure 5.5a) and are presumed to have developed in response to the infiltration or presence of hydrous fluid. Regions of plagioclase which appear unreacted in the optical microscope are found to be up to 10% (by volume) transformed when examined by transmission electron microscopy (TEM). The resulting micrographs show that small, oriented zoisites have nucleated in the least transformed regions (Figure 5.5b) and where the degree of transformation is about 10%, zoisite, kyanite and rare quartz have developed (Figure 5.5c) and the plagioclase composition has become more sodic. In highly reacted regions adjacent to cracks and cleavage planes, relict plagioclase is always present in the sample studied and the degree of transformation varies from about 50–90%. Variably-oriented zoisite, sodic plagioclase, kyanite and rare quartz are the main reaction products (Figure 5.5d), but in the most transformed regions jadeite is also present (Figure 5.5e). Plagioclase compositions in the partially pseudomorphed crystals vary from the original igneous composition of about An_{65} in the least transformed regions to An_{13} at around 90% transformation. In any given microdomain (e.g. Figures 5.5c–e), plagioclase compositions are highly variable but the most sodic composition generally corresponds closely to that predicted by the calculated equilibrium phase relations (Figure 5.4a) for the particular extent of transformation (Wayte *et al*, 1989, figure 9).

It is clear in this example that plagioclase did not react according to the equilibrium model (R5.1 and R5.2). Wayte *et al* (1989) have proposed two alternative disequilibrium models:

(a) Plagioclase remains metastably at its original igneous composition until the pressure exceeds that required for R5.2. Reaction then occurs in response to the

---

Figure 5.5 Micrographs showing plagioclase which has undergone various degrees of breakdown during HP metamorphism. (a)–(e) Partially-reacted plagioclase from the Allalin gabbro, Western Alps (from Wayte *et al*, 1989). (a) Light optical micrograph showing very fine-grained reaction products (black) which have developed in a plagioclase crystal along cracks and grain margins. Some cracks have developed along cleavage planes. (b)–(e) Bright field transmission electron micrographs. (b) Plagioclase, which appears unaltered in (a), showing small, oriented, zoisite crystals. (c) Plagioclase which is ~10% transformed to oriented Zo and Ky. Qtz, which is also present, is not seen in this field of view. (d) Plagioclase ~50% transformed to randomly oriented Zo + Ky + Qtz. Ky and Qtz are not present in this field of view. (e) An advanced stage of transformation to Zo + Ky + Qtz + Jd from the most highly altered regions shown in (a). Quartz has been found by TEM although it is not present in the micrograph shown here. (f) Backscattered electron micrograph of plagioclase completely reacted to Jd (dark grey) + Zo (white needles) + interstitial Qtz (black) in meta-quartz diorite from Monte Mucrone, Sesia Zone (see also Figure 5.1). The distribution of the larger zoisites suggests nucleation along a crack running from top left to bottom right (see text). Occasional small grains of phengite (pale grey) are present in this fractured region.

infiltration or presence of aqueous fluid in cracks and grain boundaries. Adjacent to these features, plagioclase reacts to Zo + Ky + Jd + Qtz in the presence of fluid (Figure 5.5e). Reaction products at these sites are variably oriented (Figure 5.5d, e), suggesting non-topotactic nucleation and growth in the presence of fluid. At intracrystalline sites distant from the cracks, reaction is severely impeded for kinetic reasons. The absence of jadeite (Figures 5.5b–d) and in some cases kyanite and quartz (Figure 5.5b) can be ascribed to large nucleation barriers for these minerals at intracrystalline sites in the absence of fluid. The crystallographic orientation of reaction products at such sites (Figure 5.5b, c) suggests topotactic nucleation and growth.

(b) In an alternative disequilibrium model, the reaction of metastable plagioclase occurs in response to several discrete episodes of aqueous fluid infiltration during pressure increase. During each episode, equilibration develops mostly in localised regions adjacent to the fluid-filled cracks. For example, at 600°C, 10% transformation (Figure 5.5c) corresponds to equilibration at 8–9 kbar, and 50% transformation (Figure 5.5d) corresponds to equilibration at about 12 kbar (Wayte *et al*, 1989, figure 9). Because the hydration reactions consume H_2O, the size of equilibrated regions could be restricted by a limited quantity of fluid available during each episode.

In both these models, the amount of zoisite, kyanite and quartz which can form in the absence of jadeite depends on the ability of plagioclase to change composition by NaSi–CaAl interdiffusion. Rates of this process have been determined by Yund (1986) in the temperature range 900–1100°C at 15 kbar, in the presence of H_2O. Extrapolating the results to lower T(600°C) suggests that NaSi–CaAl interdiffusion is too slow for plagioclase to break down either under equilibrium conditions (R5.1) or by disequilibrium model (b) if subduction is rapid (e.g. 10 mm/year) (Wayte *et al*, 1989).

The preservation of metastable plagioclase during HP metamorphism of the Allalin gabbro can be explained by an absence (at least on a localised scale) of the H_2O which is required as a reactant in R5.1 and R5.2. However, at the peak metamorphic conditions of around 600°C and > 20 kbar, the solid–solid reaction (R5.3) is overstepped by at least 6 kbar and it is likely that R5.4 was also overstepped by > 2 kbar (Figure 5.4b). The failure of these reactions to occur must result from unfavourable kinetics in the absence of H_2O (e.g. Hays, 1967). In particular, the absence of grossular suggests that the activation energy for nucleation of this phase by R5.3 and R5.4 is very large. Although the addition of water would enhance reaction rates, when H_2O becomes available the hydration reactions (R5.1 and R5.2) occur in preference to the solid–solid reactions.

In some examples, reaction at plagioclase sites has not been isochemical (Figures 5.1, 5.2, and 5.5f). For example, an absence of kyanite as a reaction product, growth of jadeitic pyroxene with about 10 mole% diopside component and the chemistry of adjacent coronas suggest loss of Al and Ca and gain of Fe and Mg at the plagioclase sites during reaction (Pognante, 1985; Koons *et al*, 1987). Such mass transport indicates limited chemical exchange between reactions at plagioclase sites, reactions producing adjacent coronas and therefore reactions in other nearby domains (e.g. olivine or biotite sites).

Typical textures of the products of complete plagioclase breakdown are shown in Figure 5.5f. On the basis of this discussion, metastable plagioclase (of the original igneous composition) could have reacted directly to Jd + Zo + Qtz due to the infiltration or presence of H_2O. The textures suggest that the reaction products initially nucleated along cracks containing H_2O (Figure 5.5f). The progressive widening of reaction zones outwards from such cracks (*cf.* Figure 5.5a) could occur at the interface with unaltered metastable plagioclase by two processes: (a) growth of the product phases by a dissolution–precipitation process; (b) nucleation of new grains with variable orientations. The large volume decrease of the reaction may assist fluid infiltration by producing a transiently-high permeability.

In the Rocciavre metagabbro (Western Alps), bulk rock composition appears to influence plagioclase-breakdown reactions. In olivine gabbros and norites, plagioclase initially reacted to Jd + Zo + Qtz whereas in ferrogabbros, plagioclase was initially pseudomorphed by Omp + Grt (Figure 5.2a). The latter reaction requires considerably more mass transport between the individual domains than the former. It is possible that, due to internal buffering, the chemical environment (e.g. f_{O_2}) differed between the two rock types and affected diffusion rates and the relative rates of competing reactions (Pognante, 1985).

Plagioclase breakdown in the Flemsøy gabbro (western Norway) occurred, at higher temperature than in the examples from the Western Alps, by solid–solid reactions and there is no direct evidence of fluid involvement (Figures 5.2d, 5.3b, c; Mørk, 1985a). In a closed system, calcic plagioclase, in the absence of H_2O, should break down by R5.3 to produce Na–Pl + Grs + Ky + Qtz (Figure 5.4b). The reactions which occurred produced Na–Pl + Spl at intracrystalline sites and Ca–Fe-garnet at grain boundaries. Lack of evidence for R5.3 again suggests the difficulty of nucleation of the product phases at intracrystalline sites, especially in the absence of H_2O. The solid–solid reactions which occurred must be favoured kinetically over reaction R5.3. This could be because nucleation of new phases (garnet and orthopyroxene) occurred at grain boundaries, which energetically are much more favourable sites than the intracrystalline sites required for R5.3 to occur in a closed system.

The pseudomorphic breakdown of calcic plagioclase to Na-Pl + Grt $\pm$ Ky $\pm$ Qtz $\pm$ Spl (*cf.* R5.3) has occurred in leucogabbro-norite in the Kvamsøy pyroxenite complex, western Norway, during HP metamorphism at 13–16 kbar and 750°C (Mørk, 1985b). The composition of the garnet varies from $Prp_{25}Alm_{57}Grs_{18}$ to $Prp_{21}Alm_{50}Grs_{28}$, so that Fe and Mg must have diffused into the plagioclase sites from adjacent domains. Such mass transport would increase a_{Alm} and a_{Prp} and reduce a_{Grs}, and therefore shift the equilibrium boundary for R5.3 to lower pressures (*cf.* Figure 5.4b). This would increase the overstepping of the equilibrium boundary and would therefore increase the probability of nucleation of new phases such as garnet. The development of amphibole adjacent to plagioclase sites and the coarse grain size of the garnet (Mørk, 1985b, figure 4a) suggest some fluid involvement in these reactions.

The examples given demonstrate that several factors can affect the nature of plagioclase-breakdown reactions at pressures of 14–20 kbar. Firstly, P–T conditions relative to the stability field of Jd + Qtz (Figure 5.4) are important. Secondly, a combination of temperature and fluid availability probably has a major effect. At the lower temperatures (400–600°C), H_2O is necessary as a reaction catalyst and, in the presence of this phase, hydration reactions are stable relative to solid–solid reactions in most lithologies. Solid–solid reactions occur at higher temperature (e.g. 750°C) but very small amounts of H_2O may still be involved as a catalyst. The different reactions that occur in the Norwegian examples probably result, at least partly, from differences in diffusion rates between different domains.

5.4 Breakdown of augite and olivine

Augite is commonly pseudomorphed by sodic pyroxene (Na-augite, omphacite or chloromelanite) during HP metamorphism of gabbros (Figure 5.2). This process involves the replacement of Ca, Mg and Fe in augite by Na and Al. The Na and Al are derived from reactions occurring at plagioclase sites (Section 5.3). Ca, Mg and Fe, produced by the augite-breakdown reaction, are consumed by corona-forming reactions and may also diffuse into plagioclase sites to form a diopside component in jadeitic or omphacitic pyroxene.

Diffusion of Ca and Al in clinopyroxene is extremely slow (Freer *et al*, 1982; Brady and McCallister, 1983). Thus, at typical temperatures of eclogitisation (450–750°C), the alteration of large grains of augite to Na-pyroxene by lattice diffusion alone is unlikely on the time scale of HP metamorphism. Detailed studies of partially-reacted augites show that several possible alternative reaction mechanism can operate.

In a TEM study of augite in an eclogitised ferrogabbro, from the Western Italian Alps, Otten and Buseck (1987) identified two competing reaction mechanism which operated at different sites (at 400–500°C, $>$ 10 kbar). A topotactic mechanism operated at intracrystalline sites to produce sodic pyroxene that is oriented in the host augite. Exsolution (001) lamellae of low-Ca clinopyroxene were preferentially altered to Na-pyroxene. These altered lamellae progressively recrystallized and grew into the host augite. In addition, needles and thin (100) lamellae of Na-pyroxene developed on the (001) lamellae. All transformed regions contain a high density of dislocations and stacking faults which may have facilitated the relatively large-scale diffusion necessary for the exchange reaction (Otten and Buseck, 1987). A non-topotactic mechanism operated at augite grain boundaries to produce new grains which do not have a consistent orientation relation with the relict augite. It is likely that this latter mechanism could only operate in the presence of a fluid, by a dissolution–precipitation process. Therefore, Otten and Buseck (1987) concluded that the dominance of the topotactic mechanism over the non-topotactic mechanism in the partially transformed augite is indicative of a scarcity of fluid.

Carpenter and Okay (1978) have studied the partial replacement of augite by omphacite which occurred at $<$ 350°C, during blueschist facies metamorphism of a dolerite from north-west Turkey. Omphacite developed in irregular veins in the augite, which suggests that reaction occurred in response to fluid infiltration. The replacement is invariably topotactic, in contrast to the fluid-assisted reaction in the previous example.

Differences between the reaction mechanisms in the above examples may be related to differences between the *P–T* conditions of metamorphism. There is clearly still considerable scope for further similar TEM studies of augites partially reacted to sodic pyroxene over a wider range of *P–T* conditions.

Reactions consuming olivine (Figures 5.2, 5.3) invariably operate by the inward growth of new phases from the grain boundaries. Thus, phases which eventually pseudomorph olivine initially develop as coronas, usually between olivine and plagioclase (Griffin and Heier, 1973; Chinner and Dixon, 1973; Mørk, 1985a, 1986; Pognante, 1985). Reactions at lower temperatures ($\leqslant$ 600°C) commonly involve hydration, produce phases such as talc and chlorite, and are therefore indicative of the presence of aqueous fluid. In contrast, at higher temperatures (e.g. 750°C), solid–solid reactions occur, presumably under relatively dry conditions.

5.5 Corona formation

Coronas consist characteristically of a concentric sequence of sharply-defined mineral layers which surround relicts of precursor minerals and separate domains of different chemistries (Figures 5.1–5.3, 5.6). Each individual layer in a corona can be either monominerallic or polymineralic (e.g. Figure 5.6). The mineralogy of coronas and the sequence of layers depends on the mineralogy of the adjacent domains, *P–T* conditions of reaction, and on kinetic factors. The diversity of coronas found in eclogite facies rocks (Figure 5.2) is probably a reflection of all of these factors.

It is sometimes difficult to evaluate the *P–T* conditions of corona formation. For example, orthopyroxene–garnet coronas in metagabbros of the Western Gneiss Region of Norway were originally considered to have formed at normal crustal pressures during cooling from igneous temperatures (Griffin and Heier, 1973; Griffin and Råheim, 1973). Mørk (1985a) has argued that the coronas could have formed during HP metamorphism, much later than the time of igneous intrusion, and this has been confirmed by Sm–Nd dating on igneous and metamorphic phases in the Flemsøy metagabbro which gave a Precambrian protolith age and Caledonian metamorphic ages (Mørk and Mearns, 1986). However, the possibility of a more complex history, involving both pre-metamorphic and eclogite facies corona stages, cannot be excluded (Mørk, 1986, pp. 425–426).

Reactions which produce coronas consume chemical components which are derived from adjacent domains. The process involves reactions at layer contacts (i.e. dissolution or precipitation) and multicomponent diffusion through the various layers (Joesten, 1986). For example, in the Flemsøy metagabbro, a garnet–orthopyroxene corona formed between plagioclase and olivine (Figure 5.3b). Growth of this corona requires that mafic components are released by reaction at the olivine–orthopyroxene interface, diffuse through the corona and are consumed by reaction at the plagioclase–garnet interface. Similarly, Al_2O_3 is released by reaction at the plagioclase–garnet interface and is consumed by reactions producing orthopyroxene. The geometry of coronas and the sharp contacts between the layers indicate that their growth is controlled by diffusion (Fisher, 1978; Joesten, 1986). In contrast, rates of reactions at the layer contacts are likely to be relatively fast (Rubie, 1986, figure 4) so that local equilibrium is maintained. Under these conditions, significant chemical potential gradients ($d\mu_i/dx$) will be maintained between domains and across the regions of corona growth.

Spatial variation in chemical potentials is demonstrated by the zoning patterns of some corona minerals. In the Flemsøy metagabbro, the chemical potential of Al_2O_3 ($\mu_{Al_2O_3}$) decreased from plagioclase towards olivine sites and this variation was the primary factor that controlled the composition of orthopyroxene (Figure 5.3c; Mørk, 1985a). Similarly, asymmetric zoning patterns in small garnets which grew between biotite and plagioclase sites in a meta-quartz diorite from the Sesia Zone (Figure 5.6b) indicate that values of $\mu_{CaO}/\mu_{(FeO+MgO)}$ were higher adjacent to plagioclase than adjacent to biotite. During growth, garnet compositions were buffered under different conditions of $\mu_{CaO}/\mu_{(FeO+MgO)}$ at different positions relative to the adjacent domains (Koons *et al*, 1987). The variation in composition of white micas (phengite and muscovite) from different layers in the same corona is a further example of this phenomenon (Figure 5.6).

If the kinetics of corona formation are understood, it should, in principle, be possible to estimate rates of growth and therefore the time duration of reaction (*cf*. Rubie, 1986).

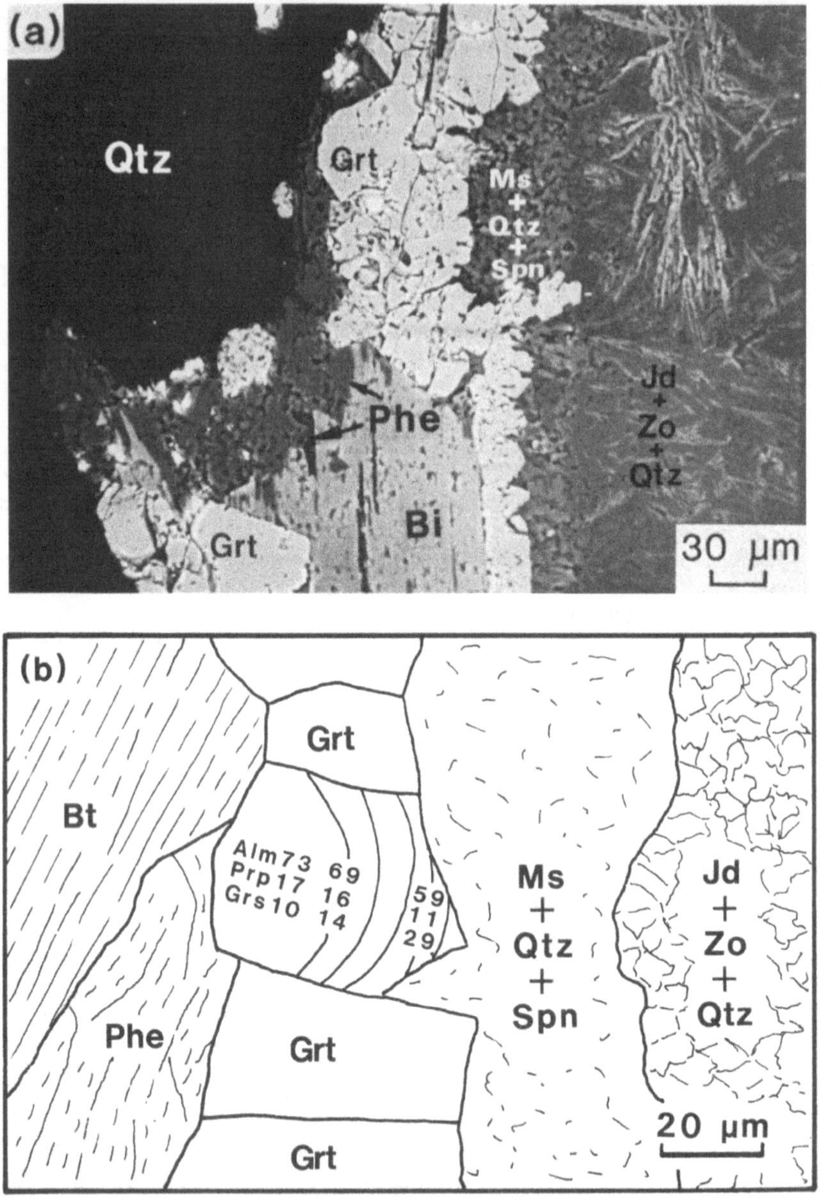

Figure 5.6 Details of reaction coronas between pseudomorphed plagioclase (Jd + Zo + Qtz) and partially-reacted biotite in meta-quartz diorite (Monte Mucrone, Sesia Zone). The layer adjacent to pseudomorphed plagioclase consists of fine-grained Ms + Qtz + Spn. The muscovite in this layer is low in celadonite and TiO_2 whereas the white mica adjacent to biotite is a high-titanium phengite. (a) Backscattered electron micrograph (phengite, situated adjacent to quartz, is dark grey.) (b) Typical zoning pattern in garnet (after Koons *et al*, 1987).

Currently, however, there is a problem in evaluating balanced chemical reactions which produce coronas, and therefore relative fluxes of components across layers, because of the assumptions which must be made (Mongkolip and Ashworth, 1983, p. 639). The problem lies in having to define arbitrarily a reference frame. For example, two frequently used reference frames, which in general will lead to conflicting results, are (*a*) the assumption of the immobility of certain cations, such as Al and/or Si (component-fixed frame), and (*b*) the assumption that the volume of reactants equals the volume of products (volume-fixed frame).

One possible approach is to use the theory of nonequilibrium thermodynamics (Katchalsky and Curran, 1967) which has been discussed in the context of a number of diffusion-controlled geological processes by Fisher (1973, 1978) and Joesten (1977). This approach, using a layer-boundary fixed reference frame, has been applied to corona formation, mainly between plagioclase and olivine, at a variety of metamorphic grades (Nishiyama, 1983; Joesten, 1986; Grant, 1988), but apparently not to eclogitic coronas (see, however, Holland and Ray, 1985). Assumptions which are normally made include (*a*) local equilibrium, (*b*) steady state diffusion, (*c*) a system which is either closed to all components or is open only to a few specifically chosen components, and (*d*) spatially-independent diffusion coefficients. On the basis of these assumptions, a set of simultaneous equations can be written, the solution of which provides the stoichiometric coefficients of reactions at each layer contact (Joesten, 1977). At each layer contact p and for each component i in an N-component system, mass balance equations

$$\sum_{i=1}^{N} \nu_{\phi}^{p} n_i^{\phi} + \nu_i^{p} = 0 \qquad \text{(E5.1)}$$

and conservation equations

$$\nu_i^{p} = J_i^{p/\text{out}} - J_i^{p/\text{in}} \qquad \text{(E5.2)}$$

can be written, where ν_{ϕ}^{p} is the number of moles of phase ϕ produced or consumed by reaction, ν_i^{p} is the number of moles of component i produced or consumed, n_i^{ϕ} is the number of moles of i in phase ϕ, $J_i^{p/\text{out}}$ is the flux of i on the low potential side of layer contact p and $J_i^{p/\text{in}}$ is the flux on the high potential side of this contact. In addition, diffusion equations can be written for each of N components for each layer:

$$J_i = \sum_{j=1}^{N} L_{ij}\, \mathrm{d}(-\mu_j)/\mathrm{d}x \qquad \text{(E5.3)}$$

where L_{ij} are phenomenological diffusion coefficients and μ_j is the chemical potential of component j. In most studies, the cross coefficients L_{ij} ($i \neq j$), which link the flux of component i to the chemical potential gradient of component j, have been assumed to be negligible so that E5.3 simplifies to

$$J_1 = L_{ii}\, \mathrm{d}(-\mu_i)/\mathrm{d}x \qquad \text{(E5.4)}$$

By substituting Gibbs-Duhem equations

$$\sum_{i=1}^{N} n_i^{\phi}\, \mathrm{d}\mu_i = 0 \qquad \text{(E5.5)}$$

written for each phase in a particular layer, the chemical potential terms in E5.4 can be eliminated if the diffusion equations are rewritten as flux ratio equations incorporating

L_{ii}/L_{jj} terms (Joesten, 1977). By solving E5.1 and E5.2 and the flux ratio equations for specified values of L_{ii}/L_{jj}, the stoichiometric reaction coefficients ν_{ϕ}^{p} are calculated, which enables the stability of a particular layer sequence as a function of L_{ii}/L_{jj} to be tested (Joesten, 1977). Mineral proportions in each layer and the relative widths of layers can be calculated from the reaction coefficients. By comparing these results with observed values, relative values of diffusion coefficients (L_{ii}/L_{jj}) during metamorphism can be constrained. If the time of reaction is known, absolute values for diffusion coefficients can also be estimated (Rubie and Brearley, 1988), or vice versa.

One problem with this model is that coronas often do not form in a closed system in eclogites (*cf.* Mongkoltip and Ashworth, 1983). Corona growth is frequently linked chemically to reactions that occur within adjacent or nearby domains rather than just at their margins. For example, in the Flemsøy metagabbro (Mørk, 1985a), the progressive enrichment of augite in the jadeite component acts as a sink for Na and Al and a source of Fe, Mg and Ca during the growth of garnet–orthopyroxene coronas between plagioclase and olivine (Figures 5.2, 5.3b). Another example is the corona shown in Figure 5.6 which may have involved a reaction such as

$$Bt + Al_2O_3 + CaO \rightarrow Phe + Grt \qquad \text{(R5.5)}$$

(Koons *et al*, 1987). The source of the Al_2O_3 and CaO required for this reaction is the plagioclase-breakdown reaction (to Jd + Zo + Qtz ± excess components) in adjacent domains. The kinetics of this latter reaction is likely to affect the variation of $\mu_{Al_2O_3}$ and μ_{CaO} with time in regions adjacent to the reacting plagioclase and could therefore influence the rate of corona formation. Under such circumstances, particularly if plagioclase breaks down rapidly (see Section 5.3), it may not be valid to assume steady state conditions and it would be necessary to replace the diffusion equation (E5.4) by an equation of the form

$$\delta\mu_i/\delta t = L_{ii}\,\nabla^2\mu_i + P_i \qquad \text{(E5.6)}$$

where P_i is a time-dependent source–sink term for component i (Koons *et al*, 1987).

5.6 Formation of metastable phases

In rocks in which there is a state of disequilibrium between textural and mineralogical domains, new phases form with compositions which are determined by the local chemical environment. Although such minerals may be stable within their localised microdomain, they are often metastable with respect to the bulk rock composition. Examples of minerals which develop with metastable compositions include pyroxenes, micas and garnet (Figure 5.6). As discussed in Section 5.7, once formed, such minerals may persist through processes which are commonly believed to be favourable to equilibration, such as deformation and fluid infiltration. Pyroxenes and garnets are particularly slow to equilibrate because of slow rates of lattice diffusion.

Bulk rock composition, as well as P and T, is one of the most important controls on the equilibrium compositions of both pyroxene and garnet during HP metamorphism. For example, in equilibrated rocks, jadeite only occurs in eclogitic schists of broadly granitic composition, which are low in FeO, MgO and CaO and high in SiO_2, whereas in rocks of more mafic composition, pyroxenes are omphacitic (Lombardo *et al*, 1977). In metagabbros and meta-quartz diorites, which show evidence of disequilibrium, it is common for jadeitic pyroxene (about $Jd_{0.9}$) to nucleate and grow within plagioclase

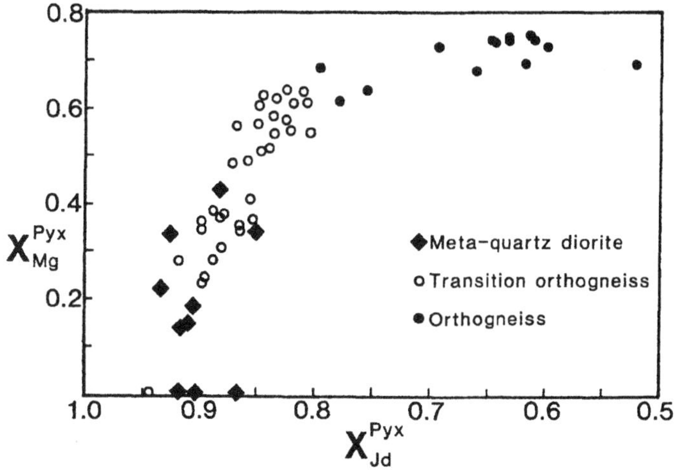

Figure 5.7 Compositions of sodic pyroxenes in undeformed meta-quartz diorite and deformed equivalents (transition orthogneiss and orthogneiss) from Monte Mucrone, Sesia Zone. $X_{Mg}^{Pyx} = Mg/(Mg + Fe^{2+})$ and $X_{Jd}^{Pyx} = Jd/(Jd + Di + Acm)$. (After Koons *et al*, 1987).

pseudomorphs (Figure 5.2). This is because of limited chemical communication with sites of mafic composition, such as olivine or biotite, which results in low μ_{FeO} and μ_{MgO} within the plagioclase sites. Compositions of jadeitic pyroxenes are often quite variable within plagioclase pseudomorphs (e.g. $Jd_{0.85-0.95}$, $Mg/(Mg + Fe) = 0–0.45$; see Figure 5.7), which demonstrates that significant chemical potential gradients can exist within individual domains.

In the Flemsøy metagabbro, orthopyroxene is present at the corona stage in rocks of all bulk compositions but it is seldom present in the equilibrated eclogites (Figure 5.2d; Mørk, 1985a). Therefore, there is the possibility that this mineral formed metastably at 15–20 kbar during the disequilibrium stage but was later resorbed during deformation and bulk rock equilibration. Orthopyroxene compositions (with low Al_2O_3) were certainly metastable (Figure 5.3c; Section 5.5). In other Norwegian eclogites, growth of orthopyroxene in a disequilibrium environment (i.e. not in equilibrium with garnet) probably explains Al_2O_3 zoning patterns which were originally interpreted as resulting from growth during a large decrease in pressure (Carswell *et al*, 1985; Mørk, 1985a).

5.7 Effects of deformation

The textural and mineralogical characteristics described so far in this chapter indicate that a state of disequilibrium can persist under eclogite facies conditions, perhaps for tens of Ma, provided the rocks remain undeformed. It is frequently found that where eclogitic rocks are deformed at high pressure, mineral assemblages have approached equilibrium owing to enhanced reaction kinetics. Deformation results in the disappearance of both relict precursor minerals and precursor (e.g. igneous) textures. An excellent example of this phenomenon is the eclogitisation of granulite facies anorthosites of the Bergen Arcs, western Norway (Austrheim and Griffin, 1985). Eclogitic mineral assemblages developed only in fine-grained, mylonitic shear zones, at

700–800°C and 16–19 kbar. Where the anorthosites remained undeformed, the granulite facies mineralogy persisted metastably at eclogite facies conditions with no occurrence of HP mineral reactions. In other examples, layers or lenses of eclogite show a zonation in which undeformed and either unequilibrated or partially equilibrated cores are surrounded by deformed and equilibrated margins (e.g. Mørk, 1985a; Koons *et al*, 1987; Morten *et al*, 1987). A relationship between the magnitude of the strain and the extent of equilibration can sometimes be demonstrated qualitatively (Ridley and Dixon, 1984; Austrheim and Griffin, 1985; Koons *et al*, 1987).

In addition to the effect that deformation can have on reaction kinetics, mineral reactions can, in turn, affect rheology (White and Knipe, 1978; Brodie and Rutter, 1985; Rubie, 1990). In some cases, at least, rocks can become significantly more deformable during eclogitisation as a consequence of mineral reactions (Rubie, 1983, 1990; Koons *et al*, 1987). The interaction between the processes of deformation and metamorphism can therefore be complex.

Equilibration of mineral assemblages during deformation results primarily from an increase in diffusion rates so that characteristic distances of mass transport increase. The original textural or mineralogical domains of local equilibrium become larger and eventually coalesce when equilibrium is achieved on a thin section scale. At intermediate stages in this process, distinct domains of local equilibrium still exist but may be difficult to recognise from textural criteria (see below Koons *et al*, 1987).

There are several mechanisms by which deformation can enhance diffusion rates (Brodie and Rutter, 1985; Rubie and Thompson, 1985; Smith, 1985), but the following are probably the most important.

(a) If aqueous fluids infiltrate into dry rocks during deformation, rates of diffusion will be greatly enhanced (e.g. Rubie, 1986). Fluid infiltration during deformation can result from an increase in permeability (Rutter and Brodie, 1985) and/or by dilatancy (Sibson *et al*, 1975; Fyfe *et al*, 1978, p. 299). On the basis of the crystallisation of hydrous phases and other criteria, it is often concluded that fluid infiltration accompanied deformation under eclogite facies conditions (e.g. Mørk 1985a; Frueh-Green, 1985; Koons *et al*, 1987).

(b) The operation of dislocation creep results in the migration of dislocations and possibly an increase in dislocation density (Nicolas and Poirier, 1976), both of which result in increased rates of lattice diffusion (Yund and Tullis, 1980; Yund *et al*, 1981).

(c) Deformation often results in a reduction in grain size, either by dynamic recrystallisation or by cataclasis (Brodie and Rutter, 1985). When grain boundary diffusion dominates over lattice diffusion, as is likely at typical metamorphic temperatures, a grain size reduction results in a larger effective diffusion coefficient because of an increase in the grain boundary area per unit volume (Brady, 1983). Grain size reduction will also increase the density of nucleation sites and may reduce the distances required for diffusion between sites of mineral growth and dissolution.

In order to understand in detail the sequence of steps involved in the processes of equilibration and deformation and their interaction (i.e. reaction pathway), samples must be studied in which the various steps are preserved. Such samples may be collected across a continuous strain gradient, between undeformed and strongly deformed lithologies, which developed under HP conditions. This approach is valid

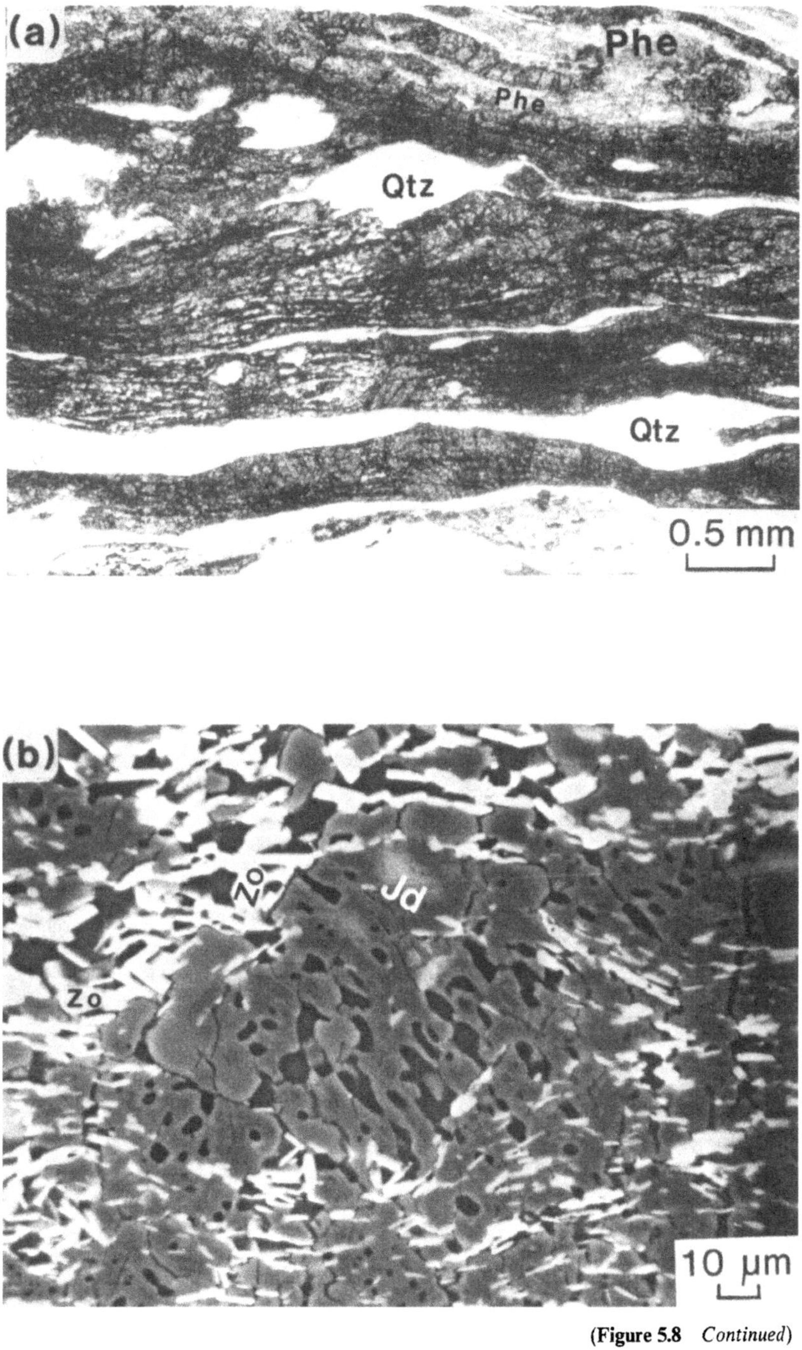

(Figure 5.8 *Continued*)

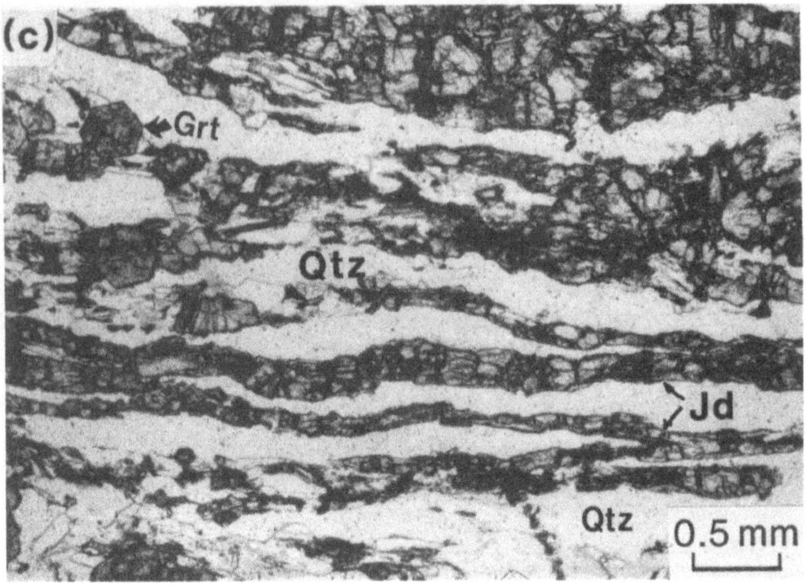
(c)
Grt
Qtz
Jd
Qtz
0.5 mm

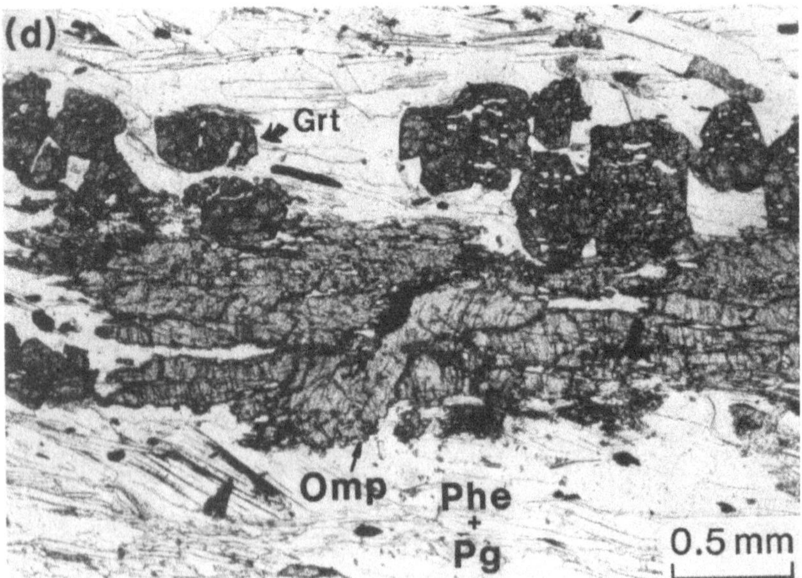
(d)
Grt
Omp
Phe
+
Pg
0.5 mm

only if it can be inferred that the time sequence of steps followed by a single sample is the same as the observed spatial sequence of steps (Ridley and Dixon, 1984; Koons *et al*, 1987; Rubie 1990). It must therefore be established that the strain gradient and mineralogical zonation developed during eclogite facies metamorphism and are not overprinted, relicts of some earlier LP metamorphic event. For example, the occurrence of sea-floor metamorphism before subduction would have a major effect on both rheology and the nature of eclogite reactions by influencing the availability and distribution of fluid and hydrous phases (Messiga and Scambelluri, 1988).

Some of the possible complications involved in the process of equilibration during deformation are illustrated by the progressive deformation of meta-quartz diorite to form eclogitic orthogneiss in the Sesia Zone, Western Alps (Rubie, 1983, 1990; Koons *et al*, 1987). In undeformed meta-quartz diorite, plagioclase is pseudomorphed by very fine-grained jadeitic pyroxene (about $Jd_{0.9}$) + zoisite + quartz (Figures 5.1, 5.5f). Biotite is partially reacted to a corona consisting of high-Ti phengite and layers of zoned garnet and fine-grained muscovite + quartz + sphene situated adjacent to pseudomorphs after plagioclase (Figures 5.1b, 5.6). As already discussed, because equilibrium was restricted to small localised domains, the compositions of some of the product phases were metastable with respect to the bulk rock composition.

Deformation in the eclogite facies may have been a continuous event but it is conveniently described in terms of two stages, producing first a transition orthogneiss and finally a relatively equilibrated orthogneiss (Koons *et al*, 1987; Rubie, 1990). During the first stage, deformation converted the jadeite-rich aggregates, which pseudomorphed plagioclase, to highly-strained, semi-continuous layers, and the original medium-grained igneous quartz to discontinuous lenses (Figure 5.8). The jadeite-rich aggregates were unusually ductile and Rubie (1983, 1990) suggested that this superplastic behaviour was due to the very small grain size of the products of the plagioclase-breakdown reaction. Deformation may have occurred primarily by diffusion-accommodated grain boundary sliding, although there are also other possibilities (Rubie, 1990). Biotite-breakdown reactions went to completion during this first deformation stage and resulted in the growth of relatively large and homogeneous garnets. White mica compositions also homogenised. In marked contrast, pyroxene compositions did not homogenise and often differ little from those of the undeformed meta-quartz diorite (Figure 5.7). These observations suggest that equilibrium was achieved among all phases except pyroxene at this stage. Two domains of local equilibrium, one consisting of Na–Pyx and the other of Qtz + Grt + Zo + Phe, can

Figure 5.8 Micrographs of deformed meta-quartz diorite (from Monte Mucrone, Sesia Zone). (a) Optical micrograph of transition orthogneiss showing semi-continuous Jd + Zo + Qtz layers (dark) which were derived by deformation of pseudomorphed plagioclase (cf. Figure 5.1a). Quartz lenses originated by deformation of igneous quartz. Fine-grained phengite probably formed during and after the deformation, during fluid infiltration. (Plane polarised light). (b) Back-scattered electron image of a layer of Jd + Zo + Qtz from (a) showing the small grain size of jadeite after deformation. The white elongate grains are zoisite and the dark grey interstitial mineral is quartz. (cf. Figure 5.5f.) (c) Transition orthogneiss with jadeite layers (dark) which have become relatively coarse grained. Garnet is also present. (Optical micrograph, plane polarised light). (d) Optical micrograph of orthogneiss. Compared with the transition orthogneiss, the pyroxene-rich layers have become disrupted by further deformation and are coarser-grained. Pyroxene compositions range from $Jd_{0.5}$ to $Jd_{0.8}$ (Figure 5.7). Sheet silicates, which surround the pyroxene and garnet in this micrograph consist of both phengite and paragonite. (Plane polarised light) therefore be distinguished on the basis of mineral compositions, but not textures. There is evidence for limited fluid infiltration during and after the deformation (Frueh-Green, 1985; Koons *et al*, 1987) which may be largely responsible for the partial equilibration. Pyroxene presumably failed to equilibrate because of slow rates of lattice diffusion: the inferred deformation mechanism of grain boundary sliding involved little or no intracrystalline plasticity (e.g. dislocation creep) which might otherwise have enhanced diffusion rates in this mineral.

During and after the first deformation stage, the grain size of jadeite coarsened from $\leqslant 30\,\mu m$ to 0.2–0.5 mm (Figure 5.8c). As a result, the deformation mechanism for jadeite changed to dislocation creep and the jadeite-rich layers became less ductile and were strongly disrupted during the second deformation stage (Figure 5.8d). The pyroxene became more omphacitic, homogenised in Mg/(Mg + Fe) (Figure 5.7) and is inferred to have approached an equilibrium composition. The kinetics of this composition change may have been enhanced by the generation and movement of dislocations. The reaction which produced the change in pyroxene composition involved the formation of paragonite and the dissolution of garnet. However, as there is no evidence for any change in garnet composition, it is concluded that pyroxene equilibrated in a domain of local equilibrium which excluded relict garnet. This conclusion is confirmed by a lack of Mg–Fe exchange equilibrium between the two phases even when they are in close proximity (Koons *et al*, 1987). Applying the garnet–clinopyroxene geothermometer of Ellis and Green (1979) gives temperature estimates about 200°C higher than generally accepted estimates for the Sesia Zone metamorphism.

The above example demonstrates the complexity of possible reaction pathways during eclogitisation. It also shows that, even when eclogites are affected by intense deformation and fluid infiltration, an equilibrium assemblage, on the scale of a thin section, will not necessarily develop. The variability of pyroxene compositions ($Jd_{0.5-0.8}$) in the orthogneiss is regarded as resulting primarily from the failure of this mineral to equilibrate completely during the enlargement of equilibrium domains during deformation. It is uncertain whether changing *P–T* conditions played any role in controlling the mineralogical evolution at high pressure.

5.8 Preservation of eclogites

The preservation of eclogite facies rocks during uplift to the Earth's surface has frequently been regarded as a significant problem. Simple thermal models for continental collision zones predict that rocks buried to depths of the order of 60 km by underthrusting should undergo significant heating during erosion-controlled uplift and pass through the granulite/amphibolite and greenschist facies (e.g. Thompson and England, 1984; Rubie, 1984, figure 5). The fact that eclogites are often little affected by LP metamorphic reactions has led to the hypothesis that uplift rates must be extremely rapid and must be controlled by tectonic rather than erosional processes. Rapid (adiabatic) uplift could favour the preservation of eclogites for kinetic reasons by eliminating the temperature increase and reducing the time available for LP reactions to occur. In evaluating this hypothesis, two factors have to be considered:

(a) Thermal–tectonic models can be formulated in which temperatures do not increase subsequent to HP conditions, even when uplift is slow.

(b) Even if the temperature does increase during uplift, retrograde reactions will not occur in some eclogites (especially mafic lithologies) provided they remain isolated from aqueous fluid.

5.8.1 *Thermal–tectonic uplift rates*

There is little or no direct evidence for sustained rapid uplift of HP rocks in collision belts. Although over short time intervals (e.g. 4 Ma), erosion-controlled uplift can be as fast as 10 mm/year (e.g. Koons, 1987), estimated average rates, over long time intervals, are generally much slower. For example, radiometric dating suggests that uplift of the Sesia Zone, Western Alps, from a depth of around 60 km to 15 km, took place at an average rate of 0.5–0.8 mm/year (Rubie, 1984). In addition, estimated P–T paths for many HP terranes indicate that uplift was near-isothermal to quite shallow levels (e.g. Heinrich, 1982; Cuthbert *et al*, 1983; Rubie, 1984; Barnicoat and Fry, 1986). It is very difficult to model isothermal uplift, with moderate average uplift rates, following the attainment of HP conditions (e.g. 16 kbar, 550°C) by a single underthrusting or thickening event (Draper and Bone, 1981; England and Thompson, 1984; Rubie, 1984). Such results are readily obtained, however, when uplift of the underthrust unit is accompanied by continuing underthrusting or subduction at lower tectonic levels. For example, a model proposed by Rubie (1984) for the Sesia Zone incorporates the following features:

(a) underthrusting of Sesia Zone continental crust to a depth of about 60 km;
(b) subduction of oceanic lithosphere below the underthrust unit for a prolonged period (e.g. 20 Ma), thus suppressing thermal recovery of the Sesia rocks;
(c) erosion-controlled uplift of the Sesia Zone began during this subduction episode and was accommodated by underplating of oceanic material (Piemonte Zone).

This type of model has been further developed, with application to collision belts in general, by Davy and Gillet (1986) and Gillet *et al* (1986) (see also Gillet and Goffé, 1988). A related model has been proposed by Platt (1986) in which uplift is controlled primarily by tectonic rather than erosional processes.

In the Western Gneiss Region of Norway, retrogression of eclogites is often extensive. This may be partly the result of initial uplift at relatively high T with the additional possibility of T_{max} being reached during uplift, after P_{max} (Jamtveit, 1987a). However, the close spatial association of retrogressed and preserved eclogites suggests that fluid availability was also an important controlling factor.

5.8.2 *Kinetics of retrograde reactions*

Mafic eclogites, consisting primarily of omphacite + garnet, have to hydrate to equilibrate in the amphibolite facies. Therefore, a major factor that controls the rate of such equilibration can be the availability and rate of supply of H_2O from an external source, during uplift. This is demonstrated by a petrological study of kyanite eclogites in the Adula nappe, Central Alps (Heinrich, 1982). Basement rocks of the Adula nappe, which include metapelites containing mafic layers and lenses, underwent an eclogite to amphibolite facies transition during the Tertiary (Lepontine) metamorphism. In the metapelites, this transition occurred by the dehydration reaction

$$\text{Phe} + \text{Pg} + \text{Grt} \rightarrow \text{Pl} + \text{Bt} + \text{Ms} + H_2O \qquad \text{(R5.6)}$$

In contrast, lenses of mafic eclogite equilibrated in the amphibolite facies by the hydration reaction

$$Grt + Omp + Ky + H_2O \rightarrow Hbl + Pl \qquad (R5.7)$$

Alteration by reaction R5.7 developed initially at the margins of the eclogite bodies and concentric zones of symplectite eclogite and amphibolite grew progressively inwards (Figure 5.9). The progress of R5.7 was evidently controlled by the infiltration of hydrous fluid from the surrounding pelites. Heinrich (1982) has suggested that dehydration of the pelites by R5.6 was the source of these fluids and has estimated that this source could provide sufficient H_2O to amphibolitise all the mafic eclogites of the Adula nappe. On the basis of the width of reaction zones and likely permeability values,

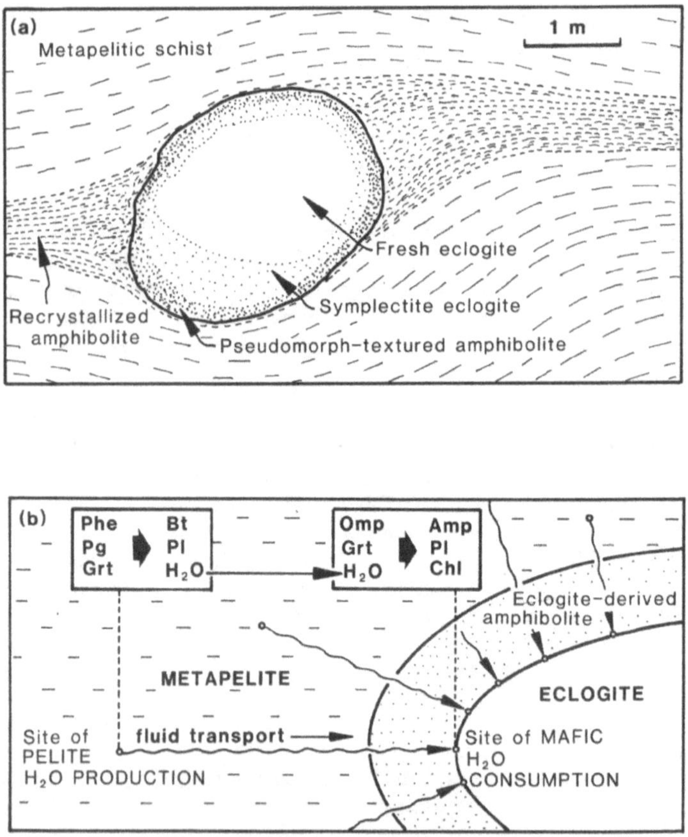

Figure 5.9 Conversion of mafic eclogite to amphibolite in the Adula nappe (Central Alps) (after Heinrich, 1982). (a) Field relations of mafic lenses and pelitic schists at Gagnone (SW Adula nappe). In the symplectite eclogite, omphacite is replaced by Pl-Di symplectite in an inner zone and by Pl-Amp symplectite in an outer zone. (b) Mafic lens being hydrated to amphibolite by the H_2O released by the dehydration of the enclosing metapelites.

Rubie (1986) estimated that fluid-present conditions in the pelites and amphibolitisation of the mafic eclogites occurred over the relatively short time interval of < 1 Ma.

Retrograde metamorphism of HP assemblages could occur in some lithologies by solid–solid reactions, such as

$$\text{Jd} + \text{Qtz} \rightarrow \text{Ab} \quad \text{(R5.8)}$$

$$\text{Omp} + \text{Qtz} \rightarrow \text{Ab} + \text{Di} \quad \text{(R5.9)}$$

The replacement of omphacite by plagioclase–diopside symplectite (Figure 5.10a) has occurred, for example, in the partially-altered eclogites of the Adula nappe (Heinrich, 1982). Concentric zones of alteration around a central core of fresh eclogite consist of (a) sequential zones in which omphacite is replaced by (i) Pl–Di symplectite and (ii) Pl–Amp symplectite, and (b) an outer zone of amphibolite (Figure 5.9a). The nature of this zonal structure suggests that it developed under a gradient in μ_{H_2O} between the anhydrous eclogite core (low μ_{H_2O}) and the surrounding dehydrating metapelites (high μ_{H_2O}) (Figure 5.9b). The formation of Pl–Di symplectite evidently required the presence of some H_2O as a catalyst but only occurred where μ_{H_2O} was low. Pl–Amp symplectite and amphibolite formed by hydration under conditions of higher μ_{H_2O}.

In the Sesia Zone, the extent of R5.8 and R5.9 is often highly variable. Perfectly preserved sodic pyroxene + quartz and partially or completely pseudomorphed sodic pyroxene can be found in adjacent outcrops (Figure 5.10b–d) or occasionally even within the same thin section. Initial alteration of pyroxene in veins, the spatial variability of alteration at all scales down to that of individual crystals, and the common occurrence of phengite as a reaction product indicate that reaction has only occurred where H_2O was present as a catalyst (and in some cases as a reactant). Rubie (1986) has substantiated this conclusion by modelling the kinetics of growth of an albite layer between jadeite and quartz, during uplift of the Sesia Zone, on the basis of available experimental data for rates of grain boundary diffusion. Results suggest that with water-saturated grain boundaries in the albite layer, 100 mm jadeite crystals could be completely pseudomorphed by albite in less than six years at 550°C. In contrast, with completely dry grain boundaries, the maximum estimated thickness of an albite layer after 40 Ma is 1–10 μm (assuming that nucleation of albite could occur in absence of fluid). A whole spectrum of intermediate reaction rates are probably possible because the rate of grain boundary diffusion may depend strongly on the exact concentration of H_2O-related species in intergranular regions (Rubie, 1986, figure 4).

The above examples show that when retrograde reactions occur by dehydration, kinetics will not be a major limiting factor. The occurrence of hydration and solid–solid reactions, however, will often be limited by the absence of H_2O. In the case of solid–solid reactions, H_2O is required as a catalyst because in the absence of this phase kinetics are extremely unfavourable.

Although the preservation of mafic eclogites depends primarily on the absence of hydrous fluid during uplift, the maintenance of low temperatures is also important. If temperatures become too high, certain lithologies may dehydrate and consequently act as a source of hydrous fluid. For example, in the Adula nappe, metapelites have dehydrated to cause extensive retrogression of mafic eclogites primarily in regions where temperatures reached 600–650°C during uplift. Where uplift temperatures were relatively low (e.g. 525°C), dehydration of the metapelites was comparatively limited, and mafic eclogites are well preserved (Heinrich, 1982).

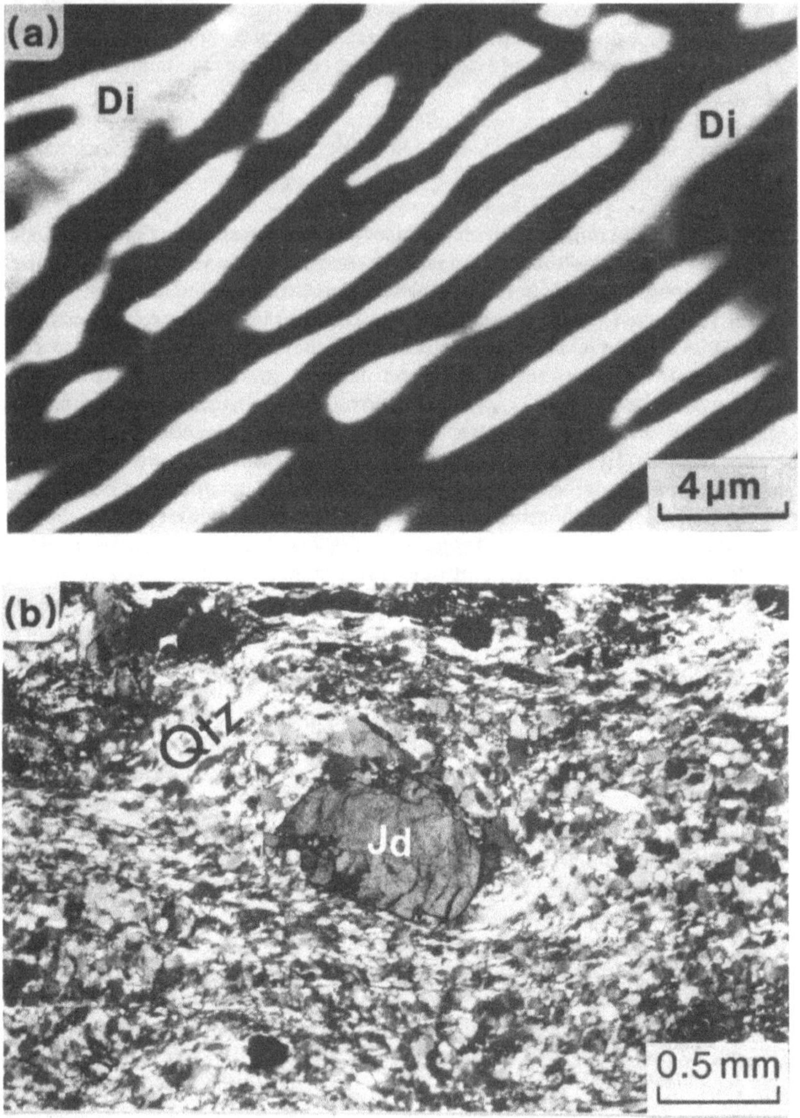

Figure 5.10 Micrographs showing various stages of retrograde reaction in eclogites. (a) Backscattered electron image of a plagioclase–diopside symplectite which has replaced omphacite in eclogites at Gagnone, Adula Nappe, Central Alps (Figure 5.9 and Heinrich, 1982). The black matrix consists of plagioclase. (b)–(d) Stages in the breakdown of sodic pyroxene (jadeite or omphacite) in samples from the central Sesia Zone which must have followed similar *P–T–t* histories during uplift. The extent of alteration in these examples was controlled primarily by the availability of H_2O. (b) Completely unaltered jadeite crystal in a quartz matrix (after Rubie, 1986). (c) An early stage of alteration showing the development of albite in veins. Very little alteration is present along Omp–Qtz grain boundaries. (d) Na-pyroxene crystals completely pseudomorphed by fine-grained albite and phengite (after Rubie, 1986).

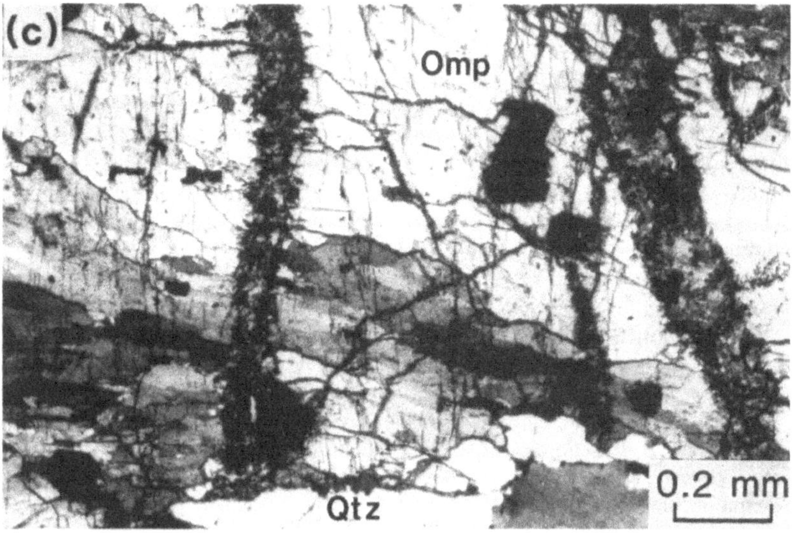
(c)
Omp
Qtz
0.2 mm

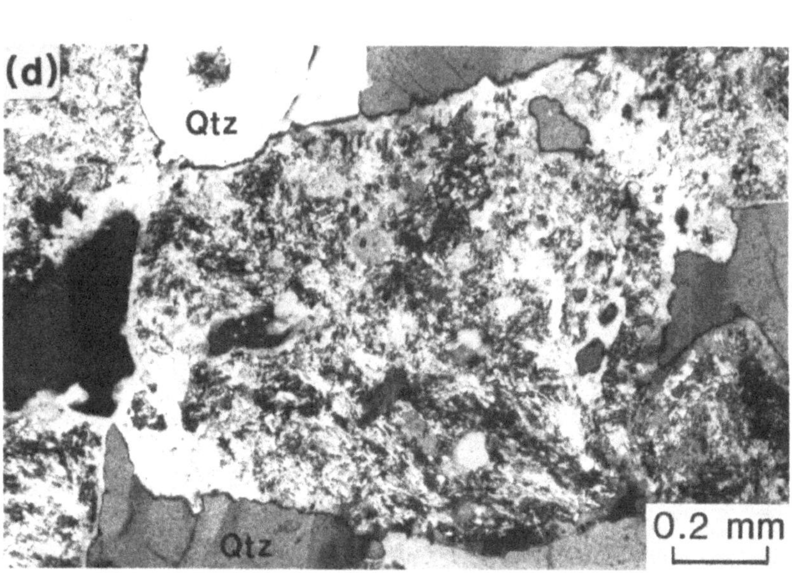
(d)
Qtz
Qtz
0.2 mm

5.9 Conclusions

The examples presented in this chapter demonstrate the importance of H_2O as a reaction catalyst during the eclogitisation of gabbros and granitoids. Even during eclogite formation at relatively high temperatures (e.g. 750°C), it is likely that small traces of H_2O were available to enhance reaction rates. These conclusions are similar to those reached by Ahrens and Schubert (1975a, b) on the basis of a theoretical model. Similar conclusions can be drawn concerning retrograde reactions during the uplift of eclogites to the Earth's surface—preservation of these rocks is favoured by the continuous absence of H_2O (Rubie, 1986).

In the absence of catalytic H_2O, gabbros and similar coarse-grained lithologies can persist metastably at eclogite facies conditions without any HP mineral reactions occurring (Austrheim and Griffin, 1985). This phenomenon can be ascribed to both the inhibition of the nucleation process (e.g. Wayte *et al*, 1989) and slow rates of diffusion. Even when small amounts of H_2O are present, disequilibrium (on a thin section scale) frequently results because of slow diffusion rates, provided simultaneous deformation does not occur. Reactions occur within small mineralogical or textural domains and also between such domains to form coronas. Phases which form during this disequilibrium stage have compositions which reflect the chemical environment of the localised domain and are often metastable with respect to the bulk rock chemistry.

It follows from these conclusions that the timing of mineral reactions will not necessarily depend on when equilibrium boundaries are crossed in P–T space but rather on when fluid becomes available at reaction sites. The concept of progressive equilibration keeping pace with changing P and T will not be valid unless fluid is continuously present in the rock. In cases where fluid infiltration occurs at a large overstepping of an equilibrium boundary (e.g. $\Delta P = 12$ kbar—see Wayte *et al*, 1989), the driving force for reaction will be large and consequently reaction rates may be fast.

In general, the effect of deformation is to enhance greatly rates of equilibration. This results largely from the associated infiltration of catalytic hydrous fluid. Also, rates of lattice diffusion in minerals, such as plagioclase and pyroxene, may be enhanced by deformation mechanisms involving the movement of dislocations. Grain size reduction during deformation will also favour equilibration by reducing diffusion distances, increasing effective diffusion coefficients (Brady, 1983) and creating new nucleation sites. However, metastable phases can survive intense deformation, and this process does not necessarily result in the development of an equilibrium assemblage (Koons *et al*, 1987).

Eclogites discussed in this chapter have formed along reaction pathways that involve intermediate disequilibrium stages (Figure 5.2) owing to slow reaction rates in the absence of fluid. Unfortunately, it is difficult to evaluate if this behaviour is the exception or the rule, because rocks which have completely equilibrated contain no record of their earlier history. Certainly it is to be expected that if rocks are deformed early in the reaction history, with simultaneous fluid infiltration, equilibrium may be more closely attained throughout the process of eclogitisation.

Acknowledgements

A review by M.B.E. Mørk and discussions with W.S. MacKenzie and G.J. Wayte are gratefully acknowledged. This Chapter was written while the author held a Research Fellowship at the Geology Department, University of Manchester, with financial support from the Natural Environment Research Council (Grant GR3/6409).

6 Isotopic systems—geochronology of eclogites

D. GEBAUER

6.1 Introduction

Absolute age dating of eclogites started only at the end of the 1970s, whereas numerous geochronological data had been produced previously for felsic rocks of the continental crust. The delay, of about 20 years, had various reasons depending on the individual techniques. In the case of the Rb-Sr method, mainly analytical difficulties caused most isotope geologists to be very reluctant even to attempt to date eclogites. As also the prerequisites to obtain isochrons were known often to be missing in both non-metamorphic and metamorphic mafic and ultramafic rocks (see references in Gebauer and Grünenfelder, 1979a) only few attempts were made to date eclogites by Rb-Sr and the results generally confirmed the negative expectations (e.g. Gebauer and Grünenfelder, 1979a or Griffin and Brueckner, 1985). The classical Rb-Sr technique therefore plays an insignificant role in the attempt to date eclogites and/or their respective protoliths.

Similarly, conventional K-Ar or $^{39}Ar/^{40}Ar$ dating can not contribute to the detection of protolith ages of eclogites and only rarely can contribute to dating the individual high pressure events. Only under favourable conditions, i.e. if amphibole or phengite can be shown to be part of the high-pressure assemblage formed at or below about 500°C, if excess Ar was not incorporated and later metamorphic overprinting under lower pressure did not affect the earlier mineral assemblage, can mineral Ar ages be expected to be close to the metamorphc peak conditions (e.g. Suppe and Armstrong, 1972 or Altherr *et al*, 1979). Unfortunately, analytical errors are generally too high to describe reliable P–T–t paths by analysing minerals formed at different times during progressive and retrogressive metamorphism by either the same or by different dating techniques.

Given the above considerations, we are left with conventional and ion microprobe U-Pb zircon dating or with the Sm-Nd technique. Of these three methods, conventional U-Pb dating using the zircon suite method has been applied since 1963. The U-Pb ion microprobe technique, the most powerful of the three methods, has been in operation only since the early 1980s, while the Sm-Nd method was developed in the early and middle 1970s. These three methods currently play the dominant role in dating eclogites.

6.2 First U-Pb zircon attempts in Norway: K-Ar as comparison

The first U-Pb analyses on zircons of an eclogite were carried out at the Carnegie Institution of Washington by Tom Krogh (Krogh *et al*, 1974). The sample was collected from one of the world's largest eclogite bodies (about 4 km^2), which occurred within high-grade gneisses of the Western Gneiss Region in Norway. Interestingly, the small rounded zircons contained inclusions of omphacite, rutile and quartz, i.e. minerals of the high-pressure mineral paragenesis, while inclusions of the post-eclogite facies mineral assemblage were not observed. Therefore, it was probable that the analysed zircons were part of the high-pressure assemblage and therefore could date the metamorphic formation of the respective eclogite body. Owing to the low U-content of about 50 ppm and the unexpectedly young age falling into the Caledonian cycle, only 12 ng and 2 ng of radiogenic lead were present in the two analysed zircon fractions. Although in the early and middle seventies such low lead loads caused considerable mass-spectrometry problems, the $^{206}Pb/^{238}U$ ages still suggested a Caledonian high-pressure formation around 400 Ma of the two analysed zircon separates. Because of a $^{207}Pb/^{206}Pb$ age of 600 ± 100 Ma, the presence of inherited zircon with pre-eclogite facies radiogenic lead was suggested. However, owing to the relatively large analytical uncertainties of the slightly discordant data points no attempt was made to define an upper intercept age and speculate on its meaning. This only became possible with the application of the zircon suite method to various other country rock eclogites of the Western Gneiss Region in Norway (Gebauer *et al*, 1982 and 1985). The Caledonian eclogite forming event as suggested by the pioneering work of Krogh *et al* (1974) strongly contradicted other interpretations on the timing of eclogite formation based on both field- and geochronological data (e.g. Pidgeon and Råheim, 1972 or Mysen and Heier, 1972). Based on Rb-Sr whole-rock work the main metamorphism of the gneisses was interpreted to be around 1.7 Ga which was considered to be a minimum age for the eclogite formation in the Western Gneiss Region.

Krogh's analyses also contradicted the interpretation of K-Ar data (McDougall and Green, 1964) obtained on different clinopyroxenes, amphiboles and phlogopites extracted from a country-rock eclogite (external eclogite) and an eclogite within a garnet-peridotite (internal eclogite). These data, ranging from 415 Ma for one phlogopite to 4.5–8.1 Ga for four different clinopyroxenes, clearly demonstrated the presence of excess radiogenic argon. Yet, based on figures around 1.8 Ga for the secondary amphiboles in two different eclogite occurrences, the eclogites were interpreted to be 'as old as 1.8 Ga' (McDougall and Green, 1964). Other amphibole- and phlogopite-ages, scattering between 950 Ma and 1170 Ma, were interpreted to reflect possible events. Obviously, the K-Ar data yielded inconsistent ages which were, at this very early stage of geochronological research, overinterpreted. Sm-Nd ages by Griffin and Brueckner (1980) and U-Pb zircon-suite data of Gebauer *et al* (1982 and 1985) on further eclogites of the Western Gneiss Region have corroborated the interpretation of a Caledonian formation of eclogites in Western Norway.

6.3 U–Pb zircon-suite method

6.3.1 *Historical background and description*

The zircon suite technique has been extensively used for zircons from (meta)-granitic and (meta)-sedimentary rocks. It was introduced by Silver and Deutsch (1963) who first found that finer and relatively more magnetic fractions of a cogenetic zircon population usually have younger apparent U-Pb ages. This is thought to be due to lead loss, facilitated by their larger surface to volume ratio during a secondary event. Additionally, the finer and more magnetic fractions of magmatic crystals usually also contain more U and Th which means that radiation damage in these fractions is larger than in the coarser relatively less magnetic fractions. Again, this facilitates loss of radiogenic lead, as it is fixed less strongly into the crystal lattice of zircon than the respective parent isotopes of U and Th. U/Pb data are usually presented graphically in a conventional Condordia diagram in which the atomic ratios of $^{206}Pb/^{238}U$ and $^{207}Pb/^{235}U$ are plotted on the ordinate and abscissa, respectively. The Concordia curve is then the locus of all points with equal $^{206}Pb/^{238}U$- and $^{207}Pb/^{235}U$-ages. U-Th-Pb dating techniques for minerals are detailed by Gebauer and Grünenfelder (1979b).

6.3.2 *How much zircon does an eclogite contain?*

In 1975 the first analyses on eclogite-zircons separated into grain-size and magnetic fractions were carried out at ETH Zürich (e.g. Gebauer and Grünenfelder, 1977, 1978, 1979a; Gebauer *et al*, 1978). This followed a systematic search for zircons in gabbros, amphibolites, eclogites, serpentinites and various peridotites. Zircon abundance levels in about half of the rocks stimulated an attempt at U-Pb dating. In rare cases the zircon content of mafic rocks was found to even exceed that of granitic rocks. Zircon abundances are not always a close reflection of the Zr-content of the rocks, since other Zr-containing phases may be present (e.g pyroxenes, amphiboles, garnet, etc.). Even in the few cases where Zr is at or below the detection limit of XRF-analysis (a few ppm), there can be enough zircon present for U-Pb dating. In future it may be possible to date a larger variety of rock-types by the U-Pb method by analysing single grains especially with the ion probe.

6.3.3 *First application: eclogites of Münchberg Gneiss Massif*

6.3.3.1 *Geological, petrological and geochemical characteristics.* Based on the above mentioned search for zircons in mafic and ultramafic rocks, one metagabbro and two eclogite samples from the Münchberg Gneiss Massif (MGM) in Northeast Bavaria (FRG) were chosen first for the zircon suite method (Gebauer and Grünenfelder, 1979a). The MGM is an oval shaped tectonic klippe of about 25 × 15 km. The samples were taken from the Weissenstein area, a classical locality for Central European eclogites which also contains the largest occurrence of eclogites within the non-Alpine part of Central Europe (about 0.6 × 1.5 km). One sample of a metagabbro was taken from a further lens situated about 300 m from the large eclogite lens. Based on field mapping, tectonic contacts of the mafic bodies with the metasedimentary country-rocks were concluded by Stettner (1960). However, it is important to note that the eclogites of the Weissenstein area are also primarily interbedded with numerous layers of paragneisses as has been found in drill-cores within the eclogite body (Matthes *et al*, 1974). Therefore, the eclogite protoliths might have been, at least in parts, basaltic lava flows and/or tuffs. Based on major and trace element data by Matthes *et al* (1975) and Puchelt *et al* (1978) as well as trace element and initial Sr- and Nd-data by Gebauer and Grünenfelder (1979a and unpublished) and Stosch and Lugmair (1987), the respective protoliths probably formed from a suboceanic mantle. The fact that within the large Weissenstein eclogite body eclogites of different geochemical characteristics occur is in agreement with findings from drill-cores in recent oceanic settings (e.g. Bougault and Cande, 1983 or Tual *et al*, 1983).

The close geological association of different eclogites and metagabbros within a small area rather favours the existence of protoliths formed over a small time range in the same geotectonic scenario. This was very probably some sort of oceanic environment. The intimate interlayering of eclogites and paragneisses precludes a true ocean floor scenario. Instead, a geotectonic environment with a near continental source seems to be probable, such as a rift- or back-arc basin. Minimum P–T conditions for the Weissenstein eclogites have been estimated as $T = 620 \pm 50$°C and $P = 13$–17 kbar (Franz *et al*, 1986). This implies that the respective protoliths were buried to more than 50 km, probably in the course of subduction processes. Earlier work by Chesnokov and Popov (1965) as well as by Smith, Thomas and Franz (in Smith, 1984) indicated the possibility of even higher pressures (> 30 kbar) based on the possible former presence of coesite in some eclogites of the MGM.

6.3.3.2 *Zircon morphologies in metagabbro and eclogites.* The abundance of exclusively primary magmatic zircons in the analysed metagabbro was quite high, about 35 mg of zircon in 1 kg of rock. All crystals were transparent, lightbrown. Only $< 1\%$ of the original crystals were completely intact: these rare zircons were bipyramidal-prismatic with smooth crystal faces and sharp edges and length/breadth ratios of about 7:1. Most zircons occur as fragments, frequently without regular crystal faces, i.e. they were derived from the inner parts of an originally much larger crystal. Such observations on zircons from gabbroic rocks are common (e.g. Gebauer *et al*, 1982; Gebauer *et al*, 1985 and unpublished data) and possibly are related to pressure release during ascent of the magma. Observation of such broken pieces of zircon in thin sections excludes an exclusive break up of zircon during crushing.

Zircons of the two eclogite samples taken from the top of the Weissenstein mountain are very similar. The abundance is around 10 mg per kilogram of rock, i.e. about a third of that of the metagabbro. The zircon crystals are mostly irregularly shaped, almost isometric. They contain numerous small crystal faces, the edges of which are frequently round. They may represent recrystallised and refaceted equivalents of zircon fragments which are commonly found in unmetamorphosed mafic protoliths. Cathodeluminescence studies by electron microprobe often show areas with multiple euhedral zoning of trace elements (similar to Figure 6.4) which indicates magmatic formation with free growth conditions for all crystal faces. However, many crystals also or only show irregular trace element distributions indicative of growth in a solid state. Some contain a euhedral core which indicates metamorphic overgrowth.

Compared with the zircons of the metagabbro, those of the two eclogites are, on average, smaller by almost a factor of two although the grain-size of the eclogites is about twice that of the metagabbro. This is compatible with the hypothesis that these eclogites formed from basaltic lava flows and/or tuffs, as suggested by the interbedding of numerous paragneisses within the eclogite body. Because of rapid cooling lavas and/or tuffs usually contain, much smaller primary zircons than the relatively slowly cooling gabbroic magmas, or none at all, and this feature can be preserved during metamorphic overprinting.

6.3.3.3 *U-Pb age results and discussion.* The data points of seven grain-size and magnetic fractions extracted from the two eclogite samples as well as the data points of two extreme size and magnetic fractions of the metagabbro plot within analytical errors on a single discordia trajectory. The intersection ages are $525 \pm 40/31$ Ma and $380 \pm 14/22$ Ma, respectively, which indicate a Cambrian, possibly even uppermost Precambrian age of the respective protoliths and a Devonian age for the eclogite facies metamorphism. Omission of the gabbro data does not result in a significant change of the intercept ages.

In contrast to the two zircon populations of the eclogites the two extreme size and magnetic fractions of the metagabbro are only slightly (about 20%) discordant although their U contents (300 and 475 ppm) are two to five times higher. This can best be explained with the significant differences in the mineralogical and structural transformation of the two rock types: zircons of the metagabbro appear typically magmatic without clear signs of metamorphic rounding, and/or over-growth. Their weak discordance demonstrates that zircons only little damaged by radiation can survive metamorphic temperatures > about 620°C without losing significant amounts of radiogenic lead, if the host rock remained simultaneously undeformed and fluid activity was low. Temperatures well in excess of even 1200°C might leave U-Pb systems in zircons unaffected if no recrystallisation of the host rock occurred. The weak degree of discordancy therefore is not surprising and is also confirmed by numerous results on zircons from similar rock types (e.g. Gebaur and Grünenfelder, 1982).

As for the eclogite zircon fractions, the most striking difference from zircon suites in granitic host rocks is that the usual positive correlation of degree of discordance versus U-content, magnetic susceptibility and decreasing grain size is not observed. This finding later turned out to be generally true for eclogite zircons with a large proportion of metamorphic zircons. In general, the degree of disordance is correlated with the amount of metamorphic mineral growth, while the U-content and its range—118–168 ppm in the case of the Weissenstein eclogites—is of lesser importance. As other criteria for separating a zircon population are often not, or not easily, applicable, separation according to grain size and magnetic susceptibilities is still the best method to produce a spread in the degree of isotopic discordances.

The Acadian U-Pb age of eclogite facies metamorphism is in good agreement with Rb-Sr mineral isochron ages and several muscovite and biotite Rb-Sr ages. These were obtained from the felsic country-rocks (Gebauer and Grünenfelder, 1979a) and other meta-igneous and metasedimentary rocks of the MGM; they fall into the range 370–395 Ma (Söllner *et al*, 1981). They indicate quick uplift following eclogite formation, as frequently observed also for other eclogite-bearing terranes. The 380 Ma eclogite facies event in the MGM was later confirmed by Sm-Nd mineral data on samples also taken from the top of the Weissenstein (Stosch and Lugmair, 1987).

6.3.4 *Further applications: Sauviat-sur-Vige (French Massif Central)*

6.3.4.1 *Petrology and geochemistry.* The largest eclogite facies mafic–ultramafic body (6 km × 2 km) of the French Massif Central occurs in its northwestern part at Sauviat-sur-Vige. According to Coffrant (1974) the primary magmatic mineral assemblages corresponded to troctolite, gabbro and peridotite. Depending on grain size, deformation and chemical composition, including fluid activity, these rocks were variably transformed into different kinds of metagabbros, eclogites and garnet-peridotites under usually static conditions. In a subsequent syntectonic stage of metamorphism most of these rocks were retrograded under amphibolite facies conditions with variable influxes of fluids. Based on major, trace element data (including REE) a model of fractional crystallisation for the genesis of the various rock types of the Sauviat Massif was proposed by Bernard-Griffiths and Jahn (1981). According to these authors, the primary magmatic crystallisation in an intraplate tectonic environment took place at $P < 9$ kbar, whereas the metamorphic transformation may have occurred at $T > 820$°C and $P > 15$ kbar as indicated by geobarometric and geothermometric data on clinopyroxene and garnet.

6.3.4.2 *Analytical results and discussion.* Nine fractions, according to size, magnetic property and colour, extracted from variously transformed and retrograded mafic and ultramafic rocks plotted linearly on a Concordia diagram with intercept ages at 496 Ma for the protolith formation and 320 Ma for the eclogite facies overprint (Gebauer *et al*, 1981). The range in degree of discordance varied from about 14% for zircons of a coronitic metagabbro to about 70% for a zircon fraction from garnet-peridotite (ultramafic cumulate). Other rock types fall between these limits. Therefore the field, petrological and, in particular, the geochemical evidence that all rock types are cogenetic is confirmed by the colinearity of the U-Pb data.

There is a poor spread in the data points obtained for zircons from the individual eclogite samples in the MGM, whilst the spread of data obtained from four individual zircon populations in the Sauviat Massif is almost negligible. This unfortunate fact has also been observed in a series of other studies where cogenetic protoliths were either not available to obtain a better data spread or not used. For the U-Pb systematics the conclusions drawn for the Weissenstein eclogites of the MGM were confirmed and extended. New- and over-growth of zircon in the course of eclogite facies mineral reactions is the major cause for the degree of discordance. The U-Pb systems of the primary magmatic zircons were only slightly disturbed even though the metamorphic end product was an equilibrated Grt-Omp rock formed above 820°C and at a depth of more than 45 km. Strong amphibolitisation of metagabbros and eclogites appeared to have no significant effect on the degree of discordance (Gebauer *et al*, 1985). However, this observation can not necessarily be generalised and, optimally, each case should be assessed separately.

Indication of a possible synchronous HP transformation of the country-rocks exists only in the immediate meta-igneous wallrocks of the Sauviat Massif. Monazites extracted from these kyanite-bearing, coarse-grained metagranites (augen-gneisses) showed a wide range of data spread from about 15 to 83% discordance. The lower intercept age is identical to that of the eclogite facies mafic and ultramafic rocks. This confirms the Carboniferous HP metamorphism which occurred, like that in the Norwegian Caledonides, relatively late in the orogenic cycle. As the regional geological and geochronological data preclude a general Carboniferous HP metamorphism in Limousin, the eclogite facies felsic to ultramafic rocks, which occur immediately along a thrust zone, probably were introduced during Carboniferous nappe emplacement. Similarly, U-Pb zircon data by Schellhorn *et al* (1982) confirmed the existence of Carboniferous eclogites for various other rocks in the southern part of Limousin.

6.3.5 *Other Phanerozoic eclogites formed from probable Phanerozoic protoliths within the European Variscides*

Like for the Münchberg Gneiss Massif and Sauviat in the French Massif Central there are a number of other eclogites and amphibolites the protoliths of which were formed within about 30 Ma around the Cambrian boundaries. Such rocks are known from the

Monts de Lyonnais and the Montagne Noire within the Massif Central. In both cases Carboniferous eclogite formation and subsequent tectonic emplacement into the existing country rocks has been inferred (Gebauer and Grünenfelder, 1982). Rather young protolith ages, between 440 and 470 Ma, are also probable for the Aiguilles Rouge Massif in the Alps (Paquette, 1987). Generally, post-Pan-African protolith formation is found frequently not only for eclogites but also for amphibolites (Gebauer and Grünenfelder, 1982) and has been interpreted in terms of successive rifting and oceanisation of different parts of Gondwana after the Pan-African multi-episodic cycle (Gebauer and Grünenfelder, 1982). Also back-arc opening can contribute to the formation of oceanic basins as young as Upper Devonian or even Lower Carboniferous which explains the fact that the oldest eclogite forming events, at about 470 Ma (Gebauer *et al*, 1988), can be older than the formation of mafic protoliths in different terranes of the European Variscides. Thus, not only successive rifting but also successive docking of microcontinental plates considerably complicates the deciphering of the geodynamic evolution of the European Variscides.

Given the fact that Precambrian zircons from either the mantle or the continental crust can be incorporated into the protoliths of eclogites without being reset isotopically in the protolithic melt, a number of discordia trajectories can not be used for the determination of a reliable protolith age (Gebauer *et al*, 1987; Paquette, 1987). Therefore, these rocks might also fall into the range of Upper Ordovician to uppermost Precambrian protoliths.

6.3.6 *Phanerozoic eclogites of unknown protolith ages*

In one category of eclogites, protolith ages could not be obtained as zircons were formed as late as in the course of eclogite transformation of the source rock. Thus, no memory of the pre-eclogite history is possible. The data points of such eclogite zircons plot on discordia lines indicate a recent or sub-recent lead loss unrelated to any significant geological event. Unfortunately, Sm-Nd whole-rock dating, which theoretically could give information on the age of protolith formation, does not seem to be promising either (Bernard-Griffiths *et al*, 1985; Quadt and Gebauer, 1988).

The first data of this kind were published by Ducrot *et al* (1983) on rocks of the western part of the French Massif Central (Haut-Allier). The authors obtained an upper intercept age of 432 + 20/ − 10 Ma and a lower, geologically insignificant intercept age at 109 + 125/ − 133 Ma. The apparent $^{206}Pb/^{238}U$ ages of seven size fractions were in the range 390–420 Ma. As the authors assume no pre-eclogite lead to be present in the analysed fractions they interpreted the upper intercept age as the age of eclogite facies metamorphism. However, previous work cited by Ducrot *et al* (Gebauer and Grünenfelder, 1979a; Gebauer *et al*, 1981 and Peucat *et al*, 1982) had unequivocally proved that magmatic zircons are only partly reset by eclogite facies metamorphism. The presence of such zircons is clearly demonstrated in figure 2 of Ducrot *et al* and also is implied by the authors in their text. So it is important to realise that the presumed age of 432 Ma for the eclogite formation is a maximum age only, as suggested also by Pin and Peucat (1986). A model, involving probable Phanerozoic protolith formation and post-432 Ma and pre-320 Ma eclogite formation followed by post-Carboniferous lead loss, probably yields a more realistic fit to the data base.

The other known example in which only indications exist for the timing of the eclogite formation was reported by Paquette (1987). The sample was an N-type MORB eclogite of the Argentera Massif, an external massif of the Western Alps about 50 km N of Nice. All six analysed fractions plot close to the upper intercept which is at 355 ± 5 Ma. The lower intercept is very close to zero Ma, i.e. all $^{207}Pb/^{206}Pb$-ages are analytically indistinguishable. There are less than 5% of slightly elongated crystals which might be magmatic. However, as with the majority of zircons which are ovoid and therefore interpreted to be of metamorphic origin, there are no signs of magmatic trace element distribution in the elongated ones as detected by cathode luminescence studies. The age of 355 Ma was therefore interpreted to reflect the age of eclogite formation, adding to the observation that eclogite formation in the European Variscides can, like the Norwegian Caledonides, occur rather late in orogenic evolution.

6.3.7 *Eclogites for which only the protolith age could be obtained*

The rare phenomenon of being able to give information only on the protolith age has been observed so far only in an N-type MORB eclogite of the Aiguilles Rouge Massif, an external massif of the Western Alps about 50 km ESE of Geneva (Paquette, 1987). Five size fractions of magmatic, euhedrally-zoned zircons plot close to the upper intercept at 454 + 6/ − 1 Ma. The lower intercept is at 102 Ma and was interpreted to be geologically meaningless as an Alpine eclogite facies metamorphism could be ruled out for the Aiguilles Rouge Massif. According to Paquette (1987) the primary magmatic zircons were resistant enough to survive the eclogite facies metamorphism of the protolith (e.g. Gebauer and Grünenfelder, 1979a or Gebauer *et al*, 1981).

As mentioned before the observation of oceanic protolith formation younger than the oldest eclogite forming events (e.g. Aiguilles Rouge versus Gotthard Massif) can be explained only with complex geodynamic models that involve formation and docking of terranes at various times.

6.3.8 *Phanerozoic eclogites formed from possible Precambrian protoliths*

6.3.8.1 *Norwegian Caledonides.* In a study on the age and origin of some Norwegian eclogites Gebauer *et al* (1982 and 1985) reported zircon suite data, as well as geochemical data, from five different eclogites and one amphibolite. All samples were taken from boudinaged lenses of metre to decimeter size, which occur within the 1.7 Ga old Basal Gneisses of Western Norway. Based on the morphology, zoning patterns, size as well as degree of discordance, protoliths of both basaltic and gabbroic origin could be distinguished. A Caledonian age of eclogite formation at about 400 Ma was detected for all five pods, in line with the early U-Pb data of Krogh *et al* (1974) and the Sm/Nd data of Griffin and Brueckner (1980).

Whereas it was not possible to obtain information on the age of the two protoliths of probable basaltic origin—because of the lack of pre-eclogite radiogenic lead—the ages of the three different gabbroic protoliths could be determined, or at least estimated, at about 1.5 Ga and > 1.76 Ga.

Based on the U-Pb systematics of primary magmatic zircons from three different eclogite boudins a Sveconorwegian overprint (ca. 1.1 Ga) is very probable. This demonstrates that these rocks can share, after an early tectonic emplacement, a long part of their history within the continental crust. This, however, does not imply that they necessarily shared the same Caledonian *P–T–t* path with the existing country-rocks as they could have been subducted to much greater depth and then brought back tectonically into the grey gneisses where they are observed today. Considering the existence of coesite and probable coesite-derived polycrystalline quartz in various eclogites of the Western Gneiss Region (Smith, 1984 and 1987), such a scenario might be the most probable, although not the only possible one, at the present stage of investigation.

6.3.8.2 *European Variscides–Brittany.* The first U-Pb zircon data on eclogites produced at Rennes University were published in 1982 by Peucat *et al.* Two eclogite samples were taken from one quarry in the Vendée close to Nantes and a total of seven zircon fractions were separated. With one exception all fractions plotted on a discordia line that intersected the Concordia at 1927 ± 60 Ma and $436 + 15/-12$ Ma, respectively. The morphologies of zircons and the variation of the degrees of discordances, unrelated to either grain size or U-content, are similar to those of the Weissenstein eclogites. The authors favoured a model in which 'the protolith may have been extracted form the upper mantle 1250–1300 Ma ago and stored in a crustal environment until it was metamorphosed under high-pressure conditions around 400 Ma ago'. The figure of around 400 Ma was chosen as zircons from two garnet-pyroxenites of the Baie d'Audierne mafic–ultramafic complex, about 250 km to the NW of the Vendée eclogites, yielded a lower intercept age of 384 ± 6 Ma. The upper intercept age (1252 ± 25 Ma) was indistinguishable from that of the Vendée eclogites. Although the two values for the individual HP metamorphisms are significantly different, the authors chose a value 'around 400 Ma' instead and did not argue for two different times of HP metamorphism. Regarding the protolith age around 1300 Ma, the authors proposed 'that a 1300 Ma old oceanic crust was tectonically incorporated into a sialic basement during the Proterozoic' and that 'this mixture was subsequently subducted during the Palaeozoic'. With current knowledge on the possibility of inherited zircons in such rocks, 1.3 Ga might far exceed the actual protolith age(s).

The third data set on eclogites of southern Brittany has been published by Paquette *et al* (1985) on an eclogite of the Champtoceaux Complex. The sample is geochemically similar to intermediate calc-alkaline rocks of island arc affinity. 100 kg of rock contained 100 mg of magmatic zircon and zircon fragments which showed typical relationships between U-content, grain size and degree of discordance ($^{206}Pb/^{238}U$ age range 840–660 Ma). The lower intercept age is 413 ± 16 Ma. The upper intercept, however, is at 2208 ± 39 Ma and can be interpreted in two ways. Firstly, that the age of 2.2 Ga is the age of the protolith and secondly that, due to contamination, no protolith age can be derived. The authors favoured the first hypothesis based on the homogeneous magmatic zircon population of the sample.

The Leon area at the northwestern tip of Brittany comprises a geological unit similar to the South Armorican domain. It contains N-type MORB eclogites with U-poor (about 10 ppm) zircons (Paquette *et al*, 1987). Three size fractions align on the lower part of a discordia intersecting the Concordia at $439 + 12/-13$ Ma and at $801 + 50/-42$ Ma. Whereas the lower intercept age agrees well with the age of eclogites of Vendée and Champtoceaux, the upper intercept is thought to represent either the age of the MORB-type protolith or be due to minor contamination. In the latter case no protolith age can be derived.

Generally, the possible presence of inherited zircons in the protoliths of eclogites can be a serious problem (Section 6.4.2). A thorough inspection of grain morphologies and cathode luminescence of zircons to be analysed is of paramount importance for an understanding of the nature of the zircons to be dated.

6.3.8.3 *European Variscides—Cabo Ortegal (Spain).* The mafic-ultramafic Cabo Ortegal Complex in Galicia (NW Spain) also contains N-type MORB eclogites. Two grain size fractions of zircon from one eclogite were analysed for U-Pb by Bernard–Griffiths *et al* (1985). The multifaceted rounded grains were very poor in U (about 7 ppm) and the two data points plot very close together and very near the Concordia curve at about 440 Ma for the $^{206}Pb/^{238}U$ age. Based on Pb-isotope studies of Kuijper (1979) that suggest a Precambrian protolith formation, the authors assumed the same protolith age as found for eclogites of NW France, i.e. 1.3 Ga. They arrived at a lower intercept age for the eclogite formation at about 420 Ma. Alternatively, they assumed lead loss due to continuous diffusion of radiogenic lead to explain the slight degree of discordance and thus arrived at an age of around 480 Ma for the eclogite forming event: the age of the eclogite facies event is bracketed between 420 and 480 Ma. However, as the possibility of pre-eclogite facies radiogenic lead can not be ruled out, an eclogite-forming event younger than 420 Ma can not be ruled out either. Such a scenario becomes realistic, if a protolith age of around 500 Ma is assumed, a typical age for many protoliths in the European Variscides. Unfortunately, Sm-Nd data on whole-rock samples did not contribute to interpretation of the protolith age(s) of Cabo Ortegal eclogites.

6.4 Ion probe dating of eclogite zircons

6.4.1 *Analytical background*

The reason for using the ion microprobe to date zircons is that ages can be obtained *in situ* on polished sections on a 20 μm (2×10^{-2} mm) scale (Figure 6.1). In this way it is possible to obtain information on the distribution of U-Pb and Pb-Pb apparent ages within single crystals and, for example, to date the formation of growth zones of

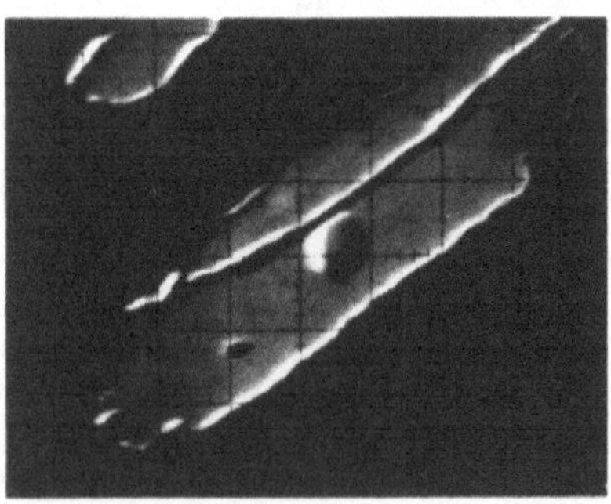

Figure 6.1 Secondary electron image of polished zircon formed co-magmatically at about 870 Ma in gabbroic protolith from a retrograded eclogite of the Gotthard Massif. Crater 20–30 μm wide and 2–3 μm deep was produced by oxygen induced sputtering using SHRIMP. For cathode-luminescence distribution of this grain see Figure 6.4. (From Gebauer *et al*, 1988)

different ages. There is only one currently operational ion microprobe that has enough sensitivity at high mass resolution to make such analyses. The development of the Sensitive High Resolution Ion Micro Probe (SHRIMP) was initiated by W. Compston at the Australian National University in Canberra in the early 1970s; design began in 1974 and the first data were obtained in 1981. Detailed information on SHRIMP is given in Compston *et al* (1986). Figure 6.1 shows the typically 20–30 μm wide, 2–3 μm deep craters produced by the oxygen-induced sputtering of a zircon after 20 minutes, the typical time for a single analysis. All ion probe Pb-U ages referred to in the text are referenced to an age of 572 Ma for the SL 3 standard zircon used at the ANU.

6.4.2 *First example: Gotthard Massif (Central Alps)*

Retrograded eclogites occur as lenses and layers up to about 100 m thick in pre-Alpine gneisses of the Central Alpine Gotthard Massif. The Alpine metamorphic overprint reached chloritoid-grade and did not cause any significant effect on the U-Pb zircon systems. Based on major-, trace- and rare earth element data (Gebauer, unpublished) the protoliths of these former high-pressure rocks could have been calc-alkaline to tholeiitic melts of island arc or even ocean floor affinity.

U-Pb analyses (Gebauer *et al*, 1988) on over a dozen spots on the rare, clear, euhedral zircons (average U about 300 ppm), which are interpreted to be formed magmatically in the protolith (Figure 6.2), plot over the whole range of a discordia trajectory which intersects Concordia at about 870 Ma and around 450 Ma respec-

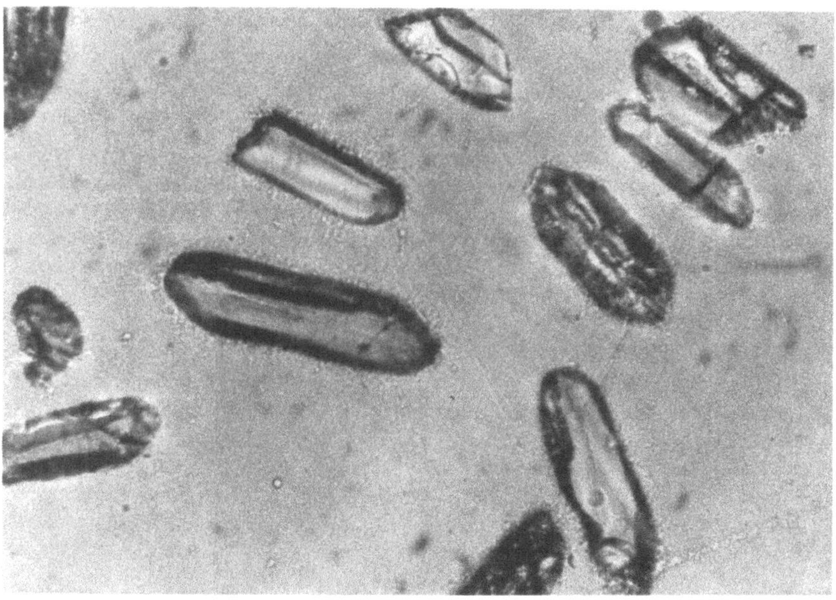

Figure 6.2 Colourless and transparent zircons interpreted to represent rare primary magmatic crystals of the protolith. Grains have a width of about 60 μm. Fracturing is a typical feature and is much more pronounced and frequent in the co-magmatic zircons than in the much more frequent inherited zircons (Figure 6.5). (From Gebauer *et al*, 1988)

tively (Figure 6.3). This wide range of discordance is found not only between grains but between different areas within single grains. Cathode-luminescence (CL) studies of the analysed zircons show areas of euhedral, oscillatory zoning of probably magmatic origin (Figure 6.4) which give an age of about 870 Ma, and irregularly shaped areas possibly resulting from metamorphic recrystallisation, which give ages at or close to the inferred age of eclogite facies metamorphism, about 468 Ma.

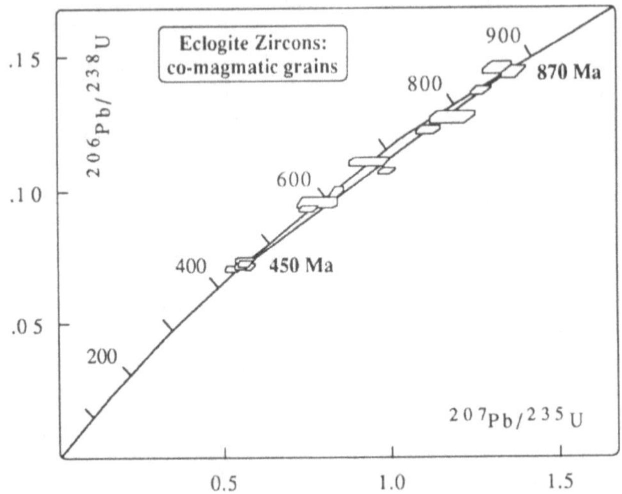

Figure 6.3 Conventional Concordia diagram with the data points obtained by single and multispot analyses (not differentiated) on euhedral zircon grains interpreted to reflect the rare co-magmatic zircon population of the protolith of the eclogite from the Gotthard Massif (Central Alps). Thus, the upper intercept age at 870 Ma probably relates to the protolith age. The lower intercept age at about 450 Ma coincides within analytical error limits with that of zircon xenocrysts as shown in Figure 6.7 and Sm-Nd data. It is interpreted to reflect the age of eclogite formation.

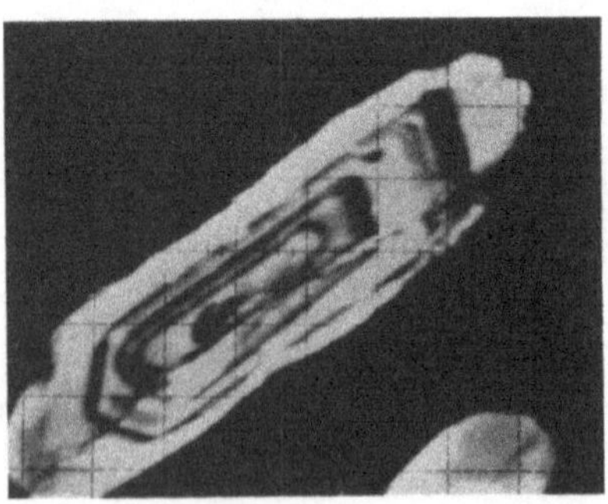

Figure 6.4 Cathode-luminescence image of the grain shown in Figure 6.1 (width about 60 μm). Multiple euhedral zoning is indicated by the different CL response which is largely based on variable sums of trace and minor elements. Probably, this grain grew under variable *P–T–X* conditions during formation and ascent of the gabbroic protolith. (From Gebauer *et al*, 1988)

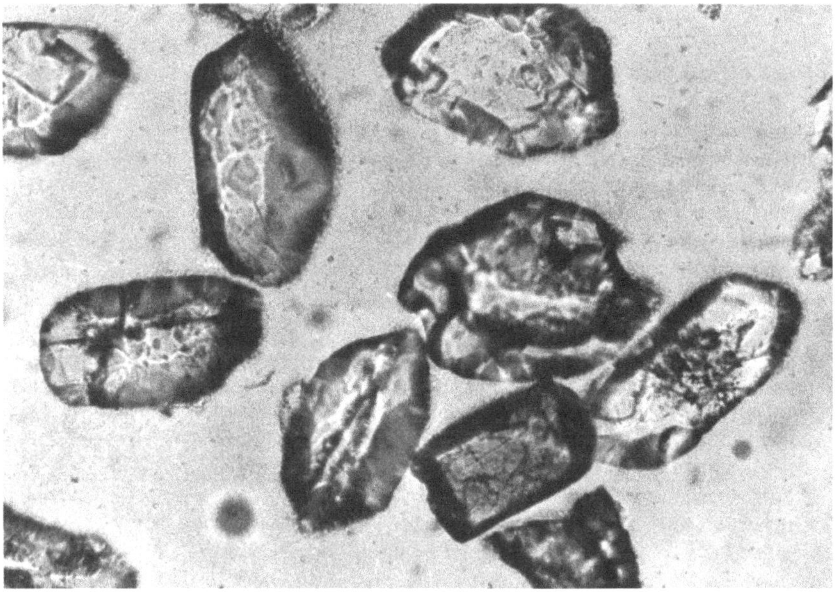

Figure 6.5 Typical, probably mantle-inherited, zircons representing almost the total zircon population of the analysed eclogite from the Gotthard Massif. Average width: about 100 μm. Further explanation in text. (From Gebauer *et al*, 1988)

More than 95% of the zircons are irregularly shaped, often rounded and partly resorbed grains with a range of colour and clarity (Figure 6.5). Unlike the rare euhedral, 870 Ma old grains, their original magmatic zoning can be cut by the present grain boundary (Figure 6.6), indicating post-crystallisation resorption. Recrystallisation, mainly starting from the grain surface, and/or new growth during the eclogite facies metamorphism is common (Figure 6.6). Most such zones and domains were almost or completely reset during the Caledonian HP metamorphism. They commonly are relatively U-rich and sometimes have lost variable amounts of radiogenic Pb during the Carboniferous overprinting. Four of the dozen grains of this type analysed still contain up to 90% of their pre-eclogite radiogenic Pb (Figure 6.7). Within-grain discordance lines defined by analyses of several areas on each of the four grains have lower Concordia intersections of around 468 Ma (Figure 6.7), the same age as indicated by the magmatic zircons. Errors in the intercept ages are between 3 Ma and 38 Ma. The upper intersection ages are at around 3.17, 2.67, 2.45 and 1.27 Ga. None of the grains appears to have been affected by protolith formation at 870 Ma, thus extending the temperature range that zircons are known to survive from that of granite melts to that of gabbroic melts. Similar conclusions have been reached previously by Compston *et al* (1985/6) and Kinny *et al* (1986).

As none of the eclogite trace element analyses shows any significant amount of crustal contamination, it is suggested that the inherited zircons were formed in the mantle. This would be consistent with their generally low U-contents. They probably formed during partial melting events and incomplete melt extraction at 3.17, 2.67, 2.45

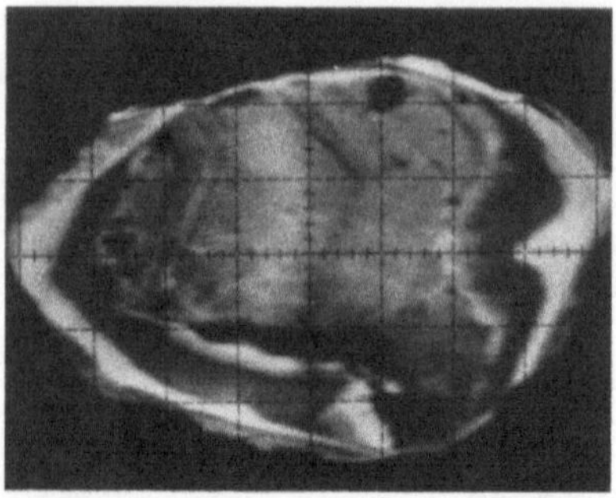

Figure 6.6 Cathode-luminescence image of magmatically formed zircon about 1.27 Ga old, probably of mantle origin. U-contents are around 250 ppm in the narrowly zoned, relict part of the crystal. The outermost, broader and more irregular zones (average U around three times the core) formed, probably by recrystallisation, during eclogite facies metamorphism at 468 Ma. (From Gebauer *et al*, 1988)

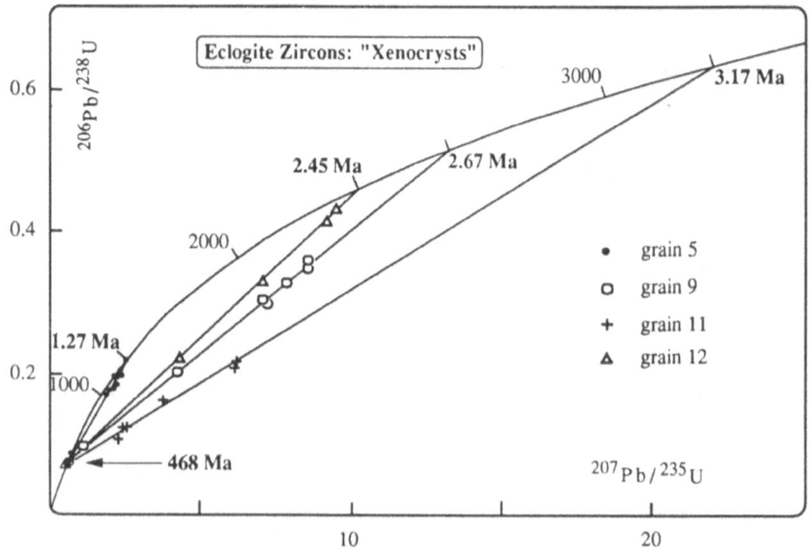

Figure 6.7 Conventional Concordia diagram with four different discordia lines obtained by multi-spot analyses on four differently old zircon xenocrysts from eclogite of the Gotthard Massif. The lower intercept ages closely scatter around 468 Ma. The latter is interpreted to reflect the age of eclogite formation, in line with Sm-Nd whole-rock and garnet data.

and 1.27 Ga and were incorporated into the mafic magma during the formation of the gabbroic protolith at 870 Ma. A second possibility might be that the irregular zircons, representing almost all of the zircon population, were brought into the mantle as a result of recycling of sediments along subduction zones. However, geochemical data indicate that such significant amounts of felsic material, unless very rich in U-poor zircon, could have played only a very minor part and therefore the first hypothesis is preferred. Of course, zircons from gabbroic rocks or even plagiogranites of the oceanic crust might also have been brought back into the mantle via subduction zones. This possibility does not, however, change the implications of the preferred mantle contamination model. If the protoliths were tuffs and/or basaltic lava flows, incorporation of zircon xenocrysts of continental origin is possible, but major and trace element data do not support this.

The age of eclogite facies metamorphism at about 468 Ma is also supported by Sm-Nd analyses on whole-rock and garnet giving 461 ± 25 Ma (Gebauer *et al*, in preparation). The initial ε_{Nd}-value at the time of emplacement of the protolith is about + 1.3, a value within the lower range for rocks which probably originated in an arc environment. However, the inherited zircons suggest the presence of significant amounts of Archaean and Proterozoic trapped liquids in the source rock which would have enriched that originally depleted source at least at 3.17, 2.67, 2.45 and 1.27 Ga, progressively decreasing the Sm/Nd ratio and lowering the rock's time-integrated ^{143}Nd content. The average age of the zircons inherited from the mantle is given by conventional multi-grain analyses as 2.23 Ga.

Assuming the deposition of the existing country-rocks to the eclogites to be syn- or post-Pan-African, as widely observed all over the European Variscides (e.g. Gebauer *et al*, 1988), it follows that the Late Riphean protoliths (at around 870 Ma) must have been emplaced tectonically, although available data do not indicate P–T conditions. If it took place at low P–T conditions, at the earliest during a Pan-African cycle, then both the eclogite precursors and the present day country-rock precursors suffered the same Caledonian HP metamorphism. On the other hand, the eclogites and associated mafic and ultramafic rocks could have been emplaced tectonically in a Caledonian or even Hercynian (Carboniferous) scenario, such as during the emplacement of pre-Alpine nappes.

An important conclusion from this ion probe study is that the conventional analyses of zircon populations from the same sample may give geologically meaningless intercept ages. Post-eclogite facies lead loss, as shown by analyses on areas frequently high in U or close to internal or external grain surfaces, clearly is responsible for the conventionally obtained lower intercept age at 400 ± 3 Ma. The Carboniferous amphibolite facies overprint is responsible for some analytically concordant but geologically meaningless ages less than 468 Ma, and in at least one case an Alpine or even more recent lead loss must have occurred. The conventionally measured upper intercept age at 2.23 ± 0.03 Ga clearly is the result of mixing inherited zircons of different ages with the few co-magmatic zircons 870 Ma old.

This study demonstrates the importance of verifying conventionally measured zircon ages by other dating techniques. The Caledonian eclogite facies metamorphism in the Gotthard Massif is now the most reliably dated HP event in continental Europe, being defined by five independent zircon discordance lines in close agreement with Sm-Nd whole-rock and mineral data. In addition, it supports the microcontinent or terrane model of Gebauer and Grünenfelder (e.g. 1982) which is based, at least in part, on the identification of HP events of different ages in different domains of the European Variscides.

6.5 Sm–Nd method

6.5.1 *First attempts in Japan; improvement of analytical techniques*

The first attempts to date rocks and minerals by the Sm-Nd method were made at the University of Tokyo on an eclogite sample taken from an eclogite inclusion within the kimberlite of the Roberts Victor Mine in South Africa (Notsu *et al*, 1973). Unfortunately, the precision in determining the $^{143}Nd/^{144}Nd$ isotopic ratio was only about 0.1%, implying a very large error in the slope through the three data points (whole-rock, Omp and Grt), thus giving a possible age range between zero and 1.3 Ga for the formation or the partial or complete resetting of Grt and Omp. Nevertheless, this work showed the feasibility of the Sm-Nd method for age dating. The Japanese group was also the first to apply a Nd-evolution diagram with ^{144}Nd as a reference isotope. The same principle had been used for the Rb-Sr isochron diagram in which ^{86}Sr is used as a reference isotope. Sm-Nd isotopic systems in Cpx, Grt and whole-rock are most frequently used for age dating of eclogites, although Sm-Nd systems in other phases like zircon, rutile or apatite can also be useful. With the work of the two Californian groups led by Lugmair and by Wasserburg and DePaolo, both analytical precision and accuracy have improved considerably. These two groups published their first Sm-Nd data in the seventies (Lugmair, 1974; Lugmair *et al*, 1975; DePaolo and Wasserburg, 1976a, b). Currently, the crucial $^{143}Nd/^{144}Nd$ ratio can typically be reproduced within 0.003% or better and the $^{147}Sm/^{144}Nd$ ratios usually have errors < about 0.2%.

In 1984 Jagoutz *et al* published Sm/Nd data on ultra-pure minerals separated from eclogite inclusions in the kimberlite of the Roberts Victor Mine. In one sample Nd-isotopes in Grt and Cpx were in equilibrium at the time of eruption about 130 Ma ago. Two further Cpx-Grt pairs from eclogite inclusions were in Nd-isotopic disequilibrium and gave internal ages at 900 Ma and 220 Ma. This, together with a 1.7 Ga Cpx-Grt age for an eclogite inclusion in a Tanzanian Tertiary kimberlite, showed that temperature alone can not be the only factor for closed or open system behaviour in Cpx and Grt. Especially under dry conditions temperature might well have to exceed 1000°C before Cpx-Grt pairs are partly or completely reset. If the eclogite inclusions in kimberlites have not formed magmatically from high pressure mantle liquids and/or cumulates, the obtained ages of Jagoutz *et al* (1984) favour the possibility of Precambrian eclogite formation, probably related to subduction. Thus, finding the oldest eclogite formed this way would give a minimum age for the onset of plate tectonic processes on our planet.

6.5.2 *Eclogites in the Caledonides and the problem of two-point isochrons*

The first data on Cpx-Grt pairs from eclogites with small analytical errors were published by Griffin and Brueckner (1980) and van Breemen and Hawkesworth (1980). Whereas the latter authors obtained a 'future age' for Omp and Grt from an eclogite in the Tauern window (Eastern Alps), Griffin and Brueckner were successful in dating various Norwegian country-rock eclogites. Five Cpx-Grt pairs showed a wide range of apparent Sm/Nd partitioning between Grt and Cpx and yielded two point isochron ages averaging about 420 Ma, while one Cpx-Grt pair yielded 887 ± 60 Ma. The five Caledonian ages were interpreted to reflect the eclogite facies transformation of the different protoliths due to Nd-isotopic equilibrium between the analysed HP minerals close to the peak of metamorphism. This interpretation has been widely applied also for later results on other eclogites from different geological units. It conforms with independent evidence described by Cohen *et al* (1988) and Jagoutz (1988) and is, probably for most cases, the most realistic interpretation of such two-point isochron data. The deduced Caledonian age of HP metamorphism is in good agreement with the interpretations of U-Pb zircon ages for various other Norwegian country rock eclogites (Krogh *et al*, 1974 or Gebauer *et al*, 1982 and 1985). Also, Mearns and

Lappin (1982) and Mearns (1986) give a rather precise Grt Cpx-whole rock Sm-Nd age of 408 ± 8 Ma for a country rock of the Western Gneiss Region which they also interpret to reflect the eclogite forming event.

Griffin and Brueckner (1985) expanded their study to lower-T eclogites (about 600°C) of the Western Gneiss Region. However, contrary to the higher-T eclogites (> 700°C) which were thought usually to give reliable metamorphic ages, no meaningful Cpx-Grt ages could be obtained. This was due to an insufficient spread in Sm/Nd ratios of the two minerals. The authors explained this as due to non-achievement of equilibrium partitioning of the REE on a mm-scale, even though Fe, Mg, Na, and so on were more mobile and therefore could be used for P–T estimates. A pegmatitic schlieren within the same eclogite contained coarse-grained Omp, apatite and rutile. The spread in Sm/Nd between Cpx and Ap was very poor, but together with the Cpx, Grt and whole-rock of the eclogite an isochron age of 547 ± 48 Ma was calculated. The authors considered this age as ambiguous and concluded that it may represent the age of the gabbroic protolith and the age of the pegmatitic segregation.

Sm-Nd data for eclogites of the Seve nappe in northern Sweden were provided by Mørk *et al* (1988). Based on petrological studies, the two different sampled eclogite-bearing terranes, separated by about 40 km, have been subjected to different P–T–t paths. The whole-rock-Cpx-Grt-ages, however, are all within analytical uncertainties at about 505 Ma, although two eclogites from the relatively lower grade terrane have formed under maximum T and P of only 500–620°C, and 12–15 kbars, respectively (e.g. Kullerud, 1987). Nevertheless, the fractionation of Sm/Nd in Grt versus Cpx corresponds in all datable examples to a frequently observed value of around 3 although cases exist in which Sm/Nd is about 8 times higher in Grt than in Cpx. In one case in the higher-T terrane Grt has a Sm/Nd ratio only 20% higher than in the Cpx which is, as typical, very close to the Sm/Nd of the whole-rock. Grt in this rock shows strong retrograde zoning and localised retrograde equilibration with Cpx. Therefore, Mørk *et al* speculated that Sm/Nd of the Grt decreased during retrograde Fe-Mg-Ca exchange between Grt and adjacent Omp and its later reaction products. Contrary to this study, however, Jagoutz (1988) found that subsolidus equilibration between Grt and Cpx considerably increased Sm/Nd in Grt of an eclogite inclusion in a kimberlite. Obviously, the partitioning of Sm/Nd between Grt and Cpx is still poorly understood. The 505 Ma eclogite forming event for the dated Seve eclogites showed that at least two different eclogite facies events can exist in different tectonic units within a mountain chain, as shown earlier for the European Variscides.

Data from garnet peridotites and included eclogites (internal eclogites) of the Western Gneiss Region corroborate this observation. The putative ages for internal eclogites and garnet peridotites are in the approximate range 1000–1700 Ma and even within a single peridotite-eclogite body such a range was obtained (Jacobsen and Wasserburg, 1980; Mearns, 1986; Brueckner, Rubenstone and Zindler, personal communication). A common Nd-isotopic composition was either never reached or, if so, not preserved in the analysed garnets and clinopyroxenes. The presence of differently old mineral generations and/or mineral zones, possibly together with partial re-equilibration during different stages of metamorphic overprinting, can account for this unfortunate observation.

In another study within the Western Gneiss Region, Mørk and Mearns (1986) separated both magmatic and metamorphic minerals from various coronitic gabbros with different degrees of transition to eclogite and from completely equilibrated and foliated country-rock eclogite. Although all samples probably have formed above 700°C and at pressures between 12 and 15 kbar, magmatic plagioclase remained closed for Sm and Nd in a coronitic gabbro. In relatively more transformed coronitic gabbros, plagioclase was disturbed while magmatic augites were not. Similarly, minerals from eclogites with pseudomorphic textures did not reach Nd-isotopic equilibrium during high-T–P metamorphism. This was only observed between whole-rock, Cpx and Grt in a foliated country-rock eclogite yielding 400 ± 16 Ma. The data of Mørk and Mearns (1986) confirmed that volume diffusion alone can not be responsible for the partial or complete resetting of many mineral systems. Instead, recrystallisation induced by deformation and/or fluid activity and/or crystallographic state (e.g. order–disorder) usually plays an overriding role. Another important conclusion is that the frequently produced two-point isochrons have to be treated with care if no metamorphic equilibrium can be demonstrated and/or if no regional distribution of these or other age data are known.

In a similar study Cohen *et al* (1988) analysed minerals from an eclogite facies shear zone developed in coronite-bearing anorthositic granulites of the Bergen Arc (West Norway). Newly formed garnets in the shear zone could not be separated from Sveconorwegian granulite garnets. Thus, the Grt data point can not be used for age calculation as it represents about equal amounts of both generations. Nonetheless, these data confirm that a simple cooling age interpretation of Grt- whole-rock ages, as suggested by Humphries and Cliff (1982), is unrealistic. This is so as their proposed blocking temperature of about 480–600°C for pyropic Grt (cooling rate 5°C Ma^{-1} and 100°C Ma^{-1}, respectively) is significantly lower than the geothermometric data of Austrheim and Griffin (1985) giving 700–800°C for the eclogitic assemblage. Although the other analysed minerals of the shear zone eclogitic assemblage (Omp and phengite) have very similar Sm/Nd ratios as the whole-rock, an imprecise Caledonian age at 507 ± 109 Ma could still be derived.

Following the Caledonian mountain belt further to the southwest into Scotland, eclogites are also known within metasediments of the eastern Glenelg Lewisian inlier, close to the Moine thrust. Two samples have been dated by Sanders *et al* (1984) who obtained Grt-Cpx-whole-rock ages between about 1.0 Ga and 1.1 Ga for the eclogite facies metamorphism. These ages are so far the oldest known for country-rock eclogites.

6.5.3 *European Variscides*

Sm-Nd data have also been obtained on eclogites of the European Variscides (e.g. Bernard-Griffiths *et al*, 1985; Stosch and Lugmair, 1987; Quadt and Gebauer, 1988).

For example, a mafic HP rock occurring as a layer within a garnet-peridotite lens of the Saxonian Granulite Massif yielded a 4 point isochron of about 380 Ma using whole-rock, Cpx, Grt and zircon (Quadt and Gebauer, 1988). Zircon in this rock is of metamorphic origin, as deduced from its morphology, and is more radiogenic (Sm/Nd = 0.55) than garnet which implies that it also can be very useful for Sm-Nd dating of eclogites.

There are a number of eclogite localities in various parts of the Moldanubicum. One of them, occurring about 50 km to the south of the German deep drilling site in the Oberpfalz, has been studied using various methods by Gebauer and Grünenfelder (1982) and Quadt and Gebauer (1988). Eclogites at this locality (Winklarn) occur within serpentinites which themselves are surrounded by cordierite-bearing metasediments that also contain orthogneisses of volcanic arc composition that preserve a granulite facies mineral assemblage. Based on data of major and trace elements including REE-data, the various mafic protoliths could have originated in an island arc environment. Whereas Rb/Sr dating on whole-rocks of various mafic and ultramafic eclogite facies rocks did not give any age information, conventional zircon dating allowed estimates on protolith ages (about 1.3 Ga?) and eclogite facies metamorphism during an Ordovician–Silurian event. Assuming the protoliths to be around 1.3 Ga, the initial ε_{Nd} of $+7.5$ is in line with formation of the protoliths from a time-integrated depleted mantle in an island arc scenario. The eclogite facies overprint is best reflected by Sm-Nd mineral- and whole-rock data using Grt, zircon and rutile. For an eclogite an age of 422 $\pm$ 8 Ma and for a felsic rock of volcanic arc composition an age of 423 $\pm$ 8 Ma were obtained. Monazite (U-Pb) of the latter rock-type reflects the general Hercynian metamorphic overprint at about 320 Ma. When compared with the existing geochronological data on the Bohemian Massif these ages might best be explained with the break up of a composite continental crust (Gondwana) after the Pan-African (Cadomian or Assyntian) orogeny and consequent convergence of the thus produced microcontinents (Gebauer *et al*, 1988). One such event must have occurred around 425 Ma, i.e. in the Silurian, probably during subduction of island arc derived rocks to a depth of at least 45 km, as derived from geobarometric data on the rocks analysed by O'Brien (O'Brien *et al*, Chapter 9). The limited extent of the island arc rocks is probably of both primary and secondary (tectonic) cause. Thus, this age signature together with its geodynamic consequences might not be a general feature for the whole Moldanubian complex although Silurian HP events are also known from other areas within the Central European Variscides. The microcontinent or terrane model readily accounts for such a complex scenario.

Carboniferous eclogite forming events, as concluded first from conventional zircon dating (e.g. Gebauer *et al*, 1981), have also been inferred by Paquette (1987) and Brueckner, Medaris and Bakun-Czubarow (personal communication). For the lower unit of the southern Brittany Champtoceaux complex Paquette (1987) found a Sm-Nd whole-rock-Grt age of 339 $\pm$ 15 Ma for an E-type MORB-type eclogite whereas high-U metamorphic zircons (about 1000 ppm) yielded 358 $\pm$ 2 Ma. The relatively younger but analytically not significantly younger Sm/Nd Grt age could be a result of recrystallisation of garnet during retrogressive blueschist facies overprinting under elevated fluid activity.

Carboniferous ages of around 330 and 345 Ma have also been found for Cpx-Grt pairs of eclogites from the Moldanubicum of Czechoslovakia and from the East Sudetes in Poland, respectively (Brueckner, personal communication). If these ages are not derived from Carboniferous recrystallisation, for example in the presence of migrating fluids, then they provide evidence for HP subduction events in the formation of the Variscan mountain chain and thus support earlier conclusions reached from conventional U-Pb zircon dating.

Acknowledgements

I gratefully acknowledge the constructive reviews of this Chapter by Bill Griffin, David Smith, Tony Carswell and Jean-Louis Paquette. Especially, the first two named colleagues guaranteed that the manuscript passed the scrutiny of scientists with currently the most divergent views on the genesis of certain famous eclogites.

7 Occurrences and stability conditions of low-temperature eclogites

M. SCHLIESTEDT

7.1 Introduction

Over the years eclogites associated with blueschists (low-T eclogites) have acted as a continuous source of interesting petrological information as well as controversial ideas. Already introduced as one principal mode of occurrence by Eskola in 1921, low-T (LT) eclogites (and blueschists) became especially important when the introduction of plate tectonic theories almost 20 years ago provided a convenient model to explain the HP–LT conditions necessary to generate the observed mineral assemblages. The concept of plate collisions involving deep burial of crustal rocks followed by a rapid uplift history has been generally accepted and most occurrences of LT eclogites are in fact related to some kind of ancient subduction process.

In this Chapter a summary of the occurrences and stability conditions of LT eclogites (group C of Coleman *et al*, 1965) will be given, concentrating on three well-documented high-pressure (HP) metamorphic terranes:

(a) the Franciscan of California, USA (e.g. Ernst *et al*, 1970; Cloos, 1986; Maruyama and Liou, 1988)
(b) New Caledonia (e.g. Yokoyama *et al*, 1986; Ghent *et al*, 1987)
(c) Sifnos, Greece (e.g. Okrusch *et al*, 1978; Schliestedt 1980, 1986).

LT eclogites from Norway and the Alps are discussed in Chapters 8 and 10 respectively.

Field evidence shows that most eclogites in the Franciscan occur as coarse grained, tectonic blocks that rest in a matrix of lower-grade blueschist facies rocks (Cloos, 1986). Both New Caledonia and Sifnos, on the other hand, exhibit conformable eclogite layers (*in-situ* eclogites) that were undoubtedly metamorphosed together with surrounding, mostly metasedimentary rocks.

There now exists general agreement that the protolith material of LT eclogite facies terranes is predominantly crustal in origin, including almost everything from typical oceanic crust (serpentinised ultrabasic rocks, gabbros, pillow basalts, cherts) to accretionary wedge rocks (mélanges, greywackes) to continental margin sediments (limestones, sandstones, shales) to typical continental crust (granitoids, gneisses). Consequently, a large variety of mineral assemblages has been described and can be used to estimate metamorphic conditions.

Particularly useful for geothermometry is the Fe^{2+}–Mg distribution between coexisting garnet and omphacite. This geothermometer has been calibrated experimentally at high temperatures (Ellis and Green, 1979; see Chapter 4) and it can be applied with success to the temperature range of LT eclogites (Newton, 1986).

Metamorphic pressures are less well constrained because of the lack of any continuous geobarometric scale: in the absence of plagioclase, jadeite contents in omphacite coexisting with quartz provide only minimum pressure values. Additional *P–T* information can be obtained from experimental phase-equilibrium data and theoretical calculations using recently developed thermodynamic data sets (Holland and Powell, 1985; Berman, 1988).

Contrary to their higher-temperature equivalents, LT eclogites are generally characterised by rather variable modal compositions. Quite often, garnet and omphacite are accompanied, in clear textural equilibrium, by significant amounts of hydrous minerals such as amphiboles (glaucophane, actinolite or hornblende), micas (phengite, paragonite) and epidote or zoisite. Even fine-scale interlayering of so-called dry garnet-omphacite assemblages with blueschists containing abundant hydrous phases has commonly been observed. This raises the following questions:

(a) Does a free fluid phase exist during HP metamorphism?
(b) If so, what is its composition?
(c) Do interlayered eclogites and blueschists represent locally heterogeneous fluids or varying bulk-rock chemical compositions (components other than H_2O)?

Many LT eclogites are polymetamorphic. Careful microtextural observations of secondary minerals replacing garnet and omphacite give important information on the uplift history. Furthermore, mineral composition zonations as well as inclusions of minerals that were formed during earlier stages sometimes record the prograde burial path of eclogites. Together with detailed geochronological work, these observations are useful in defining pressure–temperature–time (*P–T–t*) paths during metamorphism. The deduced *P–T–t* trajectories can be compared with thermal models (Oxburgh and Turcotte, 1974; England and Richardson, 1977; England and Thompson, 1984) and they may help towards a better understanding of the complex cycles of crustal thickening, HP metamorphism, rapid exhumation and erosion that characterise zones of plate collision.

7.2 Occurrences

Occurrences of LT eclogites in blueschist terranes have been described from a number of countries. Surrounding the Pacific Ocean, these include: Alaska (Forbes *et al*, 1984; Thurston, 1985; Brown and Forbes, 1986); California (Coleman and Lee, 1963; Brown and Bradshaw, 1979; Maruyama and Liou, 1988); New Caledonia (Black, 1977; Yokoyama *et al*, 1986; Ghent *et al*, 1987); Japan (Brothers and Yokoyama, 1982; Takasu, 1984; Banno, 1986); along the Alpine–Tethyan chain they include: Western and Southern Alps (Bearth, 1959; Compagnoni, 1977; Ernst and Dal Piaz, 1978; Reinsch, 1979); Eastern Alps (Miller, 1977; Holland, 1979a); Greece and Turkey (Ridley, 1984; Okay, 1986; Schliestedt, 1980, 1986); scattered occurrences mostly of Palaeozoic age include: Norway (Krogh, 1980; Cuthbert *et al*, 1983; Ohta *et al*, 1986); France (Audren and Triboulet, 1984; Peucat, 1986); and USSR (Dobretsov and Sobolev, 1984; Sobolev *et al*, 1986).

The Franciscan Complex of California has always been regarded as the classic example of HP metamorphism that produced blueschists and associated eclogites. However, owing to the exotic tectonic setting of many Franciscan eclogites it is felt that more coherent HP metamorphic terranes (with *in situ* eclogites) such as those found on

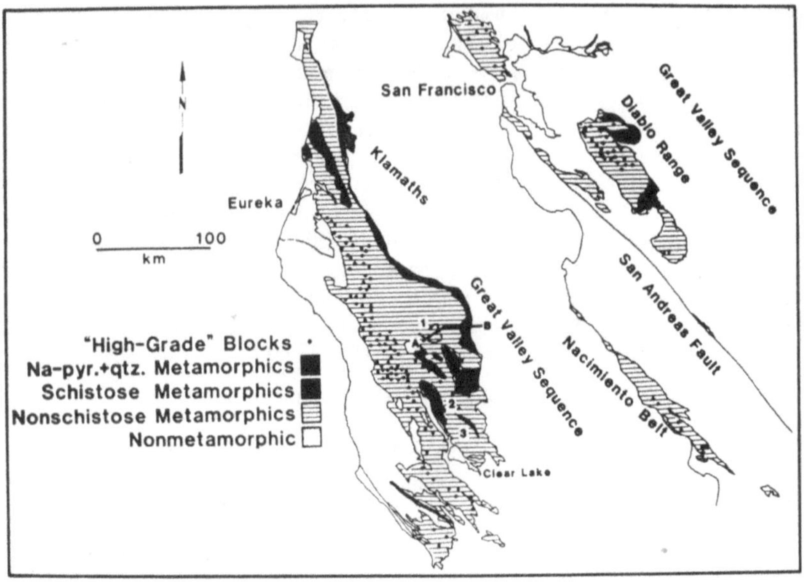

a

b

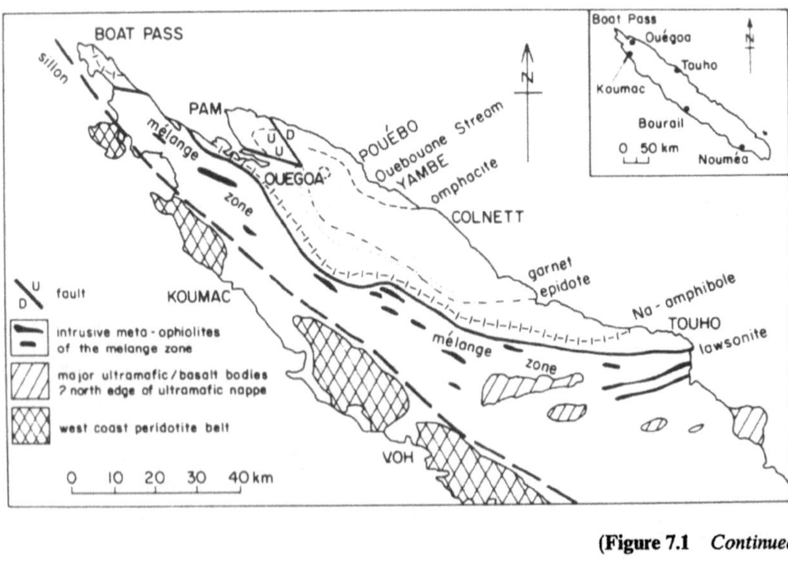

(Figure 7.1 *Continued*)

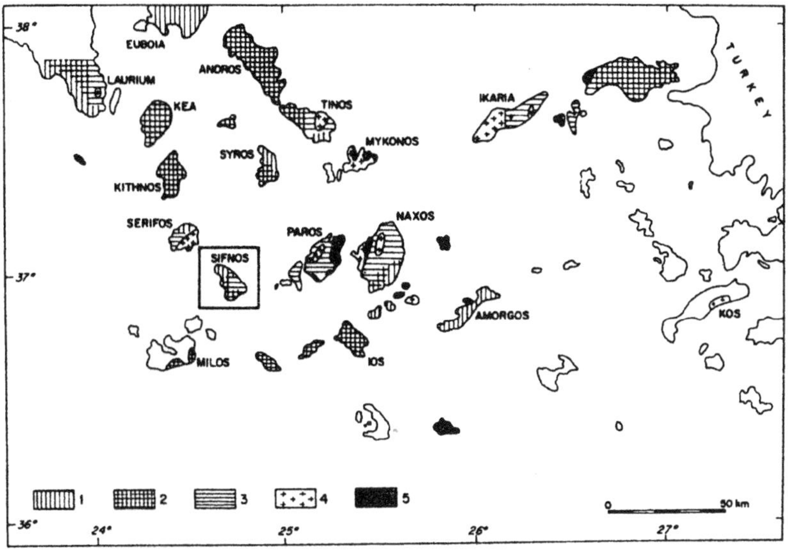

C

Figure 7.1 Regional tectonic setting of HP metamorphic terranes: (a) Franciscan, California (USA) (from Cloos, 1986); (b) New Caledonia (from Ghent *et al*, 1987); (c) Cyclades (Greece) (from Matthews and Schliestedt, 1984) with 1 = Eocene HP metamorphic rocks, 2 = Overprinted HP metamorphic rocks, 3 = Miocene MP metamorphic rocks, 4 = Miocene granitoid rocks, 5 = Rocks of the upper tectonic unit

New Caledonia and on Sifnos (Greece), are more appropriate to interpretation of the formation and preservation of LT eclogites.

7.2.1 *Franciscan Complex of California* (*USA*)

Extending from Oregon in the north all the way along the Californian coast to Mexico in the south, the Franciscan Complex probably constitutes one of the largest HP metamorphic belts (Figure 7.1a; Ernst, 1984). It consists of a large variety of rocks that were deposited, subducted and metamorphosed during convergence processes between North American and Pacific plates. Radiometric ages of 150–90 Ma (McDowell *et al*, 1984) indicate that the period of subduction and metamorphism extended from Late Jurassic to Cretaceous times, apparently close in time to the period of sedimentation.

Franciscan blueschists are primarily found in two structural settings (Cloos, 1986):

(a) as blocks, several metres in diameter, sitting in a mud-matrix mélange
(b) as variably deformed, stratigraphically coherent sequences of metasediments and metabasites that extend for several tens of kilometres.

Protolith rocks include trench-type sediments (greywacke, argillite, chert) and ocean-floor basalts.

Apart from few *in situ* eclogites in the Ward Creek sequence mentioned by Maruyama and Liou (1988), eclogite occurrences are restricted to tectonic blocks.

Other rock types found as blocks include gneissic epidote-garnet bearing blueschists and amphibolites collectively named Type IV rocks by Coleman and Lee (1963) as compared with lower-grade, Type III, schistose blueschists found in coherent sequences. Metamorphic grade is highest in blocks (LT eclogite facies) and decreases in Type III blueschists to epidote-blueschist and lawsonite-blueschist facies conditions. Franciscan eclogites and blueschists lack any significant greenschist facies overprint, as especially exemplified by the widespread occurrence of aragonite (Coleman and Lee, 1962; Cloos, 1986).

Details on the petrography and mineral compositions of eclogite and blueschist blocks given by Coleman *et al* (1965) and Brown and Bradshaw (1979) will be discussed together with New Caledonia and Sifnos data in Section 7.3.

7.2.2 *New Caledonia*

In New Caledonia, along the northeastern coast, HP metamorphic rocks form a 175 km long and 35 km wide belt that developed during the Tertiary (Figure 7.1b; Brothers and Yokoyama, 1982; Ghent *et al*, 1987). The coherent sequence consists of Cretaceous–Eocene igneous and sedimentary protolith material. Sediments, probably of shallow water origin, include carbonaceous claystones, siltstones and sandstones. They are intercalated with dolerite sills and acidic and basic tuffs. HP metamorphism occurred during the Eocene (oldest radiometric ages are 36–39 Ma) and may have lasted as long as late Oligocene (21 Ma) (Blake *et al*, 1977; Brothers, 1985). Throughout the whole belt, metamorphic grade increases from lawsonite-blueschist facies in the southwest through epidote-blueschist facies to eclogite facies in the northeast (Brothers and Yokoyama, 1982; Yokoyama *et al*, 1986). In metasediments the following progressive sequence of isograds has been identified: lawsonite, Na-amphibole, garnet (spessartine), epidote and omphacite.

Eclogites are confined to the highest-grade, omphacite zone. Depending on bulk-rock chemistry, modal compositions may vary from layers with up to 95% of omphacite + garnet (eclogite) to layers dominated by epidote and glaucophane (blueschist). Retrogression to greenschist facies assemblages occurs as a common, locally very intense phenomenon in both the epidote and omphacite zones.

7.2.3 *Sifnos (Greece)*

The Island of Sifnos is part of the Cycladic islands that, distributed across the Aegean Sea, link the crystalline massifs of mainland Greece with those in Turkey (Figure 7.1c). Igneous and metamorphic rocks of the Cycladic crystalline complex document processes of collision, subduction and exhumation during the African–Eurasian plate convergence (Dürr *et al*, 1978; Papanikolaou, 1987).

Well-preserved HP metamorphic rock units including eclogites are found on the islands of Syros (Dixon, 1969; Ridley, 1984), Tinos (Kohlmann, 1978; Bröcker and Okrusch, 1987) and Sifnos. According to radiometric dating, HP metamorphism occurred during the Eocene (40–42 Ma) and was followed in the Miocene (around 20 Ma) by a regional greenschist, locally amphibolite facies overprint that completely obliterated the earlier HP assemblages on many islands (Andriessen *et al*, 1979; Altherr *et al*, 1979). Although it is very risky to establish metamorphic field gradients when outcrops of preserved HP rocks are scattered around widely separated, small islands it seems probable that the metamorphic grade is highest along the axis Sifnos–Syros–Tinos and that it decreases, both to the NW and E (Schliestedt *et al*, 1987).

On Sifnos parent rock types of the HP sequence include Mesozoic continental margin sediments (thick horizons of marine limestone, chert and tuffaceous sandstone) interlayered with basic and minor acidic volcanites (Okrusch *et al*, 1978; Schliestedt, 1980). Eclogites are common among metabasaltic lithologies. However, large modal variations exist, sometimes on the scale of millimetres, leading to a continuous spectrum of rock types from eclogites to blueschists and actinolite-rich rocks. Retrogression to greenschist facies assemblages is restricted to the lowermost part of the section and is thought to be the result of fluid infiltration during uplift in Miocene times (Schliestedt and Matthews, 1987).

7.3 Assemblages and mineral compositional variations

7.3.1 *Mineral assemblages*

Very few LT eclogites just contain the (anhydrous) assemblage garnet + omphacite (+ quartz + rutile). Much more common are various associations of hydrous silicates together with omphacite and garnet. Although some of these phases have been shown to be clearly secondary alteration products (that can be very useful in determining the uplift path of HP rocks; see discussion in Section 7.5) others occur in textural and chemical equilibrium with omphacite and garnet. Representative examples from Sifnos eclogites are given in Figure 7.2, showing textural relations between garnet, omphacite, epidote and phengite (Figure 7.2a), omphacite and glaucophane (7.2b) and glaucophane and actinolite (7.2c). Other frequently reported, primary phases in LT eclogites are

a

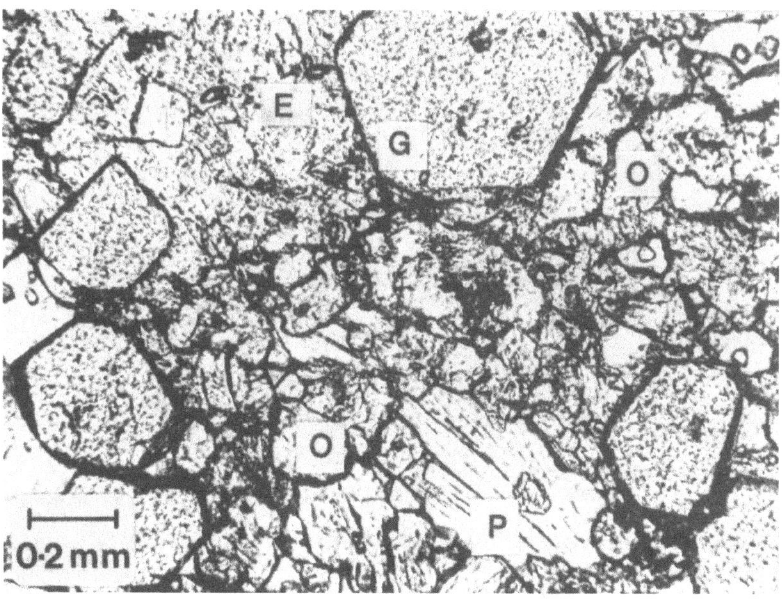

(**Figure 7.2** *Continued*)

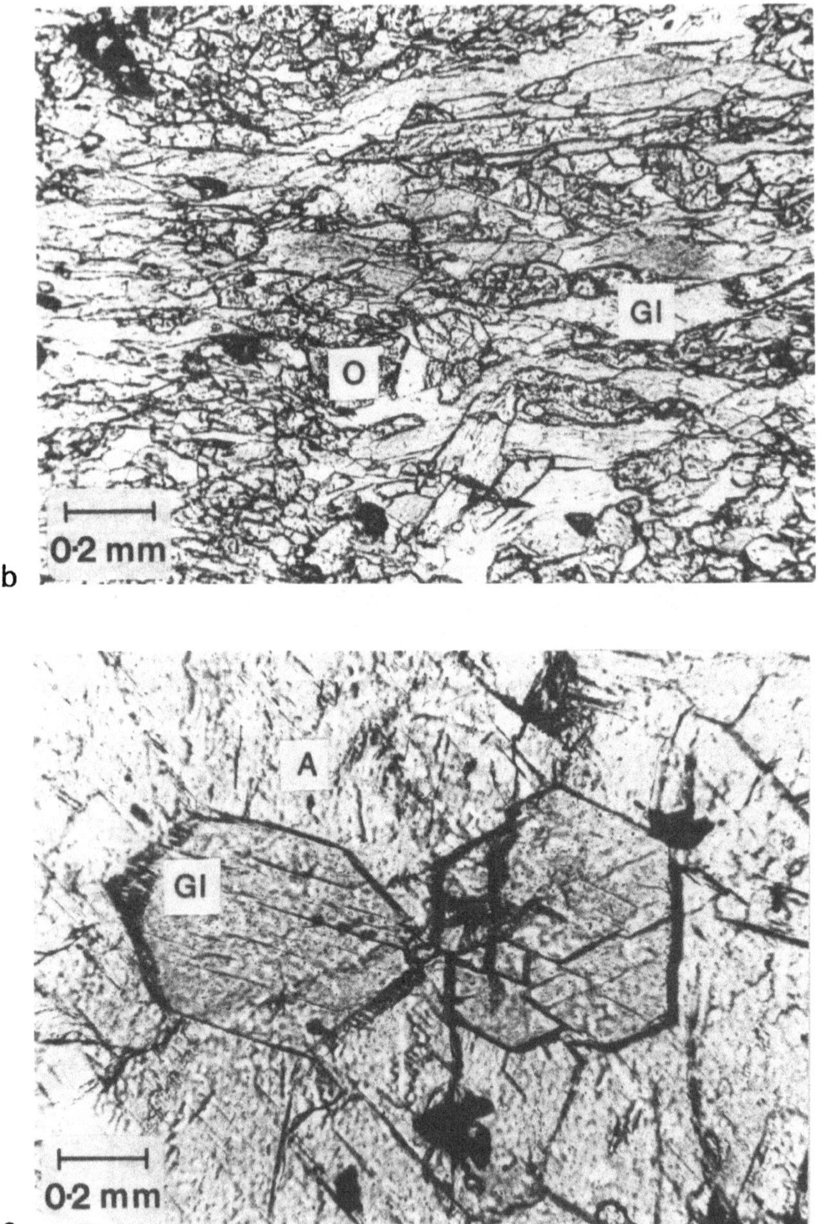

Figure 7.2 Photomicrographs of textural relations in LT eclogites from Sifnos: (a) garnet (G)–omphacite (O)–Epidote (E)–Phengite (P), specimen No. 83-Ky-31; (b) omphacite (O)–glaucophane (Gl), specimen No. 76–358; (c) glaucophane (Gl)–actinolite (A), specimen No. 76–285.

paragonite and barroistic amphibole (Takasu, 1984; Ghent *et al*, 1987), zoisite and lawsonite (mostly pseudomorphic; Green *et al*, 1968; Fry and Fyfe, 1971; Ernst and Dal Piaz, 1978).

Graphical representation of phase relations in LT eclogites can be made within the system Na_2O–CaO–MgO–FeO–Al_2O_3–SiO_2–H_2O. Assuming excess vapour, quartz, clinozoisite and garnet, eclogite assemblages are projected from these phases on to the plane $NaAlO_2$–Al_2O_3–$CaMgO_2$. The components $NaAlO_2$ and $CaMgO_2$ have been chosen because the $NaAlCa_{-1}Mg_{-1}$ exchange represents the most important chemical variation in amphibole (glaucophane–tremolite) and pyroxene (jadeite–diopside) solid solutions from HP rocks.

Resulting phase relations (Figure 7.3) display a regular arrangement of tie lines connecting omphacite with hydrous phases: Pg + Bar in case of New Caledonia and Gln + Act in Franciscan and Sifnos eclogite samples.

Paragonite has not been described from Franciscan eclogites whereas on Sifnos it is very common in blueschists associated with eclogites. The critical assemblage is Gln + Pg + Omp which has been shown to represent NaAl-richer bulk-rock chemical compositions than associated eclogites (Schliestedt, 1986). A similar interpretation has

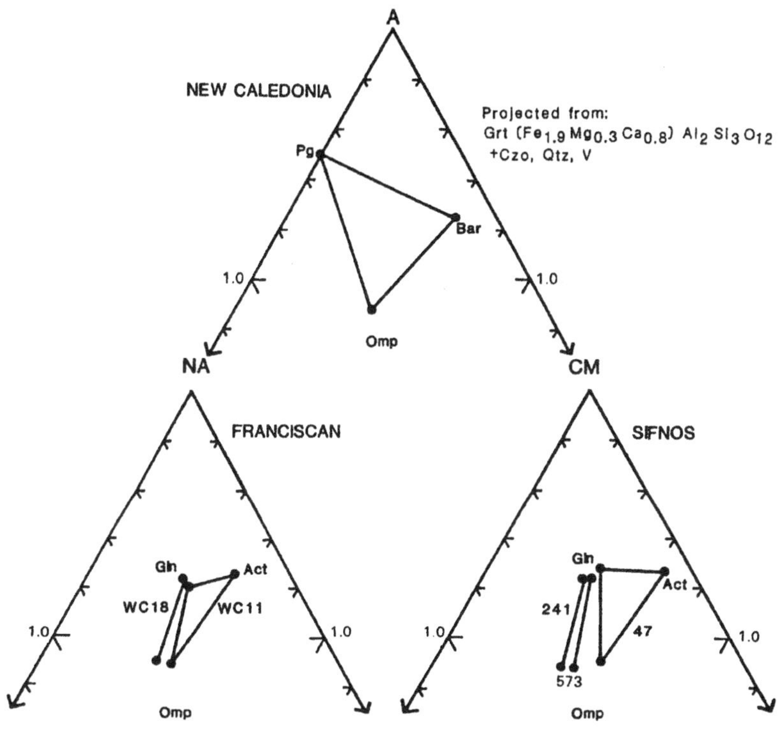

Figure 7.3 LT eclogite assemblages from New Caledonia (Yokoyama *et al*, 1986); Franciscan formation (Brown and Bradshaw, 1979); and Sifnos (Schliestedt, 1986). Projection of mineral compositions from Grt, Ep/Czo, Qtz and water on to $NaAlO_2$(NA)-Al_2O_3(A)-$CaMgO_2$ (CM)plane. Sample numbers refer to published mineral analyses.

also been given for interlayered eclogites and blueschists from the Franciscan (Bradshaw, 1978) and from New Caledonia (Yokoyama *et al*, 1986). There seems to be no longer any need to invoke differing water activities as the crucial factor. Instead appropriate bulk-rock compositions and high pressures can produce eclogite assemblages in a hydrous regime (Section 7.4.3).

Coexisting paragonite and Ca-amphibole (without glaucophane) as in New Caledonia eclogites have, however, not been observed in any Sifnos metabasic sample, suggesting that metamorphic temperatures were not high enough to cause breakdown of glaucophane in the presence of epidote (also confirmed by geothermometry; Section 7.4.1). The breakdown reaction deduced from New Caledonia metabasites is Gln + Ep = Omp + Bar + Pg (+ Qtz + H_2O) (Yokoyama *et al*, 1986) which is similar to the reaction Gln + Czo = Grt + Omp + Pg (+ Qtz + H_2O) described in metabasaltic rocks from the Sesia–Lanzo zone (Reinsch, 1979) and from the island of Syros (Ridley, 1984).

7.3.2 *Mineral compositions*

This section contains a brief summary of compositional trends observed in HP minerals from the selected LT eclogite occurrences. The data used in this comparison are given by: Franciscan: Coleman *et al* (1965); Bradshaw (1978); Brown and Bradshaw (1979); New Caledonia: Yokoyama *et al* (1986); Ghent *et al* (1987); Sifnos: Schliestedt (1980, 1986).

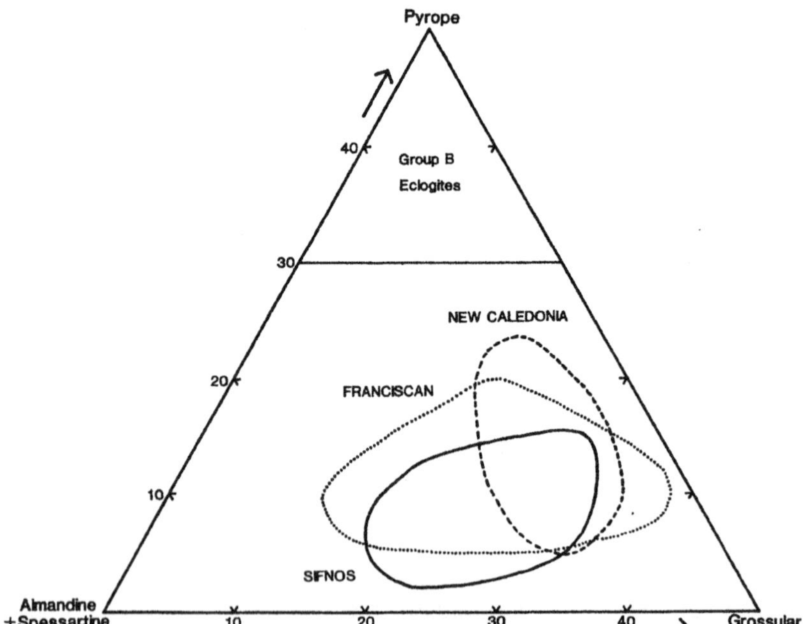

Figure 7.4 Comparison of garnet compositions from Franciscan (.......), New Caledonia (------) and Sifnos (————————). Data sources are given in the text. Dividing line between Group B and C eclogites according to Coleman *et al* (1965).

7.3.2.1 *Garnet.* Garnet compositions are similar in all three localities (Fig. 7.4) ranging in almandine (+ little spessartine) content between 50 and 80, in grossular (+ generally very little andradite) between 15 and almost 40 and in pyrope between < 5 and 25 mole%. They are clearly distinguished from MT eclogite garnets that contain $> Prp_{30}$ (Coleman *et al*, 1965). Strong zonations with spessartine-rich cores have been observed in some samples but more often only small differences (increasing grossular contents and increasing Mg/(Mg + Fe) ratios) are found and interpreted as evidence for growth during prograde conditions.

7.3.2.2 *Clinopyroxenes.* Apart from minor chloromelanites, common *sodic pyroxenes* have rather low acmite contents and plot in the omphacite compositional field (Figure 7.5). They are distinguished from pyroxenes in MT eclogites by their higher jadeite contents. Strong chemical zonation as described from lower-grade metabasic blueschists of the Franciscan (Maruyama and Liou, 1987) is absent in omphacites from LT eclogites. However, small compositional shifts from acmite-richer cores to jadeite-richer rims that have been observed in Sifnos omphacites may point to continuous growth during increasing pressure and temperature.

7.3.2.3 *Amphiboles.* Amphiboles in equilibrium with garnet and omphacite comprise both sodic and calcic (or sodic–calcic) members. Compositional variations

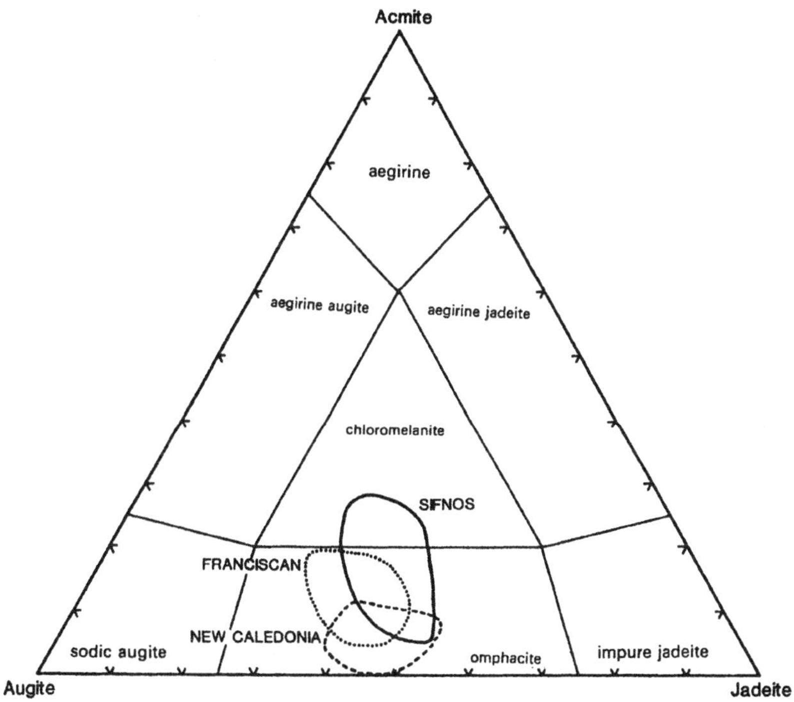

Figure 7.5 Comparison of sodic pyroxene compositions from Franciscan, New Caledonia and Sifnos eclogites. Data sources are given in the text. Compositional fields and nomenclature after Essene and Fyfe (1967).

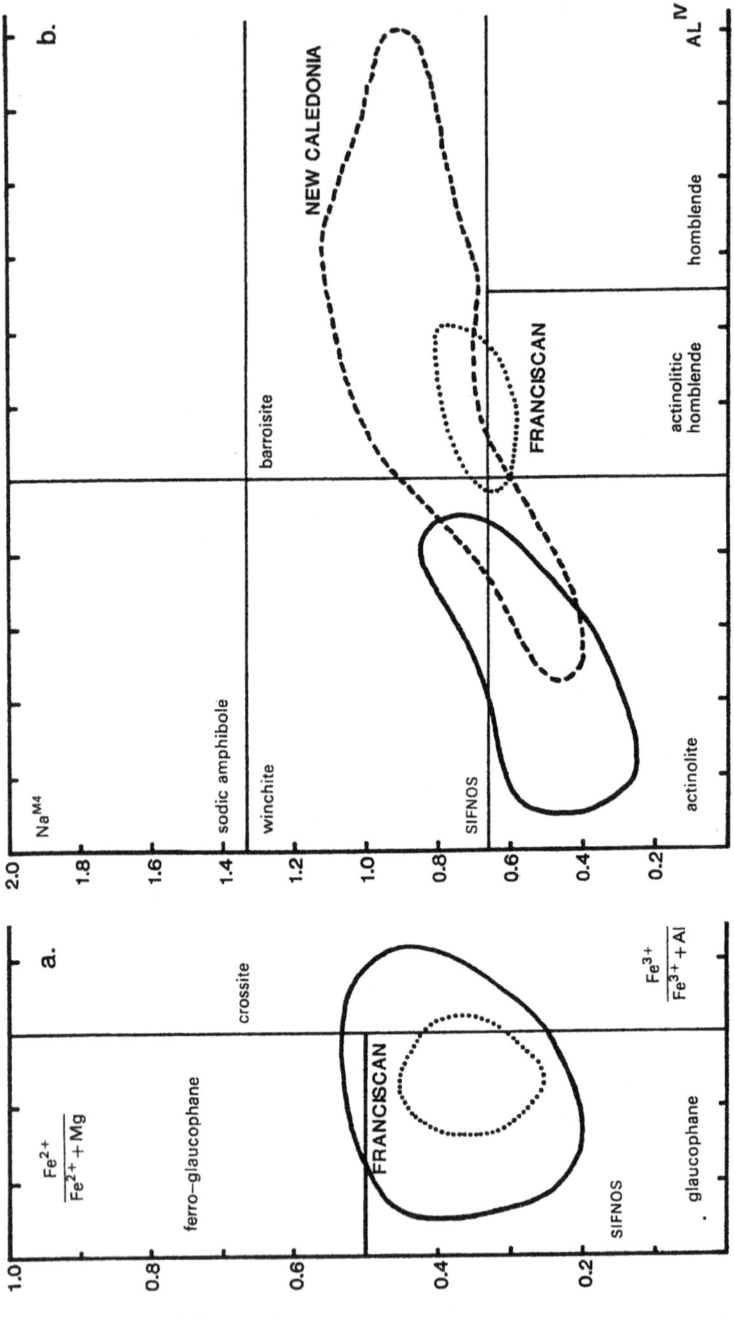

Figure 7.6 Comparison of amphibole compositions from Franciscan, New Caledonia and Sifnos eclogites: (a) sodic amphiboles in the Miyashiro diagram; (b) calcic amphiboles according to classification scheme of Leake (1978a). Data sources given in text.

in sodic amphiboles are largely restricted to the glaucophane field (Figure 7.6a); zoning is not pronounced with the exception of clearly retrograde phenomena. Calcic amphiboles coexisting with glaucophane plot mostly in the actinolite field (with a few analyses in the neighbouring winchite, barroisite and actinolitic hornblende fields; after Leake, 1978a). Calcic amphiboles from New Caledonia glaucophane-free eclogite assemblages are mostly barroisites (Figure 7.6b). The compositions of coexisting amphiboles define an asymmetric solvus that is believed to close at temperatures around 600°C (Reynard and Ballèvre, 1988).

7.3.2.4 *Paragonite and phengite.* Both *paragonite* and *phengite* have been found in LT eclogites. Paragonite compositional variations are limited to the KNa_{-1} exchange; where coexisting with phengite, muscovite end-member contents of paragonites can reach 7 (Sifnos) or 11 mole% (New Caledonia). Phengites are characterised by high celadonite (35–45 mole%) and low paragonite contents (5–15 mole%). The highest sodium contents have always been analysed in phengites coexisting with paragonite. Again, mica compositions from Sifnos eclogites systematically point to lower equilibration temperatures than New Caledonia phengites (up to 8 and 16 mole% paragonite, respectively). On Sifnos, large zonations with very celadonite-rich (up to 60%) and paragonite-poor (< 3%) cores have been observed.

7.3.2.5 *Epidote group.* One or more minerals of the *epidote group* are usually present in LT eclogites: clinozoisite/epidote (15–25 mole% pistacite) in Franciscan and Sifnos samples and clinozoisite sometimes together with zoisite (< 5% of the iron end-member) in New Caledonia. Zoning patterns are quite variable suggesting growth during different stages of metamorphism.

7.3.2.6 *Ti-bearing minerals.* The composition and distribution of the *Ti-bearing minerals* rutile and sphene in LT eclogites and associated blueschists still need to be clarified. In New Caledonia, Itaya *et al* (1985) showed that titanite (with Al_2O_3 < 4 wt%) is present in all metabasites from lowest to highest grade and that it is joined by rutile (with < 1.4 wt% FeO) from the epidote zone upwards. Apart from the retrogressive replacement of rutile by mantling titanite (a feature that is very common in HP rocks), both minerals occur in textural equilibrium. On Sifnos, on the other hand, rutile is mostly restricted to blueschist and titanite to eclogite assemblages suggesting that bulk-rock compositional variations, in addition to temperature and pressure, may also be important.

7.4 Stability conditions

Evaluation of the metamorphic conditions that prevailed during formation of LT eclogites (and associated blueschists and metasediments) has improved in recent years and has led to the recognition of higher pressure values than had been anticipated earlier (e.g. compare Newton and Fyfe, 1976 and Newton, 1986). Therefore, the need to invoke dry protoliths (Green and Ringwood, 1967a) and very low $a^{H_2O}_{fluid}$ values (Fry and Fyfe, 1971) for LT eclogite formation has largely diminished.

Table 7.1 Temperature estimates (°C) based upon garnet-omphacite and oxygen isotope thermometry ($P = 15\,\text{kbar}$) (see text for further information)

		Franciscan (Type IV)	New Caledonia	Sifnos
Fe–Mg exchange thermometry		(Coleman *et al*, 1965)	(Ghent *et al*, 1987)	(Schliestedt, 1986)
Powell (1985b)	average	530	570	500
	range	480–620	520–600	440–540
Krogh (1988)	average	505	540	475
	range	450–590	490–570	400–510
Oxygen isotope thermometry		(Taylor and Coleman, 1968)	(Black, 1974)	(Schliestedt and Matthews, 1984)
Matthews (1988)	average	500	560	490
	range	—	470–590	450–540

7.4.1 *Geothermometry*

Fe^{2+}–Mg exchange between garnet and omphacite provides the most useful way of obtaining temperature estimates from eclogites. It has been calibrated experimentally at high pressures and high temperatures by Råheim and Green (1974) and later by Ellis and Green (1979) who showed that varying grossular contents in garnet significantly affect the exchange. Recently, Ellis and Green's data have been reinterpreted by Powell (1985b) and Krogh (1988) (Chapter 4). Their revised formulations give negligible differences at high temperatures but significantly lower values at $T < 500$–600°C than the earlier Ellis and Green formulation (Krogh more so than Powell). It will be shown in this study that for LT eclogites the lower temperatures from either the Powell or the Krogh formulations are more consistent with other data (oxygen isotope thermometry, calculated phase equilibria).

In order to obtain meaningful temperatures, two additional considerations should be taken into account: (a) correction for ferric iron in omphacite (and garnet) and (b) generally, only rim compositions of coexisting garnet–omphacite pairs will be in exchange equilibrium. Temperatures obtained from core compositions of zoned minerals should be viewed with caution unless it can be established that both cores have equilibrated with each other (e.g. by analysing omphacite inclusions within garnet cores). Temperature estimates for Franciscan, New Caledonia and Sifnos eclogites are detailed in Table 7.1. Garnet-omphacite temperatures (calculated at pressures of

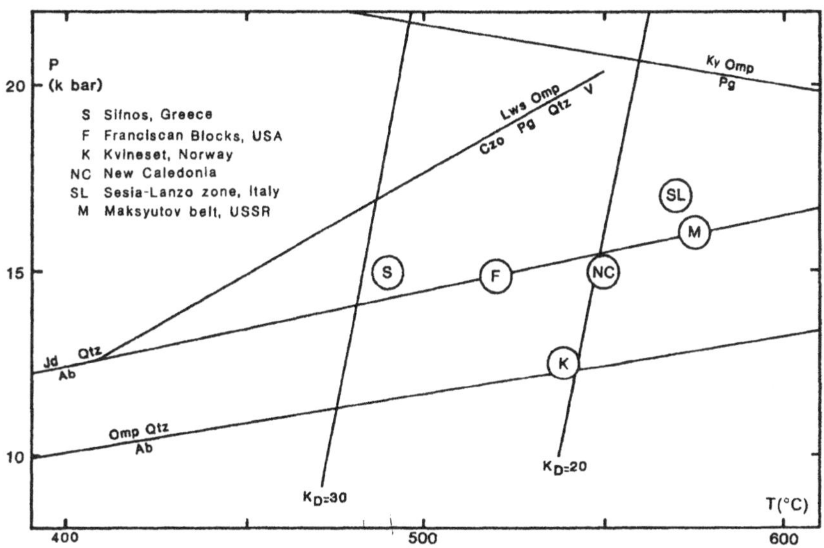

Figure 7.7 Inferred P–T conditions of various LT eclogites. For reference to Sifnos, Franciscan (Type IV eclogites) and New Caledonia data see text; Sesia–Lanzo zone (Rubie, 1984); Maksyutov belt (Sobolev *et al*, 1986), Kvineset (Krogh, 1980). $K_D = 20$ and $K_D = 30$ denote garnet–omphacite Fe^{2+}–Mg partition temperatures calculated using Powell's (1985b) equation (at $X_{Ca}^{Grt} = 0.3$). The other equilibria have been calculated with the data set of Holland and Powell (1985) taking $a_{Jd} = 0.5$ in omphacite and $a_{Czo} = 0.4$ in clinozoisite.

15 kbar) range from 475–500°C (Sifnos) to 505–530°C (Franciscan) to 540–570°C (New Caledonia) (Fig. 7.7). They agree very well with oxygen isotope thermometry (Table 7.1) based on published isotope data and a new set of mineral-pair calibration curves (Matthews, 1988). The close correspondence of Fe–Mg and oxygen isotope exchange thermometry underlines the usefulness of both methods for LT eclogites. The higher temperatures recorded for New Caledonia are in good agreement with other observations mentioned earlier: e.g. the higher sodium and potassium contents of coexisting phengites and paragonites, respectively, and the disappearance of glaucophane from New Caledonia eclogite assemblages.

A comparison with other occurrences (Sesia–Lanzo zone (Italy), Maksyutov belt (USSR), Kvineset (Norway); Figure 7.7) shows that temperatures of 450 to 550°C (possibly 600°C) are typical of LT eclogite formation. This range nearly coincides with the lower temperature limit of garnet (+ Ca-amphibole)-bearing assemblages (compared with lawsonite–chlorite) and the upper thermal limit of glaucophane (+ clinozoisite). Only eclogites with very Mg-rich glaucophanes from the Tauern Window are thought to have equilibrated above 600°C (Holland, 1979a, 1988).

7.4.2 *Geobarometry*

Metamorphic pressures are much more difficult to estimate because mineral assemblages in LT eclogites lack pressure sensitive equilibria. In the absence of plagioclase, omphacite + quartz assemblages will indicate only minimum pressure values (Newton, 1986). Maximum pressure limits are given by the general absence of lawsonite (and presence of clinozoisite + paragonite) according to the reaction Lws + Jd = Pg + Czo + Qtz + H_2O (calculated with $a_{Jd} = X_{Jd} = 0.5$ and $a_{Czo} = X_{Ca}X_{Al,M1} = 0.4$, Figure 7.7) and by the absence of any kyanite + omphacite assemblages (instead of paragonite): Pg = Jd + Ky + H_2O (calculated with $a_{Jd} = 0.5$, Figure 7.7). The broad pressure range 11–20 kbar (at T = 450–600°C) defined by these equilibria can be further restricted when almost pure jadeite + quartz parageneses from interlayered metasediments or meta-acidites (as found on Sifnos and New Caledonia) are taken into consideration. The pressure threshold of eclogite equilibration then rises to 14–16 kbar (Figure 7.7) corresponding to depths of 50 km or more. Phengites with high celadonite contents (Si = 3.4 – 3.6 pfu), although not coexisting with K-feldspar and phlogopite, support high minimum pressures (Massonne and Schreyer, 1987).

7.4.3 *Role of fluids*

The question of whether or not a free-fluid phase is present during the formation of LT eclogites has received considerable attention in the past. Influenced by straight-line extrapolation of HT–HP experiments on the gabbro–eclogite transformation by Green and Ringwood (1967a), some of the earlier pressure estimates were much lower than they are now, suggesting that eclogites must have formed at much reduced H_2O activites at P–T conditions where otherwise blueschists or amphibolites would be stable. Suggested possible mechanisms to reduce H_2O activity include:

(a) originally dry protolith material gets transformed directly into eclogites (without interim formation of any hydrous mineral) (e.g. Green and Ringwood, 1967a; Newton and Fyfe, 1976)

(b) dilution of H_2O with CO_2 (including possible H_2O–CO_2 immiscibility; e.g. Selverstone, 1986).
(c) high salt concentrations in the pore fluid (e.g. Fry and Fyfe, 1971).

It may well be that the first suggestion has been effective in certain (MT to HT) eclogites where mineral textures are indicative of solid-state reactions (e.g. corona textures, relict minerals etc.) but at temperatures below 600°C (or even 500°C) it seems quite unlikely that rates of mineral reactions are fast enough to transform completely dry basaltic rocks into eclogites. The abundance of hydrous minerals as well as textures indicative of hydrothermal growth such as monomineralic layers or veins filled with omphacite (and other HP minerals) in many LT eclogites suggests the presence of a free fluid phase (Essene and Fyfe, 1967; Brown and Bradshaw, 1979; Holland, 1979a; Yokoyama *et al*, 1986).

Recent stable isotope work on Sifnos (Schliestedt and Matthews, 1987) has shown that in a section of more than 100 m thickness interlayered rocks of different bulk-rock compositions (eclogites, blueschists, marbles, acidic gneisses) all have exchanged and equilibrated their oxygen isotopes during HP metamorphism, again providing strong evidence for the presence of a pervasive free fluid phase. However, the compositional range of such fluids remains uncertain. Fluid inclusion studies yield controversial results: whereas for example Holland (1979a) reported very H_2O-rich inclusions containing little dissolved salt (< 5 wt% NaCl equivalent) from eclogites of the Tauern Window, Luckscheiter and Morteani (1980) found high-density, pure CO_2 fluid inclusions in addition to very saline (25–50 wt% NaCl), aqueous inclusions in eclogites from the same area. More detailed investigations are needed, especially when considering that fluid inclusions trapped at peak P–T conditions are unlikely to withstand internal pressures upon isothermal decompression (Hollister *et al*, 1979). In an example from New Caledonia, Ghent *et al* (1987) argued that many of the observed H_2O-rich inclusions could have resulted from decrepitation and leakage of earlier trapped inclusions and that they might help to constrain the P–T path during uplift (Section 7.5).

Evidence for an H_2O-rich fluid is also provided by phase equilibria calculations which restrict the common presence of lawsonite (now replaced by clinozoisite/mica pseudomorphs) and the general lack of calcite + rutile + quartz assemblages in metabasites (presence of titanite instead) to $X_{CO_2} < 0.1$ (Schliestedt, 1986; Ghent *et al*, 1987).

Although discussions about the possible role of reduced H_2O activity in the formation of eclogites will continue it seems well established now that most LT eclogites (and interlayered blueschists) were formed under H_2O-rich, high pressure conditions. The concept of a LT eclogite facies characterised by a limited P–T regime remains valid.

7.5 Pressure–temperature–time paths

In contrast to peak metamorphic conditions, burial and uplift histories of LT eclogites are much more difficult to assess and, therefore, less well known. Evidence for the prograde part of the HP metamorphism is generally scarce. Mineral zonations in garnet, omphacite and phengite, as in samples from Sifnos, are compatible with growth during both increasing pressure and temperature. Inclusions of ferroglaucophane, chloromelanite and lawsonite pseudomorphs in garnet cores (Schliestedt, 1986) indicate that conditions of lawsonite-blueschist facies must have prevailed during parts of the burial process. A similar path with thermal gradients of around 10°C/km (Figure 7.8) has been suggested for early stages of burial of Franciscan (Type III blueschists; Maruyama and Liou, 1988) and New Caledonia rocks (Ghent *et al*, 1987).

In all three terranes discussed, maximum pressure and temperature conditions (about 15 kbar and 500–550°C, reached simultaneously) are similar, but uplift paths

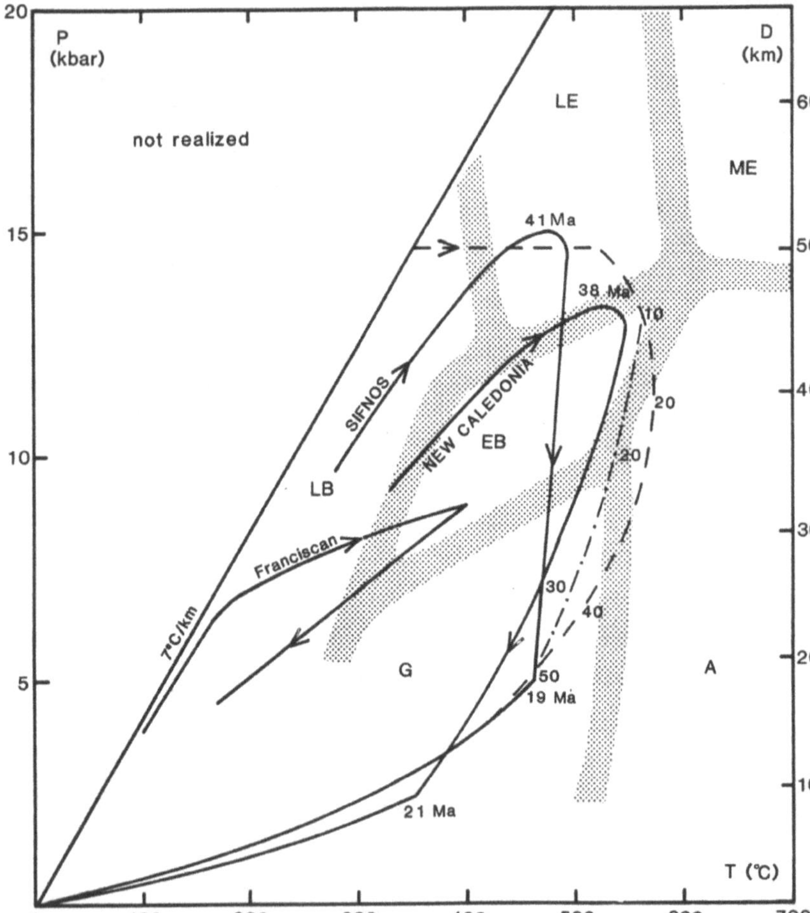

Figure 7.8 Comparative P–T–time paths for Sifnos, the Franciscan (only Type III rocks, taken from Maruyama and Liou (1988), are shown; Type IV eclogites give higher P–T conditions, Table 7.1), New Caledonia (Ghent *et al*, 1987) and a calculated thermal model for an overthrust terrane undergoing erosional uplift (dashed line) and tectonic thinning (dashed-dotted line) (Thompson and Ridley, 1987). Number on curves refer to ages in million years (Ma). Approximate stability fields of metamorphic facies: LE—LT eclogite, ME—MT eclogite; LB—lawsonite blueschist; EB—epidote blueschist; G—greenschist; A—amphibolite facies.

differ considerably: in Franciscan HP rocks retrogressive changes are generally lacking—with the exception of minor lawsonite–blueschist facies alterations along margins or veins of blocks (Cloos, 1986). The common preservation of aragonite (Coleman and Lee, 1962) also suggests that uplift must have been very rapid, without significant thermal relaxation (thermal gradient around 10°C/km), because otherwise aragonite would have reacted back to form calcite.

On New Caledonia and Sifnos, on the other hand, significant retrogression of eclogite facies assemblages (via epidote-blueschist facies assemblages, in the sense of Evans and Brown, 1987) to greenschist facies assemblages is commonly observed, which suggests a more or less isothermal uplift path (as shown in Figure 7.8). Calcite is an abundant mineral in greenschist assemblages and aragonite has not been detected. The spatial distribution of greenschist alterations is supposed to be controlled largely by the availability of a (water-rich) fluid phase which had to infiltrate eclogite facies rock units in order to generate greenschist facies assemblages. Evidence for such infiltration processes has been given by stable isotope work on Sifnos (Schliestedt and Matthews, 1987) where increased ^{18}O compositions of greenschist layers originally composed of eclogites are related to exchange with ^{18}O enriched fluids during the retrogressive event. Where fluids have not had access (probably due to impermeable rock units such as thick marble sequences) high pressure mineralogies are preserved without signs of alteration.

From detailed geochronological work on Sifnos (Altherr *et al*, 1979; Wijbrans *et al*, 1988) and New Caledonia (Blake *et al*, 1977; Brothers, 1985; also Ghent *et al*, 1987) it is clear that both terranes display a very similar history: peak metamorphic conditions were reached in Tertiary times and the time interval between the HP climax and the retrogressive, greenschist facies, overprint is only on the order of 20 Ma (Figure 7.8). Resulting uplift rates of 0.15 cm/year are at least one order of magnitude faster than average denudation rates (estimated from sediment discharge; Draper and Bone, 1981). It is therefore most likely that the preservation of LT eclogites requires more than (fast) erosional uplift. Tectonic thinning as a result of extensional deformation might be a possible process (compare calculated model P–T–t paths of Thompson and Ridley (1987) in Figure 7.8). An alternative proposition would be retarded thermal relaxation due to slow heat transfer (Rubie, 1984).

7.6 Summary: tectonic implications

The main characteristic features of LT eclogite occurrences from Franciscan, New Caledonia and Sifnos are summarised in Table 7.2. Although plate tectonic setting, protolith material and age differ considerably, peak metamorphic conditions were similar and can be interpreted in terms of rapid subduction or underthrusting of relatively cold crustal rock series to depth of more than 50 km. Mineral textures and the abundance of hydrous minerals occurring in equilibrium assemblages with garnet and omphacite suggest water-present conditions during HP metamorphism. Release of this water at greater depths will affect the thermal structure of subduction zones and, upon upward migration, it may cause hydration reactions in the hanging wall (Peacock, 1987). Significant differences in uplift paths are suggestive of different uplift mechanisms.

For the Franciscan HP terrane a process of underplating during concurrent subduction has been suggested (Platt, 1987) to account for the tectonic juxtaposition of

Table 7.2 Comparison of the characteristic features of Franciscan, New Caledonia and Sifnos low-*T* eclogite occurrences (see text for references)

	Franciscan	New Caledonia	Sifnos
Plate tectonic setting	subduction of oceanic crust and trench rocks	rapid tectonic burial underneath obducted sialic plate	subduction of volcanoclastic sequences along African–Eurasian collision zone
Protolith rocks	ocean floor basalt, greywacke, argillite	shallow water sediments, intercalated with dolerite sills, acidic and basic tuffs	continental margin sediments intercalated with acidic and basic tuffs
Age of protolith sediments	Jurassic	Cretaceous–Eocene	Mesozoic
Age of HP metamorphism	140–155 Ma	36–39 Ma	41 Ma
Age of retrogression	?	20 Ma?	19 Ma
Occurrence of eclogites	as blocks in low-grade mélange	conformable layers intercalated with metasediments	conformable layers intercalated with metasediments
Highest-grade assemblages			
basic	Omp-Grt-Ep-Gln	Omp-Grt-Ep-Pg-Bar	Omp–Grt–Gln–Ep–Act
acidic	not known	Jd_{60-100}–Qtz–Kfs–Phe–Gln	Jd_{90}–Qtz–Pg–Gln–Grt
K_D garnet–omphacite	19–24	11–21	20–29
Maximum pressure	ca. 15 kbar	15 kbar	15 kbar
Maximum temperature	520°C	550°C	490°C
Thermal gradient	10°C/km	11°C/km	9°C/km
HP metamorphic fluid	H_2O rich	H_2O rich	H_2O rich
Retrogression	to blueschist facies	to blueschist and greenschist facies	to blueschist and greenschist facies

rocks of different metamorphic grade as well as the general lack of LP alterations in high grade blueschists and eclogites.

By contrast, petrological evidence from the coherent HP units of New Caledonia and Sifnos is in favour of a rapid isothermal uplift which suggests that erosion and tectonic thinning owing to extension in upper levels may be important (Thompson and Ridley, 1987). The preservation of eclogites then becomes a question of availability or not of a fluid phase. In the case of only erosional uplift rates (England and Thompson, 1984), thermal relaxation will be fast enough to destroy HP assemblages to leave petrologists with nothing but very few eclogitic remnants.

Acknowledgements

It is my pleasure to acknowledge the help of Gabi Bartels, Heidi Forer, Ute Hahn-Schliestedt, and Petra Weise during the preparation of the manuscript. Critical reviews by Bernard Evans, Tim Holland and Alan Matthews clarified matters considerably and editorial support from Tony Carswell was most helpful.

8 Formation and exhumation of medium-temperature eclogites in the Scandinavian Caledonides

S.J. CUTHBERT and D.A. CARSWELL

8.1 Introduction

Eclogite is a lithology with widespread occurrence in the Scandinavian Caledonides, from Bergen in the south to Troms in the far north (Figure 8.1), and at several tectonostratigraphic levels. However, it is usually found in the form of relatively small bodies which, in total, constitute only a tiny proportion of the exposed area of the orogen.

A common feature of many of the eclogites is their occurrence as lenses or bands in amphibolite or granulite facies gneisses. It was this association that Coleman *et al* (1965) used to define their Type B eclogite group. Coleman *et al* found that garnet compositions in Type B eclogites were generally intermediate between those from inclusions in alkaline volcanics (Type A) and those from glaucophane schist belts (Type C), reflecting intermediate *P–T* conditions. More recent geothermometry studies support the classification of the eclogites considered here as 'medium-*T* eclogites' (see Chapter 1). Other eclogite facies rocks are commonly associated with the eclogites themselves, notably garnetiferous peridotites. The peridotites will only be dealt with briefly here, as they are discussed in some detail in Chapter 11.

Despite the great interest shown in these rocks, their origins have remained enigmatic for many years, partly because of the apparently sharp contrast in metamorphic facies between the eclogites and their enclosing country-rocks; Coleman *et al* (1965) referred to their Type B eclogites as being 'geologically out of place' in their country-rocks. The problem is exacerbated by the poor preservation of evidence for pre-metamorphic relationships between the eclogite protoliths and their host rocks. The eclogites usually take the form of pods (boudins) whose margins are retrograded to amphibolite. The amphibolite is often foliated, the foliation lying parallel to that of the adjacent gneiss, which wraps around the eclogite pod (Figure 8.2). Hence the margins of the eclogite bodies are often entirely tectonic, the tectonism being associated with a metamorphic convergence towards the facies found in the country-rocks. This has led to two fundamentally different interpretations which can be summarised as follows.

(a) The eclogites were formed at mantle-depths and subsequently tectonically transported into gneisses, which lay at a higher crustal level. Eclogite margins were amphibolitised during emplacement into the gneisses.

(b) The eclogites were formed *in situ* in their present host gneisses; differential retrogression has resulted in almost complete transformation of the gneisses to amphibolite (or granulite) facies assemblages, while kinetic factors have favoured the preservation of eclogite facies assemblages in metabasic rocks.

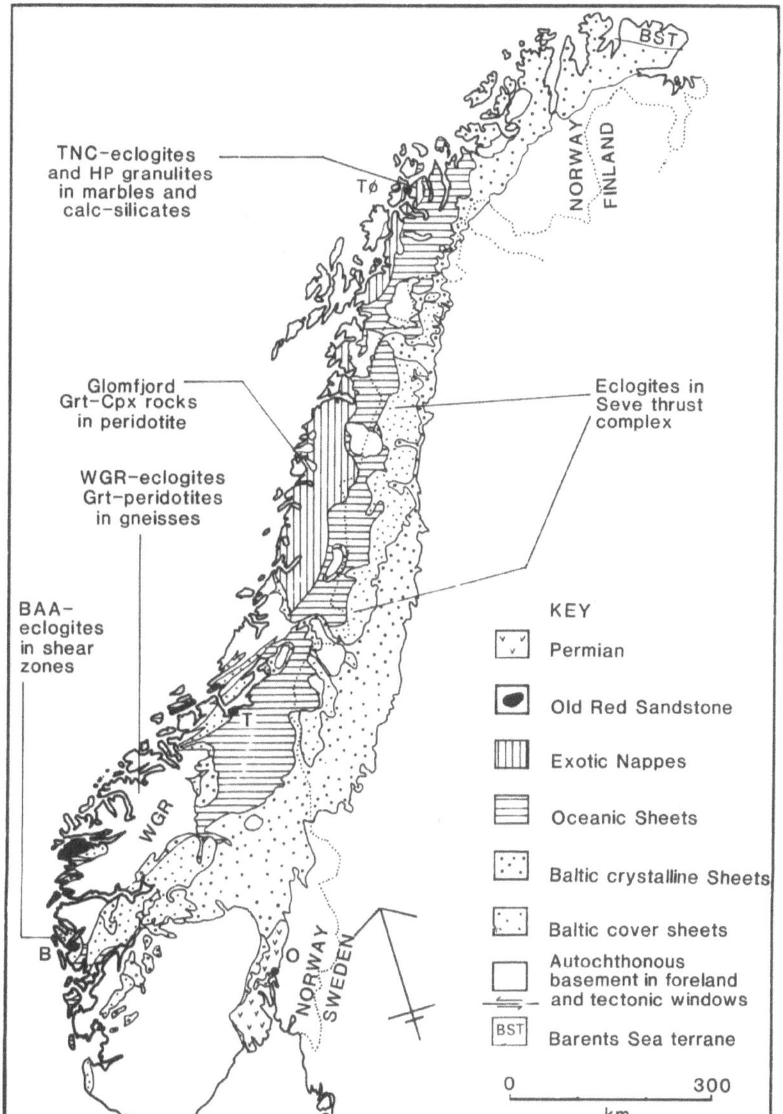

Figure 8.1 Tectonostratigraphic map of the Scandinavian Caledonides, showing the distribution of eclogites. TNC, Tromsø Nappe Complex; WGR, Western Gneiss Region; BAA, Bergen Arcs anorthosites; Tø, Tromsø; T, Trondheim; B, Bergen; O, Oslo; Compiled from Gee *et al* (1985); Roberts and Gee (1985); Griffin *et al* (1985); Hossack and Cooper (1986).

Figure 8.2 Eclogite pod from near Davik, Nordfjord. The lensoid, boudin shape with dark, sheared, amphibolitic margins against well-foliated amphibolite facies gneisses is typical of medium-*T* eclogites from the Western Gneiss Region and other parts of the Scandinavian Caledonides.

Clearly the implications of either of the above scenarios are very important not only for the geodynamic evolution of the Caledonides, but also for understanding the tectono-thermal environment of MT eclogite facies metamorphism. Comprehensive reviews of eclogite facies metamorphism in the Scandinavian Caledonides have been published by Cuthbert *et al* (1983), Griffin *et al* (1985), and Griffin (1987). These reviews have been, of necessity, preoccupied with the relevance of the eclogite facies rocks to Caledonian orogenic evolution. The aim of this article is to explore the contribution that the Scandinavian eclogites can make to a more general understanding of the processes of eclogite formation in continental crust, and the dynamics of collisional orogenesis.

8.2 Tectonostratigraphic distribution of eclogite facies rocks

The Scandinavian Caledonides (Figure 8.1) are exposed in a 1800 km long belt in Norway and western Sweden, from Stavanger (latitude 59°N) in the south to Finnmark (latitude 71°N) in the north (Gee *et al*, 1985). The orogen is constructed from a number of extensive thrust sheets which were transported up to several hundred kilometres eastwards across the margin of the Baltic Shield during the collision of two major continents, Baltica and Laurentia (Gee, 1975; Stephens, 1988). The thrust sheets are composed of metasedimentary and meta-igneous rocks of late Proterozoic to late Silurian or Devonian age, with significant involvement of Precambrian (pre-Caledonian) crystalline rocks (Roberts and Gee, 1985).

The tectonostratigraphy (Figure 8.1) has been summarised in a number of contributions in Gee and Sturt (1985). The following lithotectonic units can be defined (following Hossack and Cooper, 1986).

(a) The autochthonous Baltic basement of Proterozoic and locally Archaean high-grade gneisses and plutonic igneous rocks, with a thin, autochthonous cover of Upper Proterozoic to Lower Palaeozoic sediments. The latter act as a décollement horizon for the overlying allochthon.
(b) Baltic cover sheets: Late Proterozoic to Early Palaeozoic sediments with little or no basement involvement.
(c) Baltic crystalline sheets: a series of thrust sheets characterised by large volumes of Precambrian high-grade gneisses and igneous rocks with a cover of meta-sandstones, having affinities with the basement and cover rocks of the Baltic foreland. Examples are the great Jotun and Bergen thrust sheets. The uppermost units (Seve thrust complex) appear to represent the outermost part of the early Caledonian Baltic continental margin, largely consisting of miogeoclinal sediments and basic intrusions with only minor basement involvement (Gee, 1975).
(d) Oceanic sheets: extensive units of volcano-sedimentary rocks including ophiolites and elements of island arcs, back-arc basins, fore-arc basins and mélange. These comprise a number of allochthonous terranes which formed outboard of the Baltic continental margin, mostly during the early Ordovician (Stephens and Gee, 1985; Dunning and Pedersen, 1988). They include the Köli Nappe complex.
(e) Exotic nappe complexes: these are the highest tectonostratigraphic units, overlying the oceanic terranes. They consist of migmatitic gneisses and metasediments of uncertain affinities, and contain thrust slices of ophiolite and synorogenic granitoid batholiths (Stephens *et al*, 1985). They may represent part of the margin of Laurentia, or a microcontinental terrane outboard of Baltica (Stephens *et al*, 1985). Examples are the Helgeland Nappe Complex and the Tromsø Nappe Complex (TNC).
(f) Devonian basins: these are late-orogenic intermontane basins filled with continental sandstones and conglomerates, largely of Middle Devonian age (Steel *et al*, 1985). They lie either unconformably on, or in fault contact with, units of the Baltic crystalline sheets or the oceanic terranes.

Autochthonous or para-autochthonous basement is exposed within the orogen in a number of tectonic windows, the largest of which is the Western (or so-called Basal)

Gneiss Region (WGR) of western Norway (Figure 8.1), a large culmination in the orogen exposed between Bergen and Trondheim. These allow the Baltic basement to be traced from the foreland deep into the orogen below the allochthon.

Eclogites have been recognised in parts of all of these tectonostratigraphic units except the oceanic terranes, hence they all occur within rocks of continental affinities. No eclogites are known from the Baltic foreland basement or cover east of the Caledonide front; they are, however, well known from the para-autochthonous Baltic basement of the WGR.

Eclogites are absent in the southeast part of the WGR, but first appear northwest of Jostedalsbreen. Dietler *et al* (1985) found that Caledonian reworking of the gneisses increased in intensity westwards, and that eclogites first appeared in the zone of strongest reworking. Mineral assemblages in the country-rocks are usually of the amphibolite facies, but relicts of granulite facies and eclogite facies assemblages are occasionally found in non-mafic rocks (Bryhni, 1966; Mysen and Heier, 1972; Griffin and Mørk, 1981; Krogh, 1980b; Harvey, 1983; Griffin *et al*, 1985; Cuthbert, 1985).

Eclogites are found in two units of the Baltic crystalline sheets; a thrust sheet of granulite facies rocks in the Bergen Arcs region, and the Seve thrust complex of northern Sweden. In the Bergen arcs, eclogites are found within meta-anorthosites, metagabbros and pyroxene-syenites ('mangerites'). These rocks have had a complex history, with a protracted period of intrusion and granulite facies metamorphism during the Proterozoic at about 900 Ma, followed by Caledonian reworking at eclogite then amphibolite facies, often along discrete shear zones (Austrheim and Griffin, 1985; Cohen *et al*, 1988).

The Seve thrust complex is a very large composite unit about 800 km long (Figure 8.1) that consists of several thrust sheets with diverse lithological associations and metamorphic histories. These appear to represent fragments of the Baltic margin (miogeocline) amalgamated along high-grade ductile shear zones before eastwards thrusting into their present position (Andreasson, 1986). Eclogites are known from at least three thrust sheets, in metasediments and high grade gneisses, along with garnetiferous peridotites in one area (van Roermund, 1985). Some eclogites in psammitic metasediments are glaucophane-bearing (Stephens and van Roermund, 1984) and many of the eclogites of the Seve are considered to be metamorphosed, rift-related dolerite dykes intruded into continental-shelf sandstones at the extreme western margin of the Baltic continent (Andreasson *et al*, 1985; Dallmeyer and Gee, 1986).

Eclogites in the Exotic Nappe Complexes are found in the Tromsø Nappe Complex (Figure 8.1), which is probably the highest structural unit in the Scandinavian Caledonides (Andresen *et al*, 1985; Krogh *et al*, in press). They are found as bands and lenses within marbles, calc-silicate schists, metapelites, quartzo-feldspathic gneisses and ultramafites. Relict HP metamorphic assemblages are common in the country-rocks. Eclogites with similar lithological associations and tectonostratigraphic position occur in the Nordland region at Glomfjord (Griffin *et al*, 1985).

8.3 Protoliths of eclogite facies rocks

The problem of relating tectonised bodies of eclogite to their present non-cofacial country-rocks was touched upon in the introductory section. Such relationships have led to interpretations that involve tectonic interdigitation of upper mantle (Lappin and

Smith, 1978, 1981; Smith 1980b, 1981) or subducted oceanic crust (O'Hara, 1976) with higher-level continental crust. This point of view is still prevalent in accounts of eclogites in high-grade gneisses in some standard texts on metamorphic petrology (e.g. Turner, 1981).

Evidence for the nature of eclogite protoliths and their relationships to the country-rocks has been reviewed elsewhere (Bryhni *et al*, 1977; Griffin and Møork, 1981; Cuthbert *et al*, 1983; Griffin *et al*, 1985; Griffin, 1987). LP igneous origins are indicated by preservation of intrusive contacts with gneisses (Cuthbert, 1985; Griffin and Carswell, 1985), occurrence of eclogites within anorthosites (Brastad, 1985), geochemical trends explained by accumulation of olivine + plagioclase in layered basic intrusions (Mysen and Heier, 1972; Cuthbert, 1985; Jamtveit, 1987a), and relicts of LP igneous phases in coronites (Gjelsvik, 1952; Griffin, 1972; Griffin and Råheim, 1973; Carswell and Harvey, 1985; Mørk, 1985, 1986; Andreasson *et al*, 1985; Cuthbert, 1985). Eclogites are frequently interlayered with metasediments (Bryhni *et al*, 1977; Carswell and Harvey, 1985; Krogh *et al*, in press). Armoured mineral inclusions in eclogite garnets often show evidence that eclogites have formed by prograde metamorphism of amphibolites (Bryhni and Griffin, 1971; Krogh, 1980a, 1982; Cuthbert, 1985).

If relict eclogite facies assemblages can be found within the country-rocks it becomes unnecessary to postulate a foreign origin for the eclogites. The recognition of such assemblages is hindered by the extensive retrogression of such assemblages to lower grades. Nonetheless examples of HP assemblages from a variety of lithologies have been described in recent years. Omphacite relicts occur in quartzo-feldspathic gneisses at several localities in the WGR, usually as Cpx + Pl or Amp + Pl symplectites (Figure 8.3a), or as inclusions in garnets (Mysen and Heier, 1972; Green and Mysen, 1972; Brueckner, 1977; Krogh, 1980b; Carswell and Harvey, 1985; Cuthbert, 1985). Highly calcic garnets co-existing with kyanite, sodic plagioclase and quartz imply pressures compatible with those derived from eclogites (Griffin and Mørk, 1981; Griffin and Carswell, 1985). This is one of a number of assemblages which have been described as high-pressure granulites and may be cofacial with the eclogites (Bryhni, 1966; Krogh, 1980b; Griffin *et al*, 1985). Bt–Fsp–Ky–Qtz intergrowths (Figure 8.3b) have been increasingly widely recognised within eclogite-bearing gneisses (Griffin *et al*, 1985; Griffin, 1987), indicating that phengite was once a common phase. Fresh phengite-bearing gneisses are known from the WGR (Cuthbert, 1985 and unpublished data), and in the Seve thrust complex (Andreasson *et al*, 1985).

The association of pure jadeite + quartz is not known from the Scandinavian Caledonides. However, reconstruction of aggregates of Na-augite + albite from trondjemitic gneisses in the Tromsø Nappe Complex give highly jadeitic compositions (Jd_{70}) (Krogh *et al*, in press).

In summary, several different lines of evidence indicate that MT eclogites from the Scandinavian Caledonides have formed by metamorphism of LP protoliths, at least some of which were *in situ* in their present country-rocks before this metamorphism. This is corroborated by evidence of HP metamorphism of the country-rocks themselves.

There is, however, some danger in being dogmatic about such a hypothesis, since alpine-type peridotite bodies with relict HP garnetiferous assemblages have been widely regarded as good candidates for direct emplacement from the subcontinental mantle (Carswell, 1968, 1973, 1986; Medaris, 1980, 1984; Brueckner, 1977). Origins in sub-oceanic mantle (Griffin and Qvale, 1985; Mearns, 1986) have also been considered.

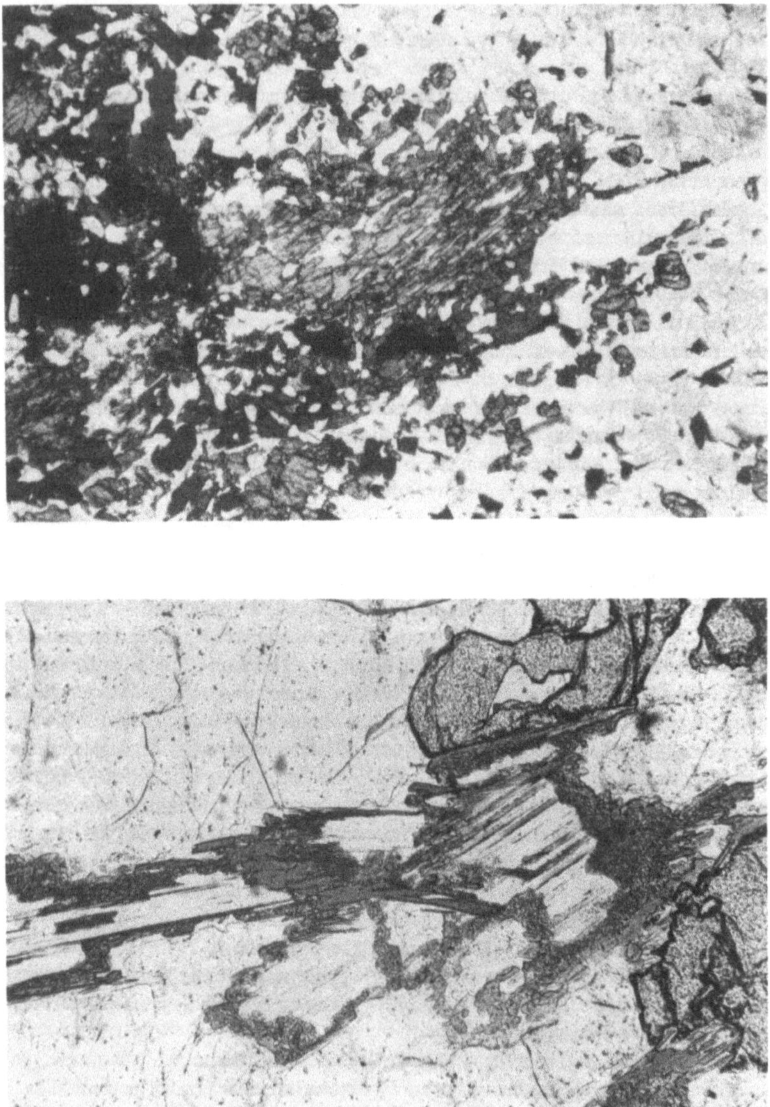

Figure 8.3 (a) Symplectite of diopsidic clinopyroxene + plagioclase (± hornblende), typical of the breakdown of pre-existing omphacite, in granodioritic orthogneiss sample U332 from Sundsbø on Otrøy in Moldefjord, Western Gneiss Region. Width of field of view about 2 mm. (b) Partial replacement of phengitic white mica by an intergrowth of biotite + feldspars (both Na-Pl and Kfs) in a HP meta-pelitic schist (Grt + Phe + Omp + Qtz) from Bryggia, Nordfjord, WGR. Width of field of view about 2 mm.

Studies have shown that these peridotites have suffered a long, polymetamorphic history (Carswell, 1986) with at least one phase of eclogite facies mineral growth during the mid-Proterozoic which predates the age of formation of some of the adjacent gneiss protoliths (Mearns, 1986). Although it is apparent from this that the early evolution of the peridotite bodies was spatially and temporally separate from that of their present enclosing gneisses, Jamtveit *et al* (1989) have presented Sm-Nd mineral-age data which demonstrate much later (437–511 Ma) recrystallisation to a second HP garnet-bearing assemblage essentially during the Caledonian HP event responsible for the widespread stabilisation of eclogite in metabasic bodies (Section 8.6).

The evidence presented in this section demonstrates that significant volumes of continental crustal material now forming the Scandinavian Caledonide orogen have been subjected to unusually high lithostatic pressures. Such evidence, which is mostly based on fieldwork and petrographic studies, shows that the Type-B eclogites of Coleman *et al* (1965) should no longer be entirely regarded as 'geologically out of place', but rather as relicts of regional HP metamorphic events. As will be shown in the following sections, this gives them a value in geodynamic analysis out of all proportion to their tiny volume.

8.4 Mineralogical evolution of metabasic eclogites

In this section the reaction pathways which led to the development of metabasic eclogites will be examined for three different situations; static growth of eclogite phases as coronas in dolerites, synkinematic growth of eclogite phases from amphibolites (based on garnet inclusion and zoning studies), and strain-related eclogite formation in shear-zones. Case-studies will be used to demonstrate the processes involved in the formation of MT eclogites, and problems of their interpretation.

8.4.1 *Coronites*

Detailed studies of metagabbros from the Nordøyane area in the extreme NW of the WGR have recently been made by Mørk (1985, 1986), and Mørk and Mearns (1986) and some of their results are discussed here (see also Chapter 5).

The eclogites locally contain relicts of medium-grained olivine-gabbros, with Ol + Pl + Aug and accessory Phl + Ilm. Pl–Aug–WR (whole rock) Sm-Nd ages of about 1200 Ma have been obtained by Mørk and Mearns (1986), who also derived a WR-Grt-Omp Sm-Nd age of 400 ± 16 Ma for a foliated eclogite from the same area. Hence metamorphism postdated intrusion by about 800 Ma. Ambient pressure during intrusion was < 8–10 kbar, while the eclogite paragenesis equilibrated at around 750°C and 15–20 kbar (Mørk, 1985).

Three stages in the transition to eclogite were defined by Mørk (1985), as illustrated in Figures 5.2d and 5.3c.

(a) Growth of thin rims of orthopyroxene and garnet at the interfaces between olivine and plagioclase grains.
(b) Encroachment of orthopyroxene into olivine grains and development of a rim of omphacite around the orthopyroxene adjacent to the garnet coronas; extensive growth of garnet within the plagioclase domains. The plagioclase becomes more sodic, with development of needles of spinel and minor omphacite. Augite

becomes compositionally inhomogeneous, trending towards omphacite, with subsequent nucleation of omphacite neoblasts.

(c) Extensive recrystallisation. Olivine, orthopyroxene and augite are entirely replaced by omphacite. Poikiloblastic masses of garnet with omphacite inclusions replace plagioclase, with further recrystallisation producing a polygonal fabric of discrete garnet and omphacite grains.

The final result is a true eclogite which pseudomorphs the original igneous fabric. Such eclogites may have no foliation, hence eclogite-formation is not uniquely correlated with deformation.

The corona phases garnet and orthopyroxene are chemically zoned. Orthopyroxene domains are marginally enriched in Al adjacent to omphacite or garnet as the result of limited Al diffusion from plagioclase domains into olivine domains. Garnet shows relative enrichment of Ca towards the igneous plagioclase domains. Ca was thought to have been released from the plagioclase sites when spinel nucleated within them, and was incorporated into the garnet corona as it grew, but limited diffusion-rates prevented its complete homogenisation. In contrast, the garnets in the completely eclogitised metagabbros are unzoned, indicating that more effective diffusive homogenisation had occurred, perhaps aided by fluids or deformation (Mørk, 1986).

Possible simplified reactions which controlled the corona development may have been:

Stage 1 $3\ \mathrm{An} + 4\ \mathrm{Fo} = 2\ \mathrm{Grt} + 2\ \mathrm{En} + \mathrm{Spl}$

Stage 2 $\mathrm{Ab} + \mathrm{Fo} = \mathrm{Jd} + \mathrm{En}$

$\mathrm{An} + \mathrm{Fo} + \mathrm{En} = \mathrm{Di} + \mathrm{Prp}$

Stage 3 $\mathrm{Spl} + \mathrm{En} + \mathrm{Ab} = \mathrm{Jd} + \mathrm{Prp}$

(in combination with those from stage 2)

The formation of orthopyroxene in stages 1 and 2 is transient, as it is subsequently consumed in the production of omphacite and garnet. Stage 3 requires extensive exchange between the mafic and felsic domains, in contrast with stages 1 and 2 where mass transfer occurs only in the vicinity of the domain interfaces. This is consistent with the flat compositional profiles in the garnets.

Mørk and Mearns (1986) showed that the Sm-Nd isotopic composition of the original plagioclase and augite domains remained little affected by these reactions until an advanced stage of textural re-equilibration, despite the extreme P–T conditions. Only well foliated eclogites had been isotopically homogenised sufficiently well to give mineral isochrons.

These coronitic rocks illustrate a number of points about the nature of the metamorphic processes and interpretation of eclogitic rocks.

Firstly, Mørk (1986) considered that the main control on the development of the coronas was the mobility and availability of elements for the growth of new phases. Early growth of orthopyroxene rather than omphacite was a result of limited mobility of Ca, Al and Na from plagioclase. Hence each stage in the coronite evolution need not represent a different point on a prograde P–T path; they simply reflect the amount of interaction between small compositional domains.

Secondly, evaluation of P–T–t paths for eclogites often involves studies of garnet zoning. In the example described here the garnet cores and their omphacite inclusions are not necessarily older than the rims. Without knowledge of the mechanism of garnet growth and the reactions involved erroneous conclusions might be drawn about the metamorphic evolution of these rocks (Møork, 1986).

Thirdly, volume diffusion, even at these extreme conditions, has not achieved chemical equilibrium on a whole-rock (about 1 litre) scale. This is particularly important for Sm-Nd dating, which is commonly used for eclogites. Two-mineral isochrons (e.g. Griffin and Brueckner, 1980) may, therefore, be unreliable for dating the HP metamorphism. Also, geothermometry based on 'closure temperatutes' for the Sm-Nd system should be evaluated carefully, as it is evident that deformation has played an important part in the achievement of isotopic equilibrium, rather than temperature alone.

8.4.2 *Garnet zoning and solid inclusions*

Solid inclusions in garnets provide an opportunity to investigate the evolution of eclogites from metamorphic precursors, frequently amphibolites. Such garnets can be a useful tool in determining the reactions which produced the eclogite parageneses. In this section examples from Krogh (1980a, 1982), Cuthbert *et al* (1983) and Cuthbert (1985) will be used to illustrate various garnet growth histories from the WGR.

Garnets in eclogites from the Sunnfjord and Nordfjord areas are typically well-zoned, commonly with inclusions of dark-green amphibole in the cores. Grt–Cpx geothermometry on these eclogites gives temperatures of 550–700°C (Krogh, 1980a, 1982; Cuthbert, 1985). Figure 8.4 illustrates features of a garnet from an eclogite at Dalsfjord, Sunnfjord (Cuthbert, 1985) which is typical of this type of zoning. The garnet core has high Mn and Fe/Mg, with inclusions of dark green, aluminous, calcic to sodic-calcic amphiboles, ilmenite and (in some other examples) epidote. Outside this zone Fe/Mg declines quite sharply, then flattens out towards the rim, where inclusions of omphacite (Jd_{35-41}), rutile and barroisitic amphibole are common. In the lowest temperature eclogites from Sunnfjord, glaucophane inclusions coexist with omphacites in the outer zones, and also in the matrix.

The zoning in such garnets has been interpreted as growth zoning (Krogh, 1980a, 1982; Cuthbert, 1985). The abrupt change in garnet composition coincident with the change from amphibole- to omphacite-dominated inclusion suites represents a frozen-in amphibolite–eclogite transition. Krogh (1982) has derived a balanced reaction equation for this transition for a kyanite-eclogite from the Nordfjord area (Verpeneset) based on measured mineral compositions:

$$60\text{Amp} + 62\text{Pg} + 60\text{Zo} + 70\text{Qtz} = 312\text{Omp} + 22\text{Grt} + 163\text{Ky} + 152\text{H}_2\text{O}$$

The zoning was probably controlled by reactions taking place during the growth of the garnet. The phase that coexisted with the garnet changed from early amphibole to later clinopyroxene. At any given temperature $K_{D_{Fe-Mg}}$ for Grt–Amp is much lower than that for Grt–Cpx (Graham and Powell, 1984), so that the Fe/Mg ratio of garnet coexisting with amphibole will have been much lower than that for garnet and omphacite. Hence the transition from amphibolite to eclogite need not have involved much temperature change, but a large increase in pressure. Cuthbert (1985), using the method of Graham and Powell (1984) on amphibole inclusions, found that temperature estimates overlapped those derived from omphacite inclusions and coexisting garnet by the Ellis and Green (1979) K_D formulation, implying a steep up-pressure path in the P–T plane. In the Verpeneset example (Krogh, 1982) such a path would be consistent

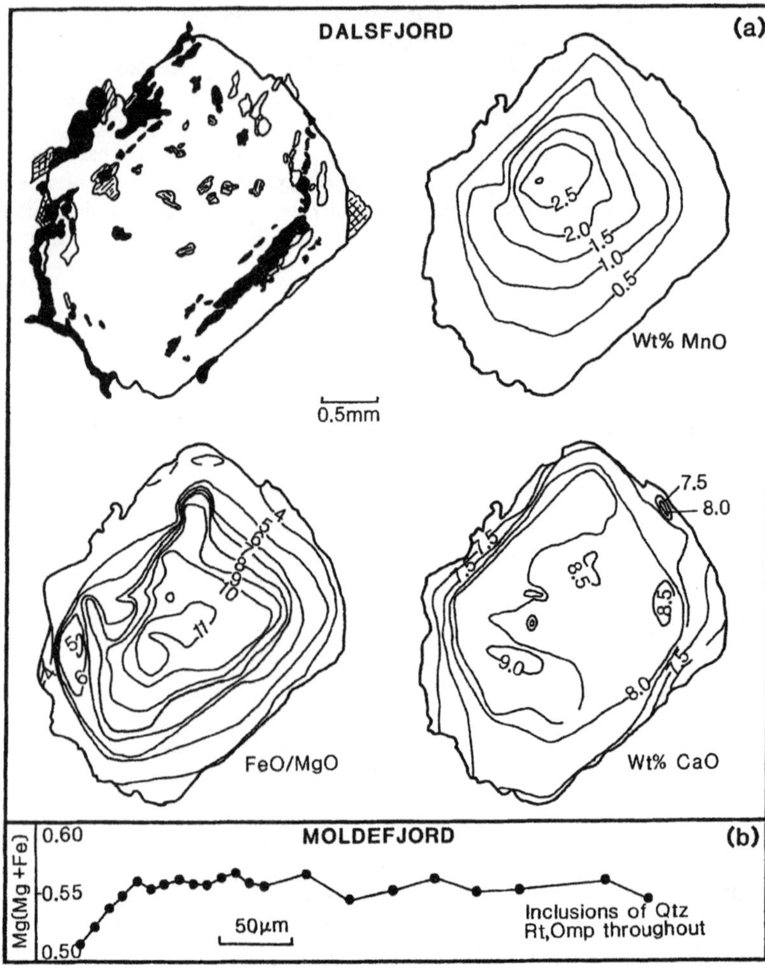

Figure 8.4 Compositional zoning of garnets from the Western Gneiss Region. (a) 'Prograde-zoned' garnet from Dalsfjord, Sunnfjord (from Cuthbert, 1985) showing distribution of inclusions (top left), and compositional maps for MnO, FeO/MgO and CaO. Ornaments for inclusions are black for rutile, unornamented for omphacite, lined for amphibole. Amphiboles near the garnet rim are barroisites, those in the core are hornblendic. (b) Compositional profile of Mg/(Mg + Fe) in garnet from Langevatnet, Moldefjord, from unpublished data of M.A. Harvey (see also Cuthbert *et al*, 1983).

with the occurrence of paragonite in the garnets and its replacement with Ky + Omp in the matrix (Holland, 1979) and available experimental evidence on the HP breakdown of amphibole (Essene *et al*, 1970; Figure 1.1).

Eclogites from the coastal NW part of the WGR have garnets with zoning profiles which contrast with those from Nordfjord and Sunnfjord. Eclogite garnets from this area have rather flat compositional zoning profiles (Krogh, 1982), and generally lack inclusions of amphibole. Equilibration temperatures from this area have been estimated as 700–800°C (Griffin *et al*, 1985).

A typical example, from Moldefjord is shown in Figure 8.4b, using unpublished data of M.A. Harvey. Mg/(Mg + Fe) has a flat profile except for a narrow zone near the rim where it declines. Inclusions of omphacite, quartz and rutile occur throughout the grain. Profiles like this have been interpreted as the result of diffusive redistribution preventing the development of zoning during growth by diffusive redistribution (Krogh, 1982), but with diffusion-controlled retrograde rims. An instructive exception to this type of profile, also from Moldefjord, is an eclogite containing very large garnets (up to 2.5 cm) which contain inclusions of hornblende, plagioclase, biotite and magnetite, and prograde compositional zoning that resembles that of garnets from the Nordfjord area (Krogh, 1982). Hence the same scenario of up-pressure evolution from amphibolites can be inferred from at least some of these eclogites, although the zoning and inclusions were only preserved in grains too large for volume diffusion (and deformation?) to be effective. Marginal retrograde zoning is interpreted as a result of diffusional re-equilibration during cooling, when the diffusional length-scale was much smaller. The form of P–T–t paths from the WGR (Section 8.7) suggests that this cooling occurred outside the eclogite stability field, making T estimates from grain-rims in such cases ambiguous.

This apparent relationship between temperature and the scale of chemical equilibrium can also be seen from Sm-Nd isotopic studies (Griffin and Brueckner, 1985). The lower-T eclogites lack equilibrium distributions of REE between garnet and omphacite, whereas in the higher-temperature region, where growth zoning is absent, isotopic equilibrium was attained and eclogites give good mineral isochrons. However, strain may also have played a part in this (Møork and Mearns, 1986).

Finally, the occurrence of growth zoning implies that P–T estimates from inclusions and grain-rim pairs will record the conditions suffered by a rock throughout its history. For unzoned garnets only T_{max} will be recorded, and this may have taken place at pressures considerably below the P_{max} if heating occurred during uplift (cf. England and Richardson, 1977).

8.4.3 *Eclogite formation in shear-zones*

The contribution of deformation and fluids to the generation of eclogite is well displayed in the anorthosite-bearing Baltic crystalline thrust sheet of the Bergen Arcs (Figure 8.1) which has been studied by Austrheim and Griffin (1985) and Austrheim (1987).

Anorthosites, gabbroic anorthosites and norites have a banding defined by granulite facies parageneses of Grenville age (900 Ma—Cohen *et al*, 1988) which give P–T estimates of 900–950°C at 10–11 kbar. The assemblage which forms the protoliths for the eclogitic rocks consists of Pl + Al – Cpx + Fe – Ti oxides + Grt I ± Opx, Scp, Hbl and Spl.

The eclogitic rocks are associated with late structures which cross-cut the granulite facies banding. Eclogitisation has occurred in mylonitic shear-zones where the granulite-banding has been obliterated and replaced by a new mylonitic foliation, in kinks where the granulite fabric is deformed but still visible, and in the neighbourhood of cracks. Eclogite facies minerals give poorly-constrained Palaeozoic ages (Cohen *et al*, 1988), so the eclogite-forming event is Caledonian, about 500 Ma after the granulite event. Eclogite phases give P–T estimates of 750 ± 50°C and 16–19 kbar.

Progressive effects of strain can be seen in the margins of shear-zones, where old Al–Cpx becomes strained and dusty, then gradually clears by exsolution of rutile needles. Sub-grains and then neoblasts developed in the clinopyroxene against adjacent plagioclase, associated with a progressive increase of jadeite content which reached Jd_{60}. Clinopyroxene also shows exsolution of epidote and grossular-rich garnet, which also appears in plagioclase. Granulite facies plagioclase becomes more sodic and clouded with kyanite, clinozoisite and phengite. These transformations are then similar to those seen in coronites (e.g. Mørk, 1985). The mylonitic parts of shear-zones consist of shape-oriented omphacite alternating with well-foliated masses of Ky + Phe + Ep + Omp ± Qtz, replacing plagioclase. These relationships demonstrate that eclogitisation has occurred progressively during deformation.

A number of features indicate the importance of fluids in the generation of eclogite (Austrheim, 1987). These include growth of epidote needles along cracks in plagioclase (these cracks contain omphacite when they pass into garnets), and the presence of microcracks in Grt I porphyroclasts filled with omphacite, kyanite and very Cl-rich hornblende.

Eclogite forms 30–45% of a number of well-exposed areas studied by Austrheim (1987). If this can be extrapolated to the entire outcrop of the thrust-sheet (about 1000 km^2) then eclogite probably forms a high proportion of its total volume (Austrheim, 1987). However, an equally significant volume has remained unaffected by the extreme P–T conditions to which the unit was subjected. Hence anorthosite may retain a metastable mineral assemblage unless recrystallisation is aided by fluid influx and deformation. The resulting alternation of dense eclogite facies and less dense granulite facies rocks on a scale of metres may be expected to have a characteristic seismic signature, and is one possible explanation for the complex, layered nature of the seismic lower crust (Austrheim, 1987).

8.5 Retrogression and preservation of eclogite facies parageneses

From the discussion in Section 8.3 it will be apparent that most eclogite facies rocks in the Scandinavian Caledonides are relicts of small volume within much larger bodies of rock in which the HP assemblages have been overprinted by later metamorphism. The eclogites themselves are the most common relicts and appear to be the most resistant to alteration (although this prejudice may be a result of ease of recognition of such striking rocks). This section addresses the problem of preservation of the eclogite facies assemblages. For other discussions on this subject see Krogh (1980b), Griffin *et al* (1985), and Griffin (1987).

Retrogressive effects in metabasic eclogites are similar in all the tectonostratigraphic units in the Scandinavian Caledonides (Eskola, 1921; Bryhni, 1966; Mysen, 1972; Krogh, 1980a, b, 1982; Stephens and van Roermund, 1984; Andreasson *et al*, 1985; Krogh *et al*, in press). Omphacites break down to symplectites of sodic augite or diopside and sodic plagioclase, which are in turn replaced by Amp + Pl symplectites and poikiloblastic hornblendes with quartz and plagioclase inclusions. Garnets are replaced by a collar of kelyphitic hornblende, commonly with magnetite, plagioclase or epidote. In areas where temperatures were relatively high, hornblende may replace earlier (granulite-facies) kelyphites of Opx + Pl ± Spl. Phengite is replaced by Bt + Pl and rutile by sphene. Where eclogites contain glaucophanitic amphibole, considered to coexist with omphacite and garnet, it is altered to barroisitic amphibole and then hornblende (Krogh, 1980a; M.B. Stephens, *pers. commun.*, 1985).

In non-mafic rocks retrogressive effects will depend on composition, but in pelitic and quartzo-feldspathic gneisses observed changes are the replacement of phengite by Bt + Kfs + Ky (Figure 8.3b) (Griffin *et al*, 1985), symplectic breakdown of omphacite to Di/Hbl + Pl (Figure 8.3a) (Mysen and Heier, 1972; Carswell and Harvey, 1985; Cuthbert, 1985) and replacement of kyanite by fibrolitic sillimanite or muscovite (Guthbert, 1985—WGR; Krogh *et al*, in press—Tromsø Nappe). Typical granodioritic gneisses in the WGR consist of Pl, Kfs, Qtz, Bt, Ep ± Ms, Grt (Bryhni, 1966).

The factor that most obviously controls retrogression is the effect of strong deformation at lower-grade conditions. In the WGR the coastal areas are characterised by the development of a blastomylonitic foliation defined by amphibolite facies mineral assemblages. Boudinage and marginal amphibolitisation of the eclogites has occurred within these mylonites (Dietler *et al*, 1985). An earlier (HP?) foliation has been preserved in strain shadows of eclogite boudins (Griffin and Carswell, 1985) in the Moldefjord area. This deformation has evidently obliterated any earlier eclogite facies mineralogy in the gneisses except for locally protected enclaves. In contrast, the eclogites have tended to act as more rigid bodies, breaking up as blocks during boudinage but not deforming internally. Hence their stiffer rheological properties have aided their preservation.

Some more subtle chemical mechanisms may have operated to cause differential retrogression of eclogite facies assemblages. The original presence of omphacite in gneisses is usually indicated only by occurrences of symplectites. A common observation in the eclogites is that omphacite tends to be more heavily altered to symplectite in the presence of quartz, and the freshest eclogites are often quartz-free. This can be ascribed to the nature of the omphacite decomposition reactions:

$$\mathrm{Jd} + \mathrm{SiO_2} = \mathrm{Ab} \qquad \text{and} \qquad 2\mathrm{Jd} = \mathrm{Ab} + \mathrm{Ne}$$

The silica-absent reaction occurs at about 7 kbar lower pressure than the silica-present reaction at any temperature (Gasparik, 1985a), so that quartz-free eclogites may still be stable when quartz-bearing systems are reacting. As nepheline has not, to the authors' knowledge, been described from retrograded eclogites, symplectite formation in quartz-free eclogites may require introduction of silica or oxidation (Mysen and Griffin, 1973). The free availability of silica in quartzo-feldspathic gneisses may, therefore, explain the rarity of omphacite preservation. The large increase in grain-surface area that results from symplectite-formation may encourage hydration and recrystallisation to amphibole, obliterating all sign of previous pyroxene.

The common observation that eclogite margins are amphibolitised, and that this involved some K-metasomatism (e.g. Krogh, 1980a) suggests that the source of water for this hydration was the adjacent gneisses. Heinrich (1982) showed that retrogression of phengite-bearing pelites involves a dehydration reaction (breakdown of phengite), which is likely to proceed rapidly, while amphibolitisation of eclogite is a rather sluggish hydration reaction. Hence during decompression, retrogression of phengite-bearing gneisses is likely to have proceeded quickly, providing a source of water to hydrate eclogites, but the slow pace of this reaction would have prevented complete alteration in all but the smallest bodies. Symplectic pseudomorphs after phengite are commonly found in the Scandinavian eclogite-bearing gneisses (Griffin *et al*, 1985;

Andreasson *et al*, 1985), so this mechanism may explain the relationships between MT eclogites and their gneissic host-rocks.

A final explanation for the lack of eclogite facies assemblages in gneisses is suggested by observations from the Bergen Arcs anorthosites (Austrheim and Griffin, 1985; Austrheim, 1987) where granulite facies rocks survive on a large scale. It is possible that some lithologies never recrystallised under ambient HP conditions at all, owing to lack of available fluids or strain to catalyse reactions, or to unfavourable rock composition.

8.6 Timing of eclogite facies metamorphism

This section summarises recent geochronological results which throw light on the age of eclogite crystallisation and the cooling histories of the eclogite-bearing terranes. These data can also be used to reconstruct the relative timing of MT eclogite facies metamorphism in different tectonic units, providing important information on the dynamics of the orogenic belt.

Eclogites from the WGR give Sm-Nd mineral isochron ages (Omp–Grt or Omp–Grt–Wr) in the range 400 ± 16–447 ± 20 Ma (Griffin and Brueckner, 1985; Mearns, 1986; Mørk and Mearns, 1986), with an average of 413 Ma. Similar ages come from U-Pb dating of metamorphic zircons from eclogites, which give 401 ± 20 Ma (T.E. Krogh *et al*, 1974) and $446 + 54/-65$ Ma (Gebauer *et al*, 1985). These dates were considered to be close to the crystallisation ages of the eclogites.

Rb-Sr and K-Ar ages of hornblendes and micas from eclogites and gneisses in the WGR tend to be slightly younger than the Sm-Nd eclogite ages, as they cluster around 385 Ma (see compilation of Kullerud *et al*, 1986). However, the error brackets of these two groups of ages often overlap, so that this difference may not necessarily be real. The dated samples come from widely-spaced localities, so it is instructive to review the results from a more restricted area in order to minimise the effects of spatial variation. In the Statlandet area on the west coast eclogites give Omp–Grt Sm-Nd isochron ages of 423 ± 12 and 447 ± 20 Ma (Griffin and Brueckner, 1985). An Omp–Grt–WR Sm-Nd isochron for an eclogite in nearby Almklovdalen gave 408 ± 8 Ma (Mearns, 1986). Lux (1985) produced a ^{40}Ar–^{39}Ar plateau age on a hornblende of 410 ± 1 Ma, and a ^{40}Ar–^{39}Ar plateau age on a biotite of 375 ± 6 Ma; both samples were amphibolitised eclogites. Rb-Sr WR–Bt ages from Stadlandet gneisses are 374 ± 2 Ma (E.W. Mearns, quoted in Kullerud *et al*, 1986) and 378 ± 22 Ma (Brueckner, 1972, recalculated by Kullerud *et al*, 1986).

The data from Stadlandet show a clear spectrum of ages defined by a range of isotopic systems (Figure 8.5). Using the closure-temperature concept, this spectrum can be used to define the cooling history of the area. Peak metamorphic conditions were about 700°C, 20 kbar (Carswell *et al*, 1985), corresponding to the Sm-Nd ages. The hornblende K-Ar age of 410 Ma corresponds to a closure temperature of about 500°C (Harrison, 1981). K-Ar and Rb-Sr biotite systems both closed at about 300°C (Dodson, 1979) at about 375 Ma.

The lowest-temperature part of the cooling curve is poorly constrained. Cuthbert *et al* (1983) considered that the Middle Devonian sandstones and conglomerates of the west coast record the surface exposure of the WGR, but the Devonian sediments show only tectonic contacts with the WGR, and the conglomerates lack any input from a WGR-type source (Cuthbert, in preparation). Post tectonic pegmatites that cut some eclogites give mineral ages as young as 345 ± 2 Ma, indicating that at least part of the

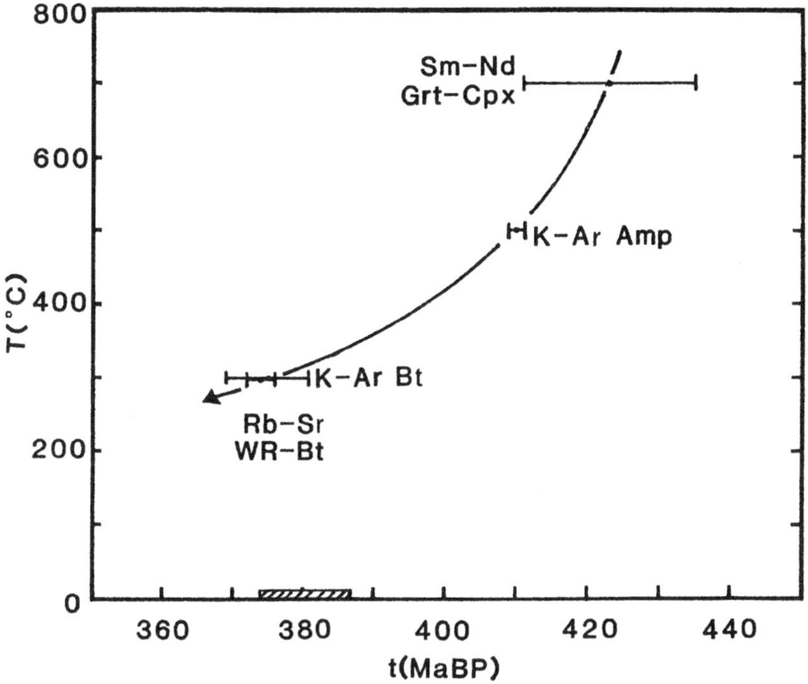

Figure 8.5 Temperature–time curve for eclogites and gneisses in the outer Nordfjord area, WGR, based on closure temperatures of mineral isotopic systems (see text for details of closure temperatures). Sm-Nd Grt-Cpx from Griffin and Brueckner (1985), K-Ar Hbl and Bt from Lux (1985), Rb-Sr Wr-Bt from Brueckner (1972) as recalculated by Kullerud *et al* (1986). Shaded area is approximate age range for Middle Devonian, when Old Red Sandstone clastics were deposited (after Harland *et al*, 1982).

WGR was still at temperatures high enough to cause partial melting in the early Carboniferous (Griffin, 1987).

Grt–Omp–WR Sm-Nd isochrons for eclogites from two terranes in the Seve thrust complex, central Sweden (Figure 8.1) give essentially identical ages of 505 ± 18 Ma and 503 ± 14 Ma respectively (Mørk *et al*, 1988) which were considered to date the peak of HP metamorphism. These eclogites are, therefore, significantly older than the WGR eclogites. ^{40}Ar–^{39}Ar dating of hornblende from amphibolitised eclogites in the same area (Dallmeyer and Gee, 1986) gives plateau ages of 491 ± 8 Ma. corresponding to temperatures of about 500°C, and phengites from host schists give plateau ages of about 440 Ma, with an estimated closure temperature of about 400°C. ^{40}Ar–^{39}Ar mica ages in other parts of the Seve and the overlying Köli terranes of the oceanic thrust sheets are in the range 425–410 Ma, recording post-metamorphic cooling that followed final thrusting of the assembled terranes over the Baltic foreland (Dallmeyer, 1988).

The Bergen Arcs anorthosite has yielded a Sm-Nd Omp–WR–Phe isochron age of 507 ± 109 Ma from an eclogitic shear zone (Cohen *et al*, 1988). This overlaps with the error brackets of Sm-Nd ages from the WGR, so these ages may not be significantly different. However, the phengite has given a K-Ar age of 443 ± 5 Ma (Cohen *et al*, 1988), which lies at the extreme oldest end of the WGR Sm-Nd ages. If the K-Ar result represents a cooling age it should postdate eclogite crystallisation, so the eclogitic shear zone development in the Bergen Arcs would, on this basis, be older than the eclogite facies metamorphism in the underlying WGR and more similar to the age of the Seve eclogites.

In the Tromsø Nappe Complex, a preliminary Sm-Nd Grt-Omp-Amp isochron gives an age of 598 ± 107 Ma (Krogh *et al*, in press). This was interpreted to correspond to T_{max} during the D1/M1 eclogite facies event. The wide error brackets mean that it is not significantly different from the Sm-Nd ages for the Seve eclogites (Mørk *et al*, 1988). However, the tectonostratigraphic position of the TNC above the outboard terranes of the oceanic thrust sheets makes a common origin for its eclogites and those of the Seve unlikely.

In summary, the lowest structural level (WGR) has the youngest eclogite ages, at about 415 Ma (late Silurian—Harland *et al*, 1982), while eclogite-bearing terranes at higher structural levels had cooled to below mica argon-closure temperatures by this time. If the 598 Ma age for the TNC is real, then it may be that eclogite formation occurred at progressively earlier times in successively higher tectonostratigraphic units. Both cooling ages for thrusted units, and eclogite Sm-Nd ages in the underlying WGR, are broadly coeval with Scandian thrusting of the Caledonian thrust sheet assembly across the Baltic foreland to the east (Gee, 1975; Hossack *et al*, 1985).

8.7 Pressure–temperature–time paths

Selected typical T and P estimates for eclogites in the Scandinavian Caledonides are given in Table 8.1.

Temperatures fall within the range 550–800°C, and pressures range from 12.5–20 kbar. In the WGR this range can be found as an apparently uninterrupted gradient across the terrane (Krogh, 1977; Griffin *et al*, 1985; figure 3b) from 550°C in the SE and E to > 800°C in the NW. In the Seve thrust complex of the Baltic crystalline sheets, however, adjacent thrust bounded lithotectonic packages may display markedly different conditions from each other.

In the classification adopted in this book all the eclogites are of medium-T type, even those bearing glaucophane. Lawsonite is unknown in eclogites on the Scandinavian mainland, and kyanite and zoisite are common, which is consistent with T > 500°C (Newton, 1986). Pressures are often minimum estimates based on the coexistence of omphacite with quartz or kyanite in the absence of plagioclase or paragonite. Unique pressures have been defined only for Opx eclogites and HP granulites with Ky + Pl + Grt + Qtz. In two cases (Eiksunddal and the TNC) an early P_{max} at $T < T_{max}$ has been inferred from relict assemblages (Jamtveit, 1987a; Krogh *et al*, in press).

Detailed reconstructions of *P–T–t* paths for Scandinavian eclogites have tended to concentrate on the WGR, and the Statlandet/outer Nordfjord area can be returned to for an example. Evidence for the prograde *P–T* path comes from zoned garnets with mineral inclusions in the Verpeneset kyanite eclogite, which show a change from Fe-Mn rich cores with inclusions of hornblendic amphibole, phengite, quartz, paragonite and zoisite, to a pyrope-rich rim with omphacite and more actinolitic amphibole inclusions (Krogh, 1982). Using Krogh's amphibole and garnet analyses (sample VP-1), the garnet– hornblende geothermometer of Graham and Powell (1984) gives temperatures of 611–652°C for inclusions in both core and rim zones of the garnet. Garnet–phengite

Table 8.1 Typical P–T estimates for Scandinavian eclogites.

Unit*	Locality	T°C	P kbar	Methods†	Source	Comments
WGR	Kvineset Sunnfjord	540 ± 35	12.5 ± 2.5	EG, H80	Krogh (1980)	Gln-eclogite P_{min}
WGR	Dalsfjord Sunnfjord	597 ± 30	16.2	EG, H79	Cuthbert (1985)	Ky-eclogite P_{min}
WGR	Verpeneset Nordfjord	650–700	14–16	EG, H79	Krogh (1982)	Ky-eclogite P_{min}
WGR	Gryttingvåg Statlandet	674–684	18–20	EG, W	Carswell *et al* (1985)	Opx eclogite
WGR	Eiksunddal Sunnmøre	ca. 750	ca. 20	EG, HL84a, b	Jamtveit (1987)	P_{max} 25 kbar at 600°C
WGR	Solholmen Nordmøre	760	18–20	EG, W	Carswell *et al* (1985)	Opx-eclogite
BA	Bergen	750 ± 50	16–19	EG, GL	Austrheim and Griffin (1985)	shear zone P_{min}
STC	Central Belt Jamtland	780 ± 50	18 ± 1	EG, GL	van Roermund (1985)	P_{min}
STC	Eastern Belt Jamtland	550 ± 70	14.0 ± 1.5	EG, GL	van Roermund (1985)	P_{min}
STC	Vaimök Lens Norbotten	735 ± 95	ca. 19	EG, H80	Santallier (1988)	P_{min}
STC	Tsakkök Lens Norbotten	610 ± 90	14.9 ± 1.5	EG, GL	Stephens and van Roermund (1984)	Gln-eclogite P_{min}
TNC	Tromsø Norway	730 ± 30	15.7 ± 0.7	K, IM, MNH H79, 80	Krogh *et al* (in press)	Eclogite and pelite. Early P_{max} 650°C, 17 kbar

*Place names; WGR, Western Gneiss Region; BA, Bergen Ares; STC, Seve thrust complex; TNC, Tromsø Nappe Complex.
†Geothermobarometry methods; EG, Ellis and Green (1979) $K_D^{Fe/Mg}$ Grt-Cpx; H80, Holland (1980) Jd in Cpx with Qtz; H79, Holland (1979) Jd in Cpx with Ky; GL, Gasparik and Lindsley (1980) Jd in Cpx with Qtz; W, Wood (1974) Al in Opx + Grt; HL84a, Harley (1984a) $K_D^{Fe/Mg}$ Grt-Opx; HL84b, Harley (1984b) Al in Opx + Grt; K, Krogh (1988) modified method of Ellis and Green (1979); IM, Indares and Martignole (1985) $K_D^{Fe/Mg}$ Grt-Bt; MNH, Newton and Haselton (1981) Grt–Pl–Ky–Qtz as modified by Krogh *et al* (in press).

thermometry (Green and Hellman, 1982) shows little change in T between a phengite inclusion in the core and a matrix phengite (688 and 669°C, $Fe^{2+} = Fe^{total}$). Pressures are poorly constrained, although the possible presence of paragonite in the garnet cores implies $P > 5$ kbar at 600°C (Holland, 1979), and Pg + Zo + Qtz inclusions in an intermediate part of the garnet grains indicate $P > 9$ kbar (Franz and Althaus, 1977).

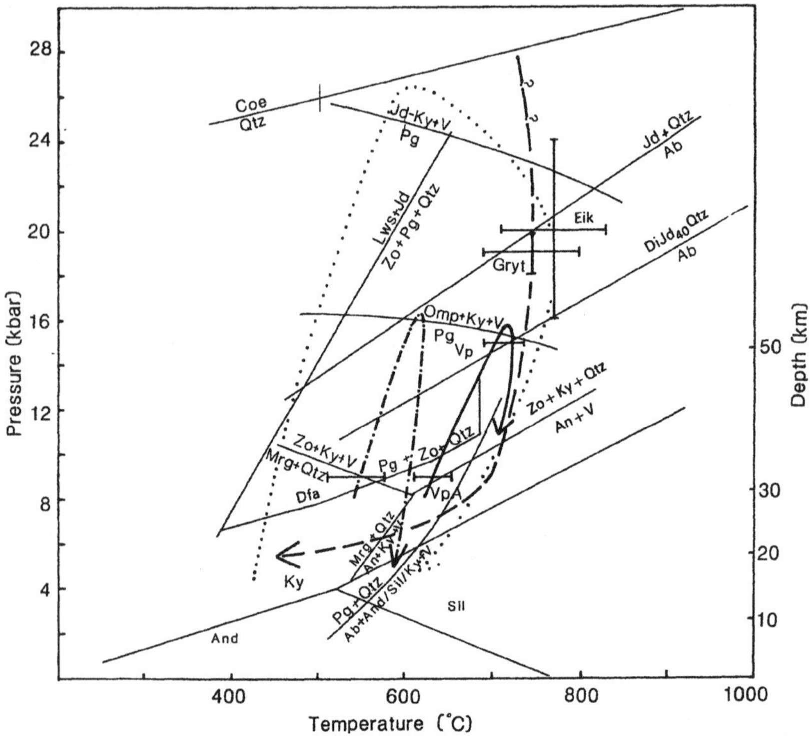

Figure 8.6 Pressure–temperature plot showing important metamorphic reaction curves for medium-T eclogites of the Scandinavian Caledonides. ———— P–T loop for the Verpeneset eclogite (from data of Krogh, 1982). — — — — retrogressive path for the Seljeneset meta-anorthosite, WGR from Cotkin *et al* (1988), extended up-pressure to account for the possible presence of coesite in the nearby eclogite at Grytting, Selje district (Smith, 1984). ·——·——·—— P–T loop for eclogites from outer Dalsfjord, Sunnfjord (from Cuthbert, 1985). Brackets marked 'Eik' define the metamorphic maximum for the Eiksunddal eclogite and ..... the deduced P–T loop (Jamtveit, 1987a). 'Gryt' is the best estimate for maximum P and T from geothermobarometry on the Gryttingvåg eclogite, from Carswell *et al* (1985). 'Vp' is the range of temperatures from Grt-Cpx geothermometry (Ellis and Green, 1979) on the Verpeneset eclogite, at an estimated $P = 15$ kbar (Krogh, 1982). 'VpA' and 'DfA' are the ranges of temperatures estimates from Grt–Hbl geothermometry (Graham and Powell, 1984) on hornblende inclusions in garnet cores in the Verpeneset and Dalsfjord eclogites, respectively, at an arbitrary $P = 9$ kbar (data from Krogh, 1982 and Cuthbert, 1985). Coesite⇌quartz curve from Bohlen and Boettcher (1982); Ab + Jd + Qtz, and Ab + $DiJd_{40}$ + Qtz from Holland (1980); Pg + Jd + Ky + V, Pg + Omp + Ky + V, Pg + Qtz + Ab + And/Sil/Ky + V and Zo + Pg + Qtz + Lws + Jd from Holland (1979); Mrg + Qtz + Zo + Ky + V, An + V + Zo + Ky + Qtz and Mrg + Qtz + An + Ky + V from Jenkins (1984); minimum stability of Pg + Zo + Qtz from Franz and Althaus (1977); Al-silicate phase boundaries from Holdaway (1971).

The matrix paragenesis of Omp + Grt + Ky + Zo + Phe + Qtz gives equilibration temperatures of 650–700°C at P_{min} 14–16 kbar (Krogh, 1982). The resulting prograde path is very steep in the *P–T* plane, as shown in Figure 8.6.

The retrograde segment of the *P–T–t* path has been resolved in detail by Cotkin *et al* (1988) from a meta-anorthosite at Seljeneset, 8 km NW of Verpeneset. Textural relationships allowed recognition of an early HP Ky + Zo + Qtz assemblage which subsequently underwent a number of reactions to produce plagioclase, quartz and margarite. Modelling these in the CASH system revealed a steep, low *T/P* trajectory from > 8.2 kbar, 800°C to 6 kbar, 610°C, with further, more rapid cooling to about 450°C at > 4.3 kbar (Figure 8.6). According to the results of Lux (1985) for nearby eclogites, cooling through 500°C occurred at about 410 Ma.

The very steep prograde and retrograde *P–T–t* paths from the outer Nordfjord area, then, define a hairpin-shaped *P–T* loop dominated by changes in P rather than T (Figure 8.6). Results from other parts of WGR show loops with similar form, including Dalsfjord in Sunnfjord, where some of the lowest-T eclogites in the WGR are found (Cuthbert, 1985).

Thermal modelling of overthrust terranes (e.g. England and Richardson, 1977; England and Thompson, 1984) has predicted a period of heating during the earliest stages of decompression. The maximum pressure recorded in the rock may then be that experienced at T_{max}, largely overprinting any mineral assemblages formed at P_{max}. Simple exchange geothermometers may be particularly sensitive to such resetting. However, some evidence can be gleaned from the WGR for the form of this critical part of the *P–T–t* path.

P–T estimates for omphacite inclusions and their host garnet can be compared with those for matrix omphacites and garnet rims. Using the garnet–clinopyroxene geothermometer of Ellis and Green (1979) and the jadeite content of omphacite in the presence of quartz (e.g. Holland, 1980), the Verpeneset eclogite shows no significant *T*-difference between inclusions and matrix (Krogh, 1982). The calculated temperature of about 700°C is close to the estimated lower limit for effective volume diffusion in garnet (Section 8.4.3), so that any large T increase following the cessation of garnet growth is ruled out. Hence in this example T_{max} probably corresponds to P_{max}, and the rock followed an essentially isothermal uplift history from its greatest depth. Unfortunately only minimum estimates of P_{max} can be made for this example. However, the hairpin-shaped Verpeneset *P–T–t* path (Figure 8.6) is at variance with the model paths derived by England and Richardson (1977) and England and Thompson (1984).

In the Eiksunddal eclogite, Jamtveit (1987a) defined a ‘metamorphic maximum’ of 750°C, 20 kbar (Table 8.1). However, a sodic amphibolite containing relicts of jadeitic pyroxene ($Di_{30}Jd_{70}$) and kyanite indicated P_{max} > 22–24 kbar at 650°C (Holland, 1979). *T*-estimates on omphacite inclusions in some garnets showed a *T*-increase from about 660–740°C from core to rim, and zoning of early (retrogressive) amphibole also suggested a period of *T*-increase, accompanied by decompression. Jamtveit (1987a) constructed a *P–T-t* loop (Figure 8.6) with a steep positive prograde trajectory to around 25 kbar, 650°C followed by decomposition with heating to 20 kbar, 750°C then further steep decomposition into the sillimanite field in the amphibolite facies.

It is not clear how the contrasting results from Verpeneset and Eiksunddal can be reconciled. It is possible that the two localities had different thermal histories (it might be surmised that decompressional heating only took place at Eiksunddal owing to a longer uplift period or closer proximity to a heat source). However, the authors consider that some uncertainties remain in determining decompressional P–T paths. For instance, Jamtveit (1987a) interpreted the zoning in the early amphiboles to be a result of breakdown of jadeitic clinopyroxene, yet his amphibole analyses show a decrease in total Na from core to rim. On the other hand, the lack of useful limiting assemblages in eclogites such as that at Verpeneset precludes precise determination of P_{max} even when an isothermal decompression can be clearly demonstrated.

The recognition of coesite in an eclogite pod from Grytting on Statlandet by Smith (1984) may have important implications for P_{max} on the P–T–t loop. The coesite was found as small inclusions in omphacite grains. Geothermobarometry by Carswell *et al* (1985) on an adjacent Opx eclogite lens (methods of Ellis and Green, 1979 and Wood, 1974) gives about 680°C and 18–20 kbar, but at the same temperature coesite is stable at $P > 27$ kbar (Figure 8.6) (Bohlen and Boettcher, 1982). This suggests that P_{max} may have been significantly higher than the recorded pressure at T_{max}.

These very high values for P_{max} require that the paragenesis jadeite + quartz should have been stable in quartzo-feldspathic rocks, but to date it has not been reported. It is possible that it has not been preserved for kinetic reasons. More importantly, perhaps, kinetic factors such as lack of deformation or fluid activity may have prevented formation of jadeite + quartz in the first place (cf. the Bergen Arcs anorthosites—Section 8.4.3). In such circumstances it might be expected that the reaction curve for Ab = Jd + Qtz may have to be grossly overstepped before the reaction could proceed. There is some petrographic evidence that significant volumes of the gneisses in the WGR were derived from dry granulite facies precursors (Bryhni, 1966: Cuthbert, 1985), so that fluid activities may well have been low at the onset of eclogite facies metamorphism.

It is evident from the foregoing discussion that an important priority for future work must be to seek out products of pressure-sensitive equilibria which can better constrain the relationships between P_{max} and T_{max}, hence providing a firmer basis for comparison with theoretical models.

8.8 Tectonic models

Tectonic models for the evolution of the MT eclogites in the Scandinavian Caledonides must also take into account the distribution of eclogites in the orogen, eclogite facies metamorphic conditions and their spatial variation, timing of eclogite facies metamorphism in different lithotectonic units, palaeogeography and palinspastic restorations of the orogen, and the present-day crustal structure.

The first attempt at a tectonic model for eclogite formation was by Krogh (1977), who proposed that the regional northwestwards increase in T across the WGR was the result of continental collision leading to its transient westwards subduction. Krogh considered that this event was of Proterozoic age, but the derivation of late Silurian mineral ages, by Griffin and Brueckner (1980) and Krogh *et al* (1974), made it apparent that eclogite formation was related to the Caledonian orogeny. Gee (1975) had independently concluded that subduction of Baltic continental crust had occurred during the Caledonian orogeny, based on regional tectonostratigraphic arguments.

Lappin and Smith (1978) also suggested a continental collision scenario for the WGR, but preferred to regard the eclogites as foreign inclusions in the gneisses. They proposed a mechanism of deep-level obduction to introduce eclogites generated at mantle depths into higher crustal levels. Smith (1980b) elaborated on this theme, describing the WGR as a 'gigantic mega-mélange' of lithologies from various crustal levels. However, the relative continuity of lithological units in the WGR is at variance with the accepted use of the term 'mélange', and the gradual change in eclogite palaeotemperatures across its outcrop indicates that it was largely a coherent body during eclogite facies metamorphism. Smith's concept may, however, be more applicable to the Seve complex, which consists of numerous thrust-bounded packages of different metamorphic grades. The Seve eclogites themselves, however, are not tectonic inclusions in their own country-rocks (Andreasson *et al*, 1985).

The continental subduction model was preferred by Griffin *et al* (1985), who were persuaded by the now extensive evidence that the eclogites have been effectively metamorphosed *in situ*. This mechanism was examined in detail by Cuthbert *et al* (1983), who incorporated refraction seismic data from Mykkeltveit *et al* (1980) which revealed a crustal low-velocity layer below the WGR. Cuthbert *et al* proposed that following subduction of the leading edge of the Baltic continental margin (now the WGR) below the Laurentian margin (Figure 8.7c, d), the locus of subduction then broke back into the Baltic foreland and Baltic crust was thrust underneath the WGR along a plane now marked by the low-velocity layer. This further period of underthrusting aided uplift and exposure of the eclogites at the surface above a normal thickness of continental crust. It was shown with this model that erosion-controlled isostatic uplift was sufficient to bring the eclogites to the surface by the middle Devonian.

Medaris and Wang (1986) have generated simple thermal models based on the Cuthbert *et al* (1983) model. For a level that corresponds to the present land surface, they predicted a period of heating after P_{max}, abruptly terminated by rapid cooling as suture progradation commences, bringing cold upper crust below the WGR. Medaris and Wang found the thermal models to be consistent with models for mineral zoning in ultramafic rocks of the WGR.

Carswell and Cuthbert (1986) developed a general tectonic model for eclogite facies metamorphism in continental crust in which multiple subduction is achieved by continental collision, followed by telescoping of the underthrust plate along low-angle crustal fractures. The model predicts that each time crust is subducted it drives the uplift of previously subducted, eclogite-bearing crust, and that the age of eclogite facies metamorphism will tend to increase with tectonostratigraphic height in the orogen.

This sequential stacking model may be broadly applicable to the Scandinavian Caledonides, as illustrated in Figure 8.7. Eclogites were generated during collision of the Baltic margin (Seve thrust complex, ? Bergen Arcs anorthosite) with an outboard arc complex during the Early Ordovician (Dallmeyer and Gee, 1986). Subsequently, subduction of more inboard Baltic crust (including the WGR) occurred below this assembly of thrust sheets in the Middle to Late Silurian (Gee, 1975; Griffin and Brueckner, 1980, 1985) and caused their final translation and uplift. Finally, transfer of thrusting to a new, more external shear-zone that cut through the Baltic foreland crust caused further crustal subduction below the WGR, driving its uplift during the Devonian (Cuthbert *et al*, 1983; Gee, 1987) and possibly into the Carboniferous. If the 600 Ma age for the TNC eclogites in the Exotic Nappes is real, this adds yet another, earlier cycle to the sequence, although strictly not related to the telescoping of the Baltic margin as it possibly took place near Laurentia (Stephens *et al*, 1985).

Crustal extension or 'collapse' may be an efficient mechanism to bring HP rocks to the surface (Platt, 1986). Some evidence now suggests that extensional tectonics may have been important during the uplift of the WGR. Hossack (1984), Seranne and Seguret (1987) and Norton (1986) demonstrated that the Middle Devonian Old Red

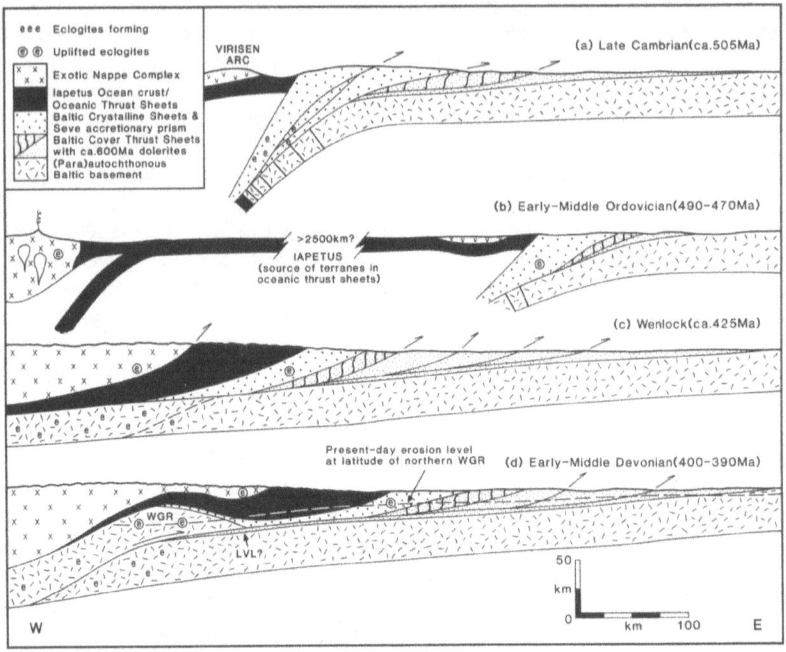

Figure 8.7 Possible environments for the formation of eclogites in the Scandinavian Caledonides, projected along strike to a NW–SE profile through the northern Western Gneiss Region. Based on Cuthbert *et al* (1983); Dallmeyer and Gee (1986); Gee (1987); and Krogh *et al* (in press). (a) Following west-directed subduction, Baltica collides with an arc complex (Virisen arc of Dallmeyer and Gee, 1986). Eclogites form in an accretionary prism of metasediments and fragments of Baltic crust which now comprise the Seve thrust complex, allocated here to the Baltic crystalline sheets. Possibly penecontemporaneous Bergen Arcs eclogites may have formed under similar circumstances. (b) Early to mid-Ordovician uplift and erosion of the thickened Baltic margin. Outboard of Baltica, subduction of Iapetus occurs below a major continental magmatic arc (Laurentia?—Stephens, 1988). The eclogite bearing Tromsø Nappe Complex may have formed part of this edifice, with eclogite formation having occurred as early as 600 Ma. These units now form the Exotic Nappe Complex. Terranes found in the oceanic thrust sheets were generated at various locations in Iapetus during this period (Stephens and Gee, 1985; Stephens, 1988). (c) Collision of Baltica with Laurentia. Assembly of the oceanic thrust sheets with the Baltic crystalline and cover sheets occurs to form the Scandian allochthon during the Wenlock. Baltica is subducted below the allochthon and the leading edge of Laurentia, generating eclogites in what now forms the Western Gneiss Region (WGR). (d) A major shear-zone forms in the Baltic crust, probably in the early to middle Devonian. Subduction transfers to this new thrust-plane below the eclogite bearing WGR, giving impetus for its rapid uplift. The shear-zone may be marked by a seismic low-velocity layer (LVL), possibly due to incorporation of part of the Baltic sedimentary cover (Mykkeltveit *et al*, 1980; Cuthbert *et al*, 1983) or the generation of mylonites.

Sandstone basins of western Norway are bounded by major low-angle normal faults or detachments associated with thick mylonites. Norton (1986) and McClay *et al* (1986) proposed that 'collapse' of the orogen along this detachment was a result of the great crustal thickening implicit in the *P*-estimates from the eclogites. In this model uplift of the eclogites was aided by removal of its overburden along the detachment, and the WGR is analogous to 'metamorphic core complexes' (Coney and Harms, 1984). Basin formation and related exhumation of the WGR may, then, be a result of the same crustal thickening processes which were responsible for the formation of the eclogites.

8.9 Concluding remarks

Eclogite formation in the Scandinavian Caledonides was widespread and took place during at least two episodes in the evolution of the orogen. It is now clear that these MT, 'type-B' eclogites are not 'geologically out of place', but shared their entire metamorphic history with the high-grade gneisses in which they lie. The deeply eroded Scandinavian Caledonides are an important example of the deep architecture of a continental subduction zone, and provide an opportunity to study a level not accessible in modern analogues such as the Himalayas. The dynamics of these deep levels are recorded in the petrology of the MT eclogites, making them especially useful sources of information for understanding the processes of continental subduction.

Acknowledgements

This chapter has benefited from a constructive review by D.G. Gee. The authors are grateful to Mai-Britt Mørk and Erling Krogh for providing manuscripts of papers before their publication. Funding for studies on Norwegian eclogites by SJC has been provided through a Studentship and a Fellowship from the Natural Environment Research Council.

9 Eclogite formation and distribution in the European Variscides

P.J. O'BRIEN, D.A. CARSWELL and D. GEBAUER

9.1 Introduction

The late Palaeozoic Variscan (Hercynian) orogeny, caused by the collision of Laurasia and Gondwana to form Pangea, resulted in the formation of an arcuate fold-belt that covered parts of Central and Southern Europe, the southern British Isles, the east coast of North America and the north-east coast of Africa. Coherent studies of the European segment of the Variscan belt have been hampered by the poor degree of exposure due to extensive Meso–Cenozoic cover, widespread post-collisional wrench faulting, disruption by the younger Alpine orogeny and, not least of all, by the presence of many national boundaries, which result in publications in many languages. In addition, the plethora of orogenic phases of varying extent—Sardic, Acadian–Ligerian, Reussic, Bretonic, Sudetic, Asturic, Saalian—is also confusing to the uninitiated. At present the belt can be followed in a number of isolated massifs which from west to east are: the Iberian Massif covering part of Spain and Portugal; the Armorican, Central and Vosges Massifs in France; the Schwarzwald in West Germany and the Bohemian Massif comprising parts of West Germany, East Germany, Poland, Czechoslovakia and Austria (Figure 9.1). Eclogites and related rocks will be reviewed from all the above named massifs. However, HP Pre-Alpine parageneses are also known in the Variscan basement of the Alpine chain, some of which are covered in Chapter 10.

A division of the Variscan belt into distinct zones that exhibit marked differences in their sedimentary, structural and metamorphic histories originated with Kossmatt (1927). This zonal subdivision into the Rhenohercynian (RZ), Saxothuringian (SZ) and Moldanubian (MZ) Zones (Figure 9.1) can be traced essentially across the whole of the Variscides (albeit with different local names), see for example Ellenberger and Tamain (1980). The tectono-metamorphic history of these zones and consequently of the Variscan belt as a whole has been a contentious issue for a number of decades. Models can be classed as either fixistic or mobilistic. To those favouring a fixistic evolution, the Variscan belt is seen as a number of fundamental blocks, separated by deep faults, and having the character of either a Proterozoic–Palaeozoic sedimentary trough or a crystalline rise (e.g. Krebs and Wachendorf, 1973; Škvor and Zeman, 1969). These models negate the existence of oceanic crust, hence excluding subduction or collision type orogenies. Instead they explain the metamorphism by means of the vertical movement of buoyant basic magmas from upper mantle partial melting causing crustal anatexis and the ascent of granitic domes. Deep-seated faults occurred at the flanks of plutonic diapirs to form sedimentary basins, within which gravity-sliding of cover rocks and fault-related ultramafic, basaltic and keratophyric magmas are a feature,

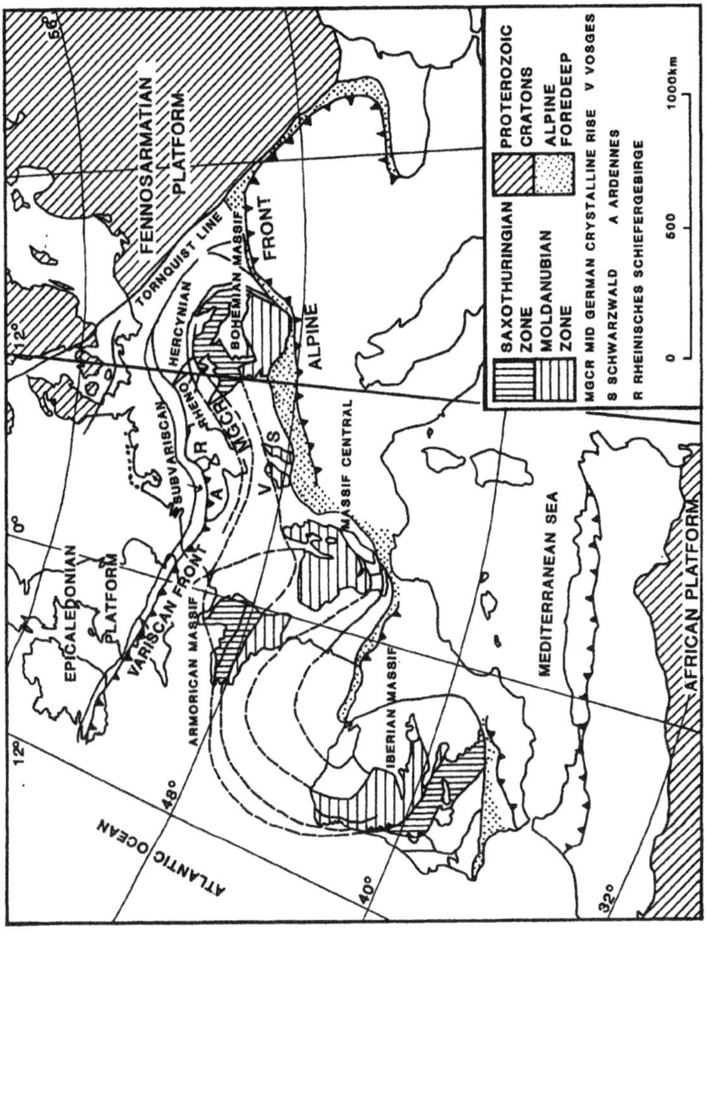

Figure 9.1 Variscan Europe (derived from Ellenberger and Tamain, 1980).

while shallow-water deposits developed above the domes. Variations in geologic history are related to differential vertical movements of the individual blocks. Metamorphism due to this tectonic scenario is of LP type (at 5–15 km depth, Zwart, 1969): the eclogites, garnet peridotites, kyanite granulites and kyanite gneisses are considered part of a pre-Variscan Assyntian (Cadomian) event (Zoubek, 1969; Suk, 1969).

In contrast, the mobilistic school presents evidence for a horizontal evolution, which typifies a classical collision belt, involving intracrustal thrusts, supracrustal nappes, wrench faults and considerable crustal shortening (e.g. Arthaud and Matte, 1977; Behr *et al*, 1980, 1984; Matte, 1986; Ries and Shackleton, 1971; Ziegler, 1984). A tectonic layering of the crust has been deduced from geophysical studies and major thrust planes at deep and shallow levels are inferred from seismic reflection profiles across all zones of the Variscan belt (Behr *et al*, 1984; KTB, 1986; Meissner, 1986; Cazes *et al*, 1985). Geochronological work has shown that the high grade metamorphic rocks generally were formed less than 480 Ma ago and hence are an integral part of the post-Pan-African development (Gebauer, 1983, 1986; Kröner *et al*, 1988; van Breemen *et al*, 1982). In addition, although the presence of true oceanic crust within the belt is disputed, rocks having the character or geochemistry of ophiolites or MORB are present at both low and high metamorphic grade (e.g. Bernard–Griffiths and Cornichet, 1985; Bodinier *et al*, 1986; Gebauer and Grünenfelder, 1982; Jelínek *et al*, 1984; Ménot *et al*, 1988; O'Brien 1989c; Paquette, 1987; Paquette *et al*, 1987; Pin *et al*, 1988; Stosch and Lugmair, 1987).

How can the study of eclogites contribute to an understanding of Variscan Europe? In the following, the petrology, geochemistry, metamorphism, geochronology and structural relations of eclogites and associated HP parageneses are presented massif by massif. In the light of this review of data—albeit selective, the significance of variations along and across the orogenic belt will be discussed in relation to possible tectono–metamorphic scenarios.

9.2 Spatial and temporal distribution and lithological associations of eclogites and associated HP mineral assemblages

9.2.1 *Bohemian Massif*

The Bohemian Massif, the easternmost block of the Variscan chain in Europe, covers parts of the MZ and SZ (Figures 9.1 and 9.2). The SZ to the north and northeast of the massif comprises a Late Proterozoic to Carboniferous volcano–sedimentary sequence only locally up to amphibolite facies grade (e.g. Mielke *et al*, 1979). The MZ is dominated by a monotonous unit of quartzo-feldspathic paragneisses—usually cordierite-bearing and often migmatitic—with minor intercalations of acid and basic orthogneisses and meta-quartzites. Localised subordinate varied units have a similar meta-psammitic and -pelitic series but with additional abundant intercalations of calcareous, graphitic, quartzitic and basic rocks (Fischer, 1968; Vejnar, 1971). The dominant metamorphism is of LP–HT type, which corresponds widely to the anatectic Crd-Sil-Kfs zone at conditions around 4.2 ± 0.4 kbar, 690 ± 50°C (e.g. Blümel, 1983). Granulites, a metamorphosed pile of acid volcanics and minor sediments (Fiala *et al*, 1987a), are closely associated with eclogites, pyroxenites and garnet peridotite and are generally considered as remnants of an allochthonous nappe (Fuchs, 1971; Tollmann,

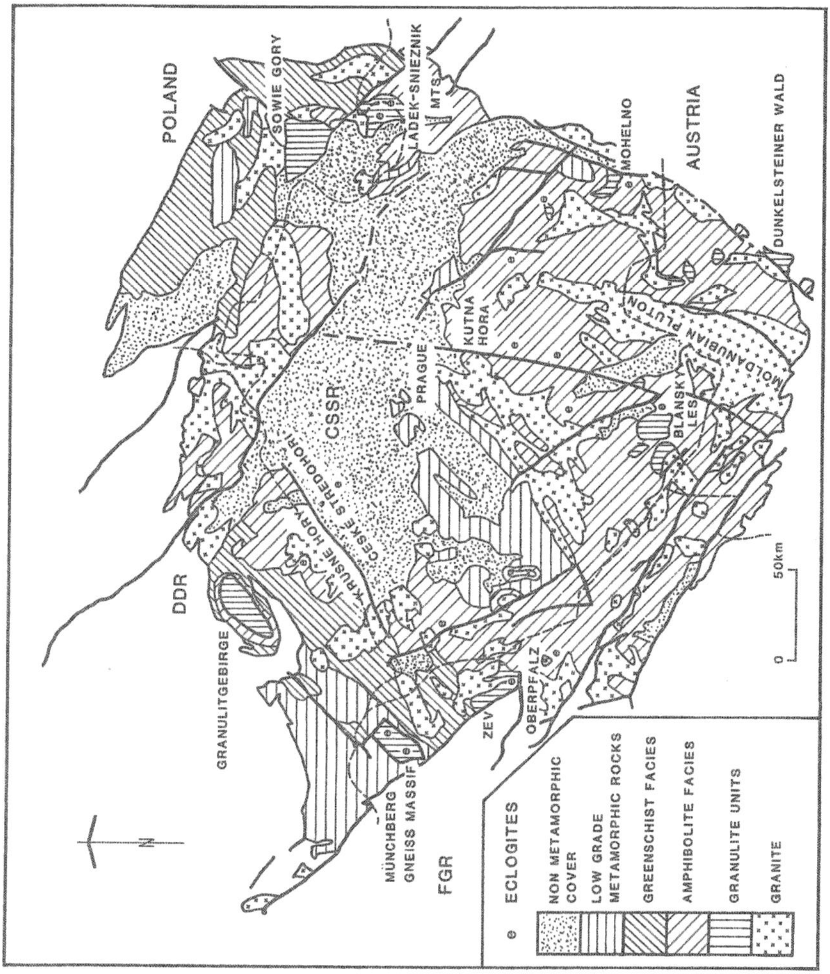

Figure 9.2 Simplified geology of the Bohemian Massif and main eclogite locations.

1982). Further klippen at the Bavarian margin are the Münchberg Gneiss Massif (MGM), the zone of Erbendorf–Vohenstrauss (ZEV) and the zone of Tepla–Taus (ZTT) (Behr *et al*, 1984; KTB, 1986).

Radiometric ages, especially from the Bavarian part of the Bohemian Massif, indicate that the detritus in the various metasedimentary sequences was largely derived from the non-Australian part of Gondwana with detrital components ranging from Pan-African (Cadomian or Assyntian) to Early Archean ages up to about 3.8 Ga (e.g. Gebauer *et al*, 1989). There are indications that Late Proterozoic rocks of oceanic (possibly island arc) origin have been tectonically emplaced into the sedimentary or metasedimentary precursors of the present high-grade metamorphic host rocks (Quadt and Gebauer, 1988). At least for the Austrian and German part of the MZ no rocks of Pan-African magmatic and metamorphic cycles have been detected unequivocally. Therefore it seems that most of this basement was covered both stratigraphically and tectonically during post-Pan-African depositional and collision cycles (Gebauer *et al*, 1989). For the Bohemian Massif the first well established data on compressional events fall into the Early Ordovician followed by events in the Silurian, Devonian and finally the Carboniferous. Eclogite-forming events are known from the Silurian and Devonian periods.

The *Münchberg Gneiss Massif* (MGM) in the NW of the Bohemian Massif, long considered as a protruding basement diapir, is now regarded as an allochthonous stratigraphically and metamorphically inverted nappe-pile resting on a very low-grade Saxothuringian Zone Palaeozoic sequence (e.g. Behr *et al*, 1980). The uppermost nappes are subdivided into the upper *Hangend Serie* of hornblendic gneisses, amphibolites and carbonate-rocks in ortho- and paragneisses and an underlying *Liegend Serie* of monotonous meta-clastics with some granulitic gneisses, metamorphosed acid and basic intrusions and intercalations of serpentinite (Stettner, 1960). Eclogites occur at the top of the *Liegend Serie* or as tectonic inclusions in the lowermost *Hangend Serie*. Eclogites from the locality Weissenstein yield ages around 380 Ma (Gebauer and Grünenfelder, 1979a; Stosch and Lugmair, 1987). Underlying units represent slivers of a basinal sequence and nappe-derived clastics telescoped beneath the advancing nappe pile. Microstructural studies have revealed a series of deformational features dominated by low-angle mylonitisation attributed to large-scale horizontal tectonics (Behr *et al*, 1980, 1984).

Eclogites, the subject of a systematic study in the 1960s and 1970s (see reviews in Matthes and Schmidt, 1974; Matthes, 1978), have been split into dark Ky-free and light Ky-bearing varieties. Recent work suggests oceanic MORB and continental crust-related cumulate sources for dark- and light-types respectively (Stosch and Lugmair, 1987) but this is inconsistent with the intimate interlayering of the two eclogite types and gneisses known from a borehole log (Matthes *et al*, 1974). All eclogites contain Omp + Grt + Qtz + Rt with varying proportions of Ky, Phe, Zo and Amp: Omp, Ky and Amp can occur in poikiloblastic forms. The derivation of some eclogites from gabbros can be deduced from honeycomb-textured garnet aggregates and inclusion patterns. Garnet aggregates show a domainal chemical zoning (Knauer, 1975) but euhedral single crystals can show high values of Ca, Fe and Mn in cores which decrease towards the rims as Mg increases (Schmidt, 1967).

A multi-stage metamorphic history is evident. LP, possibly rift-related, basic crustal protoliths were transformed to eclogites during HP metamorphism under initially anhydrous and later hydrous conditions and were then retrogaded during decompression into the amphibolite facies (Franz *et al*, 1986). Typical breakdown features are Amp + Pl kelyphites around Grt, Cpx + Amp + Pl and later Amp + Pl after Omp, Bt + Pl around Phe, Pg rims on Ky and Ttn after Rt. $K_{D_{Fe-Mg}}^{Grt-Cpx}$ values are in the range 9–13 indicating temperatures little more than 650°C for the eclogite stage at $P > 15$ kbar. Phengite geothermobarometry corroborates these values (Franz *et al*, 1986) which are significantly higher than the 5–6 kbar, 550 ± 50°C proposed by Matthes *et al* (1975) in order to match the equilibration conditions of eclogites and their St- and Ky-bearing host rock gneisses.

Another interesting rock in the MGM is a coronitic metagabbro ranging from a partially altered Opx + Cpx + Pl + Bt + Ilm assemblage to one with thick Grt coronas between Pl and mafic phases, Rt + Grt (+ later Ttn) on Ilm, Zo + Ab + white mica pseudomorphing basic Pl and Amp aggregates replacing pyroxenes (Matthes and Seidel, 1977).

In the *Oberpfalzer Wald*, retrograded eclogites associated with serpentinised peridotites have been documented from localities predominantly around Oberviechtach (Busch, 1970). At an important exposure SW of Winklarn, near Oberviechtach, serpentinites show a tectonic contact to metabasites marked by a band of Tr–Act rock and metabasites are interlayed and folded with anatectic paragneisses and Mg-rich, alkali-poor garnet pyroxenites. Thin bands of an amphibolite-facies Ol + Opx + Amp + Spl assemblage are sometimes preserved in the serpentinite but, aluminium enrichment at the rims of Cr-Spl and Opx may be indicative of the breakdown of garnet belonging to an earlier HP assemblage (O'Brien, 1989c). Typical metabasites, corresponding geochemically to N-type MORB tholeiites (O'Brien, 1989c), originally contained Grt + Omp + Rt ± Qtz ± Ky ± Zo but this is never retained intact. Garnets are prograde zoned with thin retrograde rims and are usually rimmed by a radial kelyphite of Amp + Pl ± Spl: rarer are earlier kelyphites of Opx + Pl. Omp, normally only preserved as 'armoured relicts' in garnet, breaks down to a vermicular Di + Pl symplectite later replaced by diablastic Amp + Pl aggregates. Qtz grains adjacent to Grt have reacted to give a corona sequence (Grt)–Pl–Di–Hy–(Qtz), all Ky has been replaced by Hc + An masses, Zo by An and Rt reacts to form Ilm and later Ttn. The textures indicate a multi-stage metamorphic history probably starting in the amphibolite facies (low Mg–Grt cores and Amp inclusions) and continuing through the eclogite (710°C, $P > 15$ kbar), HP granulite (800°C, $P = 12$ kbar), MP granulite (700°C, 7–9 kbar), and amphibolite facies (O'Brien, 1989a, c). Gneisses enclosing the metabasites at Winklarn are folded and tectonised stromatic metatexites. Pressure estimates of 10–14 kbar using the relictic Grt–Ky–Pl–Qtz assemblage are comparable to those from the granulite stage in the eclogites (O'Brien, 1989b, c).

Geochronological data for mafic and ultramafic rocks as well as acid country-rocks from the quarry at Winklarn are reported together with geochemical data by Gebauer (1984) and Quadt and Gebauer (1988). Based on scattered Rb–Sr whole-rock data, U–Pb data on zircons as well as Sm–Nd whole-rock data the respective magmatic protoliths might have formed around 1.25 Ga from a time-integrated depleted mantle (initial $\varepsilon_{Nd} = +7.5$). The eclogite facies overprint is best reflected by Sm–Nd mineral- and whole-rock data using Grt, Zr and Rt. For an eclogite an age of 427 ± 5 Ma and for a felsic rock of volcanic arc composition an age of 423 ± 8 Ma were obtained. Owing to the high probability that the mafic protoliths were formed before deposition of the present metasedimentary country-rocks a tectonic introduction must be concluded. Unfortunately, the geochronological data do not differentiate between a model of early obduction on to a continental margin followed by a common HP metamorphism in the Silurian and one of Silurian emplacement into LP country-rocks. Considering the latter case, some of the immediate felsic country-rocks might also have been dragged up from greater depth together with the mafic–ultramafic eclogite facies rocks.

The SZ–MZ boundary in Bavaria has been overridden by a polymetamorphic nappe pile—the *Zone of Erbendorf–Vohenstrauss* (ZEV) (KTB, 1986). Relict HP granulites—Grt + Ky + perthite + Olg + Qtz + Rt—from this unit underwent partial anatexis followed by mylonitisation and later retrogression during which Bt + Pl replaced Grt and prismatic Sil replaced Ky (Voll, 1960). Blümel (1983) estimated P–T conditions of 650–750°C, 6–8 kbar for the anatexis. Serpentinised peridotites and retrograded eclogites and garnet-pyroxenites are also known from the ZEV (Busch, 1970; O'Brien, unpublished). Geochronological data by Teufel (1988) show a very similar geological development of the ZEV and the MGM (Gebauer and Grünenfelder, 1979; Söllner *et al*, 1981a, b) supporting a common origin for the two Carboniferous nappe piles. However, this does not automatically imply that eclogites of the ZEV are of the same age as those dated in the MGM.

In the *Czech part of the Moldanubian Zone*, numerous minor eclogite bodies are known, often closely connected with serpentinised garnet peridotites or migmatitic paragneisses, within areas characterised by the presence of granulite and its retrograde product—the Gföhl gneiss (Dudek and Fediuková, 1974; Fediuková and Dudek, 1977). The eclogite-bearing units occur in distinct belts, about 160 km long, either side of the Moldanubian Pluton. The NW limb, in southern Bohemia, runs from the major granulite massifs around Blansky Les to Kutna Hora whilst the SE limb follows the SW Moravian margin of the Bohemian Massif in Czechoslovakia and continues into Austria. A coronitic metagabbro with the corona sequence (Ol–Opx–Mg Hbl–Cpx + Spl symplectite–Grt–(Pl)) between olivine and plagioclase is recorded from eastern Bohemia (Holub and Munshi, 1984).

The eclogites have been subdivided by Dudek and Fediuková (1974) on grounds of their mode of occurrence, petrography, and mineral and bulk-rock chemistry into Prp–Alm–Grs, Prp–Alm and Prp types. The Prp–Alm–Grs type are 'common' or Group-B eclogites that contain porphyroblasts of Grt in a foliated Omp matrix and can occur directly within or tectonically removed from peridotite bodies or less frequently in gneisses and migmatites unaccompanied by ultrabasites. Prp–Alm eclogites occur as lenses or boudins in peridotites or granulites and have higher bulk–rock Mg and lower alkalis and Fe than the former type whilst the Prp eclogites, found as cm-thick lenses in peridotites, are more correctly garnet pyroxenites (with little or no Opx) or garnet websterites (when both Opx and Cpx are present in substantial proportions) (Paděra and Prochazka, 1970). All three types of Grt–Cpx rock can be found within the same peridotite body.

Geochemically the bulk-rock compositions of the true eclogites correspond to those of high-alumina or alkali basalts. Bulk-rock chemistry is reflected in mineral compositions. Garnet pyroxenites and websterites have low alkalis and high MgO giving rise to Prp-rich Grt and Jd-poor Cpx whilst the broadly basaltic Prp–Alm–Grs (common) eclogites contain Omp and more Alm-rich garnet. Pyroxenes are predominantly unzoned and garnet zoning is usually confined to Fe enrichment at grain rims. However, Mg rich rims are a feature of garnets from eclogites within gneisses. The reported $K_{D_{Fe-Mg}}^{Grt-Cpx}$ values from Dudek and Fediuková (1974) (with all Fe as FeO) show a distinct regional variation for true eclogites. In the NW belt the K_D range is 3.48–5.63 with most values >4.0, whilst in the SE belt the range is 2.01–4.15 with most values <3.0. This distribution implies that equilibration temperatures were higher in the SE belt.

In the *Austrian part of the Moldanubian Zone* the characteristic association of granulite, its migmatised equivalent—the Gföhl gneiss—Grt-peridotite, pyroxenite, ortho- and paragneiss are collectively called the Gföhl unit. Granulites are mostly the felsic so called *weiss-stein* type: subordinate are dark lenses of pyroxene granulite and basic granulites (pyriclasites). Scharbert and Kurat (1974) calculated minimum conditions of 760°C, 11 kbar for the HP granulite assemblages and noted that the common retrogression—Bt, Mc and Sil growth after Grt, Or and Ky—took place during development of a strong flattening fabric at amphibolite facies conditions. Grt-pyroxenites, previously called eclogites (Kappel, 1967) though with jadeite-poor Cpx, and Grt-websterites occur within or detached from peridotites and are interpreted as cumulates of trapped mantle partial melts (Scharbert and Carswell, 1983). Mineral zoning profiles and exsolution textures indicate that these rocks suffered a decrease in T and probably also P interpreted as due to their insertion, along with their host peridotite, into granulite facies lower crust along a crust–mantle imbrication zone (ibid.). Garnet peridotites (1060 ± 26°C, 31.8 ± 2.9 kbar) probably formed by deformation-induced recrystallisation of former HT (1100–1200°C, 10–20 kbar) spinel + aluminous pyroxene-bearing peridotites before thrusting under eclogite facies (855 ± 63°C, 20.1 ± 1.8 kbar), granulite facies (775–800°C, 12–15 kbar) and amphibolite facies conditions during consecutive phases of the Variscan orogeny (Carswell, unpublished data). Recent Sm–Nd data on garnet clinopyroxenites associated with garnet peridotites (Carswell and Jamtveit, unpublished) indicate a Variscan (340–370 Ma) rather than Caledonian age for the HP assemblages.

Scattered eclogite bodies are also known from the northern part of the Bohemian Massif in Poland. Czechoslovakia and East Germany. In the *Sudetes* eclogites and associated HP rocks are known in the Sowie (Eulengebirge), Ladek and Śnieżnik Mountains of Poland and also across the border in the Rychleby Mountains of Czechoslovakia (Smulikowski, 1967; Pouba, *et al*, 1985). In the Ladek, Snieźnik and Rychleby Mountains intercalations of amphibolite, eclogite, marble, graphitic schist and calc-silicate rocks occur in migmatitic gneisses of the Orlicke-Klodzo unit (Ansilewski, 1966). The fine-scale interlayering of eclogites and gneisses (including a Grt–Ky–St gneiss at one locality) within the migmatite sequence strongly supports an *in situ* metamorphism (Smulikowski, 1967).

Apart from the obligatory Grt and Omp, primary minerals in Polish eclogites can be carinthine (Mg–Hbl), Ky, Zo, Phe and Qtz along with minor Rt, Ap, Dol, Zrn, Py, Ilm and Ttn. Textural studies reveal a multi-stage metamorphic history. Some omphacites are intergrown with relict Tr which along with inclusion suites in garnet indicate a pre-eclogitic greenschist facies stage. The presence of porphyroblasts of Omp enclosing microlaths of an earlier Omp and common poikiloblastic Amp, Phe and sometimes Ky suggest a hydrous recrystallisation stage still in the eclogite facies. Later breakdown reactions such as Bt + Pl after Phe, radial Amp + Pl around Grt, Di + Pl symplectite then diablastic Amp + Pl after Omp, white mica (Pg?) around Ky, Ilm then Ttn on Rt, Zo replaced by Czo and the growth of 'bleb-textured' Amp indicate the passage of the rocks through lower pressures into the amphibolite facies. Eclogite $K_{D_{Fe-Mg}}^{Grt-Cpx}$ values are reported as 13.2–17.7 in the porphyroblastic types whilst lower values in the range 6–11 are calculated for the other variants (Smulikowski and Smulikowski, 1985). These correspond to temperatures of 540–590°C and 650–790°C respectively ($X_{Jd}^{Cpx} = 0.4$ and $X_{Ca}^{Grt} = 0.25$). A temperature of 720 ± 30°C at a nominal 15 kbar is reported for an Ilm-bearing eclogite interpreted as a meta-lamprophyre (Bakun-Czubarow, 1986).

Granulite units in this area are Grt + Omp rocks containing a significant proportion of primary oligoclase (Smulikowski, 1967; Smulikowski and Bakun-Czubarow, 1973; Pouba *et al*, 1985). This rock has been called plagioclase-eclogite but omphacite-granulite (Smulikowski, 1980) is preferable.

North of the Orlicke-Klodzko unit is the triangular *Sowie Gory* massif which comprises a dominant monotonous paragneiss, a subordinate orthogneiss, minor lenses of amphibolite, carbonate, and graphitic rocks and also granulites closely associated with serpentinised garnet peridotites. Some amphibolites have textures that indicate their derivation from eclogites or Grt-pyroxenites (Grocholski, 1967). The serpentinised lherzolites, in which zoned Grt overgrows an earlier Spl, yield *P–T* estimates of 1030°C, 27 kbar and 800°C, 16 kbar from cores and rims respectively for the garnet peridotite stage whilst enclosed garnet-pyroxenites give 900°C at a supposed 20 kbar (Bakun-Czubarow, 1983). The Sowie Gory gneisses overlie a low-grade mafic-ultramafic ophiolite sequence with a magmatic age around 350 Ma (Pin *et al*, 1988).

Eclogites and garnet peridotites bounded by granulites are known to occur beneath the Cretaceous cover in the *Ćeské Středohoři* (Figure 9.2) from their presence as xenoliths in Tertiary volcanics and borehole investigation (Fiala and Paděra, 1984). Eclogites with Grt + Cpx + Phl + Amp ± Zo, Ky, Qtz, Rt have $K_{D_{Fe-Mg}}^{Grt-Cpx}$ values < 3.7 (as do coexisting peridotites) which indicate high temperatures. Klápová (1985) reports Ky-, Phe- and Zo-bearing Qtz-eclogites in the *Krušné Hory* comparable to those of the MGM. Features of these rocks are pre- and post-eclogite amphibolite-facies stages, Jd-rich Omp (up to Jd_{60}) and moderate maximum calculated temperatures. In the *Saxonian Granulitgebirge*, garnet peridotites, kornerupine-bearing granulites and basic granulites and pyroxenites—some probably former eclogites—occur (Mathé, 1969b; Schreyer *et al*, 1975).

9.2.2 *Schwarzwald and Vosges*

The Schwarzwald (Black Forest) and Vosges (Figure 9.3) lie east and west, respectively, of the Rhine Graben and provide a link between the Massif Central and Bohemian Massif. Both units are dominantly in the MZ (Arverno–Vosgien Zone in French literature). In the Central Vosges and main Schwarzwald areas, a multiple metamorphic sequence is recognised which comprises: a basal variegated group containing quartzites, leptynites, amphibolites, graphitic, granulitic and calcareous rocks; a lower amphibolite series; an upper group of kinzigitic garnet gneisses, granulites, leptynites and carbonates; and finally a group of Bt-Sil gneisses (see review in Fluck *et al*, 1980). Relicts of Grt + Ky in kinzigitic gneisses now bearing Crd + Sil indicate an older HP granulite event before LP metamorphism and migmatisation which increases in importance to the south. The whole sequence is cut by late Variscan granitoids.

Eclogites, retrograded to varying degrees, are found as small lenses or boudins within the monotonous gneisses and migmatites of the Central Schwarzwald along with Grt- and Spl-bearing peridotites and rocks of acid to intermediate composition which also show evidence of HP parageneses (Klein and Wimmenauer, 1984; Stenger *et al*, 1987;

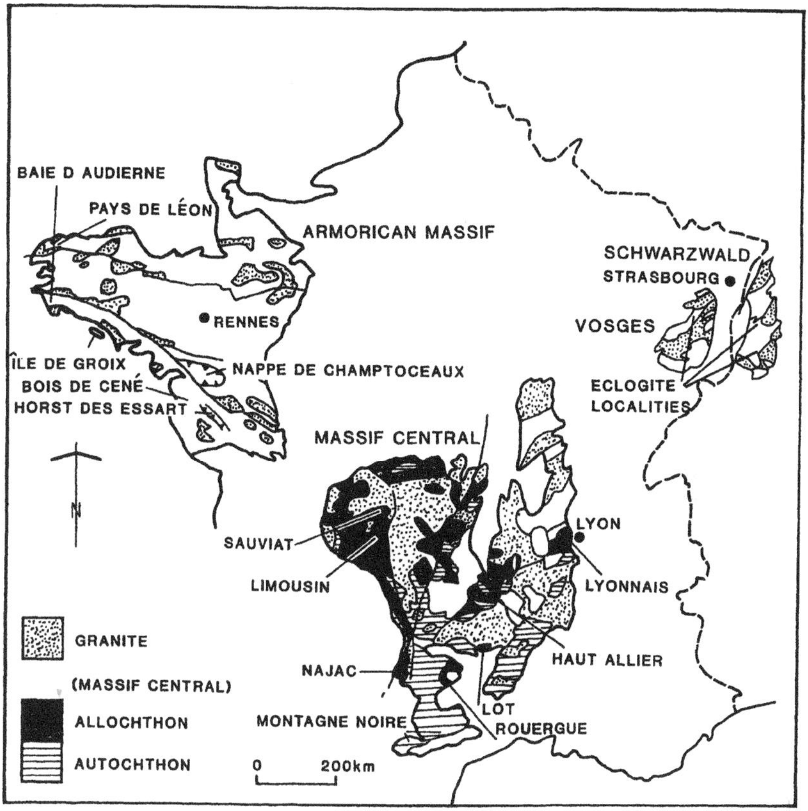

Figure 9.3 Location map for major eclogite-bearing units of the Armorican, Central, Vosges and Schwarzwald Massifs.

Wimmenauer and Stenger, 1987; Wimmenauer and Adam, 1985). The basites were Qtz-eclogites ($\pm$ Ky) but have been overprinted by subsequent granulite- and amphibolite facies conditions causing kelyphites of Amp + Pl to replace Grt, Di + Pl (later Amp + Pl) to replace Omp and Spl + Cor + An to overgrow Ky: quartz margins to garnet are rimmed by Hy (Klein and Wimmenauer, 1984). Geochemically the eclogites can be Qtz, Cor, Ne, Ol or Hy normative and correspond to island arc tholeiites: a derivation from accompanying peridotites is discounted (*ibid.*; Burgath *et al*, 1987). Recent U-Pb zircon dating (Schleicher and Kramm, 1987) gives upper intercepts of 2100 and 490 Ma and lower intercepts of 410 and 317 Ma from different samples whilst Rb–Sr systems pointed also to a 340 Ma event.

In the Vosges, eclogites have been described by Hameurt (1967). Also, leucocratic ortho-granulites (derived from calc-alkaline rhyolites) are associated with para-granulites, sometimes graphitic, and garnet peridotites (Bonhomme and Fluck, 1981). Rb–Sr systems point to a 600 Ma protolith age and HP metamorphism at 386 $\pm$ 15 Ma (*ibid.*). Polymetamorphic granulites have yielded pressures comparable with other Variscan HP granulites (Fabriès and Latouche, 1988; *see also* Pin and Vielzeuf, 1988).

9.2.3 *Massif central*

The structure and metamorphism in the French Massif Central (Figure 9.3) can most aptly be described with reference to large-scale allochthoneity (Burg *et al*, 1984; Matte, 1986; Matte and Burg, 1981). In the southern part of the massif, Late Precambrian to Visean rocks, unmetamorphosed or at low to medium grade, occur in the Montagne Noire and southern Cevennes in the inverted limbs of large recumbent folds overturned to the south (e.g. Arthaud, 1970). This series has been overthrust by a polymetamorphic unit characterised by bimodal meta-igneous rocks that appear as variable-sized boudins in metasediments (a so-called leptyno-amphibolitic group or LAG (see Santallier *et al*, 1988, for a review)): metamorphism was up to eclogite facies (e.g. Nicollet *et al*, 1979; Santallier *et al*, 1978). Above the eclogite-bearing unit occurs a thick (> 5 km) highly deformed sequence dominantly of quartzitic pelites but with minor quartzite, chert, carbonate and mafic lenses, of presumed Lower Palaeozoic age ('schistes peripheriques' of Roques, 1971). Granitic rocks and orthogneisses in this southern domain give ages between 406 and 540 Ma (Pin and Lancelot, 1978; Bernard-Griffiths *et al*, 1977). The northern part of the massif is dominated by large crystalline nappes, interpreted to be thrust towards the south, which comprise: a lower > 5 km thick LAG comparable to that in the southern domain; an upper 1–2 km thick LAG dominantly of feldspathic orthogneiss, with a considerable proportion of mafic and ultramafic eclogite facies rocks and their retrograde products; an overlying anatectic, Crd-bearing, paragneiss unit that includes relicts of HP granulite, eclogite, and garnet peridotite; and finally an unconformable epizonal, bimodal, calc-alkaline volcaniclastic (presumed Upper Devonian) sequence and an unconformably overlying unmetamorphosed post-mid Visean volcaniclastic sequence (review in Burg *et al*, 1984). Mafic and ultramafic rocks considered ophiolitic in origin (Mercier *et al*, 1982) are deformed in the basal thrust of the crystalline northern nappes: an inverted metamorphic zonation occurs beneath this thrust (Burg *et al*, 1984). Granites of various ages and tectono-metamorphic states occupy about a third of the Massif. HP metamorphic rocks are known from many localities in the Massif Central, the main ones being Limousin, Haut Allier, Lyonnais, Rouergue, Lot/Marvejols and Montagne Noire.

In the multi-nappe sequence in the Limousin area, eclogite-bearing upper and lower LAG units are separated by further allochthonous layers with highly deformed—sometimes mylonitic—mica schists and orthogneisses containing a dismembered ophiolite devoid of any HP parageneses (Friedrich *et al*, 1988; Dubuisson *et al*, 1988). The largest mafic–ultramafic body occurs at Sauviat-sur-Vige and comprises gabbros, garnet peridotites, eclogites (± Ky) and amphibolites. Equilibration conditions for the eclogite facies are estimated at $T > 800°C$, $P > 15$ kbar ($K_{D_{Fe-Mg}}^{Grt-Cpx} = 4.9–5.6$, in Bernard-Griffiths and Jahn, 1981). Chondrite-normalised REE plots from these rocks reveal a high LREE/HREE ratio and a positive Eu anomaly: a pattern interpreted as resembling continental or ocean island tholeiites. However, other Limousin basites to the south and west have MORB geochemical signatures (e.g. Schellhorn *et al*, 1982).

The metamorphic transition between various basic protoliths (mostly gabbros) and true eclogites is documented by Piboule and Coffrant (1970): Grt, Cpx, and Rt replace former Aug, Lab, Ol and Ilm and are themselves replaced, during hydrous breakdown, by Amp, Bt and Zo. Velde *et al* (1970), from a study of retrograded metabasites, put the eclogite–amphibolite transition at 650°C, 6.5 kbar. U-Pb dating of both zircon and monazite for Sauviat mafic–ultramafic rocks and metagranites respectively, gave upper intercepts, representing igneous emplacement ages, around 490 Ma and lower intercepts, corresponding to the eclogite facies metamorphism, around 320 Ma (Gebauer *et al*, 1981). A Bas Limousin eclogite sample from further south gave the same result (*ibid.*; Schellhorn *et al*, 1982). This young age for the HP metamorphism contrasts with Acadian (355 ± 2, 379 ± 19 Ma) ages from post-metamorphic granitoids south-west of Sauviat and the absence of a regional HP event at around 320 Ma (Bernard-Griffiths *et al*, 1985a). The implication is that individual nappes have had different metamorphic histories before juxtaposition.

The LAG sequence of the Haut Allier region includes eclogites, acid and basic granulites, coronitic metagabbros, Spr + Cor-bearing amphibolites and serpentinised peridotites (Forestier *et al*, 1973). A geochemical investigation of the basic rocks by Giraud *et al* (1984) revealed that although most had basaltic compositions, non-cumulate-type eclogites and amphibolitised eclogites had higher LREE/HREE ratios than common amphibolites. The complex metamorphic history in the region is exemplified by a sample from La Borie where a Ky + Phl + Cor_I eclogite passed through an initial HP granulite stage, marked by Di + Pl that replaced Omp and Ky altering to Cor_{II} + Pl $\pm$ Spr, followed by pleonaste replacement of Cor_{II} and brown Amp growth during a LP granulite stage, then finally by an amphibolite stage characterised by the presence of green Amp in symplectites and kelyphites and replacement of the Ky breakdown products by Zo (Forestier *et al*, 1973). Basic HP granulites developed Opx + Pl between Grt and Qtz and brown, then later green, Amp. The La Borie eclogite is dated at $432 + 20/-10$ Ma (Ducrot *et al*, 1983). An olivine gabbro from this area shows the corona sequence (Ol)–Opx–Prg–Cpx + Spl symplectite–Grt–(Pl) between olivine and plagioclase.

Dufour (1985) describes HP granulites (Grt + Cpx + Opx_I + Pl) from Monts du Lyonnais which he suggests are probably the result of recrystallisation of a former eclogite facies assemblage. The HP granulite stage at 800–870°C, 7.8–9.9 kbar, was succeeded by an MP granulite stage (Opx_{II} + Pl_{II}) at 710–725°C, 7.2–8.0 kbar followed by the growth of Amp and Bt in an amphibolite facies overprint, compatible with surrounding pelites, at 640–735°C, 5.2–5.5 kbar. Also in the Monts du Lyonnais are garnet peridotites developed from a former spinel-bearing peridotite (Lasnier, 1971).

Eclogites in the Montagne Noire region, geochemically corresponding to tholeiitic basalts, are found in paragneisses closely associated with orthogneisses and are generally retrograded to Amp + Pl-rich assemblages. P–T conditions of 700–800°C, 9 ± 2 kbar are calculated for the eclogite facies stage (Demange, 1985). A calc-alkaline ultramafic cumulate described by the same author underwent polyphase transformation from a spinel websterite (Cpx + Opx + pleonaste + Chr) through the eclogite field, where Grt coronas developed on the Spl, and retrogressed to Amp $\pm$ Chl-bearing assemblages. Two dated samples, from Cabardes and Airette, gave lower intercepts at 432 and 326 Ma respectively using U-Pb method on zircons (Gebauer and Grünenfelder, 1982).

The allochthonous LAG-type sequences in the Lot and Rouergue areas include relicts of eclogites, basic granulites, coronitic gabbros and even a HP trondhjemite. Geochemically the metabasic rocks have been categorised as remnants of ensialic back-arc basins rather than true oceanic crust (Briand *et al*, 1988; Bodinier *et al*, 1988). The eclogite facies conditions for Ky-eclogites of 690 ± 40°C, 15.4 ± 0.6 kbar were followed by a granulite facies overprint at around 800–850°C, 10–12 kbar (Briand *et al*, 1988). Blastomylonitic eclogites (Bouchardon, 1987) and an eclogite with chrome-rich (up to 7 wt% Cr_2O_3) kyanite (Delor and Leyreloup, 1986) gave similar P–T conditions. An interesting rock from this area is a trondhjemite from within a basite sequence interpreted as having crystallised from a melt under HP conditions around 750–840°C, 12.5–20 k bar (Nicollet *et al*, 1979).

In western Rouergue, at the margin of the Massif Central, Delor *et al* (1986, 1987) describe HP parageneses in both basic and pelitic rocks from the upper unit of the Najac klippe. Metabasite Grt contains inclusions of Clm (near the core) and Omp (near the rim) along with blue-green calcic and calcic-sodic Amp. The Grt is set in a matrix of zoned Cpx with Clm cores and Omp rims, and zoned Amp with Gln cores and calcic or calcic-sodic Amp rims. These relatively LT eclogites are estimated to have experienced conditions of 600°C, 12 kbar with a notable temperature increase before uplift (Delor *et al*, 1986): comparable P–T conditions were calculated for Grt + Phe + Chl + Cld + St + Ky metapelites (Delor *et al*, 1987). Pin and Lancelot (1982) present geochronological data for the area. They calculate ages around 480 Ma for mafic and felsic protoliths of the LAG, a 415 ± 6 Ma crystallisation age for the HP trondhjemite, and a 344 ± 13 Ma age for a syn- to post-tectonic pegmatite. Ages around 350 Ma have also been obtained from S-type granites post-dating at least one deformation event (Pin, 1981).

9.2.4. *Armorican Massif*

The Armorican Massif in NW France comprises three distinct geological provinces: the South (Ligerian = MZ), Central (Armorican = SZ) and North (Domnonean/Mancellian) Armorican domains (reviews by Autran and Cogné, 1980; Cogné and Wright, 1980). The North Armorican zone is a fragment of old (>2 Ga) Gondwana-derived basement affected by the Cadomian orogeny. The Central Armorican domain is characterised by a thick upper Proterozoic (Brioverian) shale-greywacke succession that contains bimodal marginal basin-type volcanics, only mildly affected by the Cadomian event, and non-metamorphic Lower Palaeozoic. The South Armorican Zone contains an Upper Proterozoic to Lower Palaeozoic volcano-sedimentary series with rarer graphitic, silicic and calcareous lenses and eclogites, metagabbros and peridotites. HP rocks are known from the Pays de Léon, Baie d'Audierne complex, Île de Groix, Nappe de Champtoceaux, Horst des Essart and Bois de Cené (Figure 9.3). Geochronological studies in Armorica (reviewed in Paquette *et al*, 1985) reveal detrital zircons from a 2.2 Ga continental crust, a 1200 Ma oceanic crust metamorphosed to eclogite facies in the period 380–440 Ma and Cambro–Ordovician intrusion ages for voluminous calc-alkaline orthogneisses (also Jegouzo *et al*, 1986). Post-orogenic granites were emplaced following the HP event and also after the late-Variscan LP episode. Individual HP-relicts have been dated at: 439 Ma, Léon; 384 ± 6 Ma, Baie d'Audierne; 413 Ma and 358 Ma, Champtoceaux; $436 + 15/-12$ Ma, Horst des Essart; 400–420 Ma, Groix (Paquette *et al*, 1987; Peucat *et al*, 1982; Bernard-Griffiths *et al*, 1986).

In the eastern part of the Île de Groix, metabasites are Gln-bearing LT eclogites and Grt + Gln + Ep schists (sometimes containing pseudomorphs after Lws) and metapelites are Grt + Phe + Cld schists (Triboulet, 1978). The basites correspond geochemically to ocean floor or within-plate basalt (Bernard-Griffiths *et al*, 1986). Maximum conditions of 530°C and 8.5 kbar were followed by retrogression in the greenschist facies characterised by the growth of blue-green amphibole, Chl and large porphyroblasts of Ab. Peucat and Cogné (1977) dated the HP metamorphism as occurring at 420 Ma. A mainland continuation of this blueschist belt at Bois de Cené, although devoid of eclogites, exhibits a comparable HP followed by LP evolution (Giraud *et al*, 1987). Mica schists and metabasites from the estuary of the river Vilaine experienced similar P_{max} but higher T_{max} than the blueschist rocks (Triboulet and Audren, 1985, 1988). The calculated *P–T–t* path shows a prograde HP/MT metamorphism explained by 'subduction' of the Vilaine 'unit' then a HP/HT phase because of partial thermal re-equilibration and finally retrogression during uplift caused by late underthrusting below the unit.

The main eclogite-bearing units of Armorica are the Champtoceaux Nappe and Horst des Essart of the Vendée area. Straddling the river Loire near Nantes, the Champtoceaux Nappe comprises three units (Cogné, 1966): upper and lower epizonal schist series correlated to the Brioverian and an intermediate mesozonal to catazonal unit of granulitic gneisses, leptynite gneisses, amphibolites, eclogites, serpentinites and gabbros. Coronitic basic rocks also occur (Lasnier, 1970). The eclogites correspond to N- and T-type MORB (Paquette, 1987) and crop out, along with amphibolites, as boudins within the gneisses. Metagabbros have either cumulate or N-type MORB REE features (*ibid.*; Bernard-Griffiths and Cornichet, 1985).

The eclogites are Ky + Qtz- and Qtz-bearing types and additional phases can be Amp, Pg, Phe and Rt (Godard, 1981). Garnets, sometimes in atoll form, are prograde zoned (i.e. Fe-rich core, Mg-rich rim) and contain inclusions of Omp, Rt, and white mica, whilst omphacites are Jd rich ($X_{Jd}^{Cpx} > 0.5$). Equilibration conditions of 700–750°C at a minimum of 17 kbar are estimated for the eclogite stage which was followed by a fall in *T* and *P* evidenced by Amp + Pl kelyphites and symplectites replacing Grt and Omp, Bt + Pl symplectites overgrowing Phe, fine-grained Cor + Spl + Pl reaction rims on Ky and Ilm and Ttn replacing Rt. Retrogressive growth of Gln at around 470°C, 9 kbar is documented from some samples (Godard *et al*, 1981). Similar maximum conditions are reported for eclogite facies associations of Grt + Ky + Phe + Rt in metapelites and Grt + Omp + Ky + Pg + Rt in a meta-quartzite and are inferred for coronitic transformations in an undeformed granitic body locally preserved in leptynites (Ballèvre *et al*, 1987). Lasnier *et al* (1973) document a multi-stage tectono-metamorphic history for such a 'charnockitised' granite from an inferred porphyritic granite precursor, passing through a coronitic stage and, following increased deformation, breaking down to augen-gneiss and finally blastomylonitic leptynite. A variety of crustal rocks have therefore also seen the eclogite facies conditions but have mostly retrograded during uplift. The Horst des Essart eclogites are comparable to those of the Champtoceaux unit in that they occur in acid kinzigitic gneisses, have oceanic geochemistry and suffered a lower grade re-equilibration (Godard,1981; Bernard-Griffiths and Cornichet, 1985).

In the Baie d'Audierne Complex a monometamorphic series of meta-greywackes cut by amphibolites overlies a lower unit composed of mica schist, granulitic/leptynitic orthogneiss, prasinite, amphibolite, metagabbro, garnet-pyroxenite, serpentinite and eclogite (Peucat and Cogné, 1974). Metabasic rocks have an oceanic geochemistry (Bernard-Griffiths and Cornichet, 1985). The serpentinites, devoid of any HP relicts, contain thin chromite layers, basic HP granulites (pyrigarnites) contain Grt + Ky + Qtz + Rt + Gr and Ky-eclogites are also known (Marchand, 1982). A displaced portion of the eclogite belt occurs in the Pays de Léon where migmatitic Bt + Sil gneisses are hosts to minor heavily retrograded Qtz- and Ky + Zo-eclogites with the geochemical features of N- and E-type MORB (Balé and Brun, 1986; Cabanis and Godard, 1987). P–T conditions of 650–700°C, > 13–14 kbar have been calculated by Paquette *et al* (1987).

9.2.5 *Iberian Massif*

The Variscan fold belt crops out at the western margin of the Iberian Peninsula where it forms the Iberian (or Hesperian) Massif. High-grade gneisses and mafic–ultramafic associations previously interpreted in terms of diapiric intrusion form the allochthonous massifs of Cabo Ortegal, Malpica-Tuy (blastomylonitic graben), Ordenes, Lalin, Bragança and Morais in the Middle Galicia–Trás-os-Montes sub-unit of the Central Iberian Zone (Figure 9.4) (Ribiero, 1974; Ries and Shackleton, 1971).

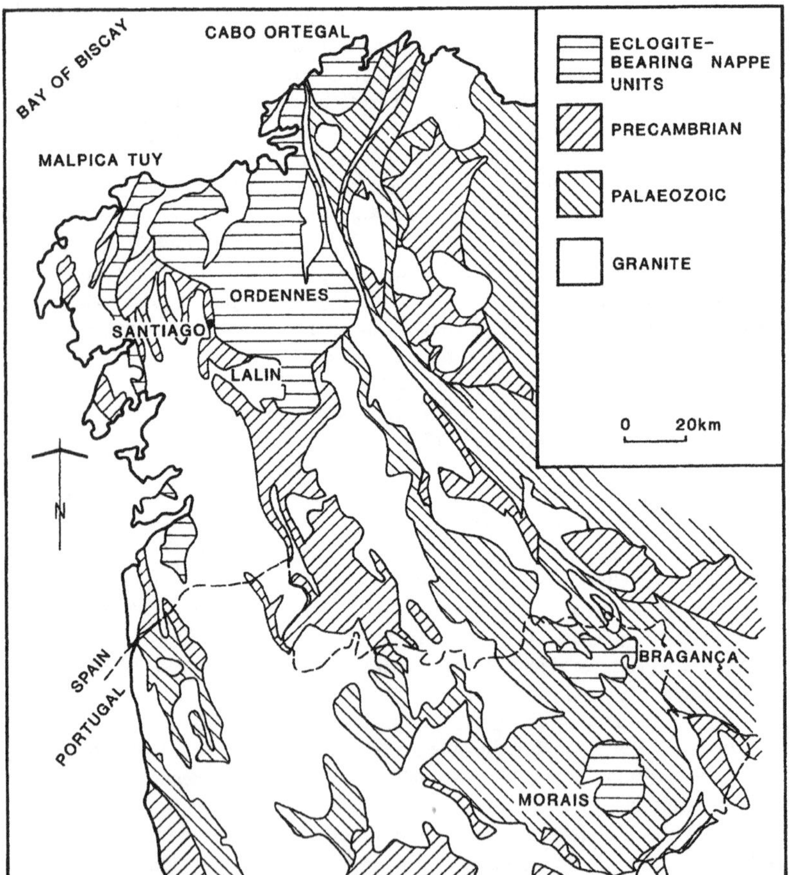

Figure 9.4 Location map for major eclogite-bearing units in NW Iberia.

At Cabo Ortegal, an upper dominantly metasedimentary unit—the Concepenido Complex—overlies an ultramafic unit and a lower, mostly metabasic, group called the Capelada Complex (Vogel, 1967). These three tectonically overlie a low-grade mafic–ultramafic 'ophiolite' complex of Siluro–Devonian age and all rest on a well dated Silurian autochthon. Within the Concepenido Complex, Ky-, Zo- and carinthine-bearing Qtz-eclogites occur in planolinear, augen or blastomylonitic—sometimes migmatitic—Grt + Bt ± Ky paragneisses of the Banded Gneiss Formation. The development of eclogites from former basic intrusions in the gneisses is confirmed by the presence of partially coronitised olivine gabbros: in one case both eclogite and coronite appear in the same layer (Vogel, 1967, p. 151). Also within the Concepenido Complex are further sub- to blastomyloniitic para- and orthogneisses enclosing mafic granulites ((plagio)pyrigarnites i.e. Grt + low Jd Cpx + Pl) rather than eclogites. In addition, a barely-deformed paragneiss in which a Ky-bearing assemblage has replaced an earlier St-bearing one contains metabasic inclusions that are common amphibolites. Both eclogites and mafic granulites suffered a polyphase retrogression that led to increasing amphibole and plagioclase growth at the expense of garnet and clinopyroxene.

Geochemically both eclogites and mafic granulites correspond to Qtz-normative tholeiites with N-type MORB REE character (Bernard-Griffiths and Cornichet, 1985) but $K_{D_{Fe-Mg}}^{Grt-Cpx}$ values of 8–10 from eclogites (where $X_{Jd}^{Cpx} = 0.3$ to 0.33) and 5–7 from mafic granulites (where $X_{Jd}^{Cpx} = 0.1$ to 0.23) correspond to vastly different P–T values of 590–670°C, 13–15 kbar and 850°C, 10 kbar respectively (Kuijper *et al*, 1985). Vogel (1984) suggests that the mafic granulites formed by local recrystallisation of former eclogite facies assemblages during a younger overprint are restricted to areas of fluid availability. Rocks of the ultramafic unit are commonly serpentinised Amp-(Chl-) peridotites—only rarely are Grt relicts found—and contain lenses and veins of garnet websterite and garnet pyroxenite. In the latter rock Grt can sometimes be seen overgrowing an earlier Spl (Vogel, 1967, p. 172). Equilibration conditions of 900°C, 18 kbar and 720°C, 10 kbar are reported for spinel-peridotites and garnet websterites respectively (Gil Ibarguchi *et al*, 1987). Bernard-Griffiths *et al* (1985b) estimate metamorphic ages between 480 and 420 Ma for near-concordant zircons from Cabo Ortegal eclogites.

In the Malpica-Tuy unit, HP parageneses are found from a variety of rock compositions which thereby strongly supports *in situ* eclogitisation. The unit comprises a sequence of mica schist, paragneiss, calc-alkaline to peralkaline orthogneiss, and Ti-rich alkaline to subalkaline basalts (Gil Ibarguchi and Gironés, 1985). Metabasites and impure sandstones contain the assemblage Grt + Omp + Gln/Bar + Rt ± Phe ± Pg ± Ky ± Zo, orthogneisses have Grt + Phe + Rt ± Zo and metasediments are Grt + Phe + Rt ± Cld ± Ky ± Bt schists. Corona development of Grt around Bt is evident in some orthogneisses. Eclogite garnets contain inclusions of lower-grade minerals and show Mg increasing towards grain rims hence indicative of prograde growth. This is reflected in P–T calculations where, using the Grt–Cpx geothermometer, Grt cores give 525 ± 20°C ($K_{D_{Fe-Mg}}^{Grt-Cpx} = 22$–24) and rims give 570 ± 40°C ($K_{D_{Fe-Mg}}^{Grt-Cpx} = 16.5$–18.5): minimum pressures of 10–12 kbar are indicated from $X_{Jd}^{Cpx} = 0.45$). Geochronological studies of rocks from the Malpica-Tuy unit reveal a 467 ± 10 Ma protolith age for a calc-alkaline orthogneiss and indicate, from K-Ar and Rb-Sr work on eclogite micas, that the HP event occurred before 360 Ma (Van Calsteren *et al*, 1979).

Blueschists are also found in the Bragança and Morais complexes of N Portugal where they occur in a volcano-sedimentary rift sequence as a lower-level nappe overlain by an ophiolite unit and an upper eclogite-, granulite- and peridotite-bearing nappe (Munhá *et al*, 1984). In this case they are LT, sometimes Lws-bearing blueschists (Schermerhorn and Kotsch, 1984). The uppermost nappe consists of felsic orthogneisses, granulites, peridotites and Ky- or Sil-bearing paragneisses that contain eclogite lenses. P–T conditions of 600 ± 30°C, 17 ± 3 kbar for Qtz + Phe + Ky-eclogite were followed by a granulite facies overprint at temperatures up to 850°C and pressures in the range 11–14 kbar before retrogression into the amphibolite facies (Munhá and Marques, 1987). Geothermobarometric studies on garnet pyroxenite lenses in peridotites indicate cooling from 900–1100°C, at 16 ± 2 kbar to 750°C at 10 kbar: HP granulites also reveal a cooling trend from 850–800°C to about 750°C at pressures of about 10–12 kbar.

9.3 Petrographic implications and geotectonic scenarios for eclogite formation

The descriptions given show that a number of clear similarities exist between the scattered occurrences of Variscan eclogites. Many of the massifs contain an acid + basic/ultrabasic association within gneisses which, owing to the presence of calcareous, graphitic and quartzitic intercalations, are obviously of supracrustal origin. In some cases the intimate interlayering of basic and pelitic material has led to the interpretation of the basic rocks as representing former lavas or mafic tuffs. In consequence, as the mafic rocks record eclogite facies conditions, then their hosts must also have endured the HP conditions. Other evidence for such a prograde evolution, i.e. from LP protoliths, is provided by chemical zoning of garnet and clinopyroxene, inclusion within these grains of lower-grade phases, and the presence of partially eclogitised acid (Lasnier *et al*, 1973) and basic (Matthes and Seidel, 1977; Piboule and Coffrant, 1970; Engels and Vogel, 1966; Lasnier, 1970; Holub and Munshi, 1984) rocks in which coronitic development of eclogite facies minerals has occurred.

Unfortunately, although the eclogites are variably retrogressed to amphibolite facies Pl + Amp-rich assemblages, their host gneisses, being more susceptible to recrystallisation than the effectively 'dry' and more structurally competent metabasites, are generally completely transformed to LP assemblages and preserve little evidence of a former HP event. The discovery of HP assemblages, relicts or texturally distinct breakdown products in rocks of non-basic composition (e.g. Nicollet *et al*, 1979; Wimmenauer and Stenger, 1987; Wimmenauer and Adam, 1985; Gil Ibarguchi and Girones, 1985; Ballèvre *et al*, 1987; Delor *et al*, 1987; Lasnier *et al*, 1973) is therefore of vital importance in deducing how much (or little) of the currently exposed gneiss terranes have undergone HP metamorphism.

Bearing this in mind, the discovery that pressures, calculated using the Grt–Ky–Pl–Qtz geobarometer, on felsic granulites from most of the Variscides, were often consistent with those of co-existing eclogites (Pin and Vielzeuf, 1988), is positive evidence for a widespread HP metamorphism. HP granulites of this type (see also Pin and Vielzeuf, 1983) are very distinctive when compared with typical granulites of old shield areas. The European HP granulites in the eclogite-bearing terranes are generally undepleted in granitophile elements and have retrogressed to lower grade by near-isothermal decompression in contrast to highly depleted (through anatexis) shield granulites that have retrogressed by near isobaric cooling (Tarney and Windley, 1977; Bohlen, 1987). LP/MP granulites of the latter type are in fact found in Central Europe—as upthrust basement in the Alpine belt and as xenoliths in young volcanics—and record the metamorphic conditions of the late Variscan lower crust (Pin and Vielzeuf, 1983).

Near-isothermal decompression paths, as revealed from selected thermobarometric results plotted in Figure 9.5, are also an integral feature of the multi-stage metamorphic development of Variscan eclogites. Rocks from Cabo Ortegal (14; Kuijper *et al*, 1985), Monts du Lyonnais (8; Dufour, 1985), Rouergue (9; Briand *et al*, 1988), Schwarzwald (16; Klein and Wimmenauer, 1984) and Oberpfalz (3; O'Brien, 1989a,b) all show evidence of complete or partial transformation to HP granulite and then later MP or LP granulite before amphibolite facies (or even lower grade) retrogression to conditions concordant with their surroundings.

Typical alteration features of these eclogites are: vermicular symplectites of Di + Pl after Omp; Opx + Pl and later Amp + Pl in kelyphites after Grt; Ky breakdown to Spl ± Cor ± Spr + An and Rt rimmed by Ilm and later Ttn. This is not the only pattern of breakdown, however. From Figure 9.5 it is also noticeable that eclogites from the MGM (1; Franz *et al*, 1986), Malpica-Tuy (17; Gil Ibarguchi and Gironés, 1985), Poland (15; Smulikowski and Smulikowski, 1985), Île de Groix (11; Triboulet, 1974) and western Rouergue (10; Delor *et al*, 1986, 1987) equilibrated at lower temperatures. Some of these examples are typical of LT (group-C) eclogites and all can contain primary amphiboles, white mica and zoisite. Once again, a multi-stage metamorphic history is evident but, because of the lower temperatures, partial re-equilibration has taken place under conditions of one or more of the eclogite-, amphibolite-, greenschist- or even blueschist-facies. Typical features of these LT eclogites are multiple evolutions of Omp, Ky, and Phe, a predominance of Amp + Pl from Grt and Omp breakdown, Bt + Pl replacing Phe and Pg overgrowing Ky.

A knowledge of the variations in P and T over time within and between rock masses of the Variscan belt is vital in deciphering the tectono-metamorphic evolution of the belt. Bearing in mind the likely LP nature of many eclogite protoliths, the low to moderate temperatures found at the depths required for the formation of the eclogite facies assemblages compared with a steady state continental geotherm must be explained. Theoretical thermal models (e.g. England and Thompson, 1984; Gillet *et al*, 1985; Thompson and England, 1984) show such low temperatures at depth for rocks taken from shallow- to deep-crustal levels during thickening of the crust by imbrication and thrusting. Post P_{max} temperatures usually increase due to thermal relaxation during uplift and erosion but can be retarded if continued underthrusting occurs. The P–T–t paths in Figure 9.5 are compatible with results from thrusting models. Eclogites could not have formed by addition of basic magma to the lower crust because such an event would cause an increase in T and not the increase in P (or decrease in T at HP) that eclogite formation requires.

In real-life situations, thrusting is probably multi-episodic (model of Cuthbert *et al*, 1983) and each increment of shortening and subsequent thermal perturbation is liable to activate or deactivate metamorphic reactions and isotopic exchanges: multi-layer thrust models are especially complex (Gillet *et al*, 1985). 'Nested' P–T loops would be produced from a horizon buried to variable depths across a region—the thermal maxima defining a piezothermic array (Spear *et al*, 1984). Such features can be deduced

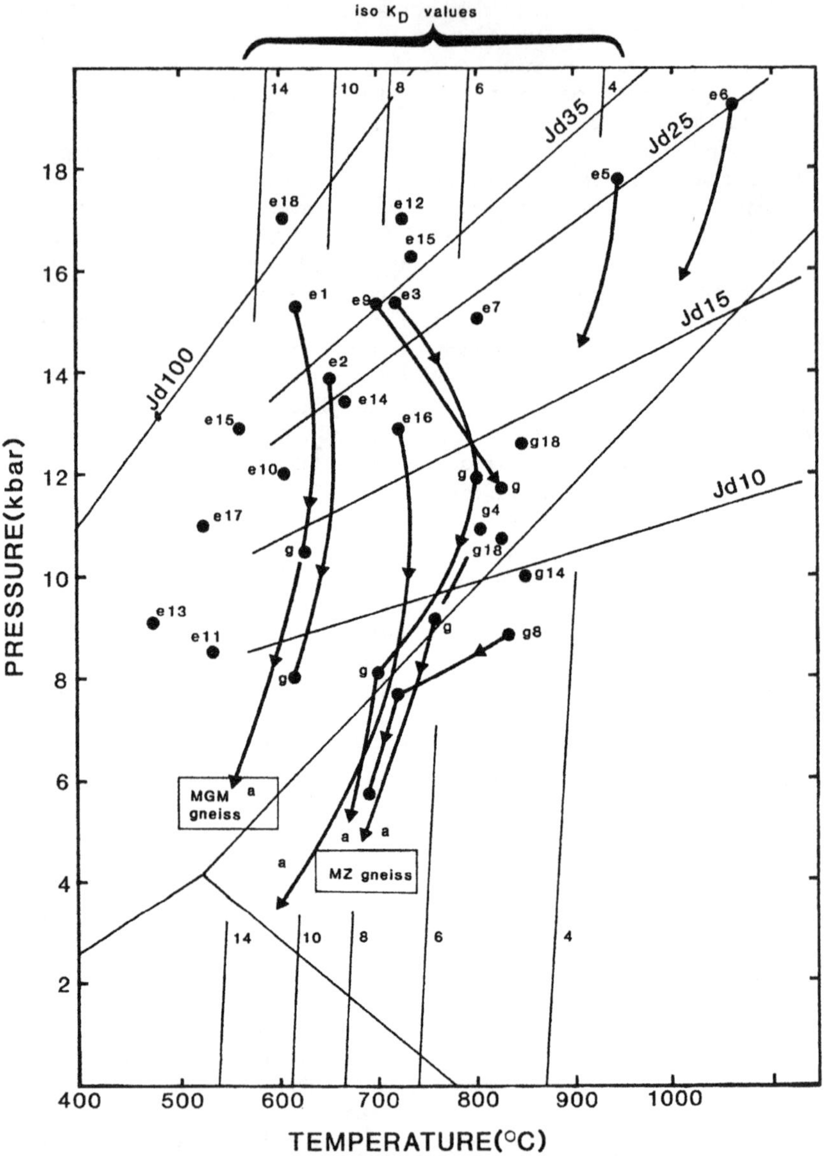

Figure 9.5 Geothermobarometric results from selected eclogites and retrograded eclogites. Jadeite curves from Holland (1980); iso-K_D curves for the Grt–Cpx geothermometer (Powell, 1985b) (assumes $X_{Ca}^{Grt} = 0.25$): alumino-silicate stability curves from Salje (1986): MM gneiss from Matthes *et al* (1975); MZ gneiss from Blümel (1983); e = eclogite; g = granulite; a = amphibolite. 1. Münchberg (Franz *et al*, 1986); 2. Münchberg (Blümel in KTB, 1986); 3. Oberpfalz (O'Brien, 1989a); 4. Oberpfalz (Blümel in KTB, 1986); 5. S. Bohemia (Dudek and Fediuková, 1974); 6. SW Moravia (ibid.); 7. Limousin (Bernard-Griffiths and Jahn, 1981); 8. Lyonnais (Dufour, 1985); 9. Rouergue (Briand *et al*, 1988); 10. W. Rouergue (Delor *et al*, 1986); 11. Île de Groix (Triboulet, 1974);

from eclogites of the Bohemian Massif (O'Brien, 1989a, Blümel in KTB, 1986) and the Armorican and Central Massifs in France (Bouchardon *et al*, 1987). In France, LT eclogites occur at the western margin of the massifs whilst equilibration temperatures and the degree of HT overprint (amphibolite- or granulite-facies) increases towards the east. In the Bohemian Massif, the temperatures at which (presumed) P_{max} occurred increase from the MGM in the northwest, through the Oberpfalz and S Bohemia, to SW Moravia in the southeast. As T would have increased with depth of burial, the direction of subduction can be expected to have been from the LT eclogite positon to that of the higher T eclogites assuming that subsequent complete tectonic inversion of the thermal profile is unlikely. Such patterns are reminiscent of the thermal variation across the Norwegian eclogite-bearing terrane (e.g. Griffin *et al*, 1985). However, the lack of stratigraphic control due to complex thrusting and wrenching and the difficulty in obtaining true pressures for eclogitisation mean that the model cannot be confirmed unequivocally. However, any tectonic model for the Variscides must explain the temperature differences in eclogites across the belt.

Tectonic incorporation of ultramafic rocks is an important feature of collision orogenies. In the Variscides there are examples of garnet peridotites—some produced by recrystallisation of former HT assemblages (Carswell, unpublished; Demange, 1985; Lasnier, 1971; Medaris *et al*, 1988a, b; Mísař *et al*, in Dudek and Mísař, 1985; Vogel, 1967) and of spinel peridotites interpreted as 'ophiolitic' (Jelínek *et al*, 1984; Mercier *et al*, 1982; Pin and Carme, 1987; Pin *et al*, 1988; Von Gehlen and Schmitt, 1987). The common low-grade recrystallisation and serpentinisation of ultrabasites often masks their former identities. However, it seems likely that various ultramafic rocks may have been (*a*) subducted from shallow levels during collision (*b*) incorporated into the crust from deep levels during imbrication or (*c*) as marginal basin ophiolites, overthrust during younger thin-skinned tectonics. Therefore all serpentinites need not have been garnet peridotites.

The combination, in the Variscan belt, of bimodal magmatism and ultramafics in a greywacke-shale sedimentary pile is interpreted by many as the product of rifting (e.g. Weber and Behr, 1983; Pin and Vielzeuf, 1988). Protolith ages for basic and acid rocks emplaced during rifting—now eclogites and granulites—are mostly in the period 500–600 Ma (Gebauer *et al*, 1981; Gebauer and Grünenfelder, 1979; Pin and Carme, 1987; Pin and Lancelot, 1978, 1982) although older material is known (e.g. Gebauer, 1986; Paquette *et al*, 1985; Peucat *et al*, 1982). A change from this extensional environment to one of compression led to closure of the basins, large scale crustal imbrication, and HP metamorphism.

Metamorphic ages for eclogites are <480 Ma (Gebauer, 1983; Bernard-Griffiths *et al*, 1977) but are variable across individual massifs (Gebauer *et al*, 1981). However, one must bear in mind the multi-stage metamorphic evolution of these high-grade rocks and ask which of the many stages on the P–T path has been recorded. In the case of the Sm–Nd method this will generally be the formation of the usually analysed

Figure 9.5 *Cont'd*
12. Vendée (Godard, 1981); 13. Vendée (Godard *et al*, 1981); 14. Cabo Ortegal (Kuijper *et al*, 1985); 15. Sudetes (Smulikowski and Smulikowski, 1985); 16. Schwarzwald (Klein and Wimmenauer, 1984); 17. Malpica-Tuy (Gil Ibarguchi and Gironés, 1985); 18. Bragança/Morais (Munhá and Marques, 1987).

minerals, garnet and omphacite. For the U–Pb zircon suite method new growth of zircon and/or loss of lead in primary magmatic zircons is the key to being able to date the HP event. After the eclogite facies stage, all rocks followed P–T paths leading to recrystallisation at lower P and higher or lower T due to uplift and erosion, thermal re-equilibration and further increments of crustal shortening (see model of Cuthbert and Carswell in Chapter 8 for eclogite formation in the Scandinavian Caledonides). The final thrusting stage, leading to the present nappe distribution, was of a much-thinned imbricated pile as evidenced by the deformation during this event, within the overthrusting nappes, of already recrystallised MP assemblages. In addition, the inverted metamorphism below the nappes is of LP type (Burg *et al*, 1984). The orientation of the final thrusting (e.g. Matte, 1986; Matte and Burg, 1981; Rajlich and Synek, 1987; Ries and Shackleton, 1971; Tollmann, 1982) was not necessarily the same as that of the earlier thrusting that led to HP metamorphism. Late Variscan geology is dominated by syn- to post-tectonic S-type granites. These were generated in the final subducted slabs in response to the nappe loading which caused granulitisation of the lowest crust (Pin and Vieleuf, 1983, 1988). Major wrenching and block-faulting at the end of the Variscan has led to the fragmentation of the belt into the present isolated blocks (e.g. Arthaud and Matte, 1977).

In conclusion it can be noted that the Variscan orogeny was one of many contrasts. Extensional features are found in areas typified by compression; thin-skinned tectonics are of as much importance as major crustal-scale imbrication; HP metamorphic rocks are found in an archetypal LP terrain; and ultramafic rocks are an integral part of a region flooded by granites. Although the European Variscides have been intensely studied for decades, problems still exist. Did oceans exist within the belt? How many rift basins were there? Where was Gondwana in relation to all this? What is the relationship between crystalline nappes of the different massifs? What is the relationship between HP rocks of the same massif? (see reviews of Behr *et al*, 1984; Matte 1986; Weber and Behr 1983; and Ziegler 1984, for further discussion). These and other problems will be solved only by integrated studies in the future. However, the fact that eclogites can keep their integrity whilst all around might have lost theirs, has been an important factor in our present understanding of the European Variscides.

10 Formation and distribution of eclogite facies rocks in the Alps

G.T.R. DROOP, B. LOMBARDO and U. POGNANTE

10.1 Introduction

Eclogite facies rocks are well represented in the European Alps (Figure 10.1). Many of the major crystalline tectonic units of the Alps contain eclogites of mafic composition. Broadly cofacial mineral assemblages have also been described from a wide range of other sedimentary and igneous bulk compositions (e.g. Fe–Al–pelites, Mg–Al–pelites, calc-schists, siliceous dolomites, Fe–Mn–cherts, granites, diorites, rodingites and peridotites), making the Alps one of the best areas in the world for demonstration of the existence of an 'eclogite facies' in the sense of Eskola (1920). As the Alps are also well exposed and have been intensively studied for over a century, they provide an excellent natural laboratory in which to study the link between orogenesis and eclogite facies metamorphism.

A large proportion, volumetrically, of the eclogite facies rocks of the Alps are of early Alpine age (i.e. Cretaceous) and are directly related to the tectonic processes responsible for the present architecture of the orogen. As indicators of the environments in which eclogite facies metamorphism can take place, these occurrences are useful because there is reasonably good control on their geological setting, and this Chapter concentrates on them for this reason. However, the review would be incomplete without some mention of the numerous occurrences (some spectacular) of Variscan and older eclogites in the pre-Mesozoic basement complexes of the Alps, but less emphasis is placed on them here because their formation was entirely unrelated to the Alpine tectonic history of the region and the pre-Alpine regional geology is poorly known in comparison.

Broadly speaking, the Alps have suffered two major phases of Alpine regional metamorphism (e.g. Frey *et al*, 1974); (*a*) an early Alpine metamorphism (the 'eo-Alpine' phase) which, in most of the units affected, was of high-pressure character, producing prehnite-pumpellyite, blueschist and eclogite facies assemblages, and (*b*) a mid-Tertiary greenschist to amphibolite facies overprint (the 'late-Alpine' phase) the effects of which are present along the whole length of the Alps but are most intense in the Ticino and Tauern culminations in the Central and Eastern Alps, respectively. In addition, some external units of the Western Alps record a very LT–HP metamorphism (the 'meso-Alpine' phase) of Palaeogene age (Caby *et al*, 1978; Pognante, 1984; Goffé and Chopin, 1986, and references therein). Whereas the late-Alpine overprint was essentially post-tectonic and produced simple concentric patterns of isograds which cut across nappe boundaries (Frey *et al*, 1974, 1980), the early- and meso-Alpine metamorphisms took place before assembly of the nappe pile was completed (e.g. Gosso *et al*, 1979) with the result that no simple regional pattern is now preserved on the erosion surface and adjacent nappes commonly have different high-pressure metamorphic signatures. In describing the eclogite facies rocks of the Alps, therefore, the Authors have chosen to group the occurrences by tectonic unit rather than by bulk composition or pure geography.

10.2 Pre-Alpine eclogites and garnet peridotites/pyroxenites

Although not particularly abundant, pre-Alpine eclogites, garnet-peridotites and their retrogressive derivatives occur widely throughout the crystalline basement complexes of both the palaeo-European continental margin (i.e. the External Massifs, Pennine Basement) and the palaeo-African plate (i.e. Middle Austro-Alpine Nappe). From a geological point of view, these occurrences are important for two reasons. Firstly, they are of potential interest to people concerned with the unravelling of the Variscan and Caledonian history of Europe, and secondly, they serve as examples of how well eclogites can survive reworking in later orogenic cycles.

The age relations of some of the eclogites described in this section are still controversial, even to the extent that their pre-Alpine status is in doubt. It used to be thought (e.g. Mottana, 1972, and references therein) that all the eclogites in the basement units of the Western Alps were of Caledonian age, but many have since been shown to be Alpine.

10.2.1 *External massifs of the Helvetic domain*

Eclogites and/or retrograded eclogites have been described from several localities in the Aiguilles Rouges, Argentera and Mont Blanc Massifs (Figure 10.1). Because these basement massifs escaped Alpine penetrative deformation and were only slightly affected by Alpine metamorphism, pre-Alpine geological features are preserved much better there than in the Penninic Domain, and a pre-Alpine age for the eclogites is more certain.

In the Argentera Massif, eclogites occur as blocks in agmatites and as layers and septa in anatectic migmatites (Franchi, 1894; Faure-Muret, 1955; Malaroda, 1970; Mottana, 1972). The typical primary assemblage is Grt (Prp_{23-30}) + Omp + brown Hbl + Qtz + Rt. At Lac Cornu in the Aiguilles Rouges Massif, the same assemblage occurs in retrograded eclogites that form the cores of thick amphibolite boudins in mica-schists and migmatitic gneisses (Brière, 1920; Bellière, 1958; Mottana, 1972; Liégeois and Duchesne, 1981). In both areas, field evidence indicates that the eclogites predate the intrusion of Variscan granites. At Lac Cornu a pre-Alpine age is also proven by the fact that, in the relatively deformed boudin margins, the primary assemblage has been overprinted by amphibolite facies minerals aligned parallel to the regional pre-Alpine foliation (Liégeois and Duchesne, 1981). As this regional amphibolite facies metamorphism is known to be pre-upper Viséan (Bellière and Streel, 1980, quoted in Liégeois and Duchesne, 1981) an early Variscan or older age for the eclogites is indicated.

Metamorphic conditions of $T \approx 780°C$ and $P \gtrsim 11$ kbar have been deduced by Liégeois and Duchesne (1981) for coexisting garnet (Prp_{24}) and omphacite (Jd_{16-25}) relics in eclogite from Lac Cornu using the garnet–clinopyroxene Fe–Mg-exchange thermometer of Ellis and Green (1979) and the jadeite content of omphacite coexisting with quartz.

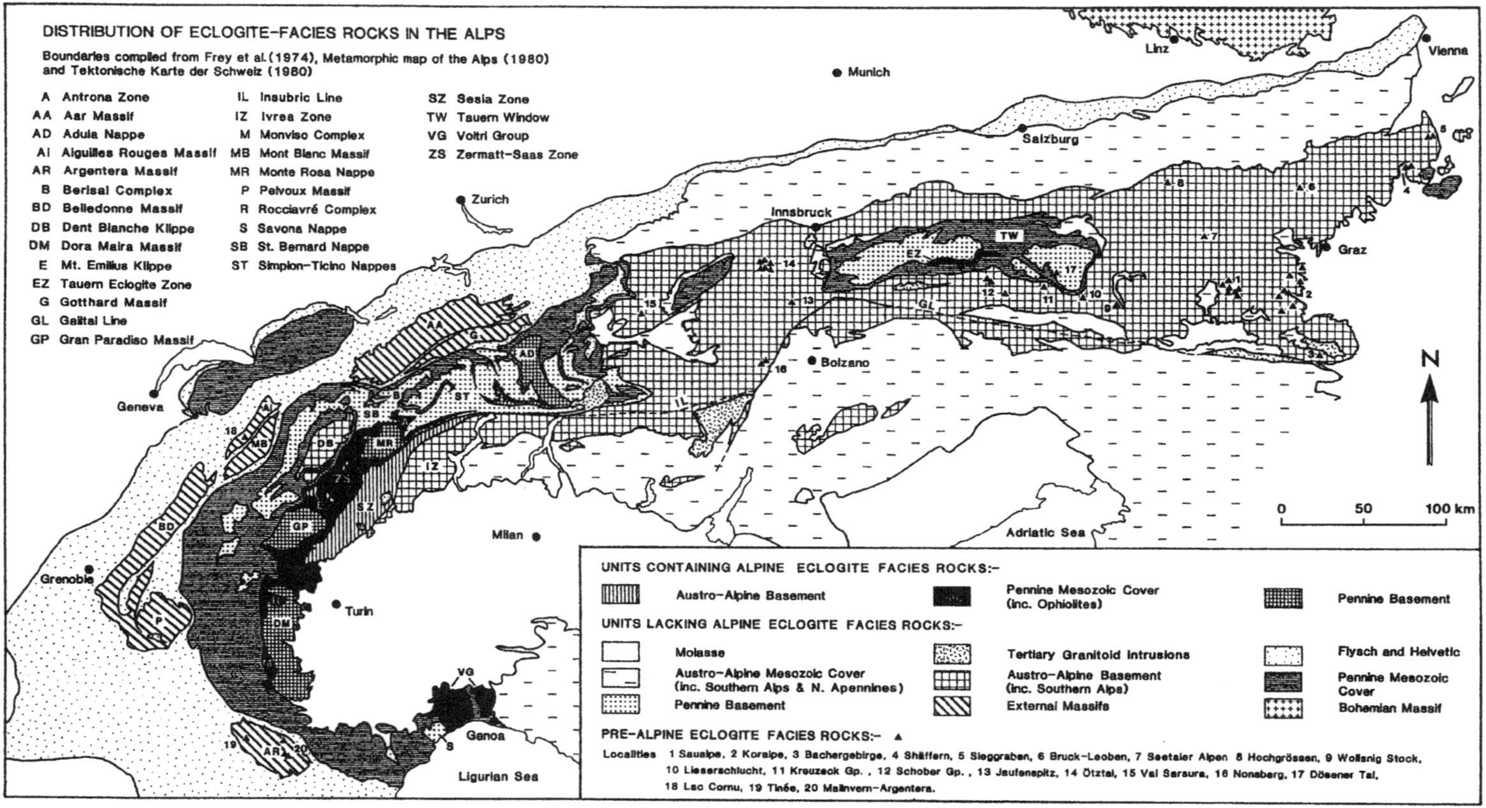

Figure 10.1 Distribution of eclogite facies rocks in the Alps.

Eclogitic rocks consisting of Hbl + Grt + Di + Pl + Qtz occur in amphibolite facies gneisses at three localities in the Mont Blanc Massif (Duparc and Mrazec, 1898; von Raumer, 1974).

10.2.2 *Pennine basement complex*

Few candidates now exist for pre-Alpine eclogites in the Penninic Domain. Rare, variably amphibolitised relics of fine-grained, iron-rich quartz-eclogites occur in concordant bands in a large garnet-amphibolite mass in the Pennine Basement of the south-east Tauern Window (Droop, 1983). As this is an area that has undergone both Variscan and Alpine deformation and amphibolite facies metamorphism, a pre-Alpine age is difficult to prove, but field evidence suggests that the primary assemblage Grt ($Prp_{\sim 12}$) + Omp (Jd_{36-46}) + Qtz + Rt predated the development of the strong pre-Alpine foliation in the associated amphibolites which is itself truncated by relatively undeformed Variscan aplite veins. Again, an early Variscan or older age for the eclogites is implied. Calculated metamorphic conditions for the primary eclogite assemblage ($T = 620 \pm 100°C$, $P \geqslant 12$ kbar) are clearly different from those recorded by both Variscan and Alpine amphibolite facies assemblages (Droop, 1983).

Garnet amphibolites with garnet–hornblende symplectite replacing omphacite have been described from the Savona Basement Nappe of the Briançonnais Zone in western Liguria. The eclogitic assemblages are believed to be of Caledonian age (Messiga, 1981).

10.2.3 *Austro-Alpine nappes*

By far the greatest concentration of pre-Alpine eclogite facies rocks occurs in the Middle Austro-Alpine nappe system of Austria (Figure 10.1). The most famous are the eclogites of the Saualpe and Koralpe Complexes (the Koriden association) which have been known since Haüy's (1822) original description of the type eclogite, and which have since been studied by numerous workers including Kieslinger (1928); Wieseneder (1934); Angel and Schaider (1950); Angel (1957); Beck-Mannagetta (1961); Machatschki and Walitzi (1962); Weissenbach (1965); Heritsch (1966, 1973); Mottana *et al* (1968); Heritsch and Bossert (1969); Heritsch and Wittmann (1969); Richter (1973); Miller and Frank (1983); Miller (1985); Miller *et al* (1988) and Manby and Thiedig (1988). These eclogites form layers and lenses within a thick sequence of polymetamorphic paragneisses which include abundant garnet–kyanite schists, blastomylonites ('Plattengneis') and minor marbles. The rocks have clearly had a long and complicated geological history and the status of the eclogites is still controversial. According to Miller (1985), the schist protoliths are of late Proterozoic to Cambrian age and have undergone at least three regional metamorphic events: (*a*) a HT–LP event, producing And/Sil + Kfs assemblages in pelites; (*b*) a Barrovian amphibolite facies event, producing Grt + Ky + St + Ms assemblages; (*c*) a HP event ($T = 600$–$620°C$, $P = 10$–12 kbar) synchronous with shearing in the Plattengneis, again with garnet and kyanite stable in pelites. Traditionally, these events have been regarded as Variscan or older, but recently Frank *et al* (1983) have obtained Cretaceous Rb–Sr and K–Ar ages from micas that crystallised during the third event. Below the Plattengneis horizon, the eclogites are demonstrably pre-Alpine as they are overprinted by amphibolite facies minerals aligned in a foliation that is cut by undeformed pegmatites dated at 270 Ma (Miller, 1985). Evidence for a Proterozoic age comes from preliminary Sm–Nd dating of the Saualpe eclogites which suggests crystallisation of garnet and omphacite prior to 700 Ma (Manby and Thiedig, 1988). In eclogite boudins within the Plattengneis, omphacite has allegedly recrystallised in the mylonitic foliation, suggesting that eclogitic assemblages were also stable during the Cretaceous shearing event (Miller and Frank, 1983).

Geochemically, the Koriden eclogites fall into two categories: (*a*) a low-Ti group with chemical characteristics indicating formation of igneous precursors as cumulates produced by pyroxene and plagioclase fractionation of basic magma, and (*b*) a high-Ti group which probably originated as basaltic liquids with normal to enriched MORB affinities (Miller *et al*, 1988).

The typical maximum phase assemblage of the unretrograded, texturally equilibrated eclogites of the Koriden association is Grt + Omp + Ky + Amp + Zo + Qtz + Rt + accessories (Table 10.1). The eclogites typically show internal banding defined by variations in grain size and modal mineralogy, and elongate omphacites, kyanites, zoisites and amphiboles are commonly aligned parallel to the schistosity in the surrounding rocks. Some unfoliated varieties preserve textural evidence of a gabbroic origin and contain pseudomorphs of omphacite after augite and of Zo + Ky + Jd + Qtz after plagioclase, separated by garnet coronas. Garnets in both texturally equilibrated and metagabbroic eclogites are magnesian (typically Prp_{34-48}) and omphacites contain 31–40% of the jadeite molecule (Miller, 1985). Primary amphiboles are barroisites and magnesio-hornblendes in the classification of Leake (1978b). Equilibration conditions for the eclogitic assemblages have been estimated by Miller (1985). Temperatures obtained by application of the Ellis and Green (1979) garnet–clinopyroxene thermometer range from 610 to 670°C. Pressures of about 18 kbar have

Table 10.1 Maximum mineral assemblages in texturally equilibrated, pre-Alpine eclogite facies rocks from the Austro-Alpine Nappes

Lithology	Locality	Eclogite facies assemblage	
Metabasites	Saualpe	Grt + Omp + Qtz + Ky + Zo + Hbl + Rt + Po	(1, 7, 9)
		Grt + Omp + Qtz + Ky + Zo + Bar + Rt	(7)
		Grt + Omp + Qtz + Ky + Zo + Hbl + Pg* + Dol + Rt	(7, 9)
		Grt + Omp + Qtz + Czo + Phe + Rt	(9)
	Koralpe	Grt + Omp + Qtz + Ky + Zo + Hbl + Rt + Py	(4, 7, 9)
		Grt + Omp + Qtz + Czo + Rt + Py + Ap + Zrn	(3, 7, 9)
		Grt + Omp + Qtz + Zo/Czo/Ep + Hbl + Rt + Ap	(7, 9)
		Grt + Omp + Qtz + Czo/Ep + Phe + Rt + Ap + Ilm/Mt	(9)
	Ötztal	Grt + Omp + Qtz + Ky + Czo + Act + Rt + Ap	(2)
	Silvretta	Grt + Omp + Rt + Py + Ilm	(5)
Peridotites	V. Ultimo	Grt + Ol + Opx + Cpx + Hbl + CrSpl	(8)
Pyroxenites	V. Ultimo	Grt + Di + Hbl + Ilm	(6)
		Grt + Opx + Cpx + Hbl + CrSpl	(6)

*Only as inclusions in garnet.
Sources: (1) Mottana *et al* (1968); (2) Miller (1970); (3) Richter (1973); (4) Miller and Frank (1983); (5) Thierrin (1983); (6) Morten and Obata (1983); (7) Miller (1985); (8) Obata and Morten (1987); (9) Miller *et al* (1988).

been calculated for coexisting omphacite, kyanite and paragonite inclusions in garnet in one sample, using the Pg = Jd + Ky + H_2O equilibrium (Holland, 1979b) assuming unit water activity, and are consistent with minimum pressures of 17 kbar implied by the coexistence of jadeite and quartz in the metagabbros.

Eclogites have also been described from the polymetamorphic terrain of the Ötztal Alps (e.g. Hezner, 1903; Hammer, 1926, 1933; Miller, 1969, 1970, 1974a) where they form concordant lenses within a 1 × 12 km outcrop of laminated amphibolite. The latter represents a metagabbroic complex containing cumulate layers of peridotitic, troctolitic and anorthositic composition; the whole complex underwent eclogite facies metamorphism but relatively few domains escaped pervasive amphibolite facies overprinting. Of the eclogite facies rocks, most are fine-grained, banded zoisite-kyanite-Ca amphibole-eclogites, but coronitic metagabbros (some kyanite-bearing) and metaperidotites are also represented. In the fine-grained eclogites, zoned garnets with Mg-rich rims ($Prp_{\sim 43}$) coexist with Jd_{30-35} omphacites (Miller, 1970). Temperatures of 640–700°C have been obtained by the Ellis and Green thermometer; minimum pressures are 13–15 kbar (Miller, personal communication). According to Söllner and Gebauer (unpublished U–Pb zircon data), the age of the eclogite facies event is 497 ± 2 Ma.

Kyanite-free eclogites and eclogitic amphibolites occur in the Silvretta Nappe (e.g. Thierrin, 1983; Maggetti *et al*, 1987) and in the Altkristallin of the Schober and Polinik Groups south of the Tauern Window (e.g. Exner and Wanderer, 1962; Hanselmeyer, 1977). The eclogites of the Schober Group are surrounded by schistose gneisses that preserve clear evidence of deformation and metamorphism of Caledonian age (Troll, 1978). Those of the Polinik Group, however, are associated with schists that have been strongly affected by mid-Cretaceous amphibolite facies metamorphism (Waters, 1976), and an early Cretaceous age for these has been tentatively proposed by Hoke (1987). Thierrin (1983) has estimated equilibration temperatures of 690–740°C for the Silvretta eclogites.

Eclogites are known from many other localities in the Austro–Alpine basement (Figure 10.1—references listed by Angel, 1957, and Richter, 1973). Few have been worked on in the later 1970s or in the 1980s and their ages are constrained even less well than those described above.

Eclogite facies ultramafic rocks occur in lenticular bodies, measuring up to 100 × 300 m, in upper amphibolite facies gneisses outcropping in the Val d'Ultimo area of northern Italy (e.g. Hammer, 1899; Andreatta, 1935; Rost and Brenneis, 1978; Obata and Morten, 1982, 1987; Morten and Obata, 1983). The lenses consist mainly of compositionally layered peridotites with subordinate pyroxenites in 10–40 cm thick concordant bands. The peridotites themselves are of two types (Obata and Morten, 1987): (*a*) coarse-grained, protogranular spinel-lherzolites, probably representing relatively unmodified upper mantle material, in which the pyroxenes show complex high-temperature exsolution textures (e.g. enstatite, spinel and garnet lamellae in clinopyroxene); (*b*) fine-grained metamorphic derivatives of these, the least retrogressed of which are garnet-hornblende-lherzolites and garnet-hornblende-spinel-harzburgites (Table 10.1). The pyroxenites are garnet-hornblende-websterites and garnet-hornblende-clinopyroxenites. Some layers contain large (~6 cm) clinopyroxene megacrysts with exsolution lamellae of orthopyroxene and garnet. The megacrysts have unusually subcalcic integrated compositions (Ca/(Ca + Mg) = 0.26–0.28) implying extremely high temperatures (> 1450°C) for their primary crystallisation in the mantle (Morten and Obata, 1983). Obata and Morten (1987) have obtained temperatures of 1100–1300°C for the spinel-lherzolites and 700–800°C for the fine-grained garnet-lherzolites using the Wells (1977) and Lindsley and Anderson (1983) two-pyroxene thermometers, and pressures for the latter rocks in the range of 16–28 kbar using the garnet–orthopyroxene geobarometer (Harley, 1984b; Wood, 1974) and the Prp–Fo–En–Spl equilibrium (O'Neill, 1981). Evolution from spinel- to garnet-lherzolite was either by isobaric cooling at great depth or by cooling accompanied by burial. Given that temperatures during eclogite facies recrystallisation exceeded 700°C, the coincidence between the 332 and 336 Ma zircon ages obtained from the pyroxenites by Gebauer and Grünenfelder (1979c) and the phlogopite K–Ar date of 330 ± 15 Ma obtained by Hunziker (quoted in Rost and Brenneis, 1978) is good evidence that this event was of Variscan age or older.

10.2.4 *Implications for pre-Alpine geotectonics*

The only firm conclusion that can be drawn from the foregoing descriptions is that pre-Cretaceous eclogite-forming events occurred widely in all three major crystalline basement domains of the Alpine region (Helvetic, Penninic and Austro-Alpine). Many of the Austro-Alpine occurrences yield P of > 17 kbar which implies metamorphism at depths of > 60 km. Some, like the Val d'Ultimo ultramafics, appear to be primary mantle rocks metamorphosed at mantle depths. Others, such as the Silvretta and Koriden eclogites, were evidently derived from upper crustal protoliths (e.g. Maggetti *et al*, 1987; Miller *et al*, 1988; Manby and Thiedig, 1988) and require models of deep underthrusting (subduction) to explain them. Unfortunately, however, crystallisation ages are not yet well enough constrained to resolve the eclogite-forming events either temporally or spatially, but it seems likely that relics of late Proterozoic, Caledonian and Variscan subduction are represented. In their elaborate evolutionary model for the Austro-Alpine basement, Frisch *et al* (1984) use evidence from eclogites and associated lithologies to argue for two periods of ocean closure in the Palaeozoic: a Caledonian phase, involving subduction towards the south, and an early Variscan phase with subduction to the north. Testing the validity of this model will be difficult until more precise age and provenance data for the eclogites are available. The relationship between the eclogite-forming event(s) in the External Massifs and those in the Bohemian Massif (see Chapter 9) also awaits clarification.

10.3 Alpine eclogites and related eclogite facies rocks

The occurrence of glaucophane eclogites in the Western Alps has been known since the second half of the nineteenth century (Bodewig, 1876; Cossa, 1879). These rocks, and especially the garnet-poor varieties known to the older authors as 'jadeitites', attracted considerable interest in connection with the debate on the autochthonous origin of the jade artifacts which are so characteristic of Neolithic sites around the Alps (Damour, 1881; Franchi, 1900; Bodmer-Beder, 1902). Most eclogite bodies occur on the internal side of the Western Alps and were found by the staff of the Regio Ufficio Geologico (S. Franchi, E. Mattirolo, V. Novarese and A. Stella) during the detailed mapping of the whole region in the years 1888–1908 (R. Ufficio Geologico, 1908; Franchi, 1900). The principal occurrences are in continental basement nappes derived from the Austroalpine (African) palaeomargin, the European palaeomargin and from parts of the intervening Mesozoic oceanic lithosphere.

Both in the ophiolite and in the continental eclogites, two types of transformations are represented: (*a*) 'static' eclogitisation, whereby the eclogite minerals developed as pseudomorphs on, or coronas around the primary minerals, thus preserving the original rock-structure; (*b*) 'dynamic' eclogitisation whereby the original mineralogy and structure were completely changed during syntectonic recrystallisation.

10.3.1 *Eclogitisation of the Mesozoic ophiolites and associated sediments*

10.3.1.1 *Western Alps.* The Jurassic ophiolites of the Western Alps formed in the small Piedmont–Ligurian ocean and, a few tens of Ma after their formation, underwent eclogitisation during the early-Alpine subduction associated with closure of the basin. Eclogitised ophiolites occur in the internal (i.e. eastern) part of the Western Alps (Bearth, 1967; Bocquet, 1974; Dal Piaz, 1974) and are widespread in the Zermatt–Saas area, in Val d'Aosta, Valli di Lanzo and Val Susa, in the Lanzo peridotite body, in the Monviso massif and in the Gruppo di Voltri of western Liguria. The meta-ophiolites include ultramafics, gabbroids, basalts, tuffites, and preserve a metasedimentary cover with detrital horizons.

Most of the large ultramafic bodies derive from tectonised lherzolites and are extensively or completely serpentinised (Ernst and Piccardo, 1979; Cimmino *et al*, 1981; Ernst, 1981; Pognante *et al*, 1986). The main early-Alpine parageneses are summarised in Table 10.2. Rodingites and ophicarbonate rocks occur at the boundary between serpentinites and adjacent rocks (Dal Piaz, 1969; Cortesogno *et al*, 1981).

The metagabbroids display a wide compositional range and comprise (mela)troctolites, Cr-diopside-bearing gabbros rich in Mg and Al, gabbro-norites, gabbros rich in Fe and Ti and rare trondhjemites (Mottana and Bocchio, 1975; Lombardo *et al*, 1978; Pognante *et al*, 1982, 1985). The spread of metagabbro bulk compositions describes a tholeiitic trend and is consistent with formation of protoliths by LP fractionation and cumulus processes under initially low but increasing f_{O_2} (Pognante *et al*, 1982; Pognante, 1985). Some of the Fe, Ti-rich metagabbros have very high bulk Fe/Mg ratios (e.g. the Vara eclogites of the Voltri Group) and have compositions similar to highly fractionated gabbros from mid-ocean ridge settings (Bocchio and Mottana, 1975; Mottana and Bocchio, 1975).

The metagabbros are characterised by the association of rock-volumes which partly escaped the Alpine syn-eclogite facies deformations with rocks strongly deformed in shear zones (e.g. the Rocciavré metagabbros; Pognante, 1985). In the poorly deformed rocks, plagioclase is transformed to omphacite or jadeite, zoisite, paragonite ± garnet; the magmatic clinopyroxene is pseudomorphed by omphacite ± glaucophane; olivine is chiefly transformed to talc ± chlorite with a garnet corona; Fe–Ti oxides are replaced by rutile ± ilmenite with a thick reaction rim of garnet. The most typical eclogitic assemblages are observed in the metagabbros rich in Fe and Ti and include abundant omphacite, garnet and rutile (Mottana and Bocchio, 1975; Cortesogno *et al*, 1977; Ernst and Dal Piaz, 1978; Lombardo *et al*, 1978; Morten *et al*, 1979; Cimmino *et al*, 1981; Baldelli *et al*, 1985; Nisio, 1985; Sandrone *et al*, 1986; Pognante and Kienast, 1987). Eclogitic parageneses in the Mg, Al-rich metagabbros are characterised by zoisite, Ca–Na amphiboles, garnet and omphacite and more rarely include Mg-chloritoid, talc, chlorite and/or kyanite (Bearth, 1967, 1973, 1974; Chinner and Dixon, 1973; Messiga *et al*, 1983b; Kienast and Messiga, 1987; Kienast and Pognante, 1988; Pognante, 1988; Table 10.2). The meta-trondhjemites display jadeite + quartz parageneses with minor ferro-omphacite (Sandrone *et al*, 1986; Pognante, 1988).

Table 10.2 Representative eclogite facies assemblages of early Alpine age in meta-ophiolites of the Western Alps

Rock type	Locality	Eclogite facies assemblage	
Ultramafic	Voltri Gp.	Ol + Atg + Di + Ti-Chu	(10)
	V. Orco	Ol + Atg + Chl + Tlc + Mag	(15)
Mg-metagabbros	Voltri Gp.	Grt + Omp + Qtz + Zo	(10)
	Monviso	Grt + Omp + Zo + Rt	(3)
	Rocciavré	Grt + Omp + Qtz + Zo + Gln + Pg + Rt	(14)
		Grt + Omp + Mg-Cld + Zo + Bar + Pg + Rt	(12)
	Zerm.-Saas	Grt + Omp + Mg-Cld + Tlc + Gln	(1)
		Grt + Omp + Mg-Cld + Tlc + Czo + Rt	(7)
		Grt + Omp + Mg-Cld + Czo + Gln + Pg + Rt + Ap	(1)
		Grt + Omp + Ky + Qtz + Tlc + Czo + Rt	(7)
		Grt + Omp + Qtz + Czo + Pg + Rt	(7)
		Grt + Omp + Qtz + Gln + Act + Phe + Rt + Py	(13)
		Grt + Omp + Qtz + Gln + Czo + Pg + Ttn	(13)
Fe, Ti-metagabbros	Voltri Gp.	Grt + Clm + Gln + Rt/Ilm	(2)
		Grt + Omp + Qtz + Rt	(10)
	Rocciavré	Grt + Omp + Qtz + Zo + Pg + Rt	(14)
	Fenestrelle	Grt + Clm + Rt + Ap	(4)
Meta-basalts	Monviso	Grt + Omp + Qtz + Czo + Gln + Pg + Rt	(3)
	Zerm.-Saas	Grt* + Omp + Qtz + Czo + Lws + Rt	(7)
		Grt + Omp + Qtz + Czo + Gln + Pg + Ank + Rt	(1, 7)
Metatrond-hjemites	V. Susa	Jd + Qtz + Gln + Ttn	(12)
	Lanzo	Jd + Qtz + Ttn	(15)
Inter-pillow sediment	Zerm.-Saas	Grt + Qtz + Ep + Gln + Pg + Dol + Rt	(11)
Pelites and calc-pelites	Voltri Gp.	Phe + Qtz + Grt + Zo + Pg	(10)
	V. Susa	Phe + Qtz + Grt + Cld + Cal + Ank + Ttn + Gr	(15)
		Phe + Qtz + Grt + Acm + Ep + Gln + Ttn	(15)
	Zerm.-Saas	Phe + Qtz + Grt + Cld + Chl + Cal	(5)
Meta-cherts	Zerm.-Saas	Qtz + Hem + Sps + Brn + Pie + Phe	(5, 8, 9)
		Qtz + Sps + Brn + Pie + Pxm + Rdn + Phe	(8)
		Pie + Mn–Cpx + Phe	(6)
		Hem + Brn + Mn–Cpx	(6)
		Qtz + Grt + Acm + Py	(15)

*Early eclogitic assemblage.

Sources: (1) Bearth (1973); (2) Mottana and Bocchio (1975); (3) Lombardo *et al* (1978); (4) Otten and Brouwer (1979); (5) Bearth and Schwander (1981); (6) Griffin and Mottana (1982); (7) Barnicoat and Fry (1986); (8) Mottana (1986b) and references therein; (9) Martin and Kienast (1987); (10) Messiga (1987); (11) Barnicoat (1988); (12) Pognante (1988); (13) Reynard and Ballèvre (1988); (14) G. Wayte, personal communication; (15) personal observation.

The extrusive ophiolitic metabasics derive from tholeiitic basalts with prevalent N-MORB affinity and include tuffites and detrital levels in the supra-ophiolitic sequences (Dal Piaz and Ernst, 1978; Lombardo *et al*, 1978; Piccardo *et al*, 1979; Dal Piaz *et al*, 1981; Beccaluva *et al*, 1984; Pognante *et al*, 1985). Metabasaltic eclogites that preserve original pillow-lava structures have been described by Bearth (1959, 1967, 1973) in the Zermatt–Saas meta-ophiolites. The pillow cores have been converted into fine-grained muscovite- and epidote-bearing eclogite, whilst pillow rims may be enriched in either epidote or omphacite and glaucophane. Inter-pillow material consists of glaucophane, epidote, white mica, garnet and omphacite with ankerite and quartz (e.g. Barnicoat, 1988). Eclogitised pillow lavas also occur in other ophiolite complexes of the Western Alps and are particularly well preserved in the Monviso complex (Lombardo *et al*, 1978). The metabasalts also occur as dykes that cross-cut gabbros and the ultramafics.

The metasediments associated with the ophiolitic rocks are mainly represented by phengite-bearing calcschists, impure marbles and carbonate-bearing micaschists (Bearth, 1967; Lombardo *et al*, 1978; Cimmino *et al*, 1981). Subordinate micaceous quartzites derive from original radiolarites or from cherts and in places are enriched in manganiferous minerals (Table 10.2; Dal Piaz *et al*, 1979; Bearth and Schwander, 1981; Mottana, 1986b), particularly in the St Marcel valley (Val d'Aosta), where manganiferous Na-pyroxene (violan) and garnet are common in impure quartzitic layers (Griffin and Mottana, 1982; Martin and Kienast, 1987).

The mineral chemistry of the eclogitised ophiolites of the Western Alps reflects the bulk-rock chemistry (Ernst, 1976; Cimmino *et al*, 1981), but in many poorly deformed rocks the composition of the eclogite facies minerals depends also on the microstructural domain of crystallisation (Pognante and Kienast, 1987). The eclogite minerals are usually zoned (Morten and Bondi, 1984) with compositions that may reflect either a growth zoning (i.e. for garnet) or a diffusion zoning imposed upon pre-existing (magmatic) compositions (i.e. for the omphacite pseudomorphs after augite). Such disequilibrium or local equilibrium features are less obvious in the blastomylonitic rocks; these show a greater degree of textural equilibration and their eclogite facies minerals display less marked chemical heterogeneities.

Na-pyroxenes are impure jadeite, omphacite, chloromelanite and sodian augite (Figure 10.2) and their distribution along the augite–jadeite join probably reflects the existence of a miscibility gap between omphacite and jadeite at low temperatures (Carpenter, 1980). Garnets are rich in almandine and display higher pyrope contents in the metagabbros rich in Mg and Al, and higher spessartine content in the metasediments (Figure 10.3). Na-amphiboles are usually glaucophane with minor ferroglaucophane and crossite (Figure 10.4); the coexistence of glaucophane and actinolite in a few poorly deformed metagabbros probably reflects a miscibility gap between Na and Ca amphiboles. Chloritoid is a magnesian variety in the Mg, Al-rich metagabbros and a ferroan variety in the calcschists and in the metagabbros rich in Fe and Ti.

Thermobarometric estimates for the eclogite event have been derived by the Fe–Mg exchange thermometer in garnet–omphacite pairs, by the coexistence of jadeite + quartz and consideration of the phase relations in the systems FMAS and CMAS (Ernst, 1976; Ernst and Dal Piaz, 1978; Oberhänsli, 1980; Messiga *et al*, 1983b; Pognante, 1985, 1988; Baldelli *et al*, 1985; Barnicoat and Fry, 1986; Lardeaux *et al*, 1986; Sandrone *et al*, 1986; Messiga, 1987; Pognante and Kienast, 1987; Kienast and Pognante, 1988). The inferred values point to higher P–T conditions of eclogitisation

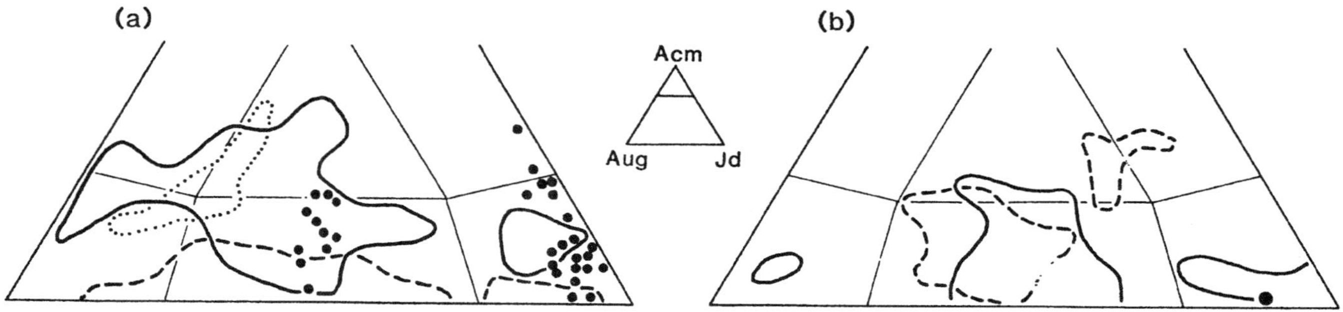

Figure 10.2 Pyroxene compositions in the eclogitic rocks of the western Alps. See text for sources of data. (a) Meta-ophiolites. Fields of pyroxenes: (————) from Fe–Ti-metagabbros; (-----) from Mg-metagabbros; (······) from metabasalts and other metabasic rocks; (● ●) Pyroxenes from metatrondhjemites. (b) Continental-type units. Fields of pyroxenes: (————) from the Sesia Zone. (-----) from the Monte Rosa, Gran Paradiso and Dora Maira massifs; (●) jadeite from the southern Dora Maira massif.

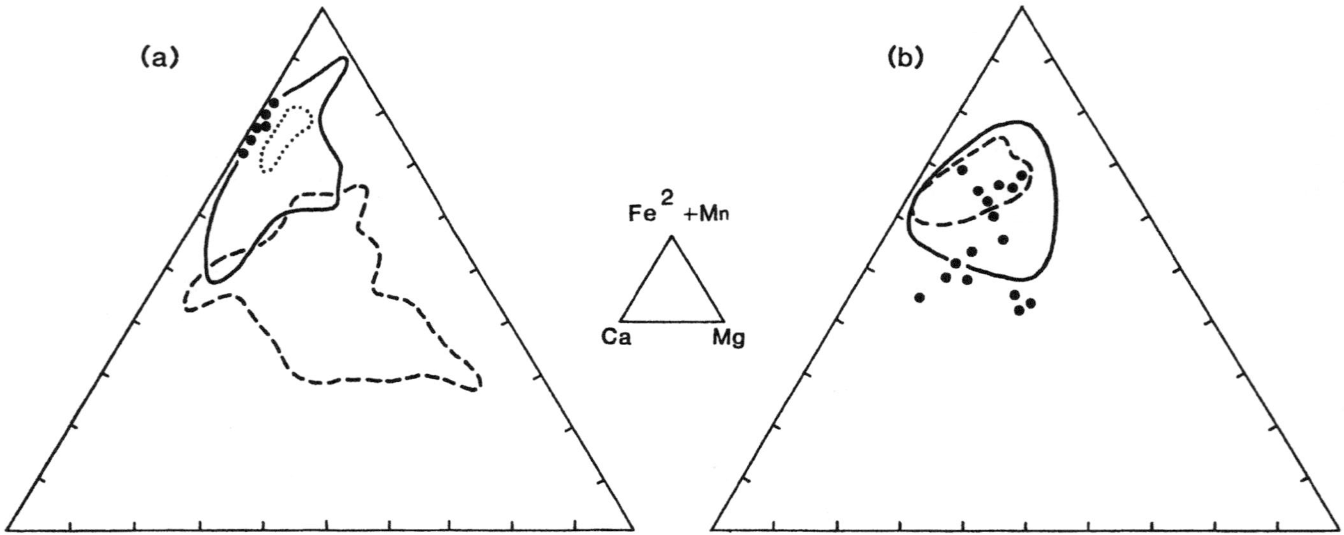

Figure 10.3 Garnet compositions in the eclogitic rocks of the Western Alps. See text for sources of data. (a) Meta-ophiolites. Fields of garnets: (————) from Fe–Ti-metagabbros; (-----) from Mg-metagabbros; (......) from metabasalts and other metabasic rocks; (• •) garnets from metatrondhjemites. (b) Continental-type units. Fields of garnets: (————) from the Sesia zone; (-----) from the Monte Rosa and Gran Paradiso massifs; (• •) garnets from the Dora Maira massif.

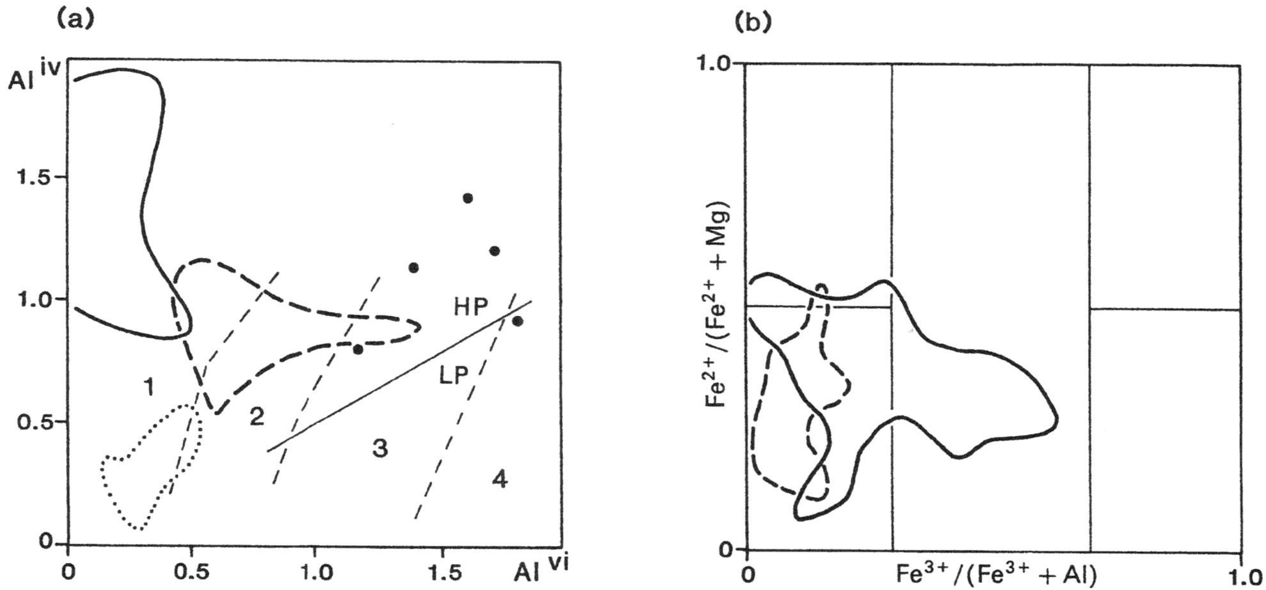

Figure 10.4 High-pressure amphiboles in the eclogitic rocks of the Western Alps. See text for sources of data. (a) Al^{VI}–Al^{IV}. (————) field of Na-amphiboles from all the eclogitised units; (-----) field of Na–Ca-amphiboles from the Sesia Zone; (······) field of syn-eclogitic actinolites from the ophiolitic metagabbros. (● ●) post-eclogitic amphiboles from the eclogites of the southern Dora Maira massif; (————) separating HP and LP (i.e. $P < 5$ kbar) amphiboles (from Raase, 1974). Compositional fields 1–4 are for greenschist, epidote–amphibolite, amphibolite and granulite facial amphiboles, respectively (Zakrutkin, 1968). (b) $Fe^{2+}/(Fe^{2+} + Mg)$ versus $Fe^{3+}/(Fe^{3+} + Al)$. Fields of Na-amphiboles: (————) from the metaophiolites, (-----) from the continental-type units.

in the Breuil–Zermatt area ($T = 550$–$600°C$ at $P > 15$–20 kbar) relative to the more southerly areas ($T = 450$–$500°C$ at $P = 10$–18 kbar). The eclogite crystallisation was followed by a decompressional path under decreasing temperatures with a syntectonic blueschist stage characterised by the growth of abundant glaucophane ($\pm$ lawsonite), and later, typically static, LP greenschist stage (late-Alpine stage).

Ages of 100–60 Ma have been derived for the HP metamorphism of the Piedmont meta-ophiolites (e.g. Bocquet *et al*, 1974; Delaloye and Desmons, 1976) though probably only the earlier part of this range corresponds to the eclogite facies event.

10.3.1.2 *Central Alps: Antrona Zone and Adula ophiolites.* Relatively few relics of eo-Alpine HP metamorphism are preserved in the Central Alps owing to the pervasive effects of the late Alpine ('Lepontine') amphibolite facies overprint in this area.

In the southern steep belt of the Central Alps, partially overprinted eclogites are known from several scattered localities in and adjacent to the Antrona and Mergoscia ophiolitic zones where they form relict pods in amphibolite (Colombi and Pfeifer, 1986, and references therein). In most samples, the primary eclogite facies assemblages have been destroyed leaving only garnet and omphacite ($X_{Jd} = 0.37$–0.45) 'islands' in a symplectic groundmass, but in a few, these phases coexist with paragonite, quartz and rutile. Colombi and Pfeifer (*op. cit.*) estimated peak metamorphic temperatures of 480–720°C using Ellis and Green's (1979) calibration of the garnet–clinopyroxene thermometer, and a minimum pressure of 14 kbar from the jadeite content of omphacite coexisting with quartz (Gasparik and Lindsley, 1980). An upper pressure limit of about 22 kbar is indicated by the stability of paragonite relative to omphacite + kyanite + fluid (Holland, 1979b).

The greatest concentration of eo-Alpine eclogite facies rocks in the Central Alps is found in the Adula Nappe. The Adula Nappe itself is a basement unit containing Variscan pelites and metabasites (Heinrich, 1986), the eo-Alpine metamorphism of which is discussed in Section 10.3.3.2). However, the Adula suite also includes a small proportion of eclogite facies ophiolitic rocks of probable Mesozoic age as slices imbricated into the frontal and marginal parts of the nappe. Lithologies in this category include garnet-lherzolite, calc-schist, meta-rodingite and associated mafic eclogites. The best evidence for an eo-Alpine age for the eclogitic metamorphism of this suite is provided by calc-mica-schists that resemble Mesozoic 'Bündnerschiefer' and contain assemblages such as Grt + Omp + Zo + Qtz (Heinrich, 1982) (Table 10.3).

The garnet-lherzolites are known from three localities, Alpe Arami (Grubenmann, 1908; Dal Vesco, 1963; O'Hara and Mercy, 1966; Möckel, 1969; Rost *et al*, 1974), Cima di Gagnone (Evans and Trommsdorff, 1978) and Monte Duria (Fumasoli, 1974, quoted in Evans and Trommsdorff, *op. cit.*), where they occur as relict layers within large lenses of otherwise foliated, garnet-free chlorite-hornblende-harzburgite. At Cima di Gagnone, the peridotites contain small discordant bodies of garnet-clinopyroxene rocks that grade in composition from tholeiitic ocean-floor basalt to rodingite (Trommsdorff *et al*, 1975; Evans *et al*, 1979, 1981). Because rodingites are produced exclusively by serpentinisation, the presence of metarodingites strongly suggests that the garnet-peridotites were derived from LP serpentinite precursors rather than primary mantle material. Further evidence that the garnet-peridotites share a common origin with lower-pressure Alpine metaserpentinites (e.g. the Malenco ultramafics) is provided by the presence of Ti-clinohumite relics in both suites (Evans and Trommsdorff, 1983). At

Table 10.3 Primary eclogite facies assemblages in ophiolitic and continental lithologies in the Adula Nappe. Subsets of the listed assemblages are common.

Rock type	Locality	Eclogite facies paragenesis	
Ultramafic	Arami, Duria	Grt + Ol + Cr-Di + Opx	(1, 2, 3)
	Gagnone	Grt + Ol + Cr-Di + Opx + Hbl	(4)
Metabasites	Arami	Grt + Omp + Ky + Qtz + Zo + Hbl + Rt + Ilm	(9)
	Gagnone	Grt + Omp + Ky + Qtz + Zo + Hbl + Rt	(5)
		Grt + Omp + Qtz + Zo + Czo + Hbl + Rt + Ilm	(9)
	Soazza	Grt + Omp + Qtz + Rt	(8)
	Trescolmen	Grt + Omp + Ky + Qtz + Czo + Hbl + Phe + Rt	(6)
		Grt + Omp + Ky + Qtz + Gln + Tlc + Pg + Rt + Py	(9)
	Confin	Grt + Omp + Ky + Qtz + Bar + Pg + Phe + Rt + Py	(9)
		Grt + Omp + Qtz + Hbl + Pg + Phe + Rt + Py	(9)
		Grt + Omp + Qtz + Hbl + Zo + Ep + Rt + Ttn + Py	(9)
	Vals	Grt + Omp + Qtz + Gln + Ep + Phe + Cal + Py	(6, 9)
		Grt + Omp + Qtz + Ab + Hbl + Czo + Phe + Rt + Py	(6, 9)
Rodingites	Gagnone	Grs + Di + Hbl + Ep + Rt	(5)
Calcschists	Trescolmen	Dol + Grt + Omp + Ky + Qtz + Zo + Rt	(6)
	Confin	Dol + Cal + Cpx + Qtz + Act + Zo + Rt + Py	(9)
	Vals	Cal + Dol + Pg + Phe + Grt + Omp + Qtz + Zo + Hbl + Rt + Py + Gr	(6)
Pelites	Trescolmen	Qtz + Pg + Phe + Grt + Rt + Py	(6)
	Vals	Qtz + Pg + Phe + Grt + Ky + Tlc + Ep + Dol + Rt + Py	(6)
Quartzites	Trescolmen	Qtz + Phe + Grt + Omp + Ky + Rt + Py	(6)

Sources; (1) O'Hara and Mercy (1966); (2) Möckel (1969); (3) Rost *et al* (1974); (4) Evans and Trommsdorff (1978); (5) Evans *et al* (1979); (6) Heinrich (1982); (7) Evans and Trommsdorff (1983); (8) Aurisicchio *et al* (1985); (9) Heinrich (1986).

Alpe Arami, the ultramafic rocks are surrounded by an envelope of hornblende-quartz-kyanite-eclogite (e.g. Bocchio, 1975, 1977; Ernst, 1977; Bocchio *et al*, 1978) whose geochemistry is also consistent with an ocean-floor provenance (Bocchio *et al*, 1985).

10.3.1.3 *Eastern Alps: Eclogite Zone of Tauern Window.* In the Penninic terrain of the Eastern Alps, eclogite facies rocks of Alpine age are restricted to a 3 km wide strip of Mesozoic cover rocks, known as the Eclogite Zone, in the south central

Tauern Window (Frank *et al*, 1987a). Most eclogite facies rocks preserved there are of mafic composition (Scharbert, 1954; Miller, 1974b, 1976, 1977a, 1977b; Holland, 1979a) but eclogite facies mineral assemblages have also been described from associated calc-schists, siliceous dolomites, pelites and aluminous quartzites (Miller, 1977b; Holland, 1979a; Franz and Spear, 1983; Spear and Franz, 1986; Dachs, 1986). The cofacial nature of the HP assemblages in metasediments and metabasites suggests that the Eclogite Zone behaved as a coherent unit during eclogite facies metamorphism (Spear and Franz, 1986) and this is supported by the apparent continuity of the stratigraphy on a scale of tens of metres to kilometres (Raith *et al*, 1980). The eclogites themselves form folded and boudinaged slab-like bodies, generally between 3 and 20 m thick and < 100 m long, enclosed in a matrix of calc-schist. Severely flattened pillow structures are locally present, suggesting that at least some eclogites originated as submarine basalts (Holland, 1979a). Most of the eclogite bodies have chloritised sheared margins, but at some contacts eclogite grades into or is finely interbanded with calc-schists that contain HP mineral assemblages (Holland, 1979a), indicating that some of the primary basalt-sediment interfaces have remained intact. Geochemically, the eclogites resemble tholeiitic to mildly alkaline basalts of within-plate character (Höck and Miller, 1987), and the presence in the Eclogite Zone of some rocks of shallow-water sedimentary origin supports the view that this rock association is not truly ophiolitic (Frank *et al*, 1987).

The mineralogy of unretrograded Tauern eclogites is characterised by an abundance of carbonates (magnesite and dolomite) and hydrous phases (white micas, zoisite, Mg-glaucophane, Mg-chloritoid and talc) (see Table 10.4 for individual assemblages). Zoisite-paragonite-quartz-kyanite-eclogites and glaucophane-paragonite-quartz-kyanite-eclogites are well represented (Holland, 1979b). Texturally, the majority are relatively fine-grained, foliated, garnet-porphyroblastic rocks with or without a millimetre-scale banding, but both coarse-grained (metagabbroic) and fine-grained, unfoliated varieties also occur. Typical eclogite facies assemblages in metasediments (Table 10.4) include Grt + Cld + Ky in pelites, Zo + Dol + Ky in calc-schists and Di + Tr + Qtz in marbles (Holland, 1979a; Franz and Spear, 1983; Spear and Franz, 1986).

The eclogites record the earliest penetrative deformation recognised so far in the Mesozoic cover of the Tauern Window. The deformation was evidently contemporaneous with eclogite facies metamorphism and resulted in (*a*) the development of the principal foliation, defined by aligned prisms of unstrained omphacite and/or Mg-glaucophane, and (*b*) widespread boudinage, with concomitant growth of coarse-grained eclogite facies minerals in boudin necks (Holland, 1979a). The early fabric (SaO) has been folded by at least two phases of intense ductile deformation (Da1, Da2), the first of which probably correlates with the main phase of overthrusting of the Austro-Alpine basement nappes during continental collision (Bickle and Hawkesworth, 1978). The structural chronology is entirely consistent with the crystallisation of the Tauern eclogites in a subduction zone operative before collision.

The eclogites also preserve relicts of pre-eclogitic parageneses in the form of hornblende, epidote, paragonite, phengite and biotite inclusions in garnet cores, suggestive of an early epidote-amphibolite stage (Miller, 1974b, 1977a, 1977b; Holland, 1979a). Clear textural evidence also exists in some samples for retrogressive overprinting of the eclogite facies mineralogy by syn/post-Da1 blueschist facies assemblages, e.g. epidote + glaucophane (a strongly coloured variety), in turn overprinted by post-

Table 10.4 Critical low-variance primary eclogite facies mineral assemblages in rocks from the Eclogite Zone in the Tauern Window, Eastern Alps. Subsets of the listed assemblages are common.

Rock type	Eclogite facies paragenesis	
Metabasalts and metagabbros	Grt + Omp + Qtz + Ky + Zo + Pg + Dol + Mag + Rt + Py	(2)
	Grt + Omp + Qtz + Ky + Zo + Pg + Phe + Rt + Py + Ap	(8)
	Grt + Omp + Qtz + Ky + MgGln + Tlc + Pg + Dol + Rt + Py + Ap	(2, 7)
	Grt + Omp + Qtz + Ky + MgGln + Tlc + Pg + Czo + Mag + Rt + Ap	(8)
	Grt + Omp + Qtz + Ky + MgGln + Pg + Mag + Rt + Py	(2)
	Grt + Omp + Qtz + Ky + Tlc + Pg + Phe + Rt + Py + Ap	(2, 3)
	Grt + Omp + Qtz + Czo/Ep + Pg + Phe + Mag + Rt + Py + Ap	(8)
	Grt + Omp + Qtz + Gln + Czo/Ep + Pg + Dol + Rt + Ap	(8)
	Grt + Omp + Ky + MgCld + Tlc + Pg + Chl + Rt + Py + Ap	(3)
Veins in metabasites	Omp + Ky + Zo + Dol + Qtz	(2)
	Omp + Ky + Pg + Ap	(8)
	Ky + MgCld + Tlc + Chl + Pg	(4)
	Ky + Pg + Qtz	(8)
	Ky + Zo + Dol + Qtz + Rt	(4)
	Ky + Tlc + Qtz + Rt	(1)
	Czo/Ep + Dol + Pg + Phe + Qtz + Ap	(8)
Calc-schists	Dol + W-Mica + Grt + Omp + Ky + Zo	(2)
	Dol + Qtz + Ky + Zo	(6)
	Dol + Cal + Qtz + Zo + Ms	(6)
Siliceous dolomites	Di + Tr + Dol + Cal + Qtz + Zo	(5)
Pelites	Ms + Qtz + Grt + Cld + Ky + Ep + Dol	(6)

Sources: (1) Miller (1977b); (2) Holland (1979a); (3) Miller (1986); (4) Frank *et al* (1987a); (5) Franz and Spear (1983); (6) Spear and Franz (1986); (7) Holland (1988); (8) Miller (personal communication).

Da2 albite-epidote-amphibolite facies assemblages of the mid-Tertiary 'late Alpine' phase. Whilst the eclogitic imprint and SaO fabric are not present in Mesozoic cover rocks overlying the Eclogite Zone (the ophiolitic 'Glockner Nappe') these units share the effects of Da1 and Da2 deformations as well as both blueschist and greenschist–amphibolite facies imprints (Holland, 1979a; Holland and Richardson, 1979; Holland and Ray, 1985; Frank *et al*, 1987a). Such observations refute the 'progressive metamorphism' models of Abraham *et al* (1974) and Raith *et al* (1977) and imply that the Eclogite Zone was tectonically juxtaposed with the Glockner Nappe between the eclogite and blueschist facies 'events'.

The eclogite facies metamorphism in the Tauern Window is presumably of early- or mid-Cretaceous age by analogy with Western Alpine occurrences. However, the existence of excess Ar^{40} in eclogitic minerals (Miller *et al*, 1980) and the insensitivity of the Nd–Sm technique to such young rocks (van Breemen and Hawkesworth, 1980)

Table 10.5 Pressure and temperature estimates for the eo-Alpine eclogite facies metamorphism in the Tauern Window. Temperatures are quoted to the nearest 5°, pressures to the nearest 0.5 kbar.

Equilibrium	Phase composition data	Calibration	Temperature (°C)		Pressure (kbar)		Source
			at 15 kbar	at 20 kbar	at 550°C	at 600°C	
FeGrt + 3MgCpx = MgGrt + 3FeCpx	$K_D = 13.4–15.8$	Råheim & Green (1974a)	535–560	560–590	—	—	Miller (1977b)
	$K_D = 9–14$	Råheim & Green (1974a)	550–635	580–665	—	—	Holland (1979a)
	$K_D = 11.5–15.3$, $X_{Ca}^{Grt} = 0.16–0.22$	Ellis & Green (1979)	560–580	575–595	—	—	Frank *et al* (1987a)
4Lws = 2Zo + Ky + Qtz + 7H_2O	$K_S = 1$ $a_{H_2O} = 1$	Newton & Kennedy (1963)	>530	>590	—	—	Holland (1979a)
Eclogite solidus	$a_{H_2O} = 1$	Hill & Boettcher (1970)	<630	<660	—	—	Holland (1979a)
Pg = Jd + Ky + H_2O	$K_S = 0.5$ $a_{H_2O} = 0.98$	Holland (1979b)	—	—	20.5	20	Holland (1979b)
		Holland (1979a)	—	—	20.5	20	Holland (1979a)

2MgChl + 5Ky = 7MgCld + Tlc	$K = 0.17$	Chopin & Schreyer (1983)	—	—	18	20.5	Frank *et al* (1987a)
Di-Tr-Dol-Cal-Qtz—fluid invariant point	$X^{Tr}_{OH} = 0.75$	Franz & Spear (1983)	—	—	13.5	24	Franz & Spear (1983)
3FeCld + 2Qtz = Alm + 2Ky + $3H_2O$	$K_S = 1.5$	Rao & Johannes (1979)	<590	<570	—	—	Spear & Franz (1986)
$^{18}O/^{16}O$ exchange Qtz-Rt	$\Delta^{18}O_{Qtz\text{-}Rt} = 7.0\text{–}7.7‰$	Matthews *et al* (1979)	545–575	545–575	—	—	Matthews *et al* (1979)
$^{18}O/^{16}O$ exchange Omp-Rt	$\Delta^{18}O_{Qtz\text{-}Omp} = 2.2‰$ Omp-Rt!	Matthews *et al* (1983)	575	575	—	—	Matthews *et al* (1983)
Simultaneous solution of 15 independent equilibria involving Prp, Di, Jd, Ky, Gln, Pg, Grs, Tlc, Qtz, Dol, H_2O, CO_2.	—	Holland & Powell (1985), Holland (1988)	—	—	—	21 ± 1	Holland (1988)

K_S = Equilibrium constant for condensed phases.

have so far obviated attempts to obtain a reliable radiometric age. The only constraint comes from K–Ar ages of 90–60 Ma from retrogressive crossites (Raith *et al*, 1980) which provide a minimum age.

The presence of an aqueous fluid phase during eclogite facies metamorphism is indicated by the abundance of eclogite facies veins (Table 10.4 gives assemblages) and low-salinity water-rich fluid inclusions in omphacites and zoisites in both veins and host rocks (Holland, 1979a). Fluid inclusions in quartz are CO_2-rich (Luckscheiter and Morteani, 1980) but probably do not preserve syn-eclogitic fluids. High water activities are also indicated by the simultaneous equilibration of the assemblages Ph + Omp (Jd_{50}) + Ky and Dol + Omp (Di_{50}) + Qtz with H_2O–CO_2 fluid at realistic temperatures (Holland, 1979a).

The predominance of low-variance assemblages in the Tauern eclogites facilitates estimation of pressures and temperatures of crystallisation. Using a variety of equilibria (Table 10.5), conditions of about 600°C and 20 kbar have been inferred for primary eclogitic assemblages (Holland, 1979b, 1988; Miller, 1986; Frank *et al*, 1987a). Franz and Spear (1983) and Spear and Franz (1986) deduced similar values for primary HP assemblages in metasediments.

10.3.2 *Continental eclogite facies rocks of the palaeo-African margin*

The effects of early-Alpine regional metamorphism are widespread in the Austro-Alpine basement. In the Eastern Alps, the dominant imprint is of a mid-Cretaceous amphibolite facies metamorphism of medium- to high-pressure character (Frank *et al*, 1987b, 1987c); eclogite facies assemblages, if represented at all (cf. Manby and Thiedig, in press), are restricted to metabasite bodies in the 'Plattengneis' horizon, which are believed by Miller and Frank (1983) to have recrystallised during Cretaceous shearing. In the Western Alps, however, Austro-Alpine outliers preserve spectacular early-Alpine eclogite facies assemblages in a wide variety of continental lithologies.

10.3.2.1 *Western Alps: Sesia Zone, Mt Emilius Klippe and Levaz Thrust Sheet.* The Sesia Zone is a composite tectonic element consisting of four sub-units which derive from a pre-Alpine continental crust: (*a*) the 'Eclogitic Micaschist Complex' (EMC); (*b*) the 'Gneiss Minuti Complex'; (*c*) tectonic slices of high-grade metamorphics partly recrystallised under blueschist facies conditions during the Alpine cycle; these slices include the 'Il Zona Diorito–Kinzingitica' to the north and the Vasario–Sparone–Cialmera element to the south; (*d*) the Rocca Canavese thrust sheets close to the Insubric line (Compagnoni *et al*, 1977a; Passchier *et al*, 1981; Lardeaux *et al*, 1982; Pognante *et al*, 1987, 1988; Pognante, 1989a).

The Mt Emilius Klippe (MEK) (a sub-unit of the Dent Blanche nappe) is a folded thrust sheet of eclogitised continental crust that preserves relics of pre-Alpine parageneses of high grade, largely overprinted by lower temperature re-equilibrations (Dal Piaz *et al*, 1983).

The Levaz Thrust Sheet (LTS) is interposed between the underlying Zermatt–Saas eclogitised meta-ophiolites and a sequence of calcschists and metabasalts (Combin unit); it consists of granulites, high-grade metamorphics and granites which show an incipient early-Alpine eclogitisation (Kienast, 1983).

This section deals with the EMC sub-unit, MEK and LTS; these display the best-developed and best-studied eclogite facies assemblages and probably derive from the palaeo-African continental margin. Scattered omphacite + garnet assemblages in metabasites (Williams and Compagnoni, 1983) and jadeite + quartz assemblages in orthogneisses (Pognante *et al*, 1987) are reported also for the 'Gneiss Minuti' sub-unit which, however, displays a pervasive LP greenschist recrystallisation. The Rocca Canavese thrust sheets, in the SW Sesia Zone, include partly serpentinised tectonite peridotites, metagranulites and orthogneisses; these rocks do not display the eclogite facies parageneses typically of the EMC and show an incipient blueschist facies recrystallisation with abundant glaucophane, jadeite + quartz, lawsonite and pumpellyite (Pognante, 1989a).

Before the early-Alpine recrystallisation, the EMC consisted of high-grade paragneisses with impure marbles, amphibolites and granulites intruded by gabbroids and Variscan granitoids. Relics of high-grade metapelites were first described in the eclogites micaschists of the Oropa region (Lago Mucrone) by Dal Piaz *et al* (1972) and have subsequently been found also in the eclogitic micaschists of Alpe Toso, Val Sesia (Lombardo and Gosso, unpublished observation; Lardeaux *et al*, 1982). At Alpe Toso a biotite-hornblende granulite was found in close association with the high-grade metapelites and with amphibolites displaying all the stages in the conversion into barroisite eclogite and amphibolite (Ungaretti *et al*, 1983). Granitoid rocks are present throughout the EMC, but are most common between the Gressoney valley and Val Sesia where they are represented by biotite granodiorites with minor diorites, tonalites and syenites (Callegari *et al*, 1976; Lardeaux *et al*, 1983). The only relict igneous minerals commonly preserved are K-feldspar and allanite, but in a few areas, notably at Monte Mucrone (Oropa) and Boccioleto (Val Sesia), the igneous structure is perfectly preserved; in these rocks biotite and/or clinopyroxene has survived the eclogite facies overprinting (Compagnoni and Maffeo, 1973; Lardeaux *et al*, 1983; Koons *et al*, 1987). Kilometric bodies of metagabbro exist in the SW EMC between Cario, Monastero and Val Chiusella (Compagnoni *et al*, 1977a; Spalla *et al*, 1983, and references therein). Serpentinised ultramafics occur near Ivozio (Val d'Aosta: Pognante *et al*, 1980) associated with lawsonite-bearing metabasites, while serpentinites rich in pre-Alpine orthopyroxene have been found near Carema (Val d'Aosta: Compagnoni and Pognante, unpublished observations).

During eclogitisation the pre-Alpine metapelites and paragneisses were converted to garnet-chloritoid-bearing micaschists or to glaucophane-jadeite/omphacite-garnet micaschists (the 'eclogitic micaschists' of the earlier authors; Table 10.6). In the micaschists kyanite is rare and forms either fine-grained pseudomorphs after pre-Alpine sillimanite (Dal Piaz *et al*, 1972) or porphyroblasts (Pognante *et al*, 1980). Impure marbles intercalated in the paraschists are characterised by the growth of omphacite, garnet and glaucophane (Castelli, 1987). Amphibolites and basic granulites were transformed to amphibole-bearing eclogites; in some of the Alpe Toso amphibolites and in the relict granulite, eclogite transformations are confined to the grain boundaries between plagioclase and the ferromagnesian phases, with extremely fine-grained aggregates of omphacite, zoisite and kyanite replacing the outer portion of the original plagioclase.

The metagranitoids from Lago Mucrone are characterised by pseudomorphs of Jd + Qtz ± Zo after plagioclase and by composite garnet–mica reaction rims bordering magmatic biotite (Compagnoni and Maffeo, 1973; Oberhänsli *et al*, 1985 and Chapter 2, this volume; Koons *et al*, 1987). Dynamically recrystallised aplites and granites were converted to garnet-free jadeite + quartz + microcline + phengite-orthoschists which in a

Table 10.6 Representative eclogite facies assemblages of early Alpine age in the eclogitic mica schist complex of the Sesia Zone

Rock type	Eclogite facies assemblage	
Metabasites	Grt + Omp + Qtz + Zo/Ep + Gln + Phe + Ank + Rt	(1)
	Grt + Omp + Qtz + Zo + Bar + Pg + Rt	(8)
	Grt + Omp + Gln/Bar + Phe + Rt	(8)
Tonalitic and granodioritic orthogneisses	Grt + Omp + Qtz + Gln + Phe + Rt + Zrn + Aln	(2)
	Grt + Omp + Qtz + Zo + Pg + Phe + Rt	(6)
	Grt + Omp + Qtz + Czo + Phe	(4)
Granitic orthogneisses and meta-aplites	Jd + Qtz + Kfs + Phe + Zo + Rt	(1)
	Jd + Qtz + Kfs + Phe + Zrn + Py	(8)
	Jd + Qtz + Phe + Grt + Gln	(2)
Paragneisses and pelitic schists	Qtz + Phe + Grt + Omp + Gln + Rt	(1)
	Qtz + Phe + Pg + Grt + Cld + Gln + Rt + Gr	(1)
	Qtz + Pg + Grt + Ky + Cld + Zo + Cal + Rt	(3)
	Qtz + Phe + Pg + Grt + Ky + Cld + Rt	(7)
	Qtz + Phe + Pg + Grt + Cld + Chl + Zo + Rt	(7)
Marbles	Cal + Dol + Phe + Py + Gr	(1)
	Cal + Dol + Qtz + Phe + Grt + Omp + Zo + Gln	(5)

Sources: (1) Compagnoni (1977); (2) Compagnoni *et al* (1977b); (3) Pognante *et al* (1980); (4) Oberhänsli *et al* (1985); (5) Castelli (1987); (6) Koons *et al* (1987); (7) Vuichard and Ballèvre (1988); (8) personal observation.

few occurrences contain centimetric megablasts of jadeite (Andreoli *et al*, 1976) whilst tonalites and quartz-diorites were converted to garnet-omphacite mica schists and felses. In the metagabbroids plagioclase is replaced by epidote ± garnet while hornblende and pyroxenes are replaced by glaucophane with a garnet reaction rim. In glaucophanitic metagabbros and in adjacent rocks from the southern part of the EMC lawsonite porphyroblasts post-date eclogitisation and overgrow a pre-existing foliation defined by glaucophane ± garnet; here lawsonite occurs also in leucocratic layers and in metamorphic veins with jadeite and quartz and is almost completely pseudomorphed by zoisite ± paragonite (Compagnoni, 1977; Pognante, 1989a).

The early-Alpine transformations in the Mt Emilius Klippe are similar to those described in the EMC. The pre-Alpine relics in the metapelites are also similar, whilst the metabasites and marbles preserve relics of pre-Alpine garnet and pyroxene.

In the Levaz Thrust Sheet, incipient eclogite facies recrystallisation of the poorly deformed metagranulites produced Jd + Zo + Ky after plagioclase, Tlc + Grt ± Omp after the pre-Alpine pyroxenes and Grt ± Tlc ± Ky ± Mg-Cld after the Al-spinel (Kienast, 1983).

The sodic pyroxenes in the EMC and the other tectonic elements described above are mainly jadeites in the metagranites, and omphacites in the metabasites and most of the paraschists (Figure 10.2); ferro-omphacite has been described in the recrystallised metasyenites of the NE sector of the EMC (Lardeaux *et al*, 1983). Garnet is rich in almandine, although in the paraschists a few cores are richer in pyrope and probably derive from the incomplete equilibration of pre-Alpine garnet (Figure 10.3). Amphiboles are mainly glaucophane, but Na–Ca amphiboles have been observed in recrystallised amphibolites and in some eclogites (Figure 10.4) (Ungaretti *et al*, 1983). A

Mg-rich variety of chloritoid has been observed coexisting with kyanite and paragonite in micaschists from Val d'Aosta, whilst more Fe-rich chloritoid is characteristic of recrystallised paragneisses and kinzigites.

P–T estimates for the peak of eclogite facies metamorphism in the EMC have been obtained from phase equilibria, oxygen isotope fractionation data and Fe–Mg exchange in garnet–omphacite and garnet–phengite pairs (Compagnoni, 1977; Desmons and Ghent, 1977; Desmons and O'Neil, 1978; Lardeaux *et al*, 1982; Ungaretti *et al*, 1983; Oberhänsli *et al*, 1985; Robert *et al*, 1985; Koons, 1986; Pognante *et al*, 1987; Pognante, 1989a). The different methods yield concordant estimates with $T=$ 500–600°C at $P > 13$–14 kbar. A metamorphic zonation in the EMC was recognised in the NE areas of this complex where albite-omphacite parageneses replace the plagioclase-free parageneses typical of the EMC (Lardeaux *et al*, 1983).

P–T conditions of the early-Alpine metamorphic peak in the MEK are less extreme than those determined in the EMC, but are similar to those of the underlying ophiolite nappe ($T=460$–480°C, $P=11$–13 kbar; Dal Piaz *et al*, 1983).

In the LTS *P–T* evaluations for the eclogitisation are $T=550$°C and $P=16$–17 kbar (Kienast, 1983).

The main eclogite stage was followed by a syntectonic blueschist facies recrystallisation which in the EMC occurred at slightly decreased *T* and *P* (Lardeaux *et al*, 1982) and produced glaucophane and clinozoisite. The blueschist stage is characterised also by the transformation of jadeite to Ab ± Agt ± Crs (Compagnoni, 1977). In the southern EMC abundant lawsonite crystallised in equilibrium with jadeite and quartz, but later than the eclogite peak, as indicated by post-tectonic porphyroblasts overgrowing a foliation defined by glaucophane, garnet and, in adjacent rocks, by omphacite. This points to a marked temperature decrease at HP that post-dates the metamorphic climax in the southern EMC (Pognante, 1989a). The blueschist stage was followed by a LP generally static, greenschist facies recrystallisation.

Rb–Sr dating of whole-rock samples of EMC metagranitoids (Oberhänsli *et al*, 1985) suggests that the eclogite facies metamorphism in the EMC unit may be as old as 130 Ma. Phengite Rb–Sr and K–Ar ages scatter in the ranges 90–70 Ma and 80–60 Ma, respectively, and are interpreted as post-eclogitic cooling ages (Oberhänsli *et al*, *op. cit.*; Stockert *et al*, 1986).

10.3.3 *Continental eclogite facies rocks of the palaeo-European margin*

Pre-Triassic basement rocks of palaeo-European provenance dominate the Penninic domain (Figure 10.1), but the distribution of eo-Alpine eclogite facies metamorphism is patchy. In the Western Alps, eclogitised basement units are restricted to the internal side of the Pennine Zone. In the Central Alps the pattern is largely obscured by the late-Alpine overprint, but eclogitic relics occur across the whole width of the zone. In both areas, the HP imprint appears to be restricted to the structurally higher, more southerly 'rooting' nappes. In the Eastern Alps (Tauern Window), there is no evidence at all for an eo-Alpine eclogitic imprint in the Penninic basement.

10.3.3.1 *Western Alps: Monte Rosa. Gran Paradiso and Dora Maira massifs.* Monte Rosa, Gran Paradiso and Dora Maira are basement nappes, each surrounded by an envelope of thrust sheets consisting of Mesozoic metasediments and dismembered ophiolite complexes.

The Monte Rosa (MR) basement complex consists of high-grade metapelites with rare basic rocks intruded by porphyritic granites of Carboniferous age, and of the younger Furgg Formation, a heterogeneous sequence including albite gneisses, metabasites and marbles. High-grade assemblages in the metapelites predate the emplacement of the Carboniferous granites and are thus of Variscan age or older. Both the metapelites and the granites were subjected to a HP metamorphism of Cretaceous age (Table 10.7) and to a pervasive MT re-equilibration in the middle Tertiary (Dal Piaz and Lombardo, 1986, and references therein).

Like the Monte Rosa Nappe, the Gran Paradiso (GP) Nappe consists of a pre-granitic complex of gneisses and metabasics, of Alpine augen gneisses derived from Variscan granites and of the younger Bonneval Gneiss and Money Complex. The relics of high-grade assemblages are less abundant than in MR, but the metamorphic history during the Alpine cycle is similar to that of MR, with an early-Alpine event producing eclogites and HP transformations, followed by extensive overprinting under greenschist facies conditions in Late Eocene time (Compagnoni *et al*, 1974; Compagnoni and Lombardo, 1974; Dal Piaz and Lombardo, 1986; Ballèvre, 1988). The metaclastic Money Complex shows no evidence of eclogite facies metamorphism, but blueschist minerals (aegirine, riebeckite and jadeite) occur in leucocratic layers of the Bonneval Gneiss (Saliot, 1979; Vearncombe, 1983).

In the Dora Maira (DM) Nappe, eclogite facies rocks occur in the 'polymetamorphic unit' (Vialon, 1966), a sub-unit sandwiched between two other continental sub-units of lower metamorphic grade (Henry *et al*, 1987). In the southern DM (Val Po and Val Varaita) the 'polymetamorphic unit' consists of Variscan granitoids and garnet

Table 10.7 Representative eclogite facies assemblages of early Alpine age in the Monte Rosa, Gran Paradiso and northern Dora Maira massifs

Rock type	Tectonic unit	Eclogite facies assemblage	
Metabasites	N. Dora Maira	Grt + Omp + Gln + Pg + Rt	(6)
	Monte Rosa and	Grt + Omp + Qtz + Zo + Pg + Rt	(5)
	Gran Paradiso	Grt + Ep + Bar + Chl + Rt	(5)
Orthogneisses	Monte Rosa, Gran Paradiso and N. Dora Maira	Qtz + Ab + Kfs + Phe + Grt + Zo (eclogite facies status not confirmed)	(5, 7)
	Gran Paradiso (Bonneval)	Qtz + Jd + *Acm + Kfs + Phe	(1,3)
Paragneisses	Monte Rosa	Phe + Qtz + Grt + Cld + Ky	(5)
	Gran Paradiso and N. Dora Maira	Phe + Qtz + Grt + Cld + Gln + Rt	(5, 7)
Mica schists	Monta Rosa	Phe + Qtz + Mg–Cld + Ky + Tlc + Chl	(4)
	Gran Paradiso	Phe + Qtz + Mg–Cld + Ky + Chl	(2, 5)
	and N. Dora Maira	Phe + Qtz + Mg–Cld + Tlc + Chl	(2, 5, 7)
		Phe + Qtz + Grt + Cld + Tlc	(2, 5)

*Equilibrium status not confirmed

Sources: (1) Saliot (1969); (2) Chopin (1981); (3) Vearncombe (1983); (4) Chopin and Monié (1984); (5) Dal Piaz and Lombardo (1986) and references therein; (6) Pognante and Sandrone (1989); (7) personal observation.

micaschists with concordant bodies and layers of eclogite and marble (Kienast and Lombardo, 1987; Henry *et al*, 1987). Relics of garnet-biotite gneiss and of a pre-Alpine metamorphic foliation have been identified in the garnet-kyanite micaschists of Isasca (Val Varaita; Lombardo and Kienast, unpublished observation). Eclogites have been described also in northern DM from Val Susa (Pognante and Sandrone, 1989) and Val Germanesca (Borghi *et al*, 1985).

The MR eclogites occur as boudins and small bodies in garnet micaschists of the southern slope of Monte Rosa, as boudinaged dykes within granitic orthogneisses (upper Valesia), or as layers and boudins in the Furgg Formation. The GP eclogites occur as boudinaged bands and sub-concordant bodies in the paragneisses of the pre-granitic complex. Eclogites derived from gabbros have been found in Val di Sea (southwestern Gran Paradiso; Pognante, unpublished observation). Bulk compositions of MR and GP eclogites are mostly basaltic, but some eclogite bodies are enriched in Fe and Ti and may derive from Fe-basaltic (or Fe-gabbro) protoliths (Dal Piaz and Lombardo, 1986). No recent data exist on the bulk composition of the DM eclogites, but relative proportions of mineral phases and their composition suggest that these rocks may derive from two protoliths, one more Fe- and Ti-rich, the other more Mg- and Al-rich.

Typical mineral parageneses are summarised in Table 10.7 and compositions of pyroxenes and garnets are shown in Figures 10.2 and 10.3. The MR and GP eclogites have rather similar omphacite and garnet compositions. In the DM, eclogites from Val Susa have chloromelanite or acmitic-jadeite instead of omphacite (Pognante and Sandrone, 1989), whereas eclogites from the 'polymetamorphic unit' of southern DM contain relatively magnesian garnets; near Brossasco (Val Varaita), thin bands rich in impure jadeite, quartz, kyanite and phengite have been found within the latter eclogites (Lombardo and Kienast, unpublished data). The amphibole is a pale coloured glaucophane in most eclogite bodies of MR and GP, but a colourless variety of sodic-calcic composition also occurs (Figure 10.4) (Dal Piaz and Lombardo, 1986). Textural evidence suggests that in most MR-GP eclogites glaucophane post-dates the crystallisation of omphacite and garnet; a later stage in the amphibole evolution is represented by Fe-rich blue-green barroisite mantling glaucophane and forming from the breakdown of garnet and omphacite. Most amphiboles in the DM eclogites are likewise post-eclogitic but, except for glaucophanes from the Val Susa and Val Germanesca eclogites, they are strongly aluminous varieties (hornblende and katophorite).

Static HP transformations in the pre-granitic complex of MR and GP and in pelitic hornfelses of the GP Nappe include: (*a*) the replacement of pre-Alpine sillimanite by pseudomorphs of fine-grained kyanite; (*b*) the replacement of an unknown Al-silicate, possibly cordierite, by Ky + Grt + Cld; (*c*) the growth of garnet rims around pre-Alpine garnet, biotite and opaques; (*d*) the replacement of biotite by white mica and rutile. Jadeite pseudomorphs, such as those described in the polymetamorphic micaschists of the Sesia zone, have not been found and the original plagioclase may still be preserved. The DM normal Fe-metapelites of the so-called polymetamorphic unit contain the HP assemblage Alm + Ky + Jd + Qtz not recorded in metapelites elsewhere in the Western Alps (Table 10.8).

The best evidence of HP transformations in metapelitic rocks of MR, GP and DM is recorded by the assemblage talc + phengite in silvery mica-schists, unusual rocks occurring as lens-shaped bodies in the granitic orthogneisses. These rocks are

Table 10.8 Early Alpine eclogite facies assemblages in the 'polymetamorphic unit' of the southern Dora Maira massif

Rock type	Locality	Eclogite facies assemblage	
Metabasites	V. Varaita	Grt + Omp + Qtz + Bar + Phe + Rt	(4)
		Grt + Jd + Qtz + Zo + Phe	(4)
Orthogneisses	V. Varaita	Qtz + Jd + Phe + Grt	(4)
Paragneisses	V. Varaita	Phe + Qtz + Alm + Ky + Jd	(4)
and quartzites	V. Po	Qtz + Alm + Ky + Jd	(1, 3)
Mg-mica schists	V. Po	Phe + Coe + Prp + Ky + Tlc + Jd + Rt	(1, 2)
Marbles	V. Varaita	Cal + Dol + Grt + Omp + Tr + Phe	(4)

Sources: (1) Chopin (1984); (2) Chopin *et al* (1986); (3) Schreyer *et al* (1987); (4) Lombardo and Kienast (unpublished observations).

characterised by magnesian mineral which mirrors the high Mg content and Mg/Fe ratios of these rocks (Chopin, 1981). In the polymetamorphic unit of the DM nappe, strongly magnesian compositions are also characteristic of the rocks described by Vialon (1966) as 'nodular pegmatites' but in reality probably meta-evaporitic pelites or metamorphosed shear-zone metasomatites, from which Chopin (1984) reported the occurrence of coesite inclusions within crystals of nearly pure pyrope in an exceptionally HP assemblage with phengite + kyanite + talc (Table 10.8). Relicts of earlier, lower temperature HP mineral parageneses, preserved in the cores of large garnets and indicative of the prograde evolution of DM metapelites, include Mg-chloritoid, Mg-staurolite and ellenbergerite (Chopin *et al*, 1986; Schreyer, 1988). The latter two minerals are not recorded elsewhere in early-Alpine HP parageneses.

Garnet-clinopyroxene geothermometry and phase compatibilities suggest early-Alpine crystallisation at $T = 440$–$530°C$ and $P > 8$–10 kbar in the MR nappe; partly similar pressures and temperatures of 400–500°C, possibly recording post-eclogitic re-equilibrations, have been derived for the GP eclogites (Dal Piaz and Lombardo, 1986; Pognante *et al*, 1987; Ballèvre, 1988) and the eclogites from the northern DM (Pognante and Sandrone, 1989). In the eclogites of the DM polymetamorphic unit, garnet-omphacite thermometry and other considerations suggest $T > 600°C$ and $P > 15$–16 kbar (Lombardo and Kienast, unpublished data), i.e. P–T conditions significantly more extreme than those reported for the other Pennidic nappes and for the Sesia zone. Such high estimates, however, are corroborated by the occurrence of minerals and parageneses diagnostic of high temperatures and very high pressures also in the metapelites and especially in the magnesian bulk-compositions, including the assemblage Prp + Phe + Ky + Tlc + Coe; for this assemblage Chopin (1984) deduced $T = 700$–$800°C$ and $P > 28$ kbar.

^{40}Ar–^{39}Ar plateau ages of 110–95 Ma on phengites from the MR and southern DM indicate a mid-Cretaceous eclogite facies metamorphism in those units (Chopin and Monié, 1984; Goffé and Chopin, 1986).

10.3.3.2 *Central Alps: Adula Nappe.* The Adula Nappe preserves evidence of

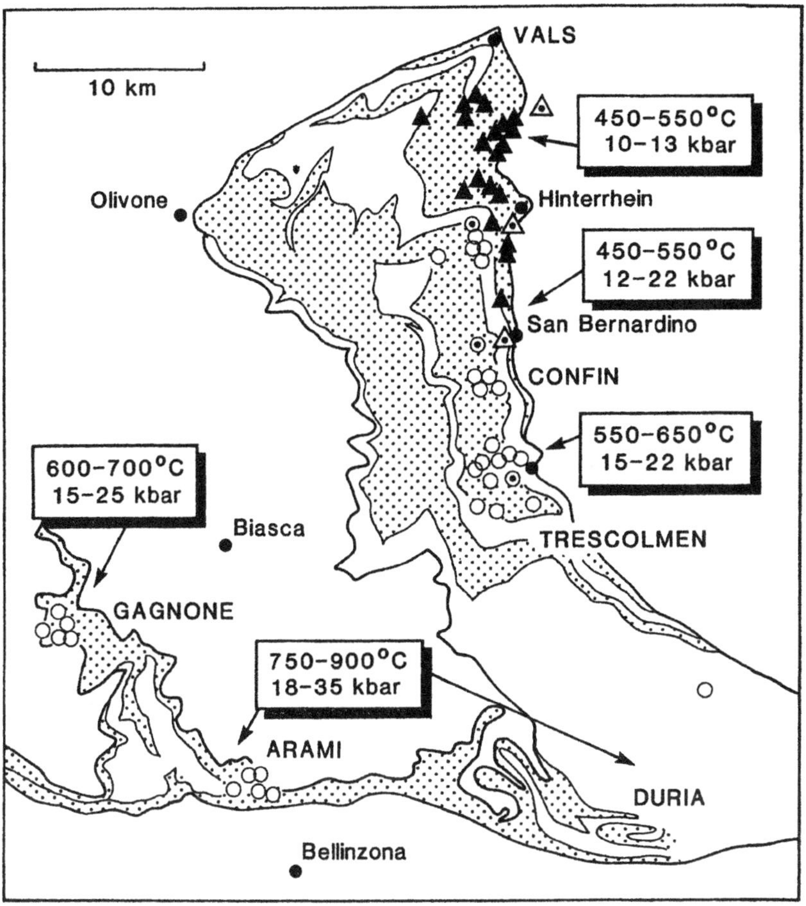

Lithological Units, Adula Nappe Basement:

☐ Metagranitoid gneisses predominant

⁙ (semi–)pelitic schists and gneisses with mafic lenses predominant

Mineral assemblages in quartz-bearing mafic rocks:

▲ Ab + Hbl + Ep (± Omp, Pg, Phe)

◬ Omp + Ep + Pg (± Amp, Phe)

⊙ Omp + Ky + Pg (± Amp, Phe)

○ Omp + Ky (± Amp, Ep, Phe)

Figure 10.5 Simplified geological map of the Adula nappe in the Central Alps, showing the distribution of eclogite facies assemblages in metabasites and estimated conditions of eclogite facies metamorphism. Adapted and redrawn from Heinrich (1986).

eo-Alpine eclogite facies metamorphism along the whole of its 50 km N–S section (Heinrich, 1982, 1986). The effects are visible both in imbricated ophiolitic lithologies and in the polymetamorphic continental basement rocks of the nappe itself, some of which also record an earlier, Variscan metamorphic imprint (e.g. Frey *et al*, 1976; Hänny *et al*, 1975). The continental suite includes pelitic schists and metabasites, which contain abundant eo-Alpine HP relics, and metagranitoid gneisses, which do not. Eclogitic assemblages are preserved better in the mafic rocks than in the pelites.

No radiometric dates are yet available for the eclogite facies metamorphism in the Adula Nappe, but according to Löw (1986), it occurred early in the deformational history. High pressures prevailed during formation and transport of the nappe, and preceded both the recumbent folding of the nappe with its Bündnerschiefer envelope and a later phase of upright refolding.

The most striking features of the eclogite facies imprint in the Adula Nappe are its regional character and its regular metamorphic zonation (Heinrich, 1982, 1986). The Adula Nappe is one of the few eclogite facies areas of the world where such a clear zonation is demonstrable, and it evidently behaved as a coherent unit during and after eo-Alpine HP metamorphism. The zonation is apparent in both the distribution of metabasic mineral assemblages (Table 10.3, Figure 10.5) and the spread of calculated metamorphic *P–T* values (Figure 10.5).

At the northern end of the nappe, quartz-eclogites contain albite (Ab_{96}), epidote and hornblende in textural equilibrium with garnet and omphacite (Jd_{40-50}) (figure 2 of Heinrich, 1986). Further south, they are characterised firstly by Ca-Amp + Pg ± Ky then by Gln + Pg ± Ky or Hbl + Ky + Zo (Table 10.3). Paragonite is absent from the most southerly occurrences. Heinrich (1986) uses phase topological arguments to show that this sequence of assemblages represents a facies series (sensu Miyashiro, 1961) and implies a gradation in metamorphic conditions in which *P* and *T* increase progressively southwards in the nappe.

Supporting evidence for such variations is provided by thermobarometry. K_D values for the garnet–clinopyroxene Fe–Mg exchange equilibrium range from about 12 in the north to 2 in the south, giving a temperature range of 450–900°C (Figure 10.5) with the Ellis and Green (1979) calibration. Corresponding pressure variations are indicated by the jadeite contents of omphacite coexisting with quartz ± albite (calibration of Gasparik and Lindsley, 1980) and the stability limit of paragonite. Estimated pressures are 10–13 kbar in the north and > 18 kbar in the south (Figure 10.5). Actual equilibration pressures of 25–35 kbar for the Adula 'root zone' are indicated by garnet–orthopyroxene barometry (Harley and Green, 1982) on coexisting phases (Prp_{67}; Opx with 1.8–2.1% MgTs) in the garnet-lherzolites of Alpe Arami and Monte Duria (Evans and Trommsdorff, 1978; Heinrich, 1986).

Pelites in the Adula Nappe contain HP assemblages such as Grt + Tlc + Ky (Table 10.3) but do not appear to be widely enough preserved to map out metamorphic zones.

No unequivocal eclogite relics have been found elsewhere in the Central Alps, but Stille and Oberhänsli (1987) interpret garnet-diopside-amphibolites in the Berisal Complex as altered eo-Alpine eclogites.

10.4 Regional patterns of Alpine eclogite facies metamorphism

Thermobarometric estimates of the early-Alpine eclogite facies climax in the Western Alps encompass pressures ranging from 8 to 30 kbar and temperatures of 400–450°C

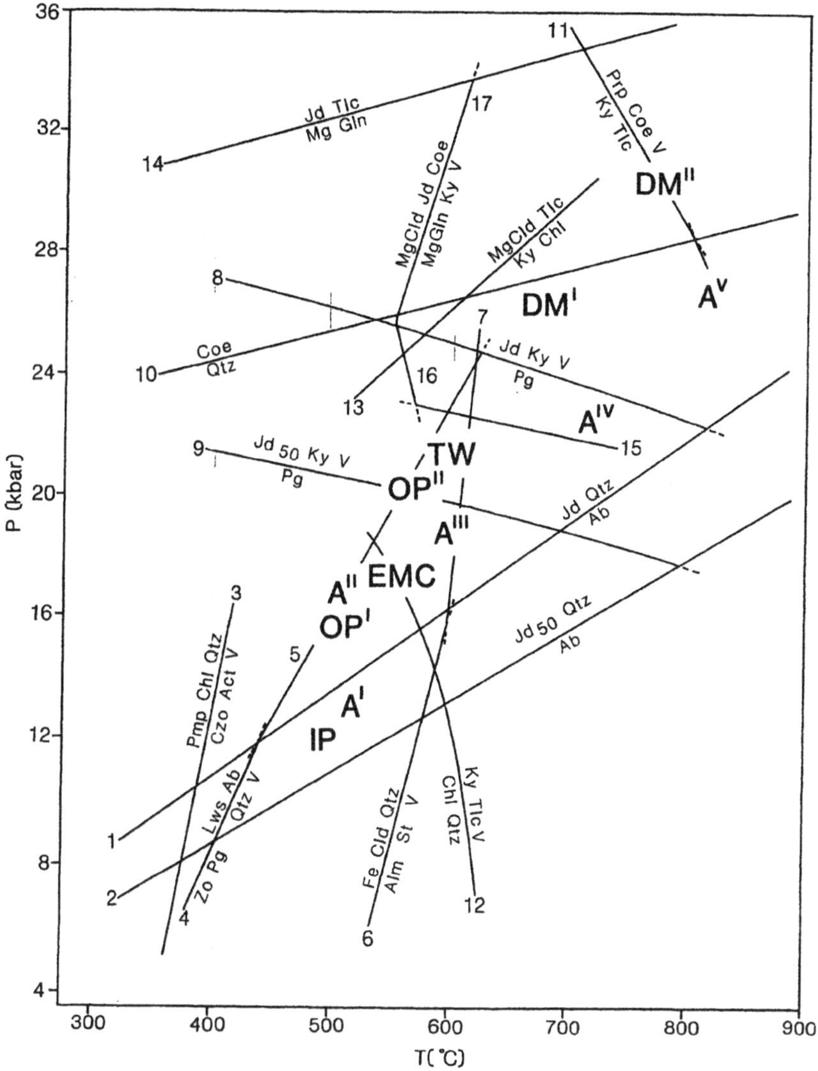

Figure 10.6 *P–T* grid showing inferred metamorphic conditions during the climax of the early-Alpine event in the eclogitised units of the Alps. IP: internal Penninic units of the Western Alps (Monte Rosa, Gran Paradiso and northern Dora Maira). A: Adula nappe (A^{I}: Vals; A^{II}: Confin; A^{III}: Trescolmen; A^{IV}: Cima di Gagnone; A^{V}: Alpe Arami and Monte Duria). EMC: Eclogitic Micaschist complex of the Sesia Zone. DM^{I}: eclogites in the southern Dora Maira massif. DM^{II}: coesite-pyrope-whiteschists of the southern Dora Maira. OP^{I}: Cottian, Graian and Voltri meta-ophiolites. OP^{II}: Pennine meta-ophiolites (Zermatt–Saas zone). TW: Tauern Window. Sources for equilibrium curves: 1. Holland (1980); 2. Holland (1983); 3. Nitsch (1971); 4 and 5. Heinrich and Althaus (1980); 6 and 7. Rao and Johannes (1979); 8 and 9. Holland (1979b); 10 and 11. Chopin (1984); 12 and 13. Chopin and Schreyer (1983); 14–17. Holland (1988). Unlabelled curves: 5. Lws + Jd = Zo + Pg + Qtz + V; 7. FeCld + Qtz = Alm + Ky + V; 15. Pg + Tlc = MgGln + Ky + V; 16. MgCld + Pg + Qtz = MgGln + Ky + V.

up to 700–800°C (Figure 10.6). This range of values is the result both of different early-Alpine histories between individual tectonic units and, less frequently, of metamorphic zonation within a structurally coherent unit. Specifically, temperatures 50–100°C higher have been obtained in the EMC sub-unit of the Sesia Zone relative to the adjacent meta-ophiolites. Temperature discontinuities related to the juxtaposition of different thrust sheets have been observed also within the Sesia Zone between the EMC and the adjacent sub-units characterised by LT (<450°C) blueschist facies metamorphism, or within the DM nappe between the rocks of the polymetamorphic unit, which apparently recrystallised at T around 700°C at $P > 27$ kbar, and the more northerly areas which were eclogitised at about 500°C and $P > 9$–13 kbar (Figure 10.7).

On the other hand, a metamorphic zonation within the same coherent tectonic unit is also recorded, notably in the meta-ophiolites, the Sesia Zone and the Adula Nappe. In the meta-ophiolites, P–T estimates are 10–18 kbar and 450–500°C in the area between Val d'Aosta and Liguria, but increase to 15–20 kbar and 550–600°C in the Pennine Alps (Breuil–Zermatt area) (Figures 10.6, 10.7). In the Sesia zone, pressures a few kilobars lower at comparable temperatures are assumed for eclogitisation in the more northerly EMC where albite and omphacite coexist, relative to the rest of the EMC where jadeite coexists with quartz. Spectacular zonation occurs in the Adula Nappe; P and T both increase dramatically towards the more internal side, and differentials of > 12 kbar and ~ 400°C exist between the front of the nappe and its root (Figure 10.5).

The regional pattern of the eclogite facies climax already discussed may in some units be obscured by the evidence of the recording by superposed parageneses of variations in time of the physical conditions attending metamorphism. The Alpine rocks recorded rare evidence of a pre-eclogitic metamorphic history of early-Alpine age and widespread, locally complete, post-eclogitic recrystallisation. Scattered relics of the pre-eclogitic Alpine history have been reported in meta-ophiolites from the Western Alps (Fry and Barnicoat, 1987) and Eastern Alps (Miller, 1974, 1977a, 1977b; Holland, 1979a), in the EMC of the Sesia Zone (Reinsch, 1979; Pognante *et al*, 1980) and in the 'polymetamorphic unit' of the DM Nappe (Chopin, 1987; Kienast and Lombardo, unpublished). Parts of the prograde trajectory are recorded in the Western Alpine meta-ophiolites and in the EMC by a (folded) foliation preserved within the eclogitic minerals. The foliation is defined by glaucophane and epidote needles which point to low thermal gradients also before the eclogite facies climax. Such a prograde trajectory, starting from very shallow depths in the crust, is supported also by the existence of upper Jurassic–Cretaceous metasedimentary cover on the meta-ophiolites and by the abundance of shallow-level intrusives of late-Palaeozoic age in the EMC and in the MR-GP-DM basement nappes. In the Eastern Alps, the pre-eclogitic epidote-amphibolite facies relics suggest a history of near-isothermal burial immediately before eclogite facies metamorphism.

The post-eclogitic trajectory is very well recorded in the Alpine rocks and began with a blueschist stage of likely early-Alpine age, followed by the LP greenschist stage which is usually attributed to the 38–40 Ma old late-Alpine stage, but which may be 60 Ma old in the internal EMC (Oberhänsli *et al*, 1985). The blueschist stage is characterised by the growth of abundant glaucophane $\pm$ clinozoisite and usually by pervasive deformation. It has been traditionally attributed to a nearly isothermal drop in P or at slightly decreased T (Ernst and Dal Piaz, 1978; Lardeaux *et al*, 1982). However, more recent data suggest an early post-eclogitic path characterised by a significant temperature decrease at high pressure; this has been proposed for the Zermatt meta-ophiolites (Fry

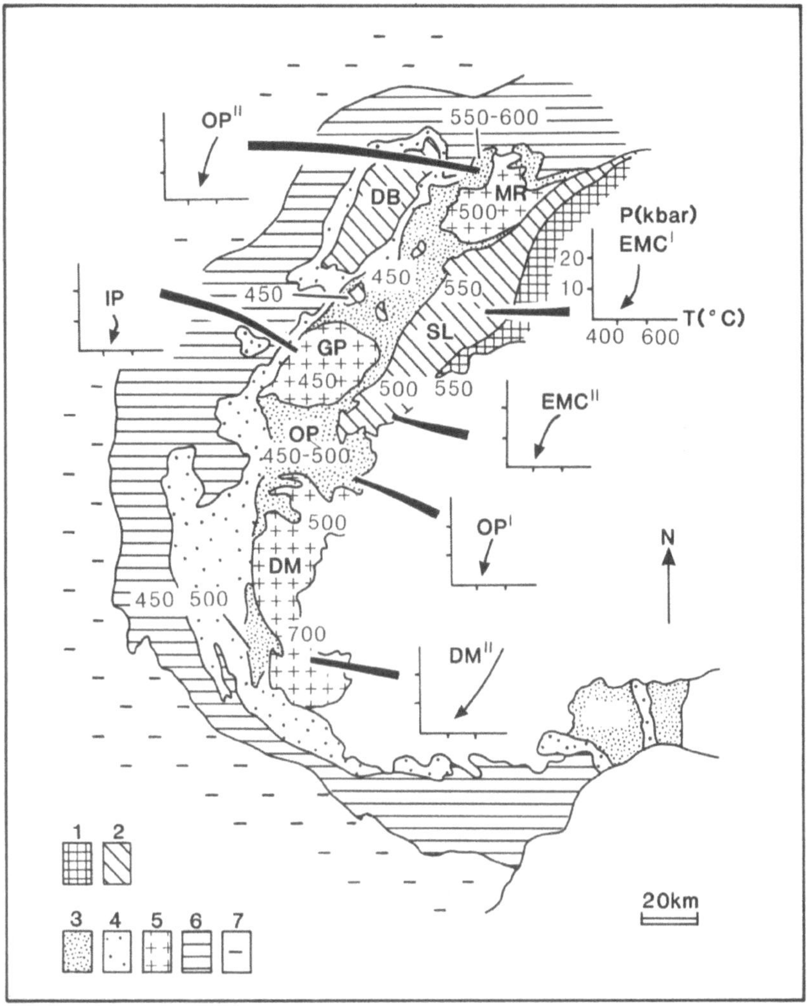

Figure 10.7 Peak metamorphic temperatures (°C) and post-eclogitic P–T paths of the eclogitised units of the Western Alps. OP^I: eclogitised ophiolites of the Cottian–Grain Alps and Voltri massif. OP^{II}: eclogitised ophiolites of the Pennine Alps (Zermatt–Saas zone). SL: Sesia Zone, including the central (EMC^I) and southern (EMC^{II}) Eclogitic Micaschist complex. IP: internal Penninic units (Monte Rosa, Gran Paradiso and northern Dora Maira). DM^{II}: pyrope-coesite-whiteschists of the southern Dora Maira. Major litho-tectonic units: 1. Southern Alps; 2. Sesia Zone and Dent Blanche nappe; 3. eclogitised ophiolites; 4. calc-schists with blueschist facies ophiolite bodies; 5. internal Penninic massifs; 6. Briançonnais zone; 7. Helvetic–Dauphinois zone, including the external massifs.

and Barnicoat, 1987), for the polymetamorphic unit of the DM Nappe (Chopin, 1987), for the southern EMC in the Sesia Zone (Pognante, 1989a), and is also implicit in the decompressive path deduced by Holland and Richardson (1979) for the blueschist to greenschist transition in the Tauern Window.

The late-Alpine LP event is chiefly post-kinematic and occurred under greenschist facies conditions in the Western Alps. It is less well developed in the easterly and structurally higher units (i.e. the EMC in the Sesia Zone) than in more westerly and structurally lower units (i.e. the internal Penninic nappes). In deeper parts of the Alpine nappe pile, such as those exposed in the Ossola–Ticino region of the Central Alps, and the Tauern Window further east, the late-Alpine event produced a pervasive MP amphibolite facies recrystallisation which has erased much of the evidence of the early-Alpine history.

10.5 Tectonic models for formation and exhumation of Alpine eclogites

Since the pioneering works of Ernst (1971, 1973) and Dal Piaz *et al* (1972), several papers have discussed the tectonic history of the Alps in terms of subduction and rapid underthrusting of ophiolitic and continental slabs (early-Alpine stage) followed by the collision between the European plate and the complex African plate (here defined as the meso-Alpine stage) (Figure 10.8). According to most of these models (Ernst, 1971; Dal Piaz *et al*, 1972; Caby *et al*, 1978; Lardeaux *et al*, 1982; Rubie, 1984; Oberhänsli *et al*, 1985; Gillet *et al*, 1986; Chopin, 1987; Fry and Barnicoat, 1987; Pognante, 1989b) eclogitisation occurred in response to southward and eastward subduction and rapid underthrusting of sheets decoupled from the margin of the South Alpine (African) plate (i.e. the composite Sesia Zone), of dismembered sections of the Jurassic oceanic lithosphere (i.e. the Piedmont–Ligurian ophiolites) and of the margin of the European plate (i.e. the Penninic units) below the Southern Alps. Geochronological data (Oberhänsli *et al*, 1985, and references therein) suggest that this process is of early to mid-Cretaceous age. Both the ophiolitic and continental slabs involved in the early-Alpine deep underthrusting stage followed a prograde trajectory reaching depths from 30–35 km up to probably 100 km under very low thermal gradients. Collision between

---

Figure 10.8 Four tectonic models for the evolution of the Western Alps simplified after: (*a*) Dal Piaz *et al* (1972); (*b*) Caby *et al* (1978); (*c*) Gillet *et al* (1986); (*d*) Platt (1986, 1987). (a1): Jurassic extension of the continental crust and opening of the Piedmont–Ligurian basin; (a2): late Cretaceous subduction of the oceanic lithosphere and continental margins; (a3): late Cretaceous building-up of the nappe pile. (Adapted and redrawn from Dal Piaz *et al*, 1972) (b1): early Cretaceous subduction of the oceanic lithosphere and of the Sesia Zone; (b2): mid-Eocene and early Oligocene westward subduction of the palaeo-European margin. (Adapted and redrawn from Caby *et al*, 1978); (c1): early Cretaceous subduction of the oceanic lithosphere and of the Sesia Zone; (c2): mid-Cretaceous eastward subduction of the palaeo-European margin. (Adapted and redrawn from Gillet *et al*, 1986); (d1): late Cretaceous subduction and shortening along the southeastern margin of the Penninic ocean system; (d2): early Eocene uplift of the HP rocks by limited buoyant uplift together with extension and thinning of the Austro–Alpine basement wedge in response to underplating. (Adapted and redrawn from Platt, 1986). Symbols are: 1. Palaeo-European margin, including the external massifs. 2. Briançonnais units and lower Penninic nappes. 3. Internal Penninic units. 4. Sub-continental mantle, Piedmont–Ligurian oceanic lithosphere and eclogitised ophiolites. 5. Calc-schists with blueschist facies ophiolite bodies, Combin unit and Valais calc-schists. 6. Austro–Alpine nappes, including the Dent Blanche nappe and Northern Calcareous Alps. 7. Sesia Zone. 8. Palaeo-African margin and Southern Alps.

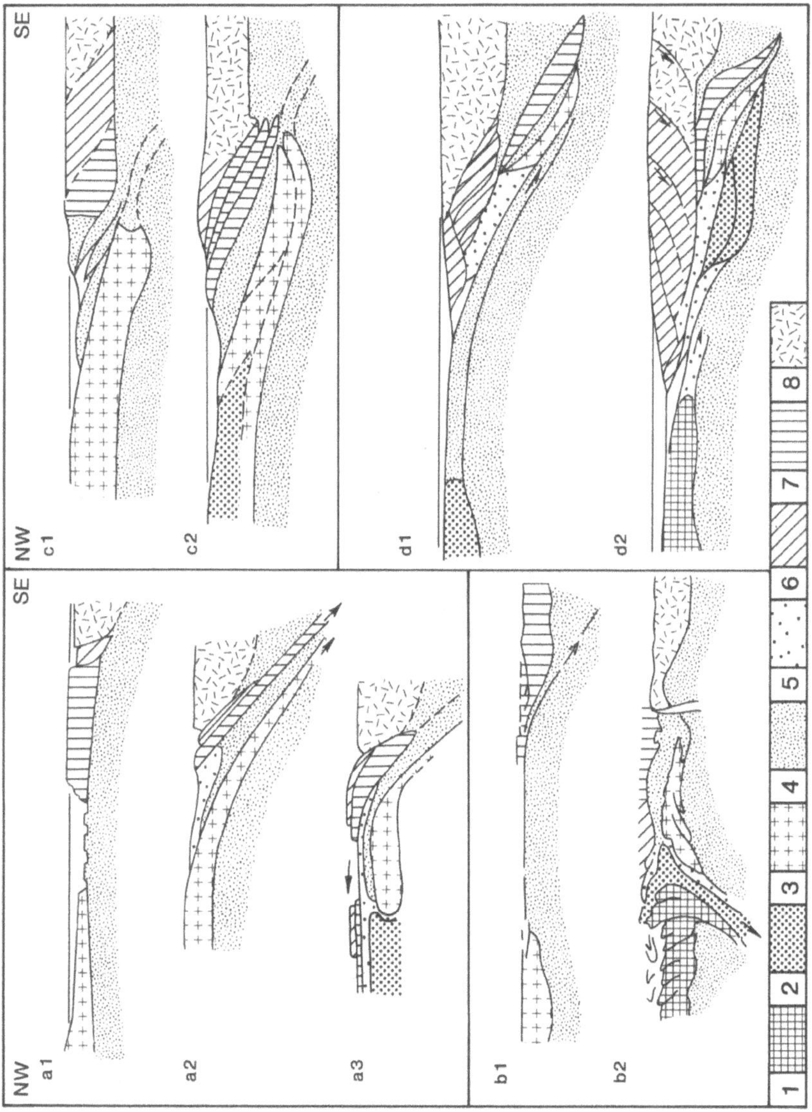
NW
SE
a1
a2
a3
b1
b2
c1
c2
d1
d2
1
2
3
4
5
6
7
8

the European and African plates during the Palaeogene determined a westward migration of the underthrusting and involved progressively more westerly and northerly sections of the European plate where a blueschist facies (non-eclogitic) metamorphism of meso-Alpine age developed (e.g. Caby *et al*, 1978; Pognante, 1984; Gillet *et al*, 1986).

Although the arguments for a southerly-dipping subduction zone are mainly structural, the eclogites themselves provide excellent corroborative evidence. Where individual nappes preserve metamorphic zonation (e.g. EMC units, Adula Nappe), rocks from internal localities record higher temperatures and depths of burial than those positioned externally. Given that the nappes now 'face' to the north or west, the metamorphic pattern is more readily explained by models in which nappes were underthrust to the south or east and were metamorphosed while inclined in those directions than by models involving north- or westwards underthrusting. If one assumes that the eclogite facies metamorphism within each thrust sheet was isochronous, then the piezothermic array may be used to give a direct estimate of the thermal gradient in the subduction zone. For the Adula Nappe, the implied gradient is about 6°C/km between depths of about 40 and 90 km. Remarkably similar values emerge when all available P–T data are plotted together (Figure 10.6; also Fry and Barnicoat, 1987; Chopin, 1987; Schreyer, 1988), which suggests that successively underthrust units experienced the same thermal gradient along the whole length of the Alps. This striking uniformity of gradient in both space and time implies the existence of a thermally mature subduction zone (Fry and Barnicoat, *op. cit.*).

The scenario presented to explain eclogite formation in the Alps is widely accepted, at least in general terms. More debated is the mechanism by which the eclogite facies rocks were returned to the earth's surface. A detailed discussion of the problem is beyond the scope of this chapter, but any proposed model must be able to explain how the eclogites were prevented from heating up at depth. Presumably the exhumation process was either very fast or took place in the presence of a heat sink or both. Credible models must also explain the following:

(a) Post-eclogite facies decompression probably occurred under more or less isothermal conditions in the internal Pennine nappes (Gillet *et al*, 1986; Ballèvre, 1988; Pognante and Sandrone, 1989) but with decreasing temperature in the meta-ophiolites (Fry and Barnicoat, 1987) and in the Sesia Zone (Pognante, 1989a, b).

(b) HP metamorphic ages are oldest in the Sesia Zone (Hunziker, 1974; Oberhänsli *et al*, 1985) and decrease discontinuously with increasing structural depth towards the external side of the orogen (Gillet *et al*, 1986).

(c) Many of the eclogite units are in direct tectonic contact with non-eclogitic ones underlying and/or overlying them. In some cases (e.g. the Tauern Eclogite Zone), tectonic mergence with overlying non-eclogitic units was completed before late Cretaceous blueschist facies metamorphism, and thus predated continental collision (see also Platt, 1986).

(d) By Tertiary times, the tectonic overburden above the majority of exposed internal units had been dramatically reduced. Only in the Ticino and Tauern culminations was the nappe pile thick enough to cause mid-Tertiary amphibolite facies metamorphism by self-heating, and for these areas, mid-Tertiary overburden thicknesses of about 20–28 km are indicated (e.g. Frey *et al*, 1980; Heinrich, 1982; Selverstone *et al*, 1984; Droop, 1985).

Mechanisms proposed (Fig. 10.8) for exhumation of the eclogitised units include: (i) 'reversal of subduction', presumably in a regional extensional régime (e.g. Dal Piaz *et al*, 1972); (ii) buoyant rise of subducted crustal material relative to the surrounding higher-density mantle (e.g. Ernst, 1971; England and Holland, 1979); (iii) isostatic uplift and erosion of the thick duplex of mainly continental lithologies above the subduction zone (Rubie, 1984; Gillet *et al*, 1986; Chopin, 1987); (iv) extensional tectonics caused by gravitational spreading of the tectonically thickened hanging wall (Platt, 1986, 1987). Several workers (e.g. Rubie, 1984; Gillet *et al*, 1986; Davy and Gillet, 1986; Chopin, 1987; Pognante, 1989b) have suggested that thermal gradients in the underthrust units were kept low by progressive downwards migration of thrust surfaces in the subduction complex; thrust slices were transferred from foot-wall to hanging-wall and thereafter overlay the still active subduction zone and any more recently underthrust slices, both of which acted as a sink for heat conducted radially from the earth's interior and heat generated within the duplex itself. This refrigeration mechanism would only have been effective so long as subduction continued, but as a large proportion of the uplift had evidently been completed before closure of the Penninic ocean, this presents no problems for any of the uplift mechanisms except the first.

Purely erosional models for exhumation cannot be reconciled with observed interleaving of eclogitic and non-eclogitic units, especially where loss of material from within (rather than from the top of) the thrust pile is implied (Platt, 1986, 1987). Models that involve buoyant uprise or gravitational spreading are satisfactory in this respect as they invoke local reversal of thrusts along S/E-dipping normal faults. Buoyancy forces are likely to have been most effective deep in the subduction zone where underthrust rocks were adjacent to mantle peridotites. At higher levels, however, reduced or reversed density contrasts would have prevented buoyancy forces from raising individual slices, though the whole nappe pile might have responded bodily to the buoyant uprise of its root.

In all probability, erosion, buoyant uplift and gravitational spreading all played a part in exhuming the eclogitic units of the Alps, but as yet the relative contribution of each process is not known.

11 The petrogenesis of Mg–Cr garnet peridotites in European metamorphic belts

L.G. MEDARIS, Jr. and D.A. CARSWELL

11.1 Introduction

Garnet-bearing alpine-type peridotites occur within continental crustal sequences at a number of localities in Europe, where, although volumetrically minor, they represent an important petrologic constituent of the Caledonian, Variscan, and Alpine orogenic belts. Garnet peridotites residing in crustal sequences are widely viewed as originally having been derived from the upper mantle, but controversy has arisen over the mechanism of crustal emplacement and the conditions of garnet stabilisation. Experiments have demonstrated that garnet is not a stable phase in peridotite at pressures attending normal crustal thickness (O'Hara *et al*, 1971), and two main hypotheses (Carswell *et al*, 1983), both of which involve collision of continental plates, have been invoked to explain the appearance of garnet peridotites in crustal sequences:

(a) garnet peridotite was tectonically emplaced from the upper mantle into thickened crust with its garnet-bearing assemblage intact (Brueckner, 1974; Carswell and Gibb, 1980a; Ernst, 1978; Medaris, 1980a, 1980b)
(b) garnet peridotite was produced by prograde HP metamorphism of spinel peridotite or serpentinite which previously had been emplaced into the crust (Evans and Trommsdorff, 1978; Griffin *et al*, 1985; Griffin and Qvale, 1985; Jamtveit, 1984; Obata and Morten, 1987).

In either case peridotites in crustal sequences are important for the information they contain on the metamorphic conditions of thickened continental crust and on the nature of interactions between mantle and crust during orogenesis.

In this Chapter the occurrences of Mg–Cr garnet peridotites in the Caledonian, Variscan, and Alpine belts of Europe are reviewed. Isotopic data are discussed that have a bearing on the age of the peridotites and on whether it is crustal or mantle signatures that are shown. Equilibration conditions of the garnet-bearing assemblages are evaluated through application of currently preferred geothermometers and barometers and, where feasible, the thermal histories of the various bodies are outlined.

11.2 Mg–Cr and Fe–Ti garnet peridotites

In western Norway an important distinction has been made between two compositionally different groups of garnet peridotites (Carswell *et al*, 1983). The Fe–Ti type of garnet peridotite is relatively ferriferous ($Mg^* = 60–75$), contains conspicuous amounts of Fe and Ti oxide phases, is poor in Cr_2O_3 (< 0.05 wt%) and rich in TiO_2 (~ 1 wt%), and is closely associated with garnet pyroxenites and eclogites. In contrast, the Mg–Cr type, which is sporadically preserved in large olivine dominated alpine-type peridotite bodies, is relatively magnesian (Mg* = 80–92), rich in Cr_2O_3 (~0.5 wt%), and poor in TiO_2 (~0.25 wt%). The contrasting rock chemistries of the two types of garnet peridotites are reflected in their mineral chemistries, with the Fe–Ti type containing more Fe-rich olivine, orthopyroxene, and garnet (Fo_{67-82}, En_{74-85} and Prp_{39-56} Alm_{29-47} Grs_{1-12}) than the Mg–Cr type (Fo_{82-92}, En_{84-93}, and Prp_{60-75} Alm_{15-30} Grs_{5-10}).

It has been demonstrated on the basis of field, textural, mineralogical, and chemical data that the Fe–Ti garnet peridotites and associated garnet pyroxenites originally formed by relatively LP ($\leqslant$ 10 kbar) crystal accumulation from layered basic intrusions and were subsequently metamorphosed under eclogite facies conditions (Carswell *et al*, 1983). A clear exposition of this type of petrogenesis is provided by the Eiksunddal complex in western Norway (Jamtveit, 1987a, 1987b). However, the petrogenesis of Mg–Cr garnet peridotites is controversial (Carswell, 1986; Griffin, 1987; Mearns, 1986), with the argument focusing on whether garnet-bearing assemblages in peridotite are allofacial or isofacial with respect to surrounding crustal gneisses. So, although Fe–Ti garnet peridotites are a significant component of orogenic belts in Europe and provide important data for evaluating the *P–T* histories of such belts, this review will concentrate on the more enigmatic Mg–Cr garnet peridotites in an attempt to elucidate their petrogenesis and mode of evolution.

11.3 Summary of Mg–Cr garnet peridotite occurrences in Europe

Mg-Cr garnet peridotites occur in crustal sequences in the Caledonian, Variscan, and Alpine orogenic belts of Europe (Figure 11.1). Although all of the garnet peridotites have certain features in common, such as bulk chemical compositions characteristic of depleted upper mantle, enough data are now available to reveal that the petrogenesis of the garnet-bearing ultramafic assemblages has not been uniform in the three orogenic belts. Rather, the garnet peridotites in Scandinavia, the Bohemian Massif, the Lepontine Alps, the Austridic Alps, and the Betic Cordillera exhibit characteristics that are distinctive to each province and that must be ascribed to a variety of processes, as will become evident in the ensuing summary.

11.3.1 *Caledonides*

11.3.1.1 *Scandinavia.* Alpine-type peridotites occur in two main settings in the Scandinavian Caledonides: in the Western Gneiss Region (WGR) of western Norway and in the Upper Allochthon (Seve and Köli Nappes) of the overlying Nappe Complex (Qvale and Stigh, 1985). All *garnet-bearing* peridotites in the Scandinavian Caledonides, however, are confined to the WGR, except for one occurrence in the Seve Nappe in Sweden (van Roermund *et al*, 1985).

The Western Gneiss Region (WGR), also referred to as the Basal Gneiss Complex, is the lowermost structural unit in western Norway and represents an autochthonous or parautochthonous basement culmination in the orogen. The WGR is a heterogeneous terrane, consisting of migmatitic orthogneisses, largely tonalite to granodiorite in composition, and supracrustal rocks, including pelitic schists, quartzites, marbles, calc-silicate rocks, and amphibolites. The supracrustal rocks contain peridotite,

Figure 11.1 Localities of Mg–Cr garnet peridotites and selected Fe–Ti garnet peridotites in Europe (map modified from Aubouin, 1980). Caledonides: 1, Scandinavia (Western Gneiss Region, Seve Nappe); 2, United Kingdom (Glenelg, Ballantrae, Outer Hebrides); Variscides: 3, Bohemian Massif (Lower Austria, Bohemia and Moravia, Sudetes, Saxony, Münchberg); 4, Vosges–Schwarzwald Massif; 5, Massif Central (Sauviat-sur-Vige, Bois des Feuilles); 6, Cabo Ortegal complex; Tethysides: 7, Lepontine Alps; 8, Austridic Alps; 9, Betic Cordillera (Ronda).

gabbro, and anorthosite and have been intruded by rapakivi textured granitoids (Bryhni, 1966; Brueckner, 1977b). The entire gneiss terrane has been extensively recrystallised under amphibolite facies conditions, but evidence for an earlier HP metamorphism is provided by the widespread occurrence of eclogites and HP granulites (Griffin *et al*, 1985).

Protolith ages for most gneisses in the WGR are Proterozoic, either Svecofennian or Sveconorwegian (Brueckner, 1972, 1979), and rapakivi granites and related augen gneisses yield ages of 1400–1600 Ma (Lappin *et al*, 1979; Harvey, 1983; Carswell and Harvey, 1985). However, some supracrustal rocks may represent infolded nappes of Palaeozoic strata (Krill, 1980). Amphibolite facies metamorphism of the WGR is Caledonian, as indicated by K–Ar and Rb–Sr mineral ages of 380–405 Ma (Brueckner, 1972). Eclogite facies metamorphism is Caledonian as well, based on a U–Pb age of 410 Ma from zircon in the Ulsteinvik eclogite (Krogh *et al*, 1974) and on Sm–Nd ages of 407–477 Ma for garnet–clinopyroxene pairs from various eclogite samples (Griffin and Brueckner, 1980; Mearns and Lappin, 1982; Mearns, 1986).

Numerous peridotites, ranging from lenses a few metres thick to large tabular bodies several kilometres in areal extent, are widely distributed in the WGR (for localities see figure 1 in Medaris, 1984, and figure 2 in Jamtveit *et al*, in press). The peridotites have been extensively recrystallised to chlorite peridotites, which are isofacial with the surrounding amphibolite facies gneisses (Medaris, 1984) and which contain the same structural elements (Brueckner, 1977b; Medaris, 1980b; Cordellier *et al*, 1981).

Locally, relict garnet peridotites and associated garnet pyroxenites have been preserved within the chlorite peridotites. Early work by Eskola (1921), O'Hara and Mercy (1963), and Mercy and O'Hara (1965a, 1965b) drew attention to the classic occurrences of garnet peridotite at Almklovdalen and Kalskaret. Subsequently, further investigations have clarified relations at Almklovdalen (Lappin, 1966; Medaris, 1980b; Carswell, 1981) and Kalskaret (Carswell, 1968a; Brueckner, 1977b) and have documented the existence of numerous other garnet peridotites in the WGR, including Aldalen (Medaris, 1984), Rødskar (Medaris, 1984), Gurskebotn (Jamtveit, 1984), Sandvik (Carswell *et al*, 1983; Medaris, 1984), Fjørtoft (Jamtveit *et al*, 1989), and Ugelvik and Raudhaugene on the island of Otrøy (Carswell, 1968b, 1973, 1986; Jamtveit *et al*, 1989). These occurrences of garnet peridotite, combined with the presence of amphibole-spinel pseudomorphs after garnet at other localities, suggest that most, and perhaps all, of the peridotites in the WGR were originally garnetiferous before extensive recrystallisation under amphibolite facies conditions.

The complex metamorphic evolution of the Norwegian peridotites has been summarised by Carswell (1986), who recognised seven stages of recrystallisation, these being from earliest to latest:

	Stage	*Interpretation*
I	$Ol_1 + Opx_1 + Cpx_1 \pm Spl_1 \pm Amp_1$	High-temperature protolith
II	$Ol_2 + Opx_2 + Cpx_2 + Grt_1$	Coarse-grained eclogite facies
III	$Ol_3 + Opx_3 + Cpx_3 + Grt_2 + Spl_2$	Fine-grained neoblast eclogite facies
IV	$Ol_4 + Opx_4 + Cpx_4 + Spl_3$	Kelyphitic granulite facies
V	$Ol_5 + Opx_5 + Amp_2 + Spl_4$	Coronitic/foliated high amphibolite facies
VI	$Ol_6 + Opx_6 + Amp_3 + Chl_1$	Foliated low amphibolite facies
VII	$Serp + Talc + Amp_4 + Chl_2$	Localised greenschist facies

This sequence of stages is based on petrographic observation of many samples; samples from a specific locality commonly contain disequilibrium assemblages of minerals from two or more, but never all seven, of the stages. The various assemblages can be viewed in terms of temperature and pressure by utilising the petrogenetic grid for the CMASH system (Figure 11.2; Chapter 2), following Obata and Thompson (1981). In Figure 11.2 stage VII has been omitted, the five-phase assemblage of stage III has been located on the boundary between the HT garnet and spinel fields, which is multivariant in natural rock compositions, and a third garnet-bearing assemblage, designated as stage IIA, has been added to represent the assemblage, Ol + Opx + Amp + Grt, which occurs in the Rødskar body (Medaris, 1984).

The prevailing pattern of post-garnet peridotite recrystallisation illustrated by the sequence of stages in Figure 11.2 is one of decompression and cooling. Mineral phases which predate the garnet-bearing assemblages are rarely preserved and only one sample of peridotite among the hundreds examined by investigators over the years exhibits the assemblage of stage I. This sample, E38 from Sandvik (Carswell *et al*, 1983; Carswell, 1986), contains aluminous pyroxenes and reaction coronas of garnet around aluminous spinel. However, the significance of sample E38 for the evolution of

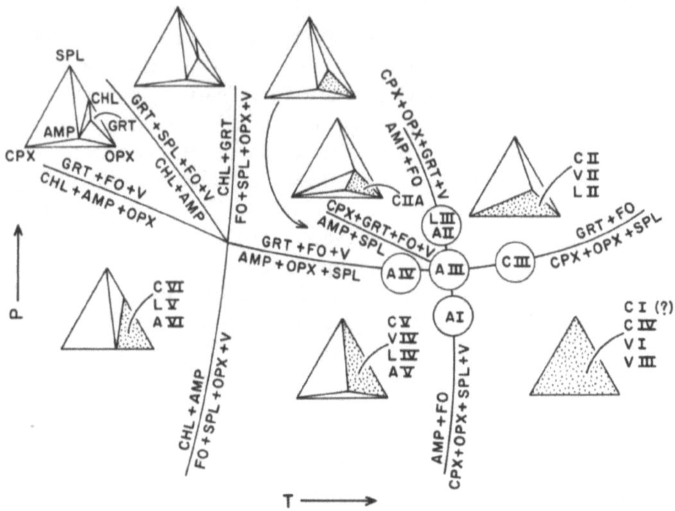

Figure 11.2 *T–P* grid for mineral assemblages in the system CaO-MgO-Al_2O_3-SiO_2-H_2O, projected from forsterite and H_2O on to the plane $MgAl_2O_4$-$CaMgSi_2O_6$-$Mg_2Si_2O_6$. Observed assemblages in peridotites of the Western Gneiss Region, Caledonides, C I–C VI; associated with Variscan granulites, V I–V IV; in the Lepontine Alps, L II–L V; in the Austridic Alps, A I–A VI.

Norwegian Mg–Cr peridotites remains unclear, because the chemical composition of sample E38 is different from that of Mg-Cr peridotites at Sandvik and elsewhere in the WGR and its relation to the surrounding rocks at Sandvik is uncertain (Carswell *et al*, 1983, p. 742).

The best petrological evidence for an early, pre-stage II assemblage in the Norwegian peridotites is provided by the occasional preservation of aluminous pyroxene megacrysts in pyroxenites and eclogites. Orthopyroxene megacrysts with exsolved garnet and clinopyroxene occur in garnet pyroxenite at Ugelvik (Carswell, 1973), clinopyroxene megacrysts with exsolved garnet and orthopyroxene are found in garnet clinopyroxenite layers at Aldalen (Medaris, unpublished), and clinopyroxene megacrysts with exsolved garnet and orthopyroxene have been described from boudinaged eclogite layers in peridotite at the Sunndal–Grubse locality in the Almklovdalen body (Lappin, 1973, 1974). The reconstituted compositions of the pyroxene megacrysts are aluminous (4.0–4.5 wt% Al_2O_3), and such pyroxenes are interpreted to have crystallised initially at high temperatures from melts produced by igneous processes in the upper mantle. Subsequent exsolution of garnet from pyroxene could have taken place during cooling in the upper mantle or could have been promoted by recrystallisation in a crustal setting related to the Caledonian orogeny.

Further evidence bearing on the early conditions of the Norwegian peridotites comes from a garnet pyroxenite at the Gurskebotn locality (Jamtveit, 1984), in which stage II garnet contains inclusions of chlorite. This textural relation, amongst other considerations, led Jamtveit to suggest that garnet peridotite may have been formed by prograde metamorphism of a LP protolith. However, stage II garnet in this sample of garnet pyroxenite also contains inclusions of pargasitic amphibole (Jamtveit, personal communication), and an alternative interpretation is that garnet host and chlorite and amphibole inclusions represent an early lower temperature, but HP, assemblage in pyroxenite (Figure 11.2).

Superferrian eclogite occurs within the Almklovdalen peridotite as boudins, which are believed to represent disrupted dykes (Griffin and Qvale, 1985). Garnet in the eclogite exhibits prograde compositional zoning and contains inclusions of Cl-rich ferropargasite, which is absent from the eclogite matrix. Griffin and Qvale have suggested that the superferrian eclogite once consisted of an assemblage of amphibole, minor plagioclase, and magnetite and that the amphibolite was subsequently transformed to eclogite by prograde HP metamorphism. By implication then, the associated garnet peridotite, which is found as scattered relicts in the surrounding ultramafic rocks, likewise is the product of HP metamorphism, having been derived from a hypothetical spinel peridotite or serpentinite protolith. However, minerals in the garnet peridotite only show retrograde compositional zoning, and there is no direct evidence in the peridotite itself for a prograde metamorphic origin from a LP ultramafic protolith.

Because petrological evidence bearing on the pre-stage II history of the Norwegian garnet peridotites is scant and equivocal, investigations utilising other methods, such as radiogenic isotopes, provide a more promising approach to evaluation of whether or not these peridotites were derived from LP protoliths. Brueckner (1974, 1977a) found that minerals separated from country rock eclogites have relatively high values of $^{87}Sr/^{86}Sr$, in the range 0.7034–0.7157, whereas clinopyroxene grains separated from garnet peridotites have low and restricted values of 0.7011–0.7029, which are consistent with values predicted for Proterozoic upper mantle. These data were interpreted as reflecting a crustal origin for the country rock eclogites and an upper mantle origin for the garnet peridotites. However, it was pointed out that the data do not preclude formation of the garnet peridotite in a crustal environment, if the peridotites were shielded from the ingress of crustal fluids.

A significant contribution to the garnet peridotite controversy has been made by Mearns (1986), who established that the stage II garnet-bearing assemblages are Proterozoic in age, rather than Caledonian. Sm–Nd data for garnet, whole rock, and clinopyroxene from a garnet lherzolite at Almklovdalen establish an isochron giving an age of 1703 ± 29 Ma. Minerals from a garnet websterite at Almklovdalen define a younger age of 1040 ± 30 Ma. Proterozoic ages have been corroborated for stage II assemblages by Jamtveit *et al* (1989), who obtained Sm–Nd ages of 1303 ± 236 Ma and 1163 ± 14 Ma for garnet peridotite and garnet pyroxenite, respectively, from Kalskaret. However, Jamtveit *et al* (1989) also analysed state III garnet-bearing assemblages and obtained an age of 437 ± 58 Ma from peridotite at Raudhaugene and 511 ± 18 Ma from Fjørtoft, thereby demonstrating that the later garnet-bearing assemblages in the Norwegian peridotites were formed during the Caledonian orogeny.

Although many questions remain to be answered about the detailed evolution of the Norwegian garnet peridotites, the simplest interpretation in view of the presently available evidence is that they were derived by tectonic emplacement of Proterozoic garnet-bearing upper mantle into the lower portions of thickened continental crust, related to collision of the Baltic and Greenland plates during the Caledonian orogeny. HP metamorphism at the base of the thickened crust was responsible for development of a second generation of garnet (stage III) in the peridotites and formation of eclogites and HP granulites in the associated crustal rocks. During decompression and cooling the peridotites were variably recrystallised to mineral assemblages of stages IV to VII.

11.3.1.2 *United Kingdom.* Garnet-bearing ultramafic rocks occur at three localities in the United Kingdom, including Glenelg in northwest Scotland (Sanders *et al*, 1984; Carswell and Smallwood, unpublished), the Ballantrae Complex of southwest Scotland (Bloxam and Allen, 1960; Smellie and Stone, 1984), and Rodel in South Harris, Outer Hebrides (Livingstone, 1967, 1976). However, no garnet peridotite of Mg-Cr type has been reported from any of these localities, so they are not described here, but merely cited for completeness.

11.3.2 *Variscides*

Although alpine-type peridotites are widely distributed in the Variscides, garnet-bearing varieties have been reported only in the Bohemian Massif, the Schwarzwald, the Vosges, the Massif Central, and the Cabo Ortegal complex (Figure 11.1), where they are commonly, but not always, located in granulite facies terranes. Two distinct groups of granulites in Variscan median Europe have been distinguished by Pin and Vielzeuf (1983). Group I, which is characterised by HP assemblages with associated eclogites, occurs as allochthonous lenses (from dm^3 to km^3) in medium- to high-grade paragneisses in internal zones of the belt. Group II, which lacks an eclogite association and contains MP granulite assemblages, represents lower crustal material that has been exhumed by Alpine tectonism or sampled by Tertiary volcanoes. The two groups of granulites are further distinguished by the mineralogy of included ultramafic rocks; the first group contains both garnet and spinel peridotites, whereas the second group only contains garnet-free peridotites, which are interpreted to be members of layered basic–ultrabasic complexes that were emplaced into the lower crust.

11.3.2.1 *Bohemian Massif.* Although ultramafic rocks are widely distributed in the Bohemian Massif, garnet peridotites have been recognised in only three settings so far: as solitary, oval-shaped bodies along major faults, within leucocratic gneiss of Gföhl type, and as lenticular masses in granulite (Dudek and Kopecký, 1966; Fiala, 1965, cf. figure 1, 1985; Machart, 1984, cf. figure 1). Of these three settings, garnet peridotites associated with granulites are most numerous and are distributed among six different granulite terranes, including the Saxonian Granulite Complex (Sachsisches Granulitgebirge) in the northwestern part of the Bohemian Massif (Mathé, 1969a), the Ohře Crystalline complex in the České Středohoři Mountains northwest of Prague, where granulite is covered by Upper Cretaceous platform sediments (Fiala and Paděra, 1977; Kopecký and Sattran, 1966. Mikhailov and Rovsha, 1966), the Góry Sowie granulites in the Sudetes of southwestern Poland (Bakun-Czubarow, 1983), the South Bohemian granulites (Fiala, 1965, 1966), the West Moravian granulites (Dobretsov *et al*, 1984; Fiala, 1965, 1966; Medaris *et al*, 1988a, 1988b; Mísař and Jelínek, 1981; Mísař *et al*, 1984; Mikhailov and Rovsha, 1966), and those in Lower Austria (Kappel, 1967; Scharbert, 1973a, 1973b; Scharbert and Carswell, 1983). Garnet peridotites in Gföhl gneiss are subordinate in number, as are the solitary bodies of garnet peridotite along major faults, most of which are located in the Kutná Hora–Svratka Complex along the northeastern border of the Moldanubian Complex (Fiala, 1965, 1985). Garnet websterite and garnet clinopyroxenite, which commonly are referred to as griquaite, pyrope eclogite, or eclogite in the Czech literature, occur as layers in garnet peridotites in all three settings (Machart, 1984; Machart and Paděra, 1982; Paděra and Prochazka, 1970; Paděra *et al*, 1971).

The granulites in southern Bohemia, western Moravia, and Austria are located in the Moldanubian Complex, which is the most internal tectonic unit in the Bohemian Massif and consists mostly of high-grade gneisses and migmatites which have been intruded by Hercynian plutons. Although the Saxony, Ohře, and Góry Sowie granulites are similar lithologically to those in the Moldanubicum, they are situated in the Saxothuringian Zone of the orogen, and only the Góry Sowie granulites have been correlated with those in the Moldanubian Complex (Bakun-Czubarow, 1983). The predominant lithostratigraphic units in the Moldanubian Complex include granulites, granitic gneisses of Gföhl type, and two groups of paragneisses (the Monotonous and Varied Groups) (Svoboda, 1966; Suk *et al*, 1984). The granulites are composed largely of banded leptynites with the assemblage, Ky + Bt + Grt + perthitic Kfs + antiperthitic Pl + Qtz, which were derived from granitic or rhyolitic protoliths; locally, more basic layers contain two-pyroxene assemblages (Jakeš, 1969a, 1969b; Vrána and Jakeš, 1982; Fiala *et al*, 1987b). Units of the Gföhl type are made up largely of migmatitic quartzofeldspathic gneiss, which is commonly referred to as orthogneiss, but which may have formed from a variety of protoliths and which may include migmatised and retrograded granulite, as well (Dudek *et al*, 1974). The paragneisses are quartzofeldspathic gneisses, probably derived from greywacke-pelite associations, with minor intercalations of metabasite and metaquartzite in the Monotonous Group and more abundant intercalations of marble, calc-silicate rocks, graphitic schists, metaquartzite, and metabasites in the Varied Group (Kröner *et al*, 1988).

Eclogites *s.s.* occur in gneisses of the Gföhl type and in paragneisses of both the Monotonous and Varied Group (Dudek and Fediuková, 1974). In addition, layers of eclogite *s.s.* occur in peridotites of the solitary type, as well as in peridotites within Gföhl gneiss (Dudek and Fediuková, 1974; Fiala, 1985; Beard *et al*, 1988). Although occurrences of eclogite *s.s.* in granulite have been cited (Dudek and Fediuková, 1974; Mísař *et al*, 1984), such occurrences are rare and may represent tectonic inclusion of disrupted eclogite facies mafic–ultramafic complexes into granulite.

The metamorphic and protolith ages of Moldanubian paragneisses and orthogneisses have long been debated (Kröner *et al*, 1988). Although some isotopic data seem to record a Caledonian thermal event in the Moldanubian Complex (Arnold and Scharbert, 1973; Grauert *et al*, 1974), more recent U–Pb, Sm–Nd, and Rb–Sr isotopic investigations have established a Hercynian age for the high-grade metamorphic assemblages and revealed Precambrian sources as old as 2680 Ma for the metasedimentary rocks (van Breemen *et al*, 1982; Kröner *et al*, 1988). U–Pb data for zircon and monazite have yielded metamorphic ages from 367 to 339 Ma for the Monotonous Group, Gföhl gneiss, and granulite. In addition, Sm–Nd ages of 324 and 338 Ma have been obtained for pairs of garnet and clinopyroxene from two samples of Moldanubian eclogites (Beard *et al*, 1988) and 344 and 370 Ma for two samples of garnet clinopyroxenite from Lower Austria (Carswell and Jamtveit, unpublished). Thus, accumulating isotopic evidence points to an Early Carboniferous age for the high-grade metamorphism in the Moldanubian Complex, proper. Two eclogites and a cogenetic metagabbro from the Münchberg Gneiss Massif in northeast Bavaria were investigated by Gebauer and Grünenfelder (1979b), who obtained a U–Pb evolution diagram for zircons with intercepts of 525 and 380 Ma, which were interpreted as reflecting Middle Devonian HP metamorphism of Cambrian protoliths; the Münchberg Gneiss Massif has been correlated with the Moldanubian Complex by Pin and Vielzeuf (1983).

Among approximately 200 occurrences of peridotites in the Bohemian part of the Moldanubian Complex, Machart (1984) found that both garnet peridotites and spinel peridotites are present in granulites and Gföhl gneisses, but that only spinel peridotites or their metamorphosed equivalents, such as tremolite schists, occur in paragneisses and migmatites of the Monotonous and Varied Groups.

Garnet peridotites associated with granulites are mineralogically complex, and although no comprehensive paragenetic scheme for these rocks has yet been published, the present authors have established a sequence of at least five stages of mineral assemblages (Figure 11.2), these being from earliest to latest:

	Stage	*Interpretation*
I	$Ol_1 + Opx_1 + Cpx_1 + Spl_1$	High temperature protolith
II	$Ol_2 + Opx_2 + Cpx_2 + Grt_1$	Formation of garnet from spinel
III	$Ol_3 + Opx_3 + Cpx_3 + Spl_2$	Kelyphitic granulite facies
IV	$Ol_4 + Opx_4 + Amp_1 + Spl_3$	Foliated amphibolite facies
V	Late serpentinisation	Final stages of uplift

The best evidence for the existence of a stage I protolith and for the relation between stages I and II is seen in western Moravia at Mohelno (Medaris *et al*, 1988a; Medaris *et al*, in preparation). The Mohelno peridotite is a large ($\sim 4\,km^2$), well-exposed, tabular body of spinel peridotite that contains a stage I assemblage characterised by aluminous pyroxenes and a porphyroclastic texture. Garnet occurs in the peridotite, but only along the margins of the body within a few metres of the contact with surrounding leptynite. The stage II assemblage has formed from stage I by the reaction:

$$\text{Al Opx}_1 + \text{Al Cpx}_1 + \text{Spl}_1 = \text{Grt}_1 + \text{Ol}_2$$

Petrographically, this reaction is demonstrated by the occurrence of spinel inclusions within garnet grains, not only in the contact zone of the peridotite, but also in garnet within boudins of peridotite in leptynite. Localisation of garnet in the contact zone suggests that development of the stage II assemblage was related to emplacement of the peridotite into leptynite (HP granulite). Contrary to the indicated phase relationships in Figure 11.2, an increase in pressure is not essential to form assemblage II from I (cf. Figure 3.1); rather, the occurrence of garnet can be interpreted as due to tectonic emplacement of a HT spinel peridotite and rapid cooling of the margins of the body into the garnet field (Medaris *et al*, 1988a; Section 11.4.4). Spinel inclusions in garnet have also been reported from peridotites in southern Bohemia (Fiala, 1965). Further evidence for the existence of a stage I HT protolith is provided by the discovery in serpentinite at Meidling-im-Tal in Austria of an aluminous orthopyroxene megacryst, which contains exsolution lamellae of garnet and clinopyroxene (Carswell, 1989). A clinopyroxene megacryst with garnet and orthopyroxene exsolution lamellae has also been described from a pyroxenite boudin in leptynite at the Bory quarry in western Moravia (Šanc and Rieder, 1983).

During decompression of peridotites and associated granulites, the stage III assemblage formed by the reaction:

$$\text{Grt}_1 + \text{Ol}_2 = \text{Al Opx}_3 + \text{Al Cpx}_3 + \text{Spl}_2$$

and is displayed in kelyphitic rims of aluminous pyroxenes and spinel around garnet. Commonly the kelyphitic rims consist of an outer corona of distinct pyroxenes and spinel and an inner zone of fine-grained filamental intergrowth of pyroxenes and spinel. Further decompression and cooling, especially promoted by deformation and associated fluid influx, led in peridotites to the development of cigar-shaped stage IV kelyphites at the expense of stage III kelyphites, according to the reaction:

$$\text{Al Opx}_3 + \text{Al Cpx}_3 + \text{Spl}_2 + H_2O = \text{Amp}_1 + \text{Ol}_4$$

and to the widespread retrogression of granulite facies leptynites to amphibolite facies assemblages. Thus, the peridotites and granulites have shared similar histories subsequent to emplacement of the ultramafic rocks; not only can granulite and amphibolite facies mineral assemblages be correlated in the two rock types, but they also share similar structural features (Dobretsov *et al*, 1984; Bowes and Misař, in preparation). Later serpentinisation (not illustrated in Figure 11.2) is widespread in peridotites in the Bohemian Massif, both in peridotites associated with granulites and in solitary bodies.

In contrast with peridotites associated with granulites, garnet peridotites in other settings in the Bohemian Massif show no evidence for an early spinel-bearing assemblage, and garnet (or kelyphite after garnet) is widespread throughout the bodies. Pyroxenes of the garnet-bearing assemblage are typically low in alumina, as opposed to the relatively aluminous pyroxenes in peridotites associated with granulites. Aside from the common replacement of garnet by fibrous, radial, amphibole-spinel kelyphite, these other bodies of garnet peridotite do not seem to show a sequence of retrograde mineral assemblages in concert with their surrounding gneisses. Such features are consistent with the interpretation that these garnet peridotites are allochthonous bodies, having been emplaced into their present positions along major faults (Dudek and Kopecký, 1966; Machart, 1984; Fiala, 1985).

To date, no age determinations have been made on garnet-bearing ultramafic assemblages in the Bohemian Massif, although such studies are essential to test the hypotheses summarised above pertaining to the origin of the different types of garnet peridotites in the Moldanubian Complex. If the hypotheses are correct, one would expect a Variscan age for the stage II garnet-bearing assemblage in peridotites associated with granulites, but perhaps an older age for garnet-bearing assemblages in peridotites that are located in other settings.

11.3.2.2 *Massif Central.* The Massif Central (Figure 11.1) is located within the Moldanubian Zone of the Variscan Orogen and is composed of southward-directed nappes and thrusts that have been divided into three main zones (Matte and Burg, 1981). Both Fe–Ti and Mg–Cr garnet peridotites occur in the most northerly and structurally uppermost of the three zones (Auvergne zone), which includes mainly pre-Lower Carboniferous high-grade rocks (mica schists, granulitic gneisses, anatexites, and eclogites).

Garnet peridotites of the Fe–Ti type occur in the well-studied Massif de Sauviat-sur-Vige (Coffrant, 1974; Bernard-Griffiths and Jahn, 1981), which is enclosed within the base of the Limousin series and is bounded on the north by coarse-grained granitic augen gneiss (Ky + Bt + Mc + Olg + Qtz) and on the south by fine-grained meta-granite (gneissic leptynite with Grt + Sil + Ms + Bt + Olg + Mc + Qtz). The 6 km by 2 km massif is composed of gabbros, microgabbros with reaction coronas, eclogites, amphibolites, and volumetrically subordinate garnet peridotites. Textural relations and geochemical data indicate that the mafic–ultramafic complex originated by LP (<9 kbar) fractional crystallisation from gabbroic magma and that the eclogites and garnet peridotites were formed by subsequent HP metamorphism (> 15 kbar). Zircon fractions from metagabbros, eclogites, and garnet peridotites of the Sauviat-sur-Vige massif define a discordia with intercepts at 496 and 320 Ma, and monazite fractions from augen gneiss to the north give the same discordia (within the limits of analytical error) with intercepts at 481 and 318 Ma (Gebauer *et al*, 1981). These data were interpreted as reflecting magmatic crystallisation of the mafic–ultramafic complex during Cambro–Ordovician time, followed by recrystallisation during Hercynian HP metamorphism. However, because no Variscan regional high-grade metamorphism is known from this part of the Massif Central, the Sauviat-sur-Vige massif and adjacent granitic gneisses are thought to have been tectonically emplaced into pre-Variscan crustal rocks.

Two Mg–Cr garnet peridotites and associated garnet pyroxenites have been described from the Massif Central, one near Artiges in the Saint-Leonard-de-Noblat area of the Haut Limousin (Bonnot and Piboule, 1980) and the other from the Bois des Feuilles, Monts du Lyonnais (Lasnier, 1971). The geological setting and petrological features of these two garnet peridotites are much like those associated with granulites in the Bohemian Massif. The Artiges body is situated in biotite-sillimanite-orthoclase gneiss, and the Bois des Feuilles massif is interlayered with biotite-sillimanite-garnet gneisses at the top of the granulite series of Monts du Lyonnais. The same sequence of ultramafic mineral assemblages has been described from both localities: an early spinel lherzolite, followed by growth of garnet (documented by spinel inclusions in garnet), in turn followed by recrystallisation under granulite facies conditions (marked by kelyphitic rims of pyroxenes and spinel around garnet), and finally, further recrystallisation leading to the growth of amphibole, chlorite, and serpentine. At both localities pyroxenites contain megacrysts of aluminous orthopyroxene and clinopyroxene that exhibit exsolution lamellae of garnet and the complementary pyroxene; as in the Bohemian Massif, this texture is related to transition from a spinel- to a garnet-lherzolite assemblage.

A dual origin for garnet peridotites in the Massif Central has been emphasised, with the Fe–Ti type representing the products of HP metamorphism of protoliths that were previously emplaced in the crust, and the Mg-Cr type being derived directly from the upper mantle by tectonic emplacement (Bonnot and Piboule, 1980).

11.3.2.3 *Other Variscan occurrences.* Elsewhere in the Variscan orogenic belt garnet peridotites have been reported in the Schwarzwald (Stenger *et al*, 1987) and in the Vosges (Bonhomme and Fluck, 1981), but to date these occurrences have not been well documented. Garnet peridotite has also been reported in the Cabo Ortegal complex of the Iberian Massif (Vogel, 1967), but a detailed investigation by Maaskant (1970) has demonstrated that garnet is stable only in pyroxenite veins. Locally, garnet occurs in peridotite due to tectonic disruption of garnet pyroxenite veins and incorporation of mechanically resistant garnet into more ductile peridotite.

11.3.3 *Tethysides*

As is the case in the Caledonides and Variscides, alpine-type peridotites are widely distributed in the European portion of the Tethysides, but garnet-bearing peridotites are restricted to only a few tectono-stratigraphic units, which are located almost entirely within the Lepontine and Austridic realms of the Alps and the Internal Zone of the Betic Cordillera in southern Spain (Figure 11.1).

11.3.3.1 *Alps.* Numerous bodies of garnet peridotite are found in the Alps between the Ivrea Zone on the west and the Adige Valley on the east (Rost *et al*, 1979). Except for one occurrence in the Ivrea Zone (Alpe Piumero) and another in the southern Alps (Arosio), all the garnet peridotites are located north of the Insubric Line and west of the Guidicarie Line in two contrasting settings. One group, including the Cima di Gagnone, Monte Duria, and Alpe Arami bodies (Evans and Trommsdorff, 1978; Ernst, 1978), is located in the Lepontine Alps within the Pennine nappes of the more northerly, external, and subducted portion of the fold belt. A second group, which includes the peridotites of the Nonsberg (Ultental) area (Rost and Brenneis, 1978; Obata and Morten, 1987), is situated in the Tyrolian Alps within the Austro-alpine nappes of the more southerly and internal segment of the fold belt.

11.3.3.2 *Lepontine Alps.* The Cima di Gagnone, Monte Duria, and Alpe Arami peridotites all occur in the Adula (Cima Lunga) Nappe of the lower Pennides (Evans and Trommsdorff, 1978). These three garnet-bearing peridotites, as well as numerous other lenses and macroboudins of garnet-free metaperidotite, are encased in semi-pelitic and quartzo-feldspathic gneiss, which commonly contains layers of amphibolite, eclogite, and impure carbonate rock. Although a Mesozoic age has been inferred for the carbonate-ultramafic suite, there is no continuous connection with documented Mesozoic sedimentary rocks elsewhere in the Lepontine Alps. The predominant amphibolite facies metamorphism of the Adula Nappe is Tertiary (Lepontine) in age and overprints an earlier, HP metamorphism that has been locally preserved in eclogite and garnet peridotite. This eclogite facies metamorphism may be early Lepontine or Eo-Alpine in age, but the presence of other garnet peridotites in nearby pre-Alpine metamorphic basement raises the possibility that the HP metamorphism might be pre-Mesozoic.

The Cima de Gagnone peridotite (Evans and Trommsdorff, 1978) is important, because among all the Mg–Cr garnet peridotites in Europe it provides the most convincing evidence for a crustal residence before eclogite facies metamorphism. The Cima di Gagnone lherzolitic peridotite is a 40 m × 50 m macroboudin with included boudins of eclogite and calc-silicate rock, which are thought to be disrupted dykes of metamorphosed basalt. The bulk of the ultramafic body is chlorite peridotite, but fine-grained garnet ($\leqslant$ 2 mm in diameter) is locally concentrated in thin layers or schlieren typically 5 mm thick. In this respect, garnet in the Cima di Gagnone peridotite differs from that in most other European peridotites, in which it typically occurs as medium- to coarse-grained porphyroblasts or megacrysts.

At least seven mineralogic stages in the polymetamorphic evolution of the Cima di Gagnone peridotite can be inferred from field, petrographic, and geochemical data (Figure 11.2):

	Stage	*Interpretation*
I	Antigorite serpentinite	Greenschist or blueschist facies
II	Ol + Opx + Cpx + Grt	Eclogite facies, early or pre-Alpine
III	Ol + Opx + Cpx + Grt + Hbl	
IV	Ol + Opx + Tr + Al-Spl	Kelyphitic rims on garnet
V	Ol + Opx + Tr + Chl + Cr-Spl	Amphibolite facies Lepontine metamorphism
VI	Ol + Tr + Cum + Chl + Cr-Spl	
VII	Late serpentinisation	Final stages of uplift

Petrographic evidence for stage I is provided by the existence of relict titanoclinohumite in one garnet peridotite sample and by olivine-ilmenite pseudomorphs after titanoclinohumite in others. Although titanoclinohumite is a relatively uncommon mineral, it is widely distributed as an accessory in Penninic antigorite serpentinites, including those in the Piedmont, Liguria, Monte Rosa, Zermatt–Saas Fee, Val Malenco, and the Tauern. The presence of titanoclinohumite and related pseudomorphs at Cima di Gagnone implies that this peridotite was serpentinised, at least partially, early in its history. Additional evidence for an early stage of serpentinisation is provided by the occurrence of a suite of eclogite-metarodingite boudins in country rocks and peridotite (Evans *et al*, 1979, 1981). If it is accepted that rodingites are formed exclusively by metasomatic processes related to serpentinisation, then metarodingites at Cima di Gagnone require an episode of serpentinisation before eclogite facies metamorphism.

The magnesiohornblende-bearing five-phase assemblage of stage III is characteristic of the eclogite facies stage at Cima di Gagnone. Although the hornblende-absent assemblage of stage II was not observed, it was inferred to have been present by analogy with the Alpe Arami and Monte Duria localities. Stage IV is based on the presence of fibrous amphibole-spinel kelyphite around garnet. The most extensively developed assemblages are those of stages V and VI, which record the pervasive overprint of Lepontine metamorphism.

The Cima di Gagnone body is therefore an example of a garnet peridotite that was formed by HP metamorphism of an ultramafic protolith with a prior crustal residence. The Cima di Gagnone peridotite cycled through at least three metamorphic facies: initially in greenschist or blueschist facies as antigorite serpentinite, followed by subduction-related early Alpine(?) eclogite facies as garnet peridotite, and finally in Lepontine amphibolite facies as chlorite peridotite.

The Alpe Arami peridotite is located about 15 km southeast of Cima di Gagnone. Like Cima di Gagnone, the Alpe Arami body is situated within the Adula Nappe, but in contrast to Cima di Gagnone, detailed investigations have failed to uncover evidence for a LP precursor to the garnetiferous assemblage (O'Hara and Mercy, 1966; Möckel, 1969; Rost *et al*, 1974; Ernst, 1978). The Alpe Arami peridotite is a 1 km × 0.5 km lensoidal body that is bordered by discontinuous selvages of eclogite, hornblendite, and clinopyroxenite against the surrounding quartzo-feldspathic gneisses. The peridotite is locally garnetiferous, with the earliest recognised assemblage consisting of Ol + Opx + Cpx + Grt. This is followed by a five-phase assemblage of Ol + Opx + Cpx + Grt + pargasitic Amp. Recrystallisation under lower pressures and temperatures is marked first by the local development of amphibole-spinel kelyphite around garnet and then by a more pervasive transformation of the body to chlorite peridotite containing the assemblage, Ol + Opx + tremolitic Amp + Chl.

On the basis of a detailed structural investigation, Möckel (1969) concluded that formation of the garnetiferous assemblage predated emplacement of the Alpe Arami body into its present position during the later stages of the Alpine orogeny. Although Evans and Trommsdorff (1978) suggest that titanoclinohumite may be present at Alpe Arami, no titanoclinohumite (or pseudomorph thereof) has yet been documented from the body itself. The only occurrence of titanoclinohumite in the surrounding area has been found in a loose block of garnet peridotite in the torrent near Gorduno, and this sample probably originated from the garnet peridotite lens near Stuello, not from Alpe Arami (Möckel, 1969, p. 80). Another distinction between Alpe Arami and

Cima di Gagnone is the apparent absence of a metarodingite suite at Arami. With the evidence in hand, Möckel (1969) and Ernst (1978, 1981) concluded that the Alpe Arami peridotite was tectonically emplaced from the upper mantle with its garnet-bearing assemblage intact and was subsequently modified by late Alpine metamorphism and deformation.

11.3.3.3 *Austridic Alps.* Numerous lenses of ultramafic rocks occur in the Nonsberg (Ultental) region of the Austrides in northern Italy (Rost and Brenneis, 1978; Obata and Morten, 1987). The Nonsberg peridotites are concentrated in a zone delineated by the Peio line on the northwest and the Rumo line on the southeast, where they occur as concordant lenses ($\leqslant 100\,m \times 300\,m$) surrounded by amphibolite to granulite facies garnet-kyanite paragneiss, which contains a characteristic mineral assemblage of Qtz + antiperthitic Pl + Grt + Ky + Bt ± Kfs ± fibrolitic Sil ± Ms, and which grades into an axial zone of migmatite with intercalations of granitic to tonalitic orthogneiss. Although the high-grade metamorphism and peridotite emplacement are definitely pre-Alpine events, their exact ages remain uncertain, despite a K–Ar age of 330 ± 15 Ma for phlogopite from the margins of an ultramafic lens (Rost and Brenneis, 1978) and a concordant U–Pb age of 332–336 Ma for zircon from three ultramafic samples (Gebauer and Grünenfelder, 1979a).

The Austridic peridotites have been divided into a coarse- and a fine-grained type, from which a sequence of six mineralogic stages has been inferred (Obata and Morten, 1987), as summarised in Figure 11.2. The coarse variety has a protogranular texture and consists predominantly of spinel lherzolite with an assemblage of:

I Ol + Opx + Cpx + Spl ± Amp

One important sample of the coarse-grained type contains garnet and has a texture varying from protogranular to porphyroclastic. The garnet surrounds spinel, appears as exsolution lamellae in both pyroxenes, and occurs in finer-grained recrystallised domains of the sample in a spinel-absent assemblage of:

II Ol + Opx + Cpx + Grt + Amp

The fine-grained type of peridotite, which has formed from the coarse type by recrystallisation, has porphyroclastic to tabular or mosaic granular textures and contains the following assemblages:

III Ol + Opx + Cpx + Grt + Amp ± Spl
IV Ol + Opx + Grt + Amp + Spl
V Ol + Opx + Amp + Spl
VI Ol + Opx + Amp + Chl

Large garnet grains typically surround vermicular grains of Cr-spinel, but smaller garnet grains appear to be in textural equilibrium with olivine, pyroxenes, amphibole, and perhaps with spinel. Amphibole is pargasitic in the higher grade assemblages (I–IV) and tremolitic in the lower (V–VI).

The textural relations and sequences of mineral assemblages in the coarse- and fine-grained types of peridotite demonstrate that the garnetiferous Austridic peridotites formed from spinel peridotite protoliths and were subsequently recrystallised in response to decreasing temperature and pressure. Evidence for an early magmatic event before recrystallisation of the Austridic peridotites is recorded in pyroxene megacrysts from garnet websterite and garnet clinopyroxenite layers. The reconstituted compositions of such megacrysts (now exsolved) require that they initially crystallised as subcalcic pyroxenes at temperatures of 1450°C or higher, presumably in the upper mantle (Morten and Obata, 1983).

Despite their geographic location in the Alps, the Nonsberg peridotites are more closely akin to the peridotite-granulite association of the middle European Variscides in terms of lithostratigraphic association, mineralogical evolution and age than they are to the peridotites of the Lepontine Alps. Although the Nonsberg garnet peridotites have formed from spinel peridotite protoliths, as have at least some of the Variscan garnet peridotites, recrystallisation temperatures during and following the growth of garnet appear to have been lower in the Nonsberg peridotites compared with the Variscan ones, as is evident in the sequence of assemblages depicted in Figure 11.2 (note the absence of a post-garnet Opx–Cpx–Spl assemblage in the Nonsberg rocks). Lower temperatures for the Nonsberg garnet peridotites compared with the Variscan ones in central Europe are corroborated by application of geothermometers (Section 11.4).

11.3.3.4 *Betic Cordillera.* The Ronda peridotite, which is located in the Internal Zone of the Betic Cordillera in southern Spain (Figure 11.1), is the largest (300 km^2), best exposed, and most thoroughly studied HT peridotite intrusion in the world. It is distinctive among garnet-bearing peridotites in having a well developed contact metamorphic aureole. Following the first modern petrologic description of the peridotite by Dickey (1970), detailed investigations have been completed on its metamorphic aureole (Loomis, 1972a), its manner of emplacement (Loomis, 1972b, 1975; Lundeen, 1978), the mineral facies and recrystallisation conditions within the peridotite (Obata, 1980), and the geochemistry and petrogenesis of the peridotite and included mafic layers (Obata *et al*, 1980; Frey *et al*, 1985; Suen and Frey, 1987). A counterpart to the Ronda massif lies across the Strait of Gibraltar in the Internal Rif of Morocco at Beni Bouchera (Kornprobst, 1969; Conquéré and Kornprobst, 1972).

The Ronda peridotite is a slab approximately 1.5 km thick; it is underlain by the Blanca Unit, over which it was thrust, and is overlain by the Casares Unit, which rode passively on the peridotite. The Blanca Unit (Palaeozoic to Triassic cordierite gneiss, marble, and quartzite) has been correlated with the deepest member of the Internal Zone of the Betic Cordillera, the Nevado-Filabride complex, and the peridotite and overlying Casares Unit (largely Palaeozoic metapelites) have been correlated with the middle member of the zone, the Alpujarride nappe complex. All of these units are overlain by the Malaguide nappe, which is the uppermost structural unit of the Internal Zone.

A distinct metamorphic aureole is developed in the Casares Unit along the northwestern margin of the peridotite, ranging from chlorite-sericite phyllite 6 km away from the contact, through mica-andalusite schist, to garnet-cordierite-sillimanite gneiss at the contact, where a temperature of about 800°C and a pressure of about 4.3 kbar were attained. A contact metamorphic zone is also present in the Blanca Unit beneath the peridotite slab, but there it is thinner than in the Casares Unit and is characterised by a slightly lower temperature, 750°C, and pressure, 3.5 kbar. Although certain ambiguities remain with respect to the detailed structural and metamorphic evolution of the Ronda peridotite and its contact aureole, the massif is generally viewed as having originated by diapiric uprise of hot upper mantle material in response to lithospheric extension that was related to counterclockwise rotation of Spain with respect to Africa. Final emplacement occurred by thrusting during late Oligocene–

early Miocene time, while the peridotite remained at an elevated temperature.

Major and trace element compositional trends for peridotites in the Ronda massif are characteristic of residues formed by partial fusion of garnet lherzolite, presumably related to penetration of a dry peridotite solidus by the massif during diapiric uprise. Approximately 5% of the body consists of concordant layers of mafic rocks, mainly pyroxenites and gabbros. Geochemical data preclude these mafic layers from being primary melts; rather, they represent cumulates (± trapped liquid) that crystallised along the walls of conduits as melts migrated from hotter to cooler portions of the diapir. This interpretation is corroborated to some extent by Nd and Sr isotope systematics, which suggest that the mafic layers formed at about 22 Ma and represent the products of interaction between mid-ocean ridge tholeiite and depleted mantle peridotite (Zindler *et al*, 1983).

Internally, the Ronda peridotite has undergone substantial recrystallisation and has been divided into four zones of metamorphic facies, based on mineral assemblages in peridotite and mafic layers. From northwest to southeast these zones are: garnet lherzolite facies, ariegite subfacies of spinel lherzolite facies, seiland subfacies of spinel lherzolite facies, and plagioclase lherzolite facies. In peridotite these facies are characterised by the assemblages Ol + Opx + Cpx + Grt ± Spl, Ol + Opx + Cpx + Spl (the ariegite and seiland subfacies are indistinguishable in peridotite compositions), and Ol + Opx + Cpx + Pl ± Spl. Spinel in the spinel lherzolite facies has a lower ratio of Cr/Al than does spinel in the garnet and plagioclase lherzolite facies. The core compositions of pyroxene grains indicate that the peridotite initially equilibrated at temperatures of 1100–1200°C and pressures of 20–25 kbar. Recrystallisation to the four metamorphic zones took place at more or less uniform temperatures of 800–900°C (however, see Section 11.4), but at pressures ranging from 5–7 kbar in the plagioclase lherzolite facies to about 15 kbar in the garnet lherzolite facies. The Ronda slab is too thin to support an in situ pressure differential of about 9 kbar, and development of the various metamorphic facies is ascribed to a process of dynamic cooling, in which various parts of the body have followed different *P–T* trajectories in response to different cooling rates during diapiric uprise and emplacement. Such a process to account for the development of garnet in the Ronda peridotite is similar to that in the Mohelno peridotite in the Bohemian Massif, except that Ronda is a much larger body than Mohelno and garnet is more widely distributed.

11.4 Thermometry and barometry

11.4.1 *General considerations*

Although numerous geothermometers and barometers are now available for estimating equilibration conditions in garnet lherzolites, application of these methods is often not a simple task. Commonly, an individual sample of garnet peridotite contains polymetamorphic and multistage mineral assemblages, and in such cases it is essential to identify mineral grains that represent a particular stage of development and to analyse adjacent grains that appear to be in textural equilibrium. Another complicating factor is introduced by the fact that minerals in garnet peridotites are frequently compositionally zoned, which requires a decision to be made as to which portions of grains, if any, represent equilibrium compositions. An additional problem arises owing to the interdependence of thermometers and barometers in calculating a *T–P* value for a particular sample, and the possibility that in some cases such values may be spurious because of different blocking temperatures for different types of mineral equilibria which are utilised in the calculation, such as Fe–Mg exchange equilibria combined with the Al content of orthopyroxene.

With these caveats in mind, the authors have calculated temperatures and pressures for garnetiferous assemblages in crustal peridotites by means of three different Fe–Mg exchange thermometers, olivine–garnet (O'Neill and Wood, 1979, 1980), orthopyroxene–garnet (Harley, 1984a), and clinopyroxene–garnet (Powell, 1985b), and one orthopyroxene barometer (Nickel and Green, 1985). These particular thermobarometric methods were selected because Carswell and Gibb (1987a) found that, among twelve different barometric and twenty different thermometric formulations, they were the ones that best reproduced the *T–P* conditions of experiments performed on 'natural' multi-component garnet lherzolites. Thermometers based on the pyroxene miscibility gap have not been used in this comparative investigation, because such thermometers are relatively insensitive at temperatures below about 900°C.

In calculating temperatures and pressures for the European garnet lherzolite assemblages, two sources of data have been drawn upon. Wherever possible, we have utilised our own published and unpublished analyses of mineral phases which we judged to be in equilibrium. This group of analyses includes samples from Norway, Czechoslovakia, Austria, and the Alps. For localities with which we have no personal experience, we employed published analyses that have been used previously for *T–P* calculations by other investigators. Temperatures and pressures for all analyses have been calculated from the preferred thermometers and barometers in a uniform manner, thereby providing a consistent basis for a comparison of temperatures and pressures for the European occurrences, as well as yielding what are hoped to be accurate *T–P* values for the garnet-bearing assemblages.

Except for olivine, the various minerals of different stages in garnet peridotites generally exhibit some degree of compositional zoning. The most common zoning pattern in garnet and pyroxenes that belongs to the four-phase garnet lherzolite assemblage is one in which grain interiors are relatively homogeneous and zoning is confined to the rims (Medaris, 1984; Carswell, 1986). Typically, the most prominent compositional variations from core to rim (and the most important ones with respect to thermobarometry) are an increase in Fe/Mg in garnet and an increase in Al in orthopyroxene. This configuration represents a retrograde zoning pattern due to homogenisation of the garnet lherzolite assemblage at elevated *T–P* conditions, followed by partial re-equilibration during cooling and decompression (Medaris and Wang, 1986). The rims of grains thus record compositions that were 'frozen in' at some point along a retrograde *T–P* path. Accordingly, in those instances where minerals are compositionally zoned, we have utilised analyses of mineral cores in an attempt to reconstruct the earliest *T–P* conditions for the garnet lherzolites.

11.4.2 *Temperature estimates*

Fe–Mg exchange among olivine, orthopyroxene, clinopyroxene, and garnet is a function predominantly of temperature, but also of pressure to a lesser degree (see Chapter 4). Consequently, as a first step in the comparison of the equilibration conditions of European garnet lherzolites and associated garnet pyroxenites, temperatures have been calculated from each of the three Fe–Mg exchange equilibria at an arbitrary pressure of 30 kbar (Table 11.1). This approach amounts essentially to a comparison of K_D values for Fe–Mg exchange among the three mineral pairs, after adjustments for compositional influences. In so doing, all Fe is treated as Fe^{2+}, owing to the difficulty in calculating Fe^{3+}/Fe^{2+} from microprobe analyses of the Mg-rich and Fe-poor minerals under consideration and in accordance with the results and recommendations of Carswell and Gibb (1987a).

On the basis of temperature there is a clear separation of most European garnet lherzolites into two groups, one at temperatures less than about 900°C, including those in the Caledonides, Lepontine Alps (except for Alpe Arami), and Austridic Alps, and another at temperatures greater than this, including all the Variscan bodies and the Ronda massif in the Betic Cordillera (Figure 11.3). Within the Variscan occurrences the garnet pyroxenites at Biskupice and Níhov, as well as the garnet peridotite at Staré, yield lower temperatures than do the remaining garnet peridotites. Among the garnet peridotites in the Lepontine Alps, the Alpe Arami body apparently equilibrated at slightly higher temperatures than did the Cima di Gagnone and Monte Duria bodies.

Generally, there is good agreement among the averaged results of the three Fe–Mg exchange thermometers at localities from which a number of samples have been analysed. This is not always the case, however, for localities from which only one analysis has been reported for each of the minerals, olivine, orthopyroxene, clinopyroxene, and garnet. In those individual samples where there is poor agreement, it may be that not all four analysed mineral grains came from the same chemical domain of the sample or represent the same stage of mineral assemblage and, therefore, that one or more of the 'representative' compositions are not in equilibrium with the others.

11.4.3 *Temperature–pressure estimates*

Specific *T–P* values for garnet lherzolites may be obtained by combining the Fe–Mg exchange thermometers with the orthopyroxene Al barometer (Nickel and Green, 1985), in which $X^{M1}_{Al(Ts)} = (Al\text{-}Cr\text{-}2Ti + Na)/2$, in an iterative calculation or by a graphical solution.

The garnet peridotite assemblages in the Western Gneiss Region of Norway in the Scandinavian Caledonides yield a *T–P* trend from about 800°C, 30 kbar to almost 1000°C, 55 kbar (Figure 11.4a, Table 11.1). The specific *T–P* values, however, are suspect for several reasons. First, it is difficult to envisage a tectonic process whereby upper mantle material from depths of 100–150 km could be injected into continental crust, even if the crust had attained a vertical dimension equal to twice that of normal thickness due to continent–continent collision. Second, the apparent *T–P* values lie along a thermal gradient that is substantially colder than that for a steady-state Precambrian shield geotherm, and such a thermal regime is difficult to explain, even considering the possible thermal perturbations that could be caused by the collision of lithospheric plates.

It has been suggested that the *T–P* values for the Norwegian garnet peridotites are artificial results arising from combination of the orthopyroxene Al barometer with Fe–Mg exchange thermometers (Carswell *et al*, 1985; Medaris and Wang, 1986). It is thought that the garnet peridotites may have experienced conditions of 600–700°C and 15–25 kbar during their early residence in thickened continental crust (Section 11.4.4). Subsequent thermal relaxation and heating to around 800°C may have caused re-equilibration and homogenisation of stage II olivine, orthopyroxene, clinopyroxene,

Table 11.1 Temperature and pressure estimates for European garnet peridotites and pyroxenites

Locality	T°C at 30 kbar Ol/Grt	Opx/Grt	Cpx/Grt	T°C	P kbar
		Caledonian (stage II)			
Kalskaret (6)[l,n,o]	821 ± 43	842 ± 12	848 ± 15	875 ± 15	38.0 ± 1.0
Aldalen (2)[n,o]	807 ± 13	819 ± 50	778 ± 29	870 ± 43	44.3 ± 3.0
Lien (12)[m,o]	808 ± 52	847 ± 21	794 ± 27	840 ± 36	34.4 ± 2.0
Rødhaugen (8)[c,o]	833 ± 77	854 ± 56	800 ± 30	860 ± 37	35.7 ± 2.0
Rødskar (6)[n,o]	849 ± 46	897 ± 33	824 ± 4	946 ± 67	48.8 ± 4.4
Sandvik (5)[n,o]	825 ± 57	888 ± 18	813 ± 27	903 ± 63	41.8 ± 3.9
Gurskebotn (1)[k]	834	960	900	974	53.5
Ugelvik (8)[d,e]	777 ± 99	833 ± 50	735 ± 34	814 ± 108	37.0 ± 6.2
Raudhaugene (4)[d,e]	730 ± 34	817 ± 23	796 ± 54	806 ± 80	35.4 ± 9.3
		Caledonian (stage III)			
Ugelvik (2)[d]	734 ± 12	780 ± 12	712 ± 33	782 ± 57	38.5 ± 4.1
Raudhaugene (2)[d,l]	740 ± 29	803 ± 12	750 ± 48	794 ± 61	36.2 ± 3.8
Fjørtoft (1)[l]	795	878	730	795 ± 79	27.7 ± 4.4
		Variscan (granulite terranes)			
Austria (8)[e]	1147 ± 195	1114 ± 58	1046 ± 16	1105 ± 77*	33.9 ± 4.1*
Czechoslovakia					
Mohelno (2)[o]	1288 ± 1	1240 ± 31	1307 ± 7	1264 ± 40	27.9 ± 1.5
Contact					
Mohelno (1)[f]	—	1194	1144	1109 ± 10	19.4 ± 0.4
Mohelno (1)[o]	1144 ± 22	1119 ± 25	1055 ± 17	1071 ± 39	23.4 ± 1.5
Boudin					
Horní Bory (1)[f]	—	—	1163	—	—
Horní Bory (1)[s]	—	885	808	766 ± 33	14.7 ± 1.5
Pyx megacryst with exsolution lamellae					
Staré (2)[j]	859 ± 1	939 ± 12	922 ± 24	920 ± 47	32.7 ± 2.1

	Variscan (granulite terranes)				
Poland					
Góry Sowie (2)[a]	1144 ± 65	1125 ± 55	972 ± 46	1044 ± 82	22.3 ± 3.2
Massif Central					
Artiges (1)[b]	997	1032	1133	974 ± 88	16.3 ± 3.4
	Variscan (Gföhl gneiss)				
Czechoslovakia					
Nové Dvory (5)[o]	1040 ± 18	1027 ± 37	1078 ± 38	1159 ± 13	51.5 ± 0.8
Biskupice (3)[o]	—	882 ± 16	870 ± 20	910 ± 21	36.8 ± 1.2
Níhov (2)[o]	—	902 ± 14	915 ± 5	936 ± 2	35.7 ± 0.1
	Variscan (Kutná Hora–Svratka Complex)				
Bečváry (1)[o]	1051 ± 40	—	1119 ± 9	—	—
Bečváry (1)[i]	788	822	931	913 ± 64	45.0 ± 4.2
Úhrov (2)[o]	1127 ± 31	1043 ± 13	968	1108 + 91	41.7 ± 4.8
	Tethyan (Lepontine Alps)				
Cima di Gagnone (1)[h]	743	797	826	806 ± 44	34.0 ± 2.7
Monte Duria (1)[h]	802	856	868	872 ± 35	36.4 ± 2.1
Alpe Arami (7)[g,o]	923 ± 69	930 ± 43	903 ± 39	1012 ± 45	45.6 ± 2.7

(*Contd.*)

Table 11.1 (*Contd.*)

Locality	T°C at 30 kbar Ol/Grt	Opx/Grt	Cpx/Grt	T°C	P kbar
	Tethyan (Austridic Alps)				
Nonsberg (6)[r]	774 ± 28	835 ± 18	737 ± 11	738 ± 43	20.3 ± 2.1
Nonsberg (2)[p] Matrix and exsolved grains	—	802 ± 12	714 ± 20	728 ± 39	19.5 ± 1.9
	Tethyan (Betic Cordillera)				
Ronda (4)[q]	1023 ± 63	1075 ± 17	1055 ± 29	1006 ± 28	21.7 ± 1.2

*These values exclude the highly variable and unsatisfactory results from Ol/Grt pairs.

Number of samples for each locality in parentheses.
Sources of data:

a Bakun–Czubarow (1983)
b Bonnot and Piboule (1980)
c Carswell (1981)
d Carswell (1986)
e Carswell (unpublished)
f Dobretsov *et al* (1984)
g Ernst (1978)
h Evans and Trommsdorff (1978)
i Fiala (1985)
j Fiala and Paděra (1977)
k Jamtveit (1984)
l Jamtveit *et al.* (1989)
m Medaris (1980)
n Medaris (1984)
o Medaris (unpublished)
p Morten and Obata (1983)
q Obata (1980)
r Obata and Morten (1987)
s Šanc and Rieder (1983)

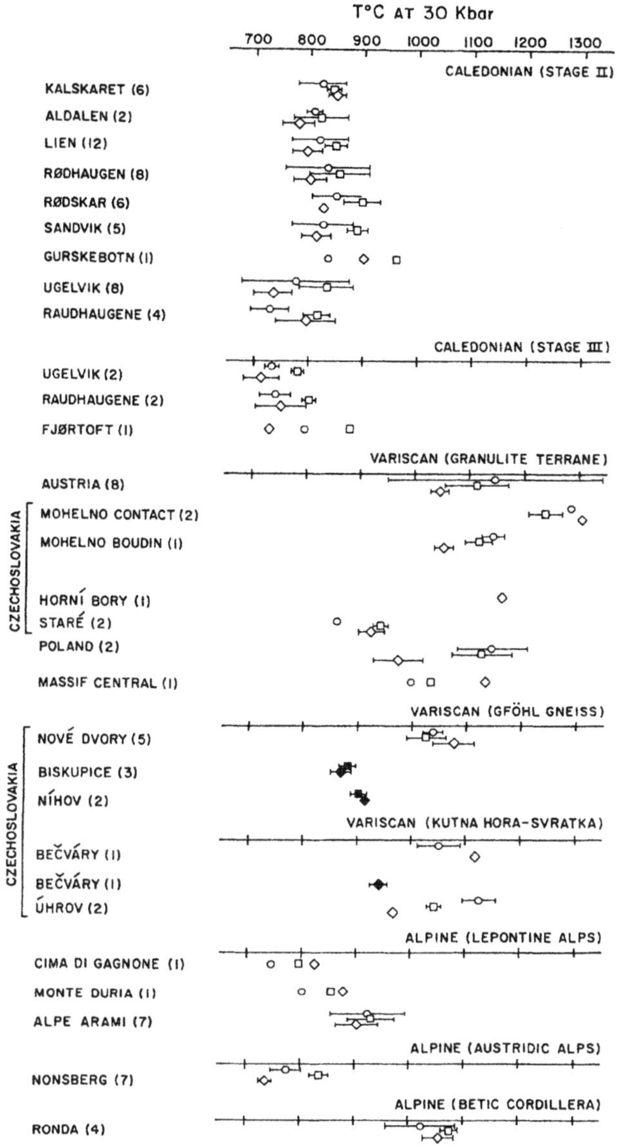

Figure 11.3 Summary of average temperatures calculated at 30 kbar for garnetiferous assemblages in European Mg–Cr peridotites. Number of analysed samples shown in parentheses following localities; error bars represent ± one standard deviation. (○ ●) temperature calculated from the olivine-garnet Fe–Mg exchange thermometer; (□ ■) orthopyroxene-garnet; (◇ ◆) clinopyroxene-garnet. Open symbols: garnet peridotite ± garnet pyroxenite; filled symbols: garnet pyroxenite.

and garnet with respect to Fe and Mg, but left the relatively immobile Al in the cores of orthopyroxene grains largely unaffected. Results from the Lien locality for the orthopyroxene barometer and olivine–garnet thermometer are plotted in Figure 11.4a. A 100°C increase in temperature along an Al 'isopleth' is accompanied by a 5 kbar increase in pressure. Thus the *T–P* values plotted in Figure 11.4a may represent spurious numbers due to the method of calculation and the particular thermal evolution that these peridotites have experienced. However, coesite inclusions have been reported in clinopyroxene from eclogite at Grytting in the WGR (Smith, 1984), which raises the possibility that the subducted continental crust of the Baltic plate and its included eclogite and peridotite bodies may have been subjected to pressures of 30 kbar or more.

Another inconsistency in the Norwegian data appears when *T–P* values for the Precambrian stage II assemblage are compared with those for the Caledonian stage III assemblage (Jamtveit *et al*, 1989). It should be expected that the stage III assemblage, which presumably formed during Caledonian crustal thickening, would yield consistently lower temperatures and pressures than the mantle-derived stage II assemblage, but this is not the case (Figure 11.4a). The puzzling results for the stage III assemblage may be due in large part to the difficulty in distinguishing among orthopyroxene grains of the various recrystallisation stages in these polymetamorphic rocks. Carswell (1986) and Jamtveit *et al* (1989) have shown that more realistic *T–P* estimates of 700–750°C and 21.0–22.6 kbar for the stage III assemblage at Ugelvik are obtained by combining the Fe–Mg exchange thermometers with a barometric calculation for a 5-phase garnet + spinel assemblage, as outlined by O'Neill (1981).

T–P estimates for most Variscan garnet peridotites in granulite terranes (with the exceptions of Austria and Staré) plot in the vicinity of the spinel-garnet lherzolite transition (Figure 11.4b), which is consistent with the widespread petrographic evidence for the formation of garnet at the expense of spinel in these rocks. The effects of Cr and Fe on calculated pressure for the spinel–garnet transition (O'Neill, 1981; Evans and Trommsdorff, 1978) are largely offsetting for the compositions of the peridotites under consideration, with a net change of about 500 bars for the Mohelno samples. A distinctive feature of this group of peridotites is their apparent HT thermal regime, as recorded by temperatures mostly above 1000°C. The *T–P* values for these peridotites are believed to be real, rather than spurious, because olivine, orthopyroxene, clinopyroxene, and garnet are thought to have equilibrated with respect to Fe, Mg, and Al during recrystallisation related to the garnet-forming reaction at elevated temperatures (in contrast to the Norwegian peridotites, which may have experienced heating

---

Figure 11.4 Temperature–pressure estimates for European Mg–Cr garnet peridotites and pyroxenites. Reactions: 3, spinel lherzolite to garnet lherzolite transition (O'Hara *et al*, 1971); 4, dry pyrolite solidus (Green and Ringwood, 1967b); 5, quartz to coesite transition (Mirwald and Massonne, 1980); Precambrian shield geotherm (Clark and Ringwood, 1964). (a) Caledonides. Stage II: A, Aldalen; D, Rødskar; G, Gurskebotn; H, Raudhaugene; K, Kalskaret; L, Lien; R, Rødhaugen; S, Sandvik; U, Ugelvik. Stage III: F, Fjørtoft; H, Raughaugene; U, Ugelvik. Curves: 1, average *T–P* locus for orthopyroxene barometer (Al isopleth) for Lien samples; 2, average *T–P* locus for olivine–garnet thermometer for Lien samples. (b) Variscides. Peridotites in granulite: A1–A2, Austria; C, Massif Central; H, Horní Bory; M1–M3, Mohelno; P, Poland; S, Staré. Peridotites in Gföhl gneiss: B, Biskupice; D, Nové Dvory; N, Níhov. Peridotites in the Kutná Hora–Svratka complex: U, Úhrov; V, Bečváry. (c) Tethysides. Lepontine Alps: A, Alpe Arami; D, Monte Duria; G, Cima di Gagnone. Austridic Alps: N1–N2, Nonsberg. Betic Cordillera: R, Ronda.

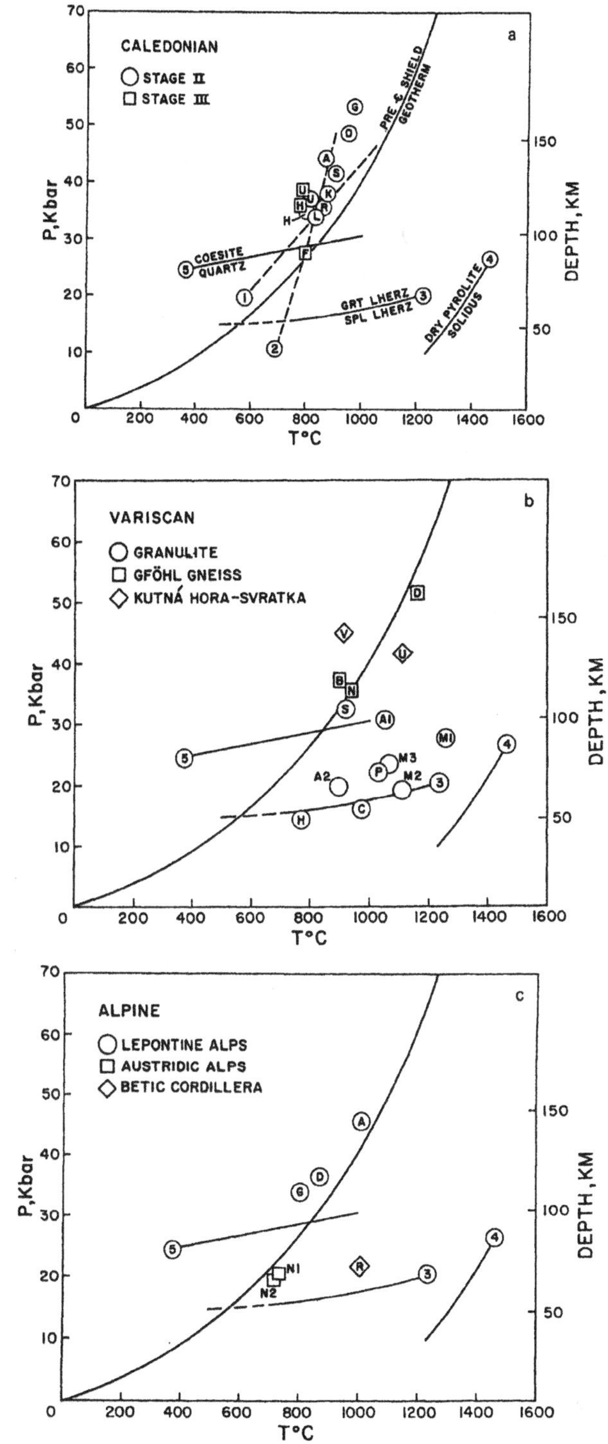
a
CALEDONIAN
STAGE II
STAGE III
PRE & SHIELD GEOTHERM
COESITE
QUARTZ
GRT LHERZ
SPL LHERZ
DRY PYROLITE SOLIDUS
b
VARISCAN
GRANULITE
GFÖHL GNEISS
KUTNÁ HORA-SVRATKA
c
ALPINE
LEPONTINE ALPS
AUSTRIDIC ALPS
BETIC CORDILLERA
P, Kbar
DEPTH, KM
T°C

after formation of the garnet-bearing assemblage). The temperature of 765°C for garnet pyroxenite from the Horní Bory quarry (Figure 11.4b) was obtained from a clinopyroxene megacryst with exsolved lamellae of garnet and orthopyroxene and reflects recrystallisation conditions following the thermal maximum.

Problems arise in estimating *T–P* conditions for the Austrian garnet peridotites owing to their heavily retrograded condition and to the difficulty in determining which of the various generations of orthopyroxene growth are in equilibrium with garnet. Ol/Grt and Opx/Grt temperature estimates are quite variable (Table 11.1), reflecting poor equilibration. Utilisation of the more consistent results from the Cpx/Grt thermometer yields a 'best estimate' for the P_{max} garnet-bearing assemblage of 1051°C and 31.4 kbar (Al in Figure 11.4b), which lies outside the *T–P* field for most other Variscan garnet peridotites that are associated with granulites. However, a sample of garnet websterite, which has a porphyroclastic texture because of deformation induced recrystallisation, but which is not amphibolitised, yields a *T–P* estimate of about 890°C and 20.8 kbar (A2 in Figure 11.4b), which lies in the vicinity of the spinel–garnet lherzolite transition in concert with the other garnet peridotites that are hosted by granulites.

The peridotite recovered from borehole T-7 near Staré in the České Středohoří Mountains of Bohemia is associated with granulite, but it differs mineralogically from other peridotites that are located in granulite terranes. Spinel occurs in garnet peridotite from the T-7 borehole, but only in kelyphitic rims around garnet and not as inclusions in garnet, and the reported mineral chemistries, as expressed in terms of a *T–P* value (Table 11.1; Figure 11.4b), more closely resemble those of solitary peridotite bodies or of peridotites in Gföhl gneiss than those in granulite. Further investigation of this problematic peridotite would be worthwhile.

T–P values for the solitary garnet peridotite bodies (Úhrov and Bečváry) and those in Gföhl gneiss (Nové Dvory, Biskupice, and Níhov) lie in the vicinity of a steady-state Precambrian shield geotherm (Figure 11.4b). Little is known so far about the thermal histories of these particular bodies, and it is difficult to assess whether their *T–P* values are real or spurious. It has been suggested that some of these peridotites have been emplaced from deep fault zones, and while this may be so, it is again implausible that mantle material was transported directly into the crust from depths of 150 km or more.

Garnet peridotites located in the Tethyan belt fall into a lower pressure group (Ronda and Austridic Alps) at around 20 kbar and a higher pressure group (Lepontine Alps) at > 30 kbar (Figure 11.4c, Table 11.1). Garnet peridotite in the Ronda massif plots close to the spinel-garnet lherzolite boundary, in accordance with the presence of spinel inclusions in garnet. *T–P* conditions for the Ronda garnet peridotite (1005°C, 21.7 kbar) are closely comparable to those for certain of the Variscan spinel-garnet peridotites, indicating a common link in thermal evolution between these HT spinel-garnet peridotites. The Nonsberg spinel-garnet peridotites, which are geologically and mineralogically similar to the Variscan spinel-garnet peridotites, plot close to the spinel-garnet lherzolite boundary, but at much lower temperatures than do the other Variscan bodies. The garnet-bearing assemblage in the Nonsberg peridotites yields an average temperature of about 740°C, which is essentially the same, about 720°C, as that for a recrystallised pyroxenite, in which pyroxene megacrysts contain garnet exsolution lamellae.

The garnet peridotites at Cima di Gagnone and Monte Duria in the Lepontine Alps both plot at pressures of about 35 kbar. If these two bodies truly had a crustal history before subduction and were transformed to garnet peridotites by prograde metamorphism, and if the pressure estimates are correct, the crust that transported the peridotites must have been depressed to depths on the order of 100 km or more. Elsewhere in the Alps, crustal rocks have been subducted to comparable depths, as documented by the occurrence of coesite inclusions in garnet from the Dora Maira massif (Chopin, 1984). Application of thermobarometric methods to the eclogite-metarodingite suite at Cima di Gagnone indicates that the eclogite facies metamorphism occurred at approximately 800°C and 25 kbar (Evans *et al*, 1979), about 10 kbar less than that apparently recorded by the garnet peridotites.

The Alpe Arami peridotite resembles the Norwegian garnet peridotites in many respects, including geologic occurrence, textures, mineral chemistry, and sequence of mineral assemblages. The *T–P* estimate for Alpe Arami falls close to the apparent *T–P* trend for the Norwegian peridotites, but at a temperature about 100°C higher. As is the case in Norway, the high *T* and *P* for the Alpe Arami garnet peridotite may be spurious, reflecting the thermal history of the body subsequent to growth of the garnetiferous assemblage.

11.4.4 *Temperature–pressure paths*

Temperature–pressure paths for garnet peridotites can be established by comparison of the observed mineral paragenesis with experimentally determined stability fields for ultramafic assemblages and by application of appropriate thermometers and barometers. It is generally agreed that European Mg-Cr garnet peridotites originate in the upper mantle, and their *T–P* paths following emplacement in the crust are well constrained and non-controversial. However, interpretation of the ages of formation and conditions of stabilisation of the garnetiferous assemblages in many of the European peridotites is subject to a degree of uncertainty, and details of the early segments of some paths remain conjectural.

The thermal evolution of the Western Gneiss Region in the Scandinavian Caledonides was a consequence of the early Palaeozoic collision of the Baltic and Greenland plates (Cuthbert *et al*, 1983). Continental crust of the subducted Baltic plate was first subjected to eclogite facies metamorphism caused by crustal thickening and then was recrystallised to granulite, amphibolite, and lower grade facies during uplift (Figure 11.5a). Two different *T–P* paths are illustrated for crustal rocks of the WGR: path 'P' is based on petrologic data (Jamtveit, 1987a; Dunn and Medaris, 1989), and path 'C' has been calculated from the tectonic model of Cuthbert *et al* (1983) for a package of rocks starting at a depth of 66 km in a column of crust that was thickened initially to 81 km (Medaris and Wang, 1986). Note that both paths display significant heating during uplift and that attainment of maximum pressures preceded achievement of maximum temperatures.

Among the seven stages in mineralogical evolution of the peridotites in the WGR, the first is the most problematic. Highly aluminous pyroxenes (now exsolved) in some pyroxenite layers provide evidence for an early HT igneous event, which is shown as stage I in Figure 11.5a. Although stage I has tentatively been located in the spinel lherzolite field (on the basis of one unusual sample at Sandvik), it could equally well have occurred at higher pressures in the garnet lherzolite field. In view of the possibility that the *T–P* values for the garnet-bearing stage II and III assemblages may be spurious, as discussed earlier, the current 'best' *T–P* estimates (Jamtveit *et al*, 1989)

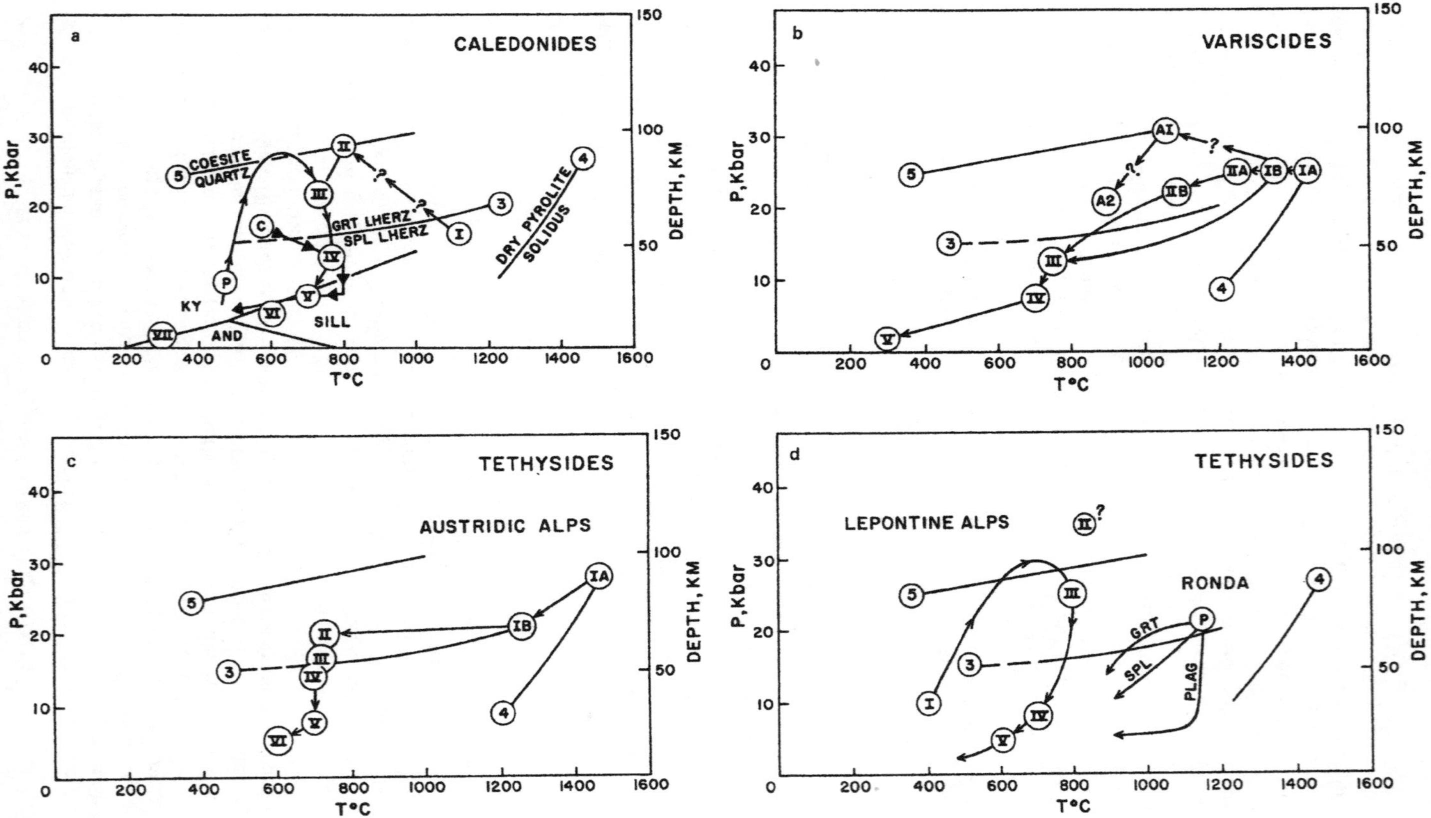
a
CALEDONIDES
COESITE
QUARTZ
GRT LHERZ
SPL LHERZ
DRY PYROLITE
SOLIDUS
KY
AND
SILL
b
VARISCIDES
c
TETHYSIDES
AUSTRIDIC ALPS
d
TETHYSIDES
LEPONTINE ALPS
RONDA
GRT
SPL
PLAG
P, Kbar
T°C
DEPTH, KM

for the Proterozoic (about 1700 Ma) stage II assemblage (around 805°C, 28.7 kbar) and the Caledonian (about 437–511 Ma) stage III assemblage (700–750°C, 21.0–22.5 kbar) have been plotted in Figure 11.5a.

The simplest interpretation of the age and petrologic relations of the garnet peridotites in the WGR is that subcontinental mantle of Proterozoic age was tectonically intercalated as thrust slices into the leading edge of the subducting Baltic plate during the Caledonian orogeny (Carswell, 1986; Medaris and Wang, 1986; Jamtveit *et al*, 1989). At this time the distinctive mosaic porphyroclastic texture of the stage II assemblage developed, and local recrystallisation produced the stage III garnetiferous assemblage. More complex *T–P* paths are possible, and it has been postulated that the peridotites may have followed a looped path, in which the peridotites resided temporarily between stages II and III at conditions of 450–500°C and 10–12 kbar with the stage II assemblage frozen in (Carswell, 1986; Jamtveit *et al*, 1989).

After emplacement in the crust, peridotites in the WGR followed the same *T–P* path as the enclosing gneisses and were extensively recrystallised to granulite and amphibolite facies assemblages (stages IV–VI). An important feature of paths 'P' and 'C' is that heating due to thermal relaxation occurred during decompression, possibly resulting in homogenisation of stage II and III minerals with respect to Fe and Mg at around 800°C and 15 kbar (Medaris and Wang, 1986), thereby erasing the record of temperature conditions during emplacement of the peridotites into the crust. Another significant feature of paths 'P' and 'C' is that cooling of the peridotites occurred at relatively low pressures outside the stability field of garnet lherzolite. Thus, the retrograde compositional zoning of Fe and Mg at garnet rims in stage II and III assemblages is the result of cation exchange among metastable phases (Medaris and Wang, 1986).

The *T–P* path for peridotites associated with granulites in the Variscides (Figure 11.5b) contrasts markedly with that for the Caledonides, the principal differences being in the higher temperature and lower pressure conditions for the Variscan garnet-bearing assemblages and the absence of any discernible temperature increase subsequent to the crystallisation of garnet. The earliest recognisable event in the Variscan bodies is recorded in pyroxenite layers, some of which contain pyroxene megacrysts with garnet exsolution lamellae (Šanc and Rieder, 1983; Scharbert and Carswell, 1983). The reconstituted compositions of such pyroxenes suggest an early HT igneous episode, which has been indicated in Figure 11.5b as stage IA, although the *T–P* conditions for this stage are poorly constrained. The spinel peridotite assemblage at Mohelno, which is shown as stage IB, contains high-temperature pyroxenes (Medaris

---

Figure 11.5 Temperature–pressure paths for European Mg–Cr peridotites. Reactions (3), (4), and (5) as in Figure 11.4. (a) Caledonides. I–VII: Stages of mineral assemblages in Norwegian peridotites. P, *T–P* path for crustal rocks in the WGR inferred from phase petrology; C, *T–P* path for WGR calculated by thermal modelling from tectonic scenario. (b) Variscides. IA–V, stages of mineral assemblages in Czech peridotites associated with granulites, based largely on the Mohelno occurrence. Al, best *T–P* estimate for Austrian garnet peridotites; A2, recrystallised Austrian garnet pyroxenite. (c) Tethysides. IA–VI, stages of mineral assemblages in the Nonsberg peridotites of the Austridic Alps. (d) Tethysides. Lepontine Alps: I–V, stages of mineral assemblages in the Cima di Gagnone and Monte Duria peridotites. Betic Cordillera: P, primary stage for the Ronda peridotite, with *T–P* trajectories shown for garnet, spinel, and plagioclase peridotites.

et al, 1988a) and is plotted near the spinel-garnet lherzolite boundary, although pressure conditions could have been lower. Two *T–P* locations are plotted for the garnet-bearing assemblages, one at higher temperature, IIA, which represents the contact zone in the Mohelno peridotite, and one at lower temperature, IIB, which represents an average for other Czech and Polish estimates.

The Czech garnet-bearing ultramafic assemblages are thought to have formed during tectonic emplacement of HT spinel peridotite into lower continental crust, related to underthrusting of a southern European lithospheric plate by a previously consolidated North European plate during early Hercynian time. The HT regime for the lower crustal and upper mantle rocks may reflect a preceding episode of lithospheric extension in the Ligerian Cordillera. As illustrated in Figure 11.5b, different portions of peridotites may have followed two different *T–P* paths in passing from stage IB to III. The interior portions of large bodies may have cooled along a trajectory at pressures below the spinel-garnet lherzolite boundary and therefore never passed through the garnet field. In contrast, the contact zones of large bodies and smaller masses, such as peridotite boudins in granulite, cooled more rapidly and passed through the garnet lherzolite field, in which spinell peridotite was recrystallised to garnet peridotite. Further cooling and decompression led to development of the granulite facies assemblage of stage III, the amphibolite facies assemblage of stage IV, and finally, late stage serpentinisation.

This evolutionary scenario harbours several obvious implications. First, if garnet developed in peridotite during tectonic emplacement into granulite, granulite itself must have been subjected to relatively high pressures of 15 kbar or more. Although the felsic granulites at Mohelno have been extensively recrystallised, one sample has yielded a pressure of 11.5 kbar (at 700°C), based on the Pl–Grt–Ky–Qtz barometer as calibrated by Koziol and Newton (1988), and granulites in southern Bohemia, some of which are less recrystallised, have yielded values of 13.7–14.3 kbar (Medaris and Peterson, unpublished). Such high pressures for felsic granulites, despite substantial recrystallisation, lend some credence to the above scenario. Second, if peridotite was emplaced into granulite at temperatures on the order of 1100–1200°C, some degree of partial fusion of granulite might be expected, even though peridotite and granulite were largely anhydrous. Tonalitic rocks, which are referred to as anatectites by Czech geologists, occur locally at the contacts of the Mohelno peridotite with granulite, possibly representing hybridisation of peridotite with granulitic melt. A third implication is that the stage II garnet peridotite assemblage should be Variscan in age, a point that remains to be fully tested, but which is supported by Sm-Nd ages of 344 ± 10 Ma and 370 ± 15 Ma for two garnet clinopyroxenite samples from within Lower Austrian garnet peridotites (Carswell and Jamtveit, unpublished).

This proposed *T–P* evolution for peridotites in Variscan granulite terranes is complicated by the higher pressures obtained for the Austrian garnet peridotites, a 'best estimate' for which is shown as point A1 in Figure 11.5b. The postulated path from stage I to A1 might conceivably represent a transition from a hot extensional environment with thin crust to a colder compressional plate collision regime with substantial tectonic thickening.

Peridotites in the Nonsberg region of the Austridic Alps have followed a *T–P* path similar to that for peridotites in the Variscan granulite terranes, except that formation of garnet-bearing assemblages occurred at much lower temperatures. An early igneous event, shown as stage IA in Figure 11.5c, is revealed by the reconstituted composition of an exsolved pyroxene megacryst in garnet websterite (Morten and Obata, 1983). Stage IB represents the HT conditions for spinel peridotite, based on the compositions of pyroxene porphyroclasts (Obata and Morten, 1987). Thermobarometric calculations demonstrate that the stage II garnet assemblage crystallised at moderate temperatures of around 750°C. Obata and Morten (1987) have proposed two possible *T–P* paths from stages IB to II, one that is isobaric, as depicted in Figure 11.5c, and another that involves an increase in pressure of 3–5 kbar; in either case they argue that transformation from spinel to garnet peridotite occurred within the lower crust. Further recrystallisation to the transitional assemblages of stages III and IV, the granulite facies assemblage of stage V, and the amphibolite facies assemblage of stage VI occurred mainly in response to decompression. As is the case in the Variscides of central Europe, Obata and Morten failed to find any evidence for post-garnet heating in the Nonsberg peridotites.

The Cima di Gagnone and Monte Duria bodies in the Lepontine Alps are exceptional among European garnet peridotites for the evidence they provide of a low *T–P*, crustally enclosed, pre-garnet protolith. The presence of titanoclinohumite (and pseudomorphs thereof) and an eclogite-metarodingite suite suggest that the Cima di Gagnone and Monte Duria bodies were emplaced into the crust and serpentinised under greenschist or blueschist facies conditions early in their histories (Evans and Trommsdorff, 1978), as represented by stage I in Figure 11.5d. An even earlier upper mantle stage (not shown) must have preceded stage I. The stage I assemblage was transformed to stage II garnet peridotite by eclogite facies metamorphism that was related to early Alpine subduction. Stage II is located in Figure 11.5d according to the results obtained by thermobarometric calculations, but as discussed previously, these results may be spurious due to the influence of possible later thermal relaxation; in comparison, eclogites from these localities yield values of approximately 800°C and 25 kbar (Evans *et al*, 1979). The peridotites may have followed a path like the one illustrated in Figure 11.5d, which represents a generalised *T–P* path for subducted crustal rocks. With decreasing pressure, amphibole was stabilised in the stage III assemblage. A further decrease in pressure, accompanied by a decrease in temperature, led to development of the stage IV assemblage in the forms of kelyphitic rims on garnet, and, eventually, Lepontine metamorphism induced extensive recrystallisation to the amphibolite facies assemblage of stage V.

In the Alpe Arami body there is little evidence for a low-pressure precursor to garnet peridotite, and it is suggested that this body may have followed a *T–P* path similar to that postulated for the Norwegian peridotites.

The different mineral facies in the Ronda peridotite of the Betic Cordillera have been explained in terms of a dynamic cooling model (Obata, 1980). The 'primary condition' for the Ronda peridotite, which is shown as point P in Figure 11.5d, represents the earliest set of conditions that can be discerned in this peridotite body and is based on the compositions of pyroxene porphyroclasts. The early HT conditions in the Ronda massif have been ascribed to mid-Tertiary lithospheric extension in the Alboran Basin. During ascent of the Ronda diapir from its primary stage, the exterior portions of the body cooled more quickly than did the interior, owing to heat loss to surrounding mantle. As a result, different parts of the ascending diapir followed different *T–P* trajectories (Figure 11.5d), thereby imparting a mineralogical zonation to the body. The outer zone of the Ronda massif cooled to garnet peridotite, the intermediate zone, to spinel peridotite, and the innermost portion, to plagioclase peridotite. The dynamic cooling model for the Ronda peridotite is analogous to the thermal evolution envisaged for certain HT Variscan spinel-garnet peridotites, such as at Mohelno, except for differences in scale and site of development.

11.5 Concluding remarks

An accumulating body of evidence, which is based on a combination of field, petrographic, phase petrologic, thermobarometric, and isotopic investigations, indicates that European crustal garnet peridotites originated in at least three different ways, including:

(a) direct emplacement of garnetiferous upper mantle into continental crust (Caledonian bodies in Norway; certain Variscan bodies in Austria and Czechoslovakia; Alpe Arami (?) in the Lepontine Alps);
(b) prograde HP metamorphism of spinel peridotites ($\pm$ their serpentinised equivalents) that were previously emplaced into the crust (Cima di Gagnone and Monte Duria, Lepontine Alps);
(c) tectonic emplacement of HT spinel peridotites, portions of which passed into the garnet stability field during dynamic cooling (the Variscan Mohelno peridotite, Czechoslovakia; Ronda massif, Betic Cordillera).

Although European crustal garnet peridotites exhibit diverse origins, it appears that all have experienced the effects of continent–continent collision at some stage in their evolution.

Further investigations, including refinements of thermometers and barometers that are applicable to peridotites, evaluation of cooling histories by diffusion modelling of compositionally zoned garnets, and isotopic studies of carefully identified mineral assemblages, should reveal additional details about the evolution of crustal garnet peridotites and thereby enhance our knowledge pertaining to the interaction of crust and upper mantle during orogenesis.

Acknowledgements

Medaris would like to thank Herb Wang for his contributions to the thermal and diffusion analysis of European garnet peridotites; the National Science Foundation for financial support through Grant EAR-8606998; and the International Research & Exchanges Board for sponsorship of a two-month research visit to Czechoslovakia in 1985. D. A. Carswell wishes to acknowledge grants from NERC supporting studies of Norwegian and Austrian garnet peridotite occurrences.

12 Eclogite stability near the crust-mantle boundary

W.L. GRIFFIN, SUZANNE Y. O'REILLY and N.J. PEARSON

12.1 Introduction

Experimental studies of phase equilibria suggest that eclogite mineral assemblages should be stable in the lower crust and uppermost mantle of cratonic areas (see summary by Green and Ringwood, 1972, and Chapter 3). These studies predict that the lower crust, in areas with normal continental heat flow (about 40 mW m^{-2}), should consist of a mixture of eclogites and granulites. The exact mixture would be dependent on bulk composition (more felsic compositions occurring as granulites) and temperature, with fluid composition playing a minor role.

Such predictions are difficult to verify through direct observation. Many uplifted terranes regarded as deep crustal sections (e.g. the Ivrea Zone) contain mafic granulites but not eclogites (Fountain and Salisbury, 1981; Fountain, 1976). However, few of these so-called deep crustal sections extend to the equivalent of > 25 km depth, and most appear to represent mid-crustal, rather than deep-crustal, levels (Fountain, 1987). A world-wide study of possible lower-crustal xenolith suites (Griffin and O'Reilly, 1987a) showed that these are strongly dominated by mafic granulites, and few of these are associated with eclogites. Xenolith suites from young fold belts commonly contain garnet clinopyroxenites, but these pyroxenes are rich in Ca-Tschermaks molecule rather than jadeite. Many of these pyroxenites have clearly formed by exsolution and recrystallisation of clinopyroxene ± spinel cumulates (Lovering and White, 1969; Griffin *et al*, 1984).

Xenoliths of eclogite occur in many basaltic and ultramafic diatremes, but their depth of origin can seldom be determined directly. Many have equilibration temperatures ⩾ 800°C, which implies derivation from depths ⩾ 100 km, assuming an ordinary cratonic conductive geotherm (Eggler *et al*, 1987; Chapter 13). However, probable lower-crust/uppermost mantle eclogites can be recognised in xenolith suites from several localities. In all of these, eclogites are intimately associated (chemically and petrographically) with granulite xenoliths, or have equilibration temperatures implying depths < 100 km, or both. Localities known to the authors are: Lesotho (Griffin *et al*, 1979; Jackson and Harte, 1977); Chino Valley, Arizona (Krieger, 1965; Arculus and Smith, 1979; Schulze and Helmstaedt, 1979); Camp Creek, Arizona (Esperanca and Holloway, 1984); Sample Creek, Liberia (Toft *et al*, 1988); Australia (Edwards *et al*, 1979; Arculus *et al*, 1988; O'Reilly *et al*, in press); Taryat Depression, Mongolia (Kepezhinkas, 1979). All of these localities are on cratons or craton margins, mainly of Proterozoic age.

Xenoliths of unusual Na-rich, lawsonite-bearing eclogites similar to the eclogites of blueschist terranes occur in diatremes of the Northern Navajo province of the American Southwest (e.g. Watson and Morton, 1969). These xenoliths have been interpreted as pieces of a subducted slab of ocean floor trapped within the uppermost part of the mantle (Helmstaedt and Doig, 1975; Helmstaedt and Schulze, 1979), and are not considered in detail here.

Heat flow studies, theoretical modelling and the geothermobarometry of xenolith suites all indicate that young fold belts and rift zones have elevated geotherms relative to cratonic areas (Sass and Lachenbruch, 1979; O'Reilly and Griffin, 1985). This implies large lateral variations in the temperature at the base of the crust (depending on tectonic environment), and thus in the relative stability of the eclogite and granulite mineral assemblages. The present authors regard the restriction of lower-crustal eclogite suites to craton or craton-margin areas, and their apparent absence from younger terranes, to be primarily a consequence of this temperature difference.

Geothermal gradients can, however, change with time, and the effects of such changes can be followed in the microstructures and mineral chemistry of some xenolith suites. In this Chapter we will illustrate the factors that control the distribution of eclogite in the lower crust and uppermost mantle (depths < 100 km) by comparing three xenolith suites, each of which shows a transition from granulite to plagioclase-free garnet + clinopyroxene ($\pm$ orthopyroxene) assemblages, and each of which represents a different thermal regime. In the following, we use the term garnet pyroxenite to include all Grt $\pm$ Cpx $\pm$ Opx assemblages but denote more specific assemblages as garnet websterite (Grt + Opx + Cpx), garnet clinopyroxenite (Grt + Cpx, not omphacitic) or eclogite (Grt + Cpx, omphacitic).

12.2 Description of xenolith suites

12.2.1 *Anakie*

The Anakie granulite-pyroxenite xenolith suite occurs in a Holocene nepheline mugearite cinder cone in south-central Victoria, and was described in detail by Wass and Hollis (1983). The Anakie xenoliths (excluding the abundant spinel lherzolites) range from mafic to ultramafic in bulk composition. Microstructures range from igneous (gabbros, orthocumulates, poikilitic cumulates) to granoblastic mosaic. Composite xenoliths show contacts between igneous and metamorphic rock types and indicate multiple deep-seated intrusive events separated by significant time intervals.

Gabbros and pyroxenites contain clinopyroxene $\pm$ orthopyroxene, plagioclase, olivine, amphibole, mica, apatite, Al-spinel, Fe-Ti oxides and Fe–Ni sulphides. Granulites and metapyroxenites ($\pm$ garnet) contain clinopyroxene $\pm$ orthopyroxene, plagioclase, garnet, anorthoclase, amphibole, mica, ilmenite, rutile, pseudobrookite, maghemite, Al-spinel, Fe–Ni sulphides, rare sphene and apatite. Fluid inclusions, filled with a CO_2-dominant fluid, are ubiquitous in both microstructural types and commonly reach the size of vugs up to 1 cm across, usually lined with euhedral crystals of amphibole or pyroxene (Wass and Hollis, 1983; Andersen *et al*, 1984).

Re-equilibration to lower temperatures is evident in a variety of microstructural and mineralogical changes. Grain-boundary readjustment produced rounded terminations on feldspar laths and an incipient granoblastic microstructure in the early stages. Exsolution of complementary pyroxenes created a spectrum of microstructures including spectacular vermicular pyroxene intergrowths and completely recrystallised polygonal aggregates. No evidence has been found for exsolution of garnet from clinopyroxene, but reaction of pyroxenes and plagioclase has resulted in garnet formation. In some xenoliths this arrested reaction is preserved, with garnet porphyroblasts evident in feldspar grains. Thus the Anakie xenoliths record the progressive change of gabbroic rocks to granulites or garnet pyroxenites; this is interpreted as the result of isobaric cooling from igneous to ambient temperatures in the upper mantle/lowermost crust.

12.2.2 *Eastern margin of the Australian craton (EMAC)*

Xenoliths from three areas near the eastern margin of the Australian craton (EMAC) have been examined in detail. These are the kimberlites at El Alamein (Port Augusta locality of Stracke *et al*, 1979) and Terowie (Calcutteroo and Pine Creek diatremes) in South Australia and the Kayrunnera basaltic diatreme in NSW (e.g. Edwards *et al*, 1979; Arculus *et al*, 1988; McCulloch *et al*, 1982; Scott-Smith *et al*, 1984; O'Reilly *et al*, in press).

Extremely altered spinel and garnet lherzolites and quartzo-feldspathic granulite xenoliths (Ky + Grt + Kfs + Pl + Qtz + Rt ± Bt) occur sparsely, but most xenoliths are mafic. These mafic xenoliths can be subdivided into four groups:

(a) garnet websterites
(b) mafic granulites (no kyanite)
(c) kyanite-bearing mafic granulites
(d) eclogites.

The mafic xenoliths have diverse microstructures which record transitional stages in the transformation of igneous assemblages and microstructures to granoblastic granulites and eclogites. Microstructural evidence for these reactions includes (i) relict igneous phases; (ii) coronas and symplectites; (iii) solid inclusions; (iv) pseudomorphism of relict igneous phases.

12.2.2.1 *Garnet websterites.* The garnet websterite xenoliths (Cpx + Opx + Grt ± Spl) are characterised by coarse (up to 15 mm) clinopyroxene and orthopyroxene, both of which have exsolved the complementary pyroxene. Garnet occurs as exsolved blebs in clinopyroxene, as narrow coronas surrounding spinel, and as strings along pyroxene–pyroxene grain boundaries. Further recrystallisation produces finer-grained granoblastic aggregates of Cpx + Opx + Grt.

12.2.2.2 *Mafic granulites.* The mafic granulite xenolith suite can be further divided into three types:

(i) two-pyroxene granulites (Cpx + Opx + Pl ± Qtz ± Kfs ± Ilm ± Amp)
(ii) garnet two-pyroxene granulites (Grt + Cpx + Opx + Pl ± Qtz ± Kfs ± Ilm ± Amp)
(iii) garnet-clinopyroxene granulites (Grt + Cpx + Pl ± Qtz ± Amp ± Rt ± Bt ± Scp ± Ttn)

Microstructures of the two-pyroxene granulites are characteristically equigranular granoblastic. Exsolution within both pyroxenes and the antiperthitic plagioclase provides the only evidence for re-equilibration or disequilibrium.

In contrast to the two-pyroxene granulites, the garnet-bearing two-pyroxene granulites display a number of disequilibrium microstructures. Rare coarse, high-Al

(a)

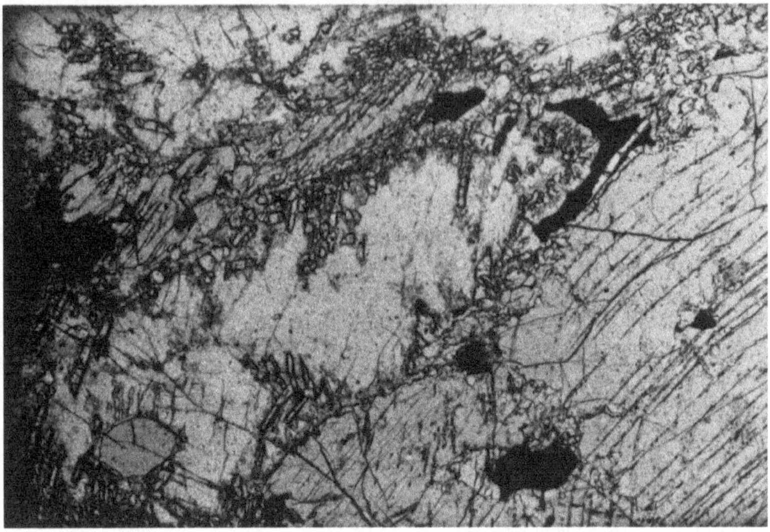

(b)

Figure 12.1 (*Continued*)

(c)

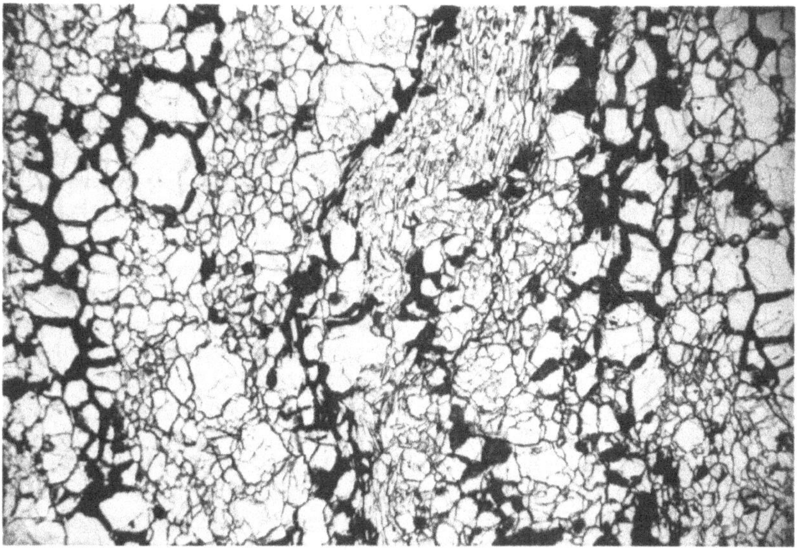

(d)

Figure 12.1 Microstructures in EMAC xenoliths. (a) Garnet growing in plagioclase at Cpx–Pl interface. Note tabular form of Grt grains, also planes of Grt exsolution in Cpx. Black grains are spinel. Plane polarised light. Width of view 2.4 mm. (b) Corona structure. Granular central portion is Opx surrounded by coarse radial Cpx, then by garnet (black). Felted mass outside garnet corona is plagioclase pseudomorph consisting of Ky and Grt. Ky needles occur in outer part of Grt shell. Crossed polars. Width of view 7 mm. (c) Layering of Cpx, Grt and Ky in foliated and folded granulite. Plane polarised light. Width of view 7 mm. (d) Rotated Grt in strongly foliated Ky, Cpx and Pl. Crossed polars. Width of view 4.8 mm.

clinopyroxenes with exsolved garnet and orthopyroxene are similar to those seen in the garnet pyroxenites and probably represent partially re-equilibrated relict igneous pyroxenes. Homogeneous garnet coronas separate these relict clinopyroxenes from plagioclase. More complex, multiple coronas are developed between orthopyroxene and plagioclase. The inner shell of the corona (adjacent to the orthopyroxene) comprises granular clinopyroxene which is in turn surrounded by a symplectic intergrowth of garnet and clinopyroxene. This intergrowth is separated from plagioclase by a homogeneous garnet corona. Isolated, euhedral grains of garnet occur within the plagioclase adjacent to the coronas.

The garnet-clinopyroxene granulite xenoliths are the most abundant of the mafic granulite suite. No relict pyroxenes are preserved, although polygonal granoblastic aggregates of clinopyroxene may represent recrystallised coarse igneous clinopyroxene. The initial growth of garnet as subhedral grains is within plagioclase, adjacent to clinopyroxene (Figure 12.1a). In more re-equilibrated rocks these grains connect to form trains or continuous coronas; in such fish-net textures aggregates of clinopyroxene and plagioclase are separated by garnet.

The kyanite-bearing mafic granulites (Grt + Cpx + Pl + Qtz + Ky ± Opx ± Rt ± Scp) contain no relict igneous clinopyroxene and orthopyroxene is rare. The breakdown of plagioclase is more extensive with the development of complex coronas between plagioclase and pyroxene, and the pseudomorphic replacement of plagioclase. As in the garnet two-pyroxene granulites, orthopyroxene is rimmed by granular clinopyroxene which is in turn separated from plagioclase (or its breakdown products) by continuous garnet coronas.

In least recrystallised samples, plagioclase grains in the centres of plagioclase-rich aggregates have tabular shapes suggestive of igneous microstructures. These grains may be partially pseudomorphed by extremely fine-grained aggregates of Ky + Grt + Qtz or Ky + Grt + Spl ± Cor. Discontinuous trains of garnet develop along plagioclase–clinopyroxene grain boundaries. With the progressive breakdown of plagioclase the garnet forms continuous coronas, with euhedral grains extending out into the relict plagioclase, and kyanite increases in abundance. Rare plagioclase relics may remain in the centres of original plagioclase-rich aggregates. The complete breakdown of plagioclase results in kyanite-rich aggregates which are separated from clinopyroxene by double coronas: a homogeneous garnet corona occurs adjacent to the clinopyroxene, and garnet with abundant kyanite inclusions adjacent to the kyanite-rich patches (Figure 12.1b).

A strong layering is developed on a 1–5 mm scale in more deformed kyanite granulite xenoliths (Figure 12.1c). Monomineralic layers or lenses of granoblastic clinopyroxene are separated by garnet-rich layers and layers of Ky + Qtz + Grt ± Pl. Parallel alignment of kyanite grains defines a strong lineation parallel to the mineralogical layering.

12.2.2.3 *Eclogites.* The eclogite xenoliths contain minor rutile, quartz, amphibole and biotite. Their microstructures range from coarse (up to 1 cm), equigranular granoblastic to those in which finer aggregates of clinopyroxene are surrounded by garnet, similar to the fish-net textures described in the garnet-clinopyroxene granulites (Figure 12.1d). Garnet-rich aggregates may represent pseudomorphed plagioclase, especially in the rare cases where the garnet contains kyanite inclusions. The clinopyroxene in the eclogite xenoliths shows no exsolution. The hydrous phases, in particular amphibole, appear to be in microstructural equilibrium with both garnet and clinopyroxene.

12.2.3 *Lesotho*

The Lesotho granulite-eclogite-garnet websterite xenolith suite was collected from a number of small kimberlite pipes in northern Lesotho. Detailed descriptions of the mineralogy, microstructures and bulk composition are given by Griffin *et al* (1979) and Jackson and Harte (1979). Trace element and isotopic data are given by Rogers (1977) and Rogers and Hawkesworth (1982). The bulk compositions are overwhelmingly mafic, but Griffin *et al* (1979) described one intermediate composition xenolith and suggested that the rarity of such rocks reflects the ease with which they are altered in the kimberlites. The suite shows a complete modal and chemical gradation from garnet granulite to eclogite and to garnet websterite. Additional phases are scapolite (S-bearing), quartz, rutile and apatite; secondary kyanite has grown within the plagioclase of some granulites. Typical grain-sizes are 1–5 mm.

Microstructures are in general very well-equilibrated granoblastic mosaics, although foliations are defined by mineral layering and shape orientation in some samples. However, several samples retain evidence for reaction and re-equilibration. (*a*) Exsolved garnet is present in both clinopyroxene and orthopyroxene in sample PHN 2508. This garnet websterite also contains amphibole both as poikiloblasts and as thin lamellae within clinopyroxene. (*b*) The occurrence of orthopyroxene in sample LT2 is restricted to small rounded inclusions within clinopyroxene. (*c*) Relict primary Al-rich pyroxenes (Cpx and Opx) in sample L6 are rimmed by zones of granular garnet (with inclusions of spinel and corundum) which produces a coronite microstructure, similar to that described from the EMAC garnet granulites. (*d*) Sample PHN 1442j contains coarse poikilitic clinopyroxene (inclusions of garnet and kyanite) which is surrounded by coarse-grained garnet. Elsewhere sheaths of kyanite are intergrown with secondary clinopyroxene, having higher Jd and lower Ts than the primary pyroxene. These microstructures indicate reactions which involve the breakdown of plagioclase and a primary Al-rich (Ts-rich) clinopyroxene to produce kyanite, a less aluminous secondary clinopyroxene and a Ca–Fe rich garnet.

The rare disequilibrium microstructures in the Lesotho suite are comparable to those seen in the EMAC suite, and suggest a similar history of chemical and microstructural re-equilibration, carried to a greater degree of completeness.

12.3 Geothermobarometry

P–T estimates relevant to these three suites are shown in Figure 12.2. Unless otherwise noted, the estimates shown are based on the Ellis and Green (1979) thermometer with the Wood (1974) or Newton and Perkins (1982) geobarometers used where applicable. Alternative approaches, which generally give concordant results, are discussed by Carswell and Griffin (1981), and Griffin *et al* (1984). Detailed discussion of the EMAC geothermobarometry will be given elsewhere (Pearson *et al*, in preparation).

The data points for the Anakie suite are, for clarity, not shown here; they lie along the southeastern Australian (SEA) geotherm of O'Reilly and Griffin (1985) (O'Reilly *et al*, 1988, figure 3). The SEA geotherm was defined using garnet websterites from the Holocene Bullenmerri–Gnotuk locality, and fits most garnet websterite and garnet

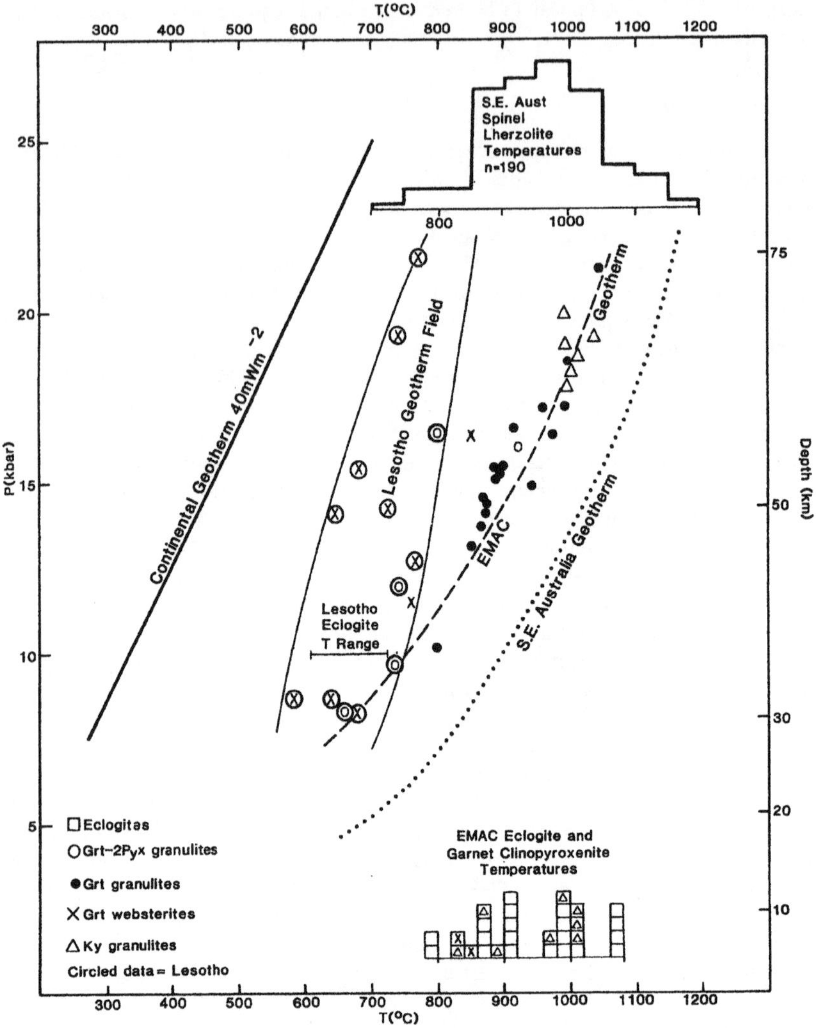

Figure 12.2 *P–T* estimates for xenoliths of the EMAC and Lesotho suites. SE Australia geotherm is from O'Reilly and Griffin (1985). Histograms show *T* estimates for spinel lherzolites from SE Australia, and Cpx-Grt pairs from EMAC xenoliths. *T* range for Lesotho eclogites is given by horizontal bar at 10 kbar.

lherzolite data from other eastern Australian localities, ranging in age from 0–200 Ma. It is interpreted as representing the present geotherm for western Victoria; this is consistent with heat-flow and MAGSAT data (Mayhew and Johnson, 1987). In other areas, the *P–T* array is regarded as a paleogeotherm, reflecting the ambient *T* at depth when the xenoliths were entrained. The spinel-lherzolite *T* histogram in Figure 12.2

shows a pronounced edge at 850°C, corresponding to a depth of about 30 km on the SEA geotherm. This depth is interpreted as the crust-mantle boundary (CMB; O'Reilly *et al*, 1988), which implies that most of the garnet granulites and garnet pyroxenites from Anakie are derived from the uppermost mantle or lowermost crust.

The data points for the EMAC suite define a curve lying about 150°C below the SEA geotherm. The depth to the CMB in this area is poorly known; the seismic Moho lies at 35–40 km (Finlayson, 1982). Spinel lherzolite xenoliths have been found, but are too altered to give T information; their presence suggests mantle material at depths $\leqslant 45$ km (O'Neill, 1981). If so, then most of the granulite and eclogite xenoliths studied here come from within the upper mantle.

The Lesotho P–T data are much more scattered, as noted by Griffin *et al* (1979) and Carswell and Griffin (1981). Most of the two-pyroxene + garnet granulites and garnet websterites have been re-analysed by the present authors, but this scatter remains. T estimates for the eclogites are in the range 600–725°C. The depth to the CMB in Lesotho is poorly defined; the seismic Moho lies at about 35 km depth (Hales and Sacks, 1959). Al-spinel lherzolites occur in some kimberlite pipes (Carswell *et al*, 1984), suggesting a CMB at <45 km. On this basis, most of the granulites and eclogites appear to come from the lower part of the crust or the uppermost mantle.

Taken at face value, these data suggest that the three suites represent different thermal regimes. Fraser and Lawless (1978) have argued that P–T arrays such as the SEA geotherm are meaningless, and reflect only freezing-in of exchange equilibria during cooling after some heating event. While such effects must occur in some cases, they cannot be assumed to apply to all xenolith suites, and especially not to those recording high T. For example, Wilson and Smith (1984) showed that diffusion rates are sufficient to maintain continuous equilibrium between garnet and pyroxene at $T > 700$°C.

Freezing-in of spurious T and P requires a decrease in diffusion rates over time, gradually limiting the distance over which equilibrium can be maintained. This must result in zoning of the coexisting phases and should be easily detectable by electron microprobe. Griffin *et al* (1984) demonstrated that the phases in SEA garnet websterites are very homogeneous, as are those in the recrystallised rocks from Anakie. This is compelling evidence for chemical equilibrium at a constant T; the SEA geotherm is therefore interpreted to reflect the ambient geotherm at the time of xenolith entrainment.

However, the new analyses of Lesotho xenoliths show measurable zoning of clinopyroxenes, even though orthopyroxenes and garnets are homogeneous. In all cases the rim compositions yield lower $K_{Fe^{2+}-Mg}^{Grt-Cpx}$ temperatures than cores, although the rim values in themselves must be meaningless. This implies that the T values for the Lesotho xenoliths are maximum values. Griffin *et al* (1979) noted that the P–T estimates for Lesotho granulites and garnet websterites lie well above the cratonic geotherm defined by the deeper garnet lherzolite xenoliths. They suggested this was a relic of Karoo heating; it might also be a relic of the formation of the granulites in Proterozoic time (Rogers and Hawkesworth, 1982).

The phases in the EMAC xenoliths are generally homogeneous, but some zoning is observed in the clinopyroxenes. The T estimates may therefore lie above the real ambient T. However, the limited scatter in the data, compared to the Lesotho suite, suggests that these xenoliths were not far out of equilibrium at the time of their entrainment.

In summary: the geothermobarometry data indicate that the three xenolith suites equilibrated under different thermal regimes. The Anakie suite reflects an elevated geotherm typical of young active areas (O'Reilly and Griffin, 1985). The Lesotho samples may represent arrested equilibration toward a cratonic geotherm, following an underplating episode. The EMAC suite appears to have been essentially in equilibrium at an intermediate state, which reflects a geotherm about 150°C cooler than the Anakie suite, but still about 400°C above a cratonic geotherm at lower crustal depths.

12.4 Mineral chemistry

12.4.1 *Analytical methods*

All of the mineral chemical data from the EMAC and Anakie xenolith suites were obtained on an ETEC electron microprobe at Macquarie University. Details of operating conditions are discussed by Griffin *et al* (1984). Most data for the Lesotho suite are taken from Griffin *et al* (1979), but a number of two-pyroxene + garnet granulites and websterites have been re-analysed at Macquarie University; these new analyses are incorporated into this study. Representative analyses are given in Table 12.1.

12.4.2 *Clinopyroxene*

Clinopyroxenes from granulite and eclogite facies outcrops differ significantly in their relative proportions of the jadeite (Jd) and Tschermaks (Ts) molecules relative to diopside (Di) and hedenbergite (Hd) (White, 1964). Calculation of these components requires estimation of the Fe^{3+} content which has been done in this study assuming stoichiometry and charge balance. Calculation of end-member components was done in the following order: acmite, jadeite, Ca-Tschermaks, Ca-Ferritschermaks, diopside + hedenbergite. End-member molecules including Ti, Cr and Mn, were also taken into account before calculation of diopside and hedenbergite. This method effectively maximises CaTs relative to Jd. Irrespective of rock type, the majority of clinopyroxenes analysed from all three suites are characterised by high Jd + Ts, Acm and Di + Hd. The relative proportions of these components vary widely within, and systematically between, the xenolith suites. The emphasis in this section is to examine the compositional changes in clinopyroxene associated with the transition from garnet granulite/pyroxenite to eclogite.

On the Jd:(Di + Hd):Ts plot (Figure 12.3) the compositional field for the Anakie clinopyroxenes stretches from the (Di + Hd) – Ts edge to a maximum Jd value of about 20%. It thus straddles the Jd:Ts (1:2) reference line, which was used by White (1964) to distinguish pyroxenes equilibrated under granulite and eclogite facies conditions. The clinopyroxenes from the EMAC xenolith suite generally show little variation in Ts content, and an increase in Jd correlated with a decrease in Di + Hd.

Clinopyroxenes in the EMAC garnet websterites and two-pyroxene garnet granulites have negligible Jd and thus plot along or close to the (Di + Hd) – Ts edge. The granular clinopyroxenes and symplectic clinopyroxenes in the coronas in the two-pyroxene garnet granulites have slightly higher Jd contents than the cores of the primary clinopyroxenes. There is a considerable overlap between the compositional fields for the garnet granulite and eclogite clinopyroxenes, with most analyses from

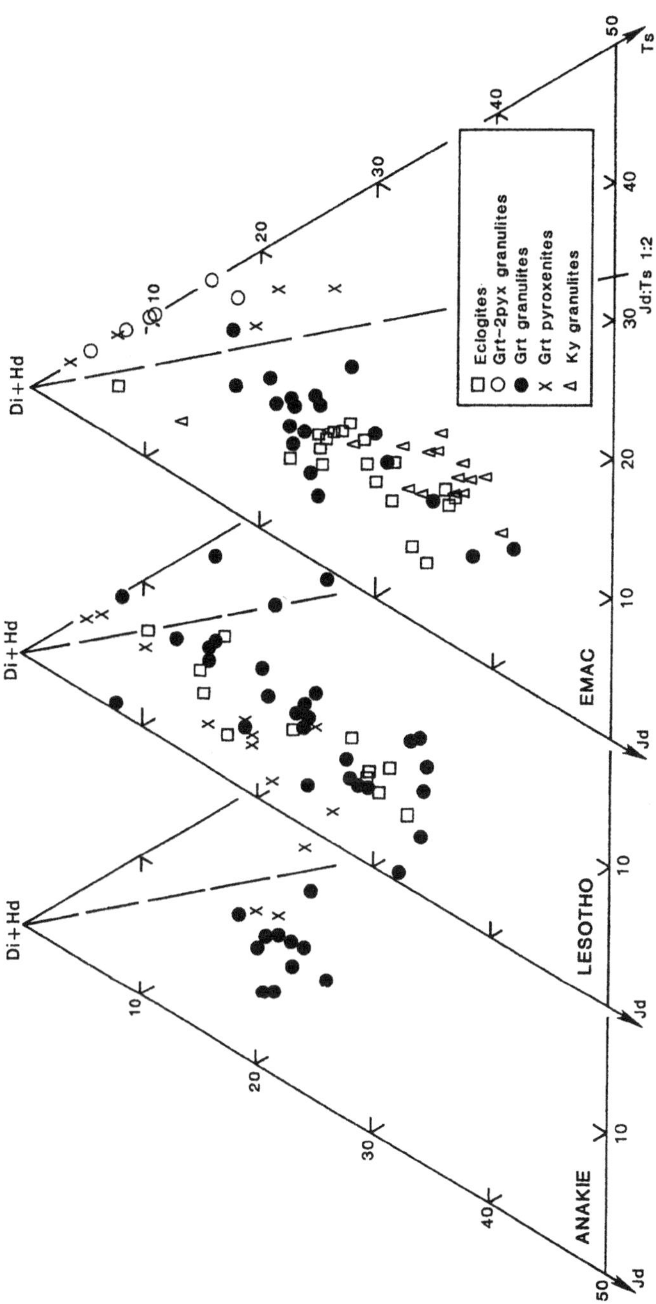

Figure 12.3 Clinopyroxene compositions in Jd-Ts-(Di + Hd). The Jd:Ts = 1:2 line was proposed by White (1964) to separate eclogite- from granulite-facies clinopyroxenes.

Table 12.1 Representative mineral analyses for EMAC and Lesotho xenoliths

	EMAC		EMAC		EMAC		EMAC	
Sample No:	WC5		78210330		78210028		78210837	
Rock type:	ec		ec		gg		gg	
X_{An}					0.15		0.20	
T(°C)	1079 (a)		982 (a)		1045 (b)		890 (b)	
P(kbar)					21.3		15.2	
	Cpx	Grt	Cpx	Grt	Cpx	Grt	Cpx	Grt
SiO_2	51.37	38.74	52.08	39.15	52.13	39.79	51.90	39.16
TiO_2	0.86	0.29	0.49	0.12	0.62	0.12	0.42	
Al_2O_3	10.04	21.69	6.82	22.26	11.44	22.37	6.82	22.12
Cr_2O_3								
FeO	8.82	22.03	10.06	24.46	6.11	19.37	7.25	22.24
MnO	0.09	0.53	0.10	0.54	0.07	0.43		0.45
MgO	8.35	6.87	9.93	7.77	8.89	9.41	10.90	7.39
CaO	16.22	10.77	17.57	7.63	16.16	9.08	19.32	9.26
Na_2O	4.60		3.25		4.51		3.03	
Total	100.35	100.92	100.30	101.93	100.03	100.57	99.64	100.62
X_{Mg}	0.628	0.357	0.638	0.361	0.724	0.464	0.728	0.372
X_{Acm}	0.121		0.092		0.030		0.097	
X_{Jd}	0.228		0.163		0.312		0.127	
X_{Ts}	0.092		0.066		0.090		0.078	
X_{Di+Hd}	0.559		0.678		0.568		0.699	
X_{Alm}		0.458		0.509		0.405		0.470
X_{Prp}		0.255		0.288		0.351		0.279
X_{Grs}		0.287		0.203		0.243		0.251

Abbreviations: ec = eclogite; Grt–2Pyx–g = Garnet 2-pyroxene granulite; gg = garnet granulite; kg = kyanite granulite; gw- = garnet websterite (a) Ellis & Green (1979) Grt–Cpx T at 15 kbar; (b) Ellis & Green (1979) Grt–Cpx T and Wood (1974) Grt–Opx P; (c) Ellis & Green (1979) Grt–Cpx T and Newton & Perkins (1982) Grt–Cpx–Pl–Qtz P; Sample numbers commencing with 78210 are Australian Bureau of Mineral Resources numbers.

both xenolith types plotted on the Jd side of the reference line. Thus according to this criterion the clinopyroxenes from both groups are eclogitic, and cannot be distinguished on the basis of the Jd versus Ts contents of the clinopyroxenes. The garnet granulites show a wider spread of Jd contents (0–31%) than the eclogite xenoliths, which are essentially restricted to the range Jd_{15-23}.

The microstructural disequilibrium preserved in many of the kyanite granulites is accompanied by chemical heterogeneity, evident both as chemical zonation and distinct compositional domains. Clinopyroxene occurring in coarse-grained aggregates is typically zoned to more Ts- and Jd-rich rims. The rim compositions of these zoned pyroxenes plot within the field defined by the clinopyroxenes coexisting with Ky + Grt + Qtz ± Pl. These clinopyroxenes are characterised by high Jd contents (22–30%); they are similar to the high-Jd clinopyroxenes from the garnet granulite and eclogite xenoliths but are distinguished from the latter by their higher Ts contents. The clinopyroxenes from kyanite granulites have low total Fe and negligible Fe^{3+} and are distinct from those in eclogites and garnet granulites (Figure 12.4). The highest acmite

Table 12.1 (*Continued*)

EMAC		EMAC		LESOTHO			LESOTHO		
78210036		78210043		PHN2495			KN206		
kg		kg		grt-2pyx-g			gw		
0.22		0.23		0.16					
995(b)		995(b)		739(c)			744(c)		
19.0		17.8		12.1			19.3		
Cpx	Grt	Cpx	Grt	Cpx	Opx	Grt	Cpx	Opx	Grt
52.79	41.67	52.10	41.24	53.45	54.94	40.29	53.93	54.20	39.44
0.56	0.11	0.57	0.11	0.51			0.34		0.13
12.01	23.57	11.23	23.49	6.54	1.80	22.66	5.93	0.86	21.97
0.08				0.45	0.11	0.40			
2.42	11.09	2.75	11.86	6.24	16.02	22.10	9.18	21.10	26.00
	0.23		0.28		0.11	0.48		0.19	0.65
10.75	16.65	11.07	15.21	11.59	27.82	11.73	10.30	24.53	9.34
17.40	7.02	18.70	8.22	18.17	0.23	4.57	15.98	0.27	4.12
4.41		3.59		3.50	0.05		4.41	0.08	
100.42	100.34	100.01	100.41	100.45	101.08	102.23	100.07	101.23	101.65
0.888	0.728	0.878	0.696	0.768	0.756	0.486	0.667	0.674	0.390
0.022		0.009		0.067			0.121		
0.298		0.253		0.199			0.216		
0.099		0.106		0.029			0.020		
0.581		0.632		0.705			0.644		
	0.223		0.240			0.452			0.543
	0.596		0.548			0.428			0.347
	0.181		0.213			0.120			0.110

contents are found in the eclogites (up to 18% Acm), although there is a substantial overlap between the eclogites and garnet granulites.

The general chemical characteristics of the clinopyroxene in the Lesotho xenoliths have been discussed by Griffin *et al* (1979). They show a compositional range similar to the EMAC clinopyroxenes but with lower Ts contents (Figure 12.3). This difference in Ts contents of the clinopyroxenes in the two suites may reflect their different thermal regimes. The great majority of analysed clinopyroxenes from the Lesotho eclogite and garnet granulite xenoliths plot on the Jd-rich side of the Jd:Ts reference line. The pyroxenes of garnet pyroxenites have the lowest Ts contents and those of the garnet granulites the highest Ts contents.

Care was taken in the re-analysis of the selected Lesotho samples to establish the extent of chemical zonation. Clinopyroxene rims adjacent to garnet are lower in Ts, and to a lesser degree Jd, than the core compositions, although the variation is restricted to narrow zones near the grain edge. Unlike the EMAC xenoliths, plots of acmite contents for the Lesotho clinopyroxenes show no distinctions between the garnet granulites, garnet websterites and eclogites (Figure 12.4).

Granulites in all three suites show a broad correlation between the Na content of the plagioclase and the Jd/(Jd + Ts) ratio of the clinopyroxene (Figure 12.5); this relationship was first noted in the Lesotho suite by Griffin *et al* (1979). Although such a relationship is predictable from the experimental work of Kushiro (1969), it has not, to the authors' knowledge, been reported from outcrop samples of granulites.

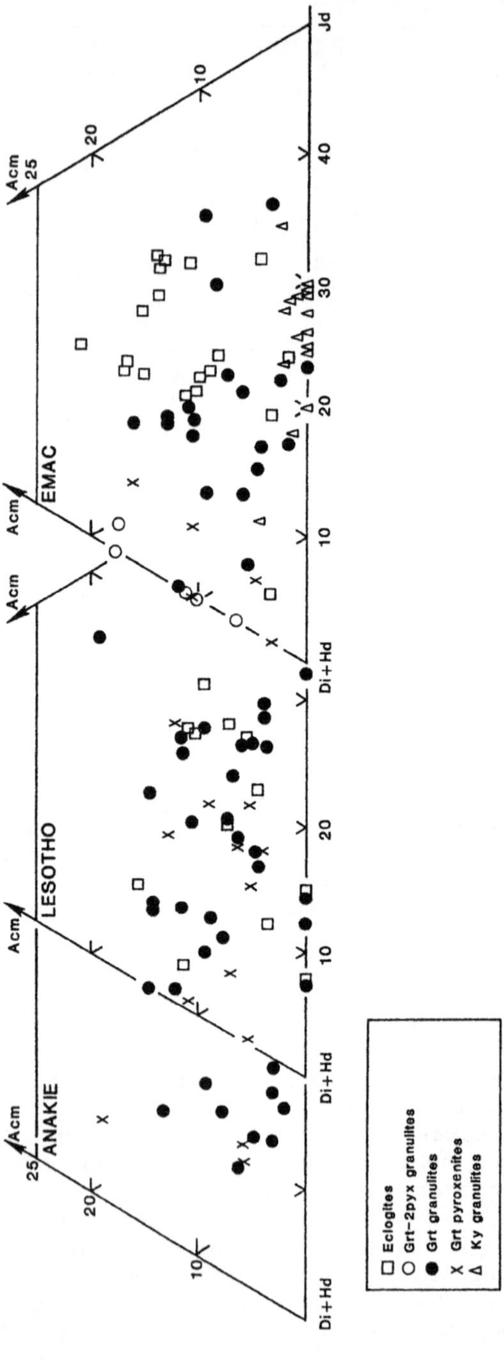

Figure 12.4 Clinopyroxene compositions in Jd-Acm-(Di + eHd).

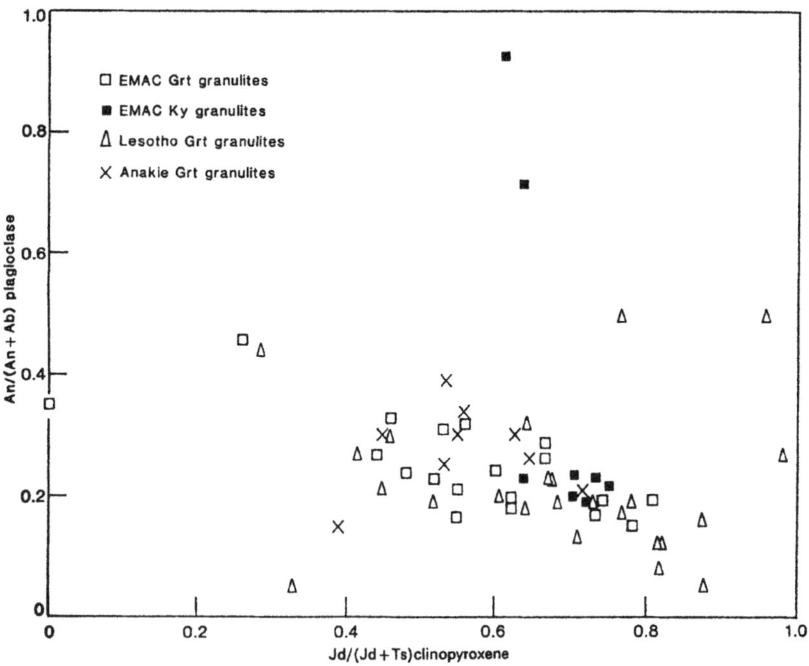

Figure 12.5 Jd/(Jd + Ts) in clinopyroxene versus An/(An + Ab) in plagioclase.

12.4.3 *Garnet*

Garnets from all rock types in each of the three suites are solid solutions of pyrope (Prp), almandine (Alm) and grossular (Grs) with minor spessartine and negligible uvarovite. There is a wide spread in Mg/Fe and a restricted range in Ca (Figure 12.6). This reflects both variations in whole rock composition and the P–T-controlled partitioning between garnet and clinopyroxene. The garnets in the Anakie xenoliths show a very restricted range in composition compared with the other suites.

The EMAC garnets show a spread of Mg/Fe but only small variation in Grs contents; the individual xenolith types form distinct clusters or fields. Garnets from two-pyroxene garnet granulites plot at the Fe-rich end of the trend. The majority of garnets from the eclogite xenoliths also cluster at the Fe-rich end but are displaced to higher Grs values. Exceptions include two samples with kyanite inclusions in the garnet, which suggests a relation with the kyanite granulite group. Both the kyanite granulites and garnet pyroxenites have higher Prp contents. The kyanite granulite garnets define a linear trend with Grs content inversely correlated with Mg/Fe. This same trend is shown by single coronas; the garnet corona adjacent to clinopyroxene has lower Ca and higher Mg/Fe (Prp:Alm:Grs = 0.61:0.21:0.18) than the garnet that coexists with kyanite in the relict plagioclase (Prp:Alm:Grs = 0.40:0.23:0.37). The garnet granulites display more scatter than the other groups of xenoliths, with Grs contents similar to both the eclogite and kyanite granulite garnets and intermediate Mg/Fe.

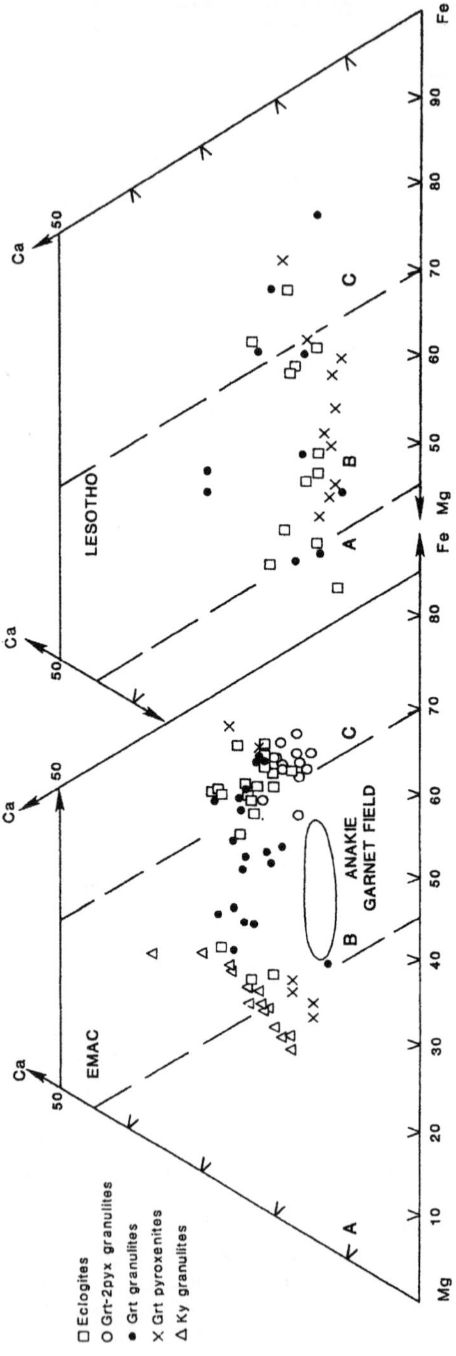

Figure 12.6 Garnet compositions.

The Lesotho garnets show a spread in Mg/Fe values similar to the EMAC garnets but displaced to lower Grs. The ranges in Prp:Alm:Grs are similar for the granulites, eclogites and websterites although garnets in the latter group have lower Grs than most of the eclogite and garnet granulite garnets. The garnet with the highest Grs content from the Lesotho suite (PHN 1442j: 23% Grs) coexists with kyanite and plots on the extension of the trend defined by the garnets from the EMAC kyanite granulites.

12.5 Bulk composition effects

Analyses of Lesotho and Anakie xenoliths are given by Griffin *et al* (1979) and O'Reilly *et al* (1988), respectively. The EMAC analyses will be presented elsewhere. Normative compositions are summarised in Figures 12.7 and 12.8. The appearance of garnet and

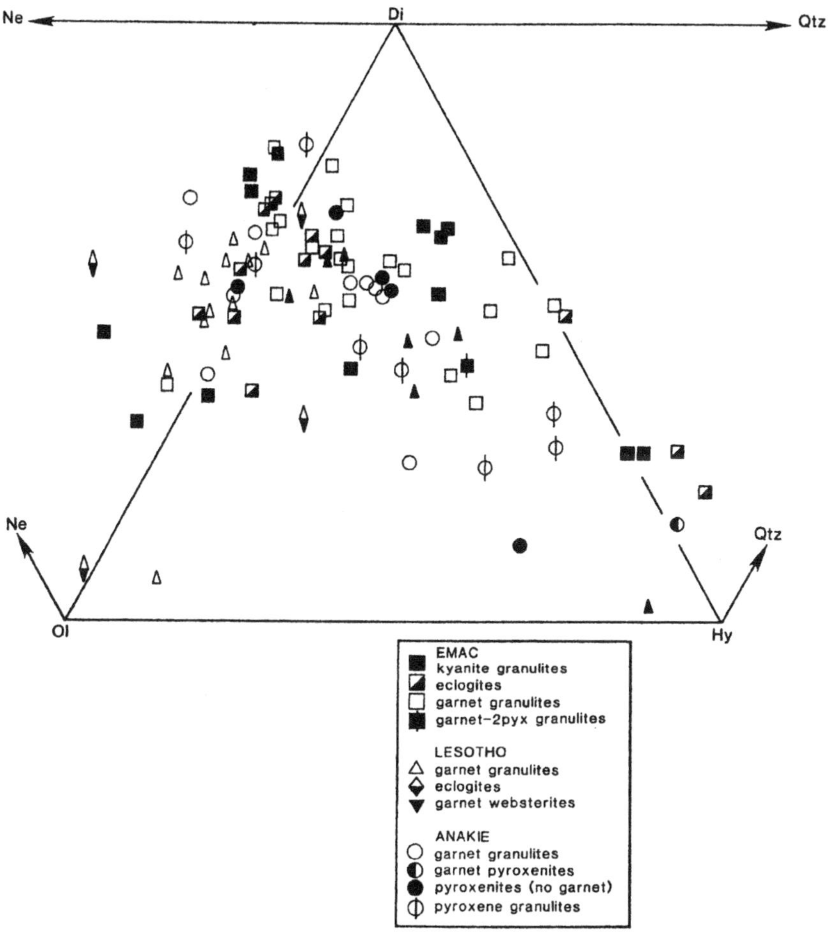

Figure 12.7 Xenolith compositions (normative) in the expanded basalt tetrahedron, projected from feldspar.

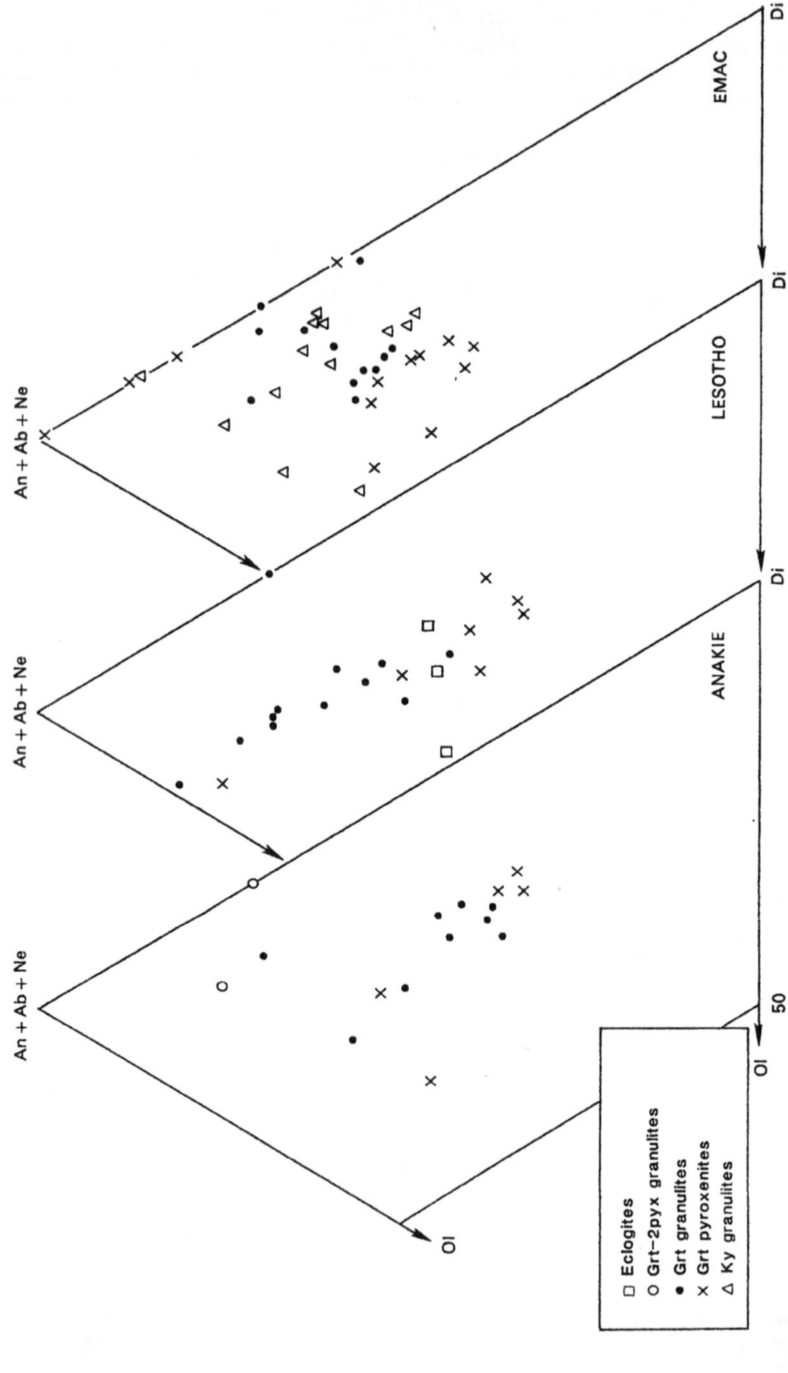

Figure 12.8 Xenolith compositions (normative) in the Diopside-Olivine-(Feldspar + Nepheline) plane of the basalt tetrahedron.

the disappearance of plagioclase from a rock with increasing *P* or decreasing *T* will be controlled by several compositional parameters, including Mg/Fe, degree of silica saturation, and total normative feldspar content (Green and Ringwood, 1972; Chapter 3). In general, feldspar-poor, nepheline-normative compositions will be converted to eclogite at lower *P* than feldspar-rich, quartz-normative compositions. These effects are demonstrated by the three suites described here.

In the Lesotho suite, the granulites are generally plagioclase-rich and slightly Ne-normative, whereas the eclogites and websterites are Hy(± Qtz)-normative,

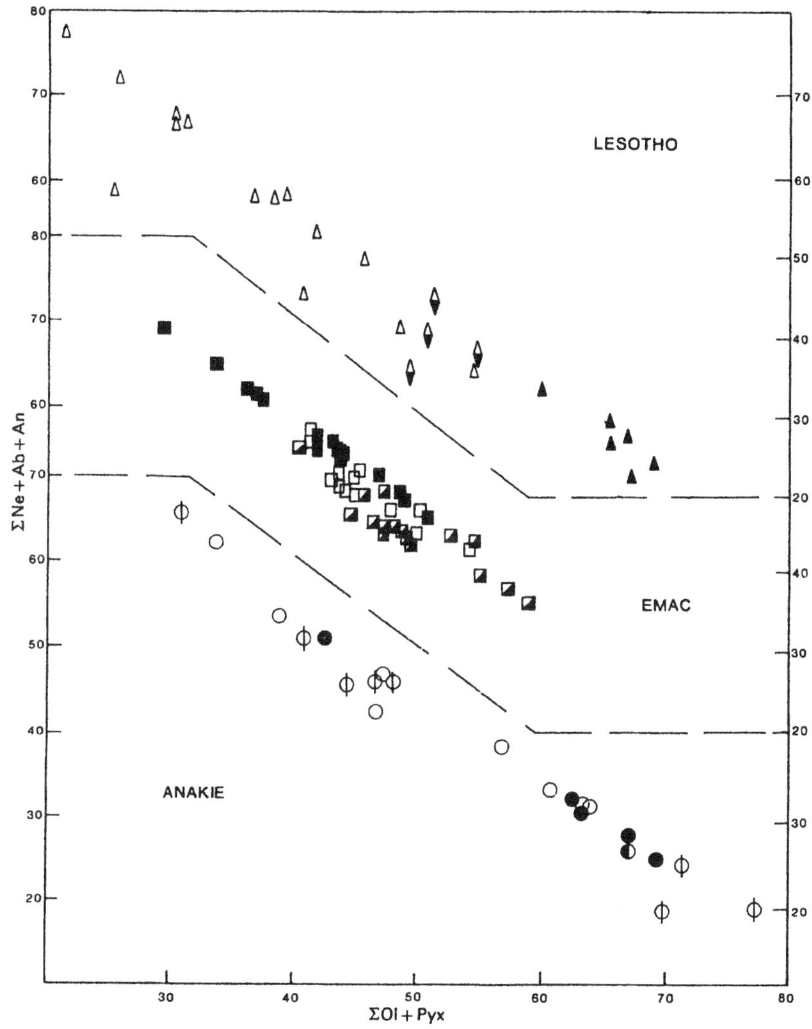

Figure 12.9 Normative mafic versus felsic components in the garnet-bearing xenoliths, illustrating the effect of bulk composition on the presence or absence of plagioclase. Symbols as in Figure 12.7.

but feldspar-poor (Figure 12.7). The rocks show a range of normative feldspar/clinopyroxene ratios at essentially constant olivine (Figure 12.8), ranging from troctolites to gabbros. The higher normative Hy in the more Cpx-rich rocks may originally have resided in subcalcic Cpx. The range of compositions is similar to that seen in some large layered intrusions. Rogers and Hawkesworth (1982) have presented trace-element data consistent with a cumulate origin for the Lesotho granulite xenoliths.

In the Anakie suite, each of the rock types spreads across the whole basalt tetrahedron (Figure 12.7). However, relict phases indicate that primary Opx was rare in this suite, and most normative Opx was in subcalcic Cpx. Figure 12.8 shows a trend similar to the Lesotho suite, but with a smaller range, more scatter and no troctolitic compositions. Wass and Hollis (1983) suggested that many Anakie rocks represent frozen magmas or cumulates. The Anakie rocks may have evolved in smaller, less fractionated intrusions than the Lesotho suite. The complete compositional overlap between the garnet-bearing and garnet-free granulites implies that P–T conditions have been the major control on the appearance of garnet. Geothermobarometry suggests that the garnet-free granulites are derived from shallower depths.

The EMAC suite is dominated by Hy($\pm$Qtz)-normative rocks (Figure 12.7). In Figure 12.8 this suite shows much more scatter than the other two. It is a very heterogeneous suite and genetic relations among the different groups are not obvious. While the compositional variations in the Lesotho and Anakie suites are consistent with relatively well-defined episodes of underplating by basaltic magmas, the processes that produced the EMAC suite appear to be more complex.

Figure 12.9 emphasises the interaction between P–T regime and bulk composition in controlling the disappearance of plagioclase in these suites. Each suite plots in a linear band on this figure, with granulites toward the Ne + Fsp axis and eclogites and pyroxenites toward the Ol + Pyx axis. In the Anakie suite feldspar is found in rocks with normative (Ol + Pyx) as high as 63%; the corresponding figure for Lesotho is 48%, reflecting the lower T at any P of the Lesotho geotherm (Figure 12.2). In both of these suites there is little overlap between the feldspar-bearing and feldspar-free compositions. The EMAC suite, however, shows more overlap, with feldspar-bearing compositions down to 45% (Ne + Ab + An). This probably reflects both the greater range in depth of the EMAC rocks, and their greater range in normative Hy.

12.6 Discussion

12.6.1 *Granulite–eclogite transitions and eclogite facies*

The three xenolith suites discussed here illustrate stages in the transformation of gabbroic rocks (*s.l.*) to granulites and then to eclogites. The early microstructural and mineralogical changes are shown in the Anakie suite, as gabbros are converted to garnet granulites. The EMAC suite shows the progressive growth of garnet at the expense of plagioclase and the transformation of granulites to eclogites. These changes are interpreted as reflecting progressive cooling, following episodes of igneous intrusion in the lower crust and uppermost mantle. The Lesotho suite contains little microstructural evidence of its prehistory, but is linked to the other two by similarities in bulk composition and mineral-chemistry relations. It is interpreted as the end result of similar processes, frozen at still lower T.

For a given P–T regime, the appearance of garnet and the disappearance of plagioclase is controlled by bulk composition. This is illustrated for the three suites described here, which represent three distinct geothermal gradients.

At higher temperatures for a given P, plagioclase is stable to greater depths and pyroxenes are rich in CaTs. At lower temperatures and comparable pressures, a series of sliding reactions take place with CaTs + An → Grt and Ab → Jd. Therefore the reactions leading from granulite to garnet pyroxenite (or eclogite) result in different mineral compositions for different P and T conditions. Under the range of P and T conditions represented by the three xenolith suites described, the normal range of basaltic compositions will result in a mix of varying proportions of granulites and garnet clinopyroxenites (or eclogites). Garnet clinopyroxenites will characterise the higher-T environments and eclogites the lower-T ones. Garnet websterites will form from the more magnesian compositions, poor in normative feldspar components.

It is important to note the equivalence in mineral chemistry of the eclogites and the granulites from the EMAC and Lesotho suites. The two rock types contain a similar range of garnet compositions; more significantly, both eclogites and granulites contain a similar range of clinopyroxenes, from sodic augite to omphacite; the only real difference is the occurrence of plagioclase, by definition, in the granulites. This difference is, in turn, related to relatively small differences in bulk composition (Figure 12.9).

White (1964) showed that the Jd:Ts = 2:1 line in Figure 12.3 effectively distinguishes clinopyroxenes from eclogite-and granulite-facies rocks sampled in surface outcrops. However, the occurrence of omphacite-bearing granulites in xenolith suites shows that this relationship does not hold at depth. Furthermore, the correlation of Jd in clinopyroxene with Ab in plagioclase, predicted by experimental work, is seen in xenoliths (Figure 12.5) but apparently not in outcropping granulites. The Cpx compositions of granulites may have been altered by re-equilibration with plagioclase during uplift of granulite terranes, whereas re-equilibration of eclogite pyroxenes was delayed by the problems of nucleating plagioclase. Xenoliths, on the other hand, represent quenched samples, transported from depth on a time scale of hours, and provide a clearer picture of equilibria at depth.

The similarities in mineral chemistry and in geothermobarometry leave little doubt that eclogite and granulite ($\pm$ garnet websterite) coexist side-by-side in the lower crust and upper mantle beneath at least some cratonic areas. Therefore although the absence of plagioclase may be essential for the definition of eclogite as a petrographic term, it has no intrinsic significance in defining an eclogite facies, unless that definition includes reference to very specific bulk compositions.

12.6.2 *Origin of protoliths of eclogite xenoliths*

Several studies of mafic granulite xenoliths have concluded that the protoliths of these granulites were formed by magmatic underplating, which involved intrusion of basaltic magma near the crust–mantle boundary (review by Griffin and O'Reilly, 1987b; Rudnick *et al*, 1986). The suites described in this Chapter demonstrate how such granulites may be converted to eclogite during thermal relaxation. We therefore suggest that magmatic underplating is the probable ultimate source of most mafic rocks in the lower crust and uppermost mantle.

A number of recent studies have proposed a subduction-related origin for eclogites derived from the deeper mantle (Chapter 13). A shallow-subduction origin has been proposed for lawsonite-bearing eclogite xenoliths from the Navajo province of the southwestern USA (Helmstaedt and Schultze, 1979), and Ater *et al* (1984) describe possible subduction-related eclogites from depths as shallow as 80 km. Ancient subducted material may thus be more common than previously recognised. However, if such material is held at depth for a time scale of the order of 100 Ma, internal heating of the slab should promote metamorphic reactions which will ultimately destroy mineralogical and microstructural evidence for a subduction origin. Examples of such reactions are described by Harley and Green (1981) in eclogites from Moses Rock. In this case, the subduction-related origin of the protoliths may only be detectable by isotopic trace-element geochemical studies.

12.6.3 *Stability of eclogites*

The stability of eclogite in the lower crust and uppermost (< 100 km) mantle can be understood by modelling the thermal evolution of a lithospheric column which has undergone magmatic underplating. During underplating the advective transport of heat by magmas produces a high, strongly convex geotherm comparable to that established for eastern Australia (Figure 12.2). Other factors, such as crustal thinning and proximity to a spreading rift, may enhance heat flow even more (e.g. Baldridge, 1979; Padovani and Carter, 1977; Amundsen *et al*, 1987). Under such P–T conditions, all mafic compositions in the lower crust and upper mantle lie in the stability field of granulite rather than eclogite.

When underplating ceases, thermal relaxation will move the ambient T at each depth toward the steady-state conductive geotherm. A geotherm typical of cratonic areas is shown in Figure 12.2. Cooling from the SEA geotherm to the cratonic geotherm will require in the order of 100 Ma as the time constant for such cooling is believed to be about 10 Ma (Sass and Lachenbruch, 1979). As this cooling takes place, an increasing proportion of mafic compositions will enter the eclogite stability field (Griffin and O'Reilly, 1987b).

Estimates of the position of the granulite–eclogite transition depend strongly on the bulk compositions considered, and on the extrapolation of experimental studies to lower temperature. Wood (1987) used thermochemical calculations to argue for a non-linear extrapolation, and concluded that a quartz-tholeiite composition would retain the granulite-facies assemblage at lower-crustal pressures, even on a $40\,mW\,m^{-2}$ geotherm. However, the great bulk of lower-crustal mafic xenoliths have olivine- or olivine + nepheline-normative compositions (Figure 12.7). In these compositions the eclogite assemblage will be stable at lower pressures, as demonstrated by the granulite–eclogite transitions described here.

If thermal stability alone were the controlling factor, then eclogite would be abundant wherever mafic rocks occur in the lower crust of cratonic regions. However, kinetic factors are obviously important, as shown by the arrested reactions seen in many of the EMAC xenoliths. Reactions that occur with declining T will tend to be sluggish (Chapter 5), and the growth of corona phases tends to shield reactants from one another. Both factors could prevent reaction rates from keeping pace with cooling rates. Austrheim and Griffin (1985) and Austrheim (1987) have demonstrated the metastable persistence of dry granulite assemblages under conditions where eclogite was the stable assemblage. In these cases the eclogite-forming reactions only took place in domains which underwent deformation and/or fluid infiltration during the eclogite-forming metamorphism. Austrheim (1987) has, on this basis, suggested that large volumes of the lower crust could retain granulite-facies mineralogy while eclogite is restricted to narrow zones.

12.6.4 *Occurrence of eclogites*

Griffin and O'Reilly (1987a) showed in a compilation of data on lower crustal xenoliths that the composition of lower crustal xenoliths is dominantly mafic. Therefore, it would be predicted that eclogites would constitute a major proportion of the lower crust in areas with cratonic geotherms and a significant proportion in regions with intermediate geotherms such as that for the EMAC suite. A dominance of mafic granulites should be characteristic of regions with high heat flow. The latter is true, but xenoliths of lower crust/uppermost mantle eclogites are rare even in cratonic regions. In some cratons, mainly Proterozoic, mafic granulite xenolith suites are known but do not carry eclogites (Griffin and O'Reilly, 1987a). In such cases, the absence of eclogites may reflect kinetic factors. However, in other regions—notably the Kaapvaal craton of southern Africa, both granulites and lower crust/uppermost mantle eclogite xenoliths are very rare (Griffin *et al*, 1979; cf. Dawson and Smith, 1987), even though xenoliths of other crustal and mantle rock types are abundant in kimberlitic and basaltic diatremes. Does this mean that mafic rocks are rare in the lower crust of such cratons, or have they simply not been sampled?

The erratic sampling of deep-seated rocks by ascending magmas is a ubiquitous problem in xenolith studies (Griffin and O'Reilly, 1987a). It is impossible to be confident about just how representative the entrained xenoliths are of the vertical rock columns beneath the eruptions. However, in the absence of data to the contrary, it seems reasonable to assume that overall, a statistically meaningful sampling of deep-seated rocks is provided by xenolith suites. The lower crust probably has distinct rheologic properties with very low yield strengths (e.g., Kirby, 1985; Carter and Tsenn, 1987) compared with stronger mantle wall-rocks. The presence of mechanically weak zones in the lower crust (due to ductile behaviour and semi-brittle fracture) is indicated by rock-mechanical measurements and calculations (Carter and Tsenn, 1987) and by marked deformation features in lower crustal xenoliths (O'Reilly *et al.*, in press; Amundsen *et al*, 1987; Griffin and O'Reilly, 1987a). However, this should not affect entrainment of fragments in ascending magmas, as brittle fracture would prevail on the short time scales involved (minutes to hours).

Several lines of geophysical evidence suggest that mafic rocks may be relatively uncommon in the lower crust of some Archean cratons. Wasilewski and Warner (1988) and Wasilewski and Weidner (in press) have shown that mafic granulites are highly magnetic, while eclogites, amphibolites, and most mantle-derived ultramafic rocks, lacking magnetite, are essentially non-magnetic. Satellite-recorded long-wave length magnetic data (MAGSAT) show that, for example, the Kaapvaal craton of southern Africa, and the Yilgarn block of Western Australia, are characterised by low magnetisation (Mayhew and Johnson, 1987; von Frese *et al*, 1987). This could in principle be explained by high heat flow (and hence a shallow Curie isotherm), but measured surface heat flows in both areas are low ($\leqslant 40\,mW\,m^{-2}$; e.g. Cull, 1982; Ballard and Pollack, 1988), so that the Curie-point isotherm lies below the Moho. Therefore, the crustal column must consist largely of non-magnetic rock types; these could be either felsic rocks (± amphibolite) or eclogites. O'Reilly and Griffin (1985) and Griffin and O'Reilly (1987b) have noted that an eclogitic lower crust would be seismically, as well as magnetically indistinguishable from the upper mantle. However, both the Kaapvaal craton and the Yilgarn block have negative Bouguer gravity anomalies (Fraser and Pettifer, 1980; F. Arnott, personal communication, 1988) which are not consistent with the presence of an eclogitic lower crust. The geophysical evidence—low magnetisation, low heat flow, and a low-density crustal column—therefore suggests that at least in these Archean cratons, the lower crust is not dominated by mafic rocks.

If then there are truly few eclogites in the lower crust of Archean cratons, perhaps

(a) Mafic rocks were never abundant in these regions because underplating by basaltic magmas was not widespread in the Archean, as distinct from the or Proterozoic and Phanerozoic,

or

(b) Eclogites were indeed present in the lower crust but became delaminated and sank due to their great density relative to the underlying mantle (e.g. McKenzie and O'Nions, 1982; Toft *et al*, 1988). Delamination would be accelerated by tectonic thrusting because of the distinct rheological properties of the lowermost crust (e.g. Passchier, 1986; Carter and Tsenn, 1987).

12.7 Conclusions

Several xenolith suites show complete chemical and petrographic gradation between granulites (Cpx ± Opx + Grt + Pl) and eclogite. Geothermobarometry indicates that both granulites and eclogites in these suites were derived over a depth range that encompasses the lower crust and uppermost mantle ($\leqslant$ 100 km). Petrographic evidence of arrested reactions in xenoliths confirms the evolution of gabbros to granulites to eclogites with declining temperature. Eclogite and granulite xenoliths in each suite have essentially identical ranges of clinopyroxene and garnet compositions, and the presence or absence of plagioclase is controlled primarily by bulk composition.

In areas of elevated geothermal gradient, such as rifts or zones of active magmatic underplating, all mafic rocks in the lower crust and uppermost mantle will have granulite-facies mineral assemblages. When such areas have cooled toward a cratonic geotherm, the same mafic rocks will consist of a mixture of granulites and eclogites, with the proportions determined by the range of bulk compositions and the ambient geothermal gradient. Complete conversion of granulites to eclogites during cooling to a cratonic geotherm may be hindered in many cases by kinetic factors.

Acknowledgements

The authors thank the BMR, Canberra for access to some of the EMAC thin sections, and Richard Arculus for a copy of his (and nine co-authors') 1982 preprint of their much delayed paper from the Third International Eclogite Conference. Stockdale Prospecting generously provided field support. S.Y. O'Reilly had financial support from ARC grant no. A38831447, the Macquarie University Research Grant scheme and Stockdale Prospecting Pty Ltd.

13 High temperature and ultra-high pressure eclogites

J.B. DAWSON and D.A. CARSWELL

13.1 Introduction

Following the classification outlined in Chapter 1, high temperature (HT) eclogites are taken as those that have equilibrated at $T > 900°C$. Observed terrestrial heat flow values effectively limit formation of HT eclogites to an upper mantle origin. This being the case, equilibration also takes place at higher pressures than for LT and MT eclogites. For example, under the ambient T conditions pertaining to sub-cratonic upper mantle (40 m W m^{-2} continental geotherm), stabilisation of HT eclogites is only to be expected at $P > 36$ kbar, equivalent to depths > 115 km. Such ultra-high P eclogite facies conditions are evidenced by mineral parageneses which may contain such HP polymorphs as diamond and coesite in rocks of appropriate bulk chemistry. Some rare occurrences of HT eclogite are also known in association with garnet peridotites in tectonically emplaced 'alpine-type' peridotite bodies of upper mantle derivation in the HP crustal terrains that represent old plate suture zones, notably in the Bohemian Massif (Chapter 11). However, most available samples have been transported to the surface by kimberlite, lamproite and alkalic basic igneous activity. In the context of older classifications, they are essentially the Group A eclogites of Coleman *et al* (1965).

By the latter half of the nineteenth century, the eclogite occurrences of Central Europe had been recognised as being the products of high-pressure metamorphism, but the realisation of the extreme conditions of formation for some bimineralic garnet-clinopyroxene rocks only followed from the finding of diamond in eclogite blocks in the kimberlite of the Newlands pipe in South Africa (Bonney, 1899). Beck (1899) recognised that these kimberlite eclogites were sufficiently different from those found in Europe as to suggest that they should be termed 'griquaites'. The differences recognised by Beck were re-emphasised in later eclogite classifications (Coleman *et al*, 1965; Smulikowski, 1972) in which it was purported that, compared with the phases in the eclogite lenses occurring within granulite terrains or accompanying blueschist facies metasediments in Alpine-type fold belts, the garnets in the kimberlite eclogites are generally richer in the pyrope molecule, and the clinopyroxenes have higher diopside: jadeite ratios. However, the range in chemistry of the phases in the kimberlite eclogites is greater than suggested in these earlier classifications. More importantly, on the basis of Fe–Mg partitioning between coexisting garnet and pyroxene, Banno (1970) showed that the kimberlite eclogites have equilibrated at higher temperatures, a suggestion reinforced by many subsequent studies.

The nomenclature of garnet-pyroxene rocks in kimberlites is confused. Some authors have used the strictly mineralogical term 'garnet pyroxenite' instead of eclogite, whereas others (e.g. Smulikowski, 1972) use the term griquaite as synonymous with 'kimberlite eclogite'. Nixon and Boyd (1973) also use 'griquaite', but in its original sense in that they recognise certain bimineralic rocks, comprising aggregates of megacrysts of chemically distinctive garnet and diopside of the megacryst suite (see Dawson 1980, Chapter 8), that are cognate with the host kimberlite; their chemistry is unlike that of the phases in eclogites. The term 'eclogite' is retained here although, as will become apparent, the term covers a group of rocks in which the fabrics, textures, mineralogy, phase chemistry, isotope chemistry and temperatures of equilibration are all variable; in particular, the bulk composition of some eclogites departs from the broadly basaltic composition referred to in Chapter 1.

In addition to eclogites themselves, garnets and pyroxenes compositionally similar to those in HT eclogites occur as inclusions in diamonds (reviewed by Meyer, 1987); they occur most often singly and, less commonly, in pairs. Other phases included with 'eclogite' garnet or pyroxene within single diamonds are graphite, kyanite, ruby corundum, sanidine, coesite (quartz), ilmenite, rutile, and sulphides (Meyer, 1987), moissanite (SiC) (Moore *et al*, 1986), and two rare potassium silicates (Bulanova *et al*, 1988). This group of chemically-distinctive HP phases is referred to as the eclogite-inclusion paragenesis and is distinct from phases of the ultrabasic-inclusion parageneses which are generally higher in both magnesium and chromium (Meyer, 1987). It was once believed that individual diamonds contain inclusions, exclusively, of either the eclogite or the ultrabasic paragenesis but there is now some evidence for mixed paragenesis inclusions (Prinz *et al*, 1975; Hall and Smith, 1984; Griffin *et al*, 1988; Moore and Gurney, 1989; Otter and Gurney, 1989). Inclusions of the eclogite paragenesis predominate in diamonds from the Premier, Orapa and Dokolwayo kimberlites, southern Africa (Tsai *et al*, 1979a; Gurney *et al*, 1984a, 1985; Daniels and Gurney, 1989), in the Sloan diatremes of Colorado (Otter and Gurney, 1989), and also in alluvial diamonds from the Urals (Sobolev, 1977). A variant of the eclogite paragenesis is the so-called calc-silicate paragenesis of grossular-rich garnets, diopside and coesite found in alluvial diamonds from SE Australia (Sobolev, 1984). A further association of diamond with eclogite minerals is found in polycrystalline aggregates from the Orapa and Star kimberlites (Gurney and Boyd, 1982; Gurney and Hatton, 1989).

The purpose of this Chapter is to review the occurrence, mineralogy and possible mode of origin of these very-high *P–T* eclogite parageneses from kimberlite and lamproite and, finally, to draw comparisons between these exotic eclogites and other high *PT* rocks occurring *in situ* in metamorphic terrains.

13.2 Petrography and textures

The eclogite xenoliths are rounded or discoidal blocks measuring up to 50 cm long. The inclusions in diamond are of the order of 100 μm in size, and display a wide variety of shapes from equant cubo-octahedral to tabular and flat (Meyer, 1987).

In many kimberlites eclogite xenoliths are rare but in kimberlites such as the Roberts Victor (RV), Bellsbank (Bobbejaan) and Orapa (southern Africa), Lake Ellen (USA) and Zagadochnaya (USSR) they form a high proportion of the xenolith population—98% in the case of RV (Hatton and Gurney, 1987). Well-studied eclogite suites are from RV (many studies—references at end), Orapa, Botswana (Shee and Gurney, 1979;

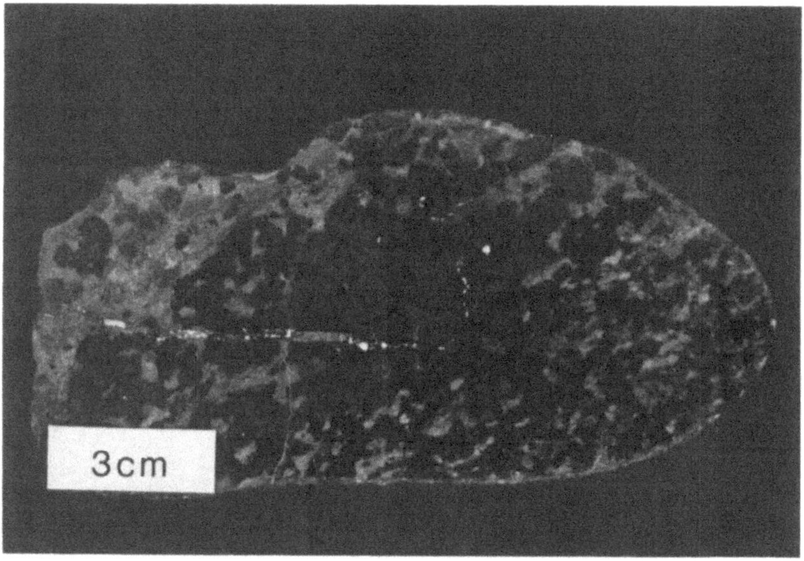

(a)

(b)

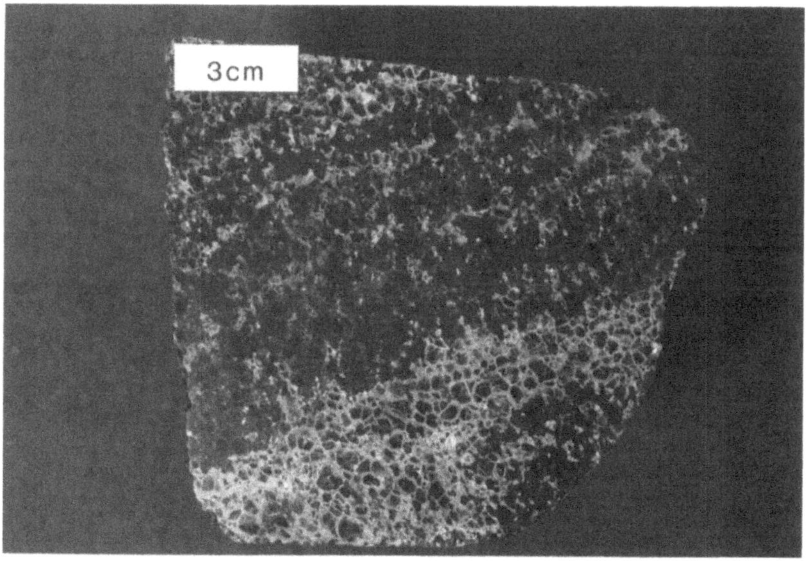

Figure 13.1 (*Continued*)

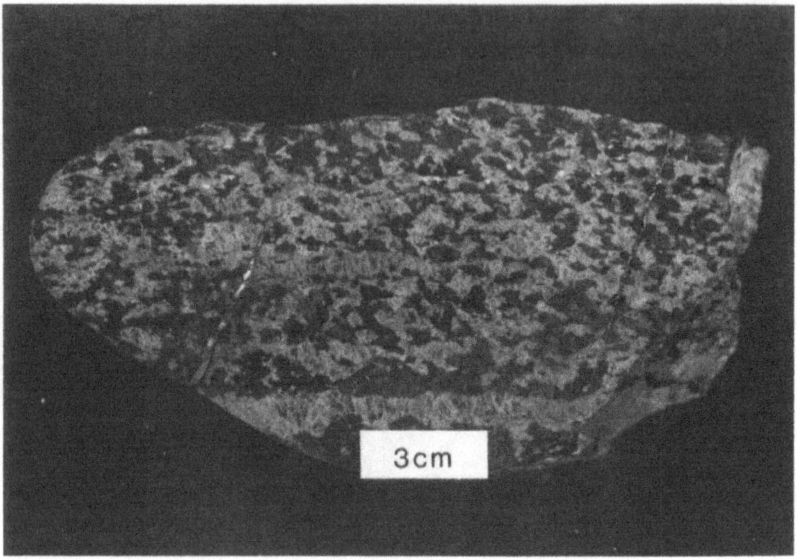

(c)

(d)

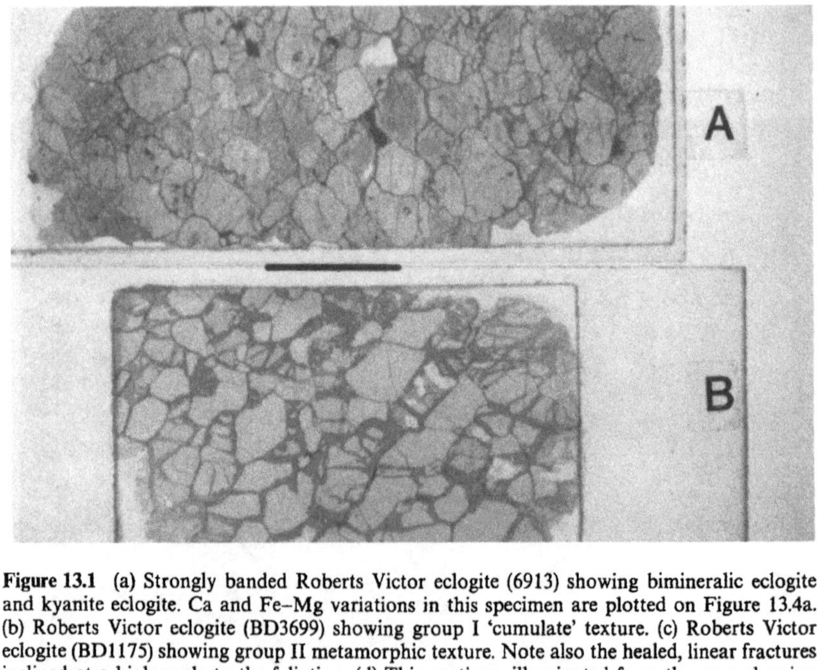

Figure 13.1 (a) Strongly banded Roberts Victor eclogite (6913) showing bimineralic eclogite and kyanite eclogite. Ca and Fe–Mg variations in this specimen are plotted on Figure 13.4a. (b) Roberts Victor eclogite (BD3699) showing group I 'cumulate' texture. (c) Roberts Victor eclogite (BD1175) showing group II metamorphic texture. Note also the healed, linear fractures inclined at a high angle to the foliation. (d) Thin sections, illuminated from the rear, showing (A) group I texture (Specimen BD1180), and (B) group II texture (BD1191). Scale bar is 10 mm.

Robinson *et al*, 1984), Bellsbank/Bobbejaan, South Africa (Bishop *et al*, 1978; Carswell *et al*, 1981; Smyth *et al*, 1984; Smyth and Caporuscio, 1984), Lake Ellen, USA (McGee and Hearn, 1984) and in the State Line and Iron Mountain Districts of northern Colorado and southern Wyoming, USA (Ater *et al*, 1984). Eclogite paragenesis diamond inclusions predominate at Orapa and Sloan but, paradoxically, they are very subordinate to ultrabasic paragenesis inclusions at RV.

Although eclogites are essentially bimineralic garnet-pyroxene rocks, many variants can be recognised in hand-specimen due to (i) presence or absence of mineral banding, (ii) differences in the proportions and colour of the garnets and pyroxenes, the colours reflecting compositional differences, (iii) variations in the textures (including different degrees of metasomatic alteration), and (iv) the presence (in some cases) of distinctive accessory minerals.

In addition to the garnet and omphacitic pyroxene, other primary phases in eclogites are diamond, graphite, orthopyroxene, phlogopite, amphibole, rutile, sphene, kyanite, corundum, sulphides, sanidine and coesite (the HP polymorph of SiO_2). Two additional very rare K-bearing phases—the potassium sulphide djerfisherite and a calcic variety of yagiite (a rare Na, K, Mg aluminosilicate first found in the Kolomero meteorite) have been found in eclogite xenoliths from the East Udachnaya kimberlite in Siberia (Govorov *et al*, 1986). Eclogite diamonds are generally sharp-edged octahedra (Robinson, 1979), contrasting with the generally more rounded diamonds that occur in kimberlite.

The most obvious fabric in many eclogite hand specimens is mineralogical banding, which results from differing proportions of phases within individual bands; for example, kyanite eclogite may be interbanded with kyanite-free eclogite (Figure 13.1a). This interbanding, which may be on a 3–5 cm scale, has been interpreted by several authors (e.g. MacGregor and Carter, 1970; Hatton and Gurney, 1987) as suggestive of an original igneous cumulate origin. In many cases extensive subsolidus recrystallisation has modified the original textures, and Dawson (1980) has suggested that the mineral banding in eclogites is equally compatible with metamorphic banding as with igneous cumulate layering. Another form of inhomogeneity in RV eclogites arises from ovoids of kyanite eclogite occurring within bimineralic eclogite bands (Hatton, 1978); the origin of these ovoids is obscure, but their shape and occurrence is not inconsistent with the boudins found in metamorphic formations. A further 'fabric' found in some eclogite xenoliths is the presence of planar fractures that are generally at a high angle to observed banding or foliation. Xenoliths bounded by angular faces, as opposed to the usual rounded surface, result from breakage along these fractures.

Replacement of garnet and inferred former orthopyroxene by metasomatic mica and amphibole is a common feature, and the pyroxene, particularly in the peraluminous eclogites, is replaced by fine aggregates in some of which secondary Ca-Al pyroxene, albite, zoisite and zeolites have been identified.

Coesite was first reported in a sanidine-bearing RV grospydite (Smyth and Hatton, 1977). The coesite is partially replaced by quartz, and the surrounding garnet has radial cracks owing to volume expansion that results from the HP $\rightarrow$ LP inversion. Coesite has also been identified, and its former presence inferred from quartz pseudomorphs, in eclogites, kyanite eclogites and grospydites from the Udachnaya kimberlite (Ponomarenko and Spetsius, 1978) and the RV, Lace, Jagersfontein and Zagadochnaya kimberlites (Schulze and Helmstaedt, 1988); in addition to simple coesite grains, some coesite is reported as occurring in graphic or vermicular intergrowths with garnet.

Diamond, when present, occurs mainly as high-temperature octahedra, and a relatively common pseudo-dodecahedral form is the result of superposition of octahedral growth layers of progressively decreasing areal extent (Robinson *et al*, 1984). Graphite, whether forming thin veneers on diamond in diamond-graphite eclogites or as single grains in graphite eclogites has rounded hexagonal, occasionally embayed, outlines and tabular habit (Robinson *et al*, 1984).

Another fairly common feature is the presence of fresh or partially-devitrified intergranular glass veinlets (Switzer and Melson, 1969; Dawson, 1980; Smyth *et al*, 1984). Except for elevated K_2O contents, the composition of the glasses is similar to that of adjacent clinopyroxenes, which have partially-melted, 'spongy' margins. Plagioclase, augite and spinel microlites are present in some glasses, and spindle-shaped secondary corundum replaces partially-melted kyanite, if present. Some veins contain rounded vesicles now filled with calcite or zeolites. The glasses are attributed to decompression melting during rapid ascent within the kimberlite.

13.3 Classifications of kimberlite eclogites

Using the mineralogy alone, Dawson (1980) subdivided eclogites into (a) eclogites i.e. those containing no distinctive accessory minerals and therefore requiring no adjectival tag, although such eclogites have been named 'metaluminous' by Ater *et al*, (1984); (b) peraluminous eclogites containing kyanite and/or corundum, of which the grospydites, in which the garnet contains $> 50\%$ of the grossular molecule, are a sub-group; and (c) carbonaceous eclogites containing diamond and/or graphite. It will be appreciated that there is an overlap in that some peraluminous eclogites also contain diamond. Additional mineral names can be added in the rocks containing distinctive phases e.g. diamond-kyanite eclogite, coesite-sanidine grospydite. Recently, Hatton and Gurney (1987) observed that a significant number of RV eclogites contain discrete patches of phlogopite between the garnets and pyroxenes; they interpret these mica patches to represent metasomatically-altered former orthopyroxene. If this interpretation is accepted, then their category of garnet websterite (equivalent to orthopyroxene-bearing eclogite) is much more common in the RV kimberlite than hitherto appreciated; by inference these garnet websterites would be relatively rich in MgO. Further support for garnet websterite as an Mg-rich member of the eclogite suite comes from diamond inclusions in the Monastery kimberlite, where highly-aluminous enstatites (Al_2O_3 4.1–11.3 wt%, Mg/(Mg + Fe) 0.83–0.85) occur together with eclogite garnets and pyroxenes (Moore and Gurney, 1989). Furthermore, diamondiferous garnet websterites, again representing the Mg-rich end of the eclogite spectrum, occur in the Udachnaya kimberlite, Siberia (Ponomarenko *et al*, 1980).

Another rock type that has been linked to eclogite is alkremite; it consists mainly of pale-pink Cr-poor pyrope-rich garnet, together with lesser amounts of dark-green spinel and, in some specimens, dark-violet corundum. First found in the Udachnaya pipe, Yakutia (Ponomarenko, 1975), variants have subsequently been found in the Bellsbank and Jagersfontein kimberlites, South Africa (Nixon *et al*, 1978; Exley *et al*, 1983; Mazzone and Haggerty 1989), and from the Moses Rock 'kimberlite', Utah, USA (Padovani and Tracy, 1981). The name is the Russian mnemonic for Al, Si and Mg which dominate in bulk analyses of the original samples (Ponomarenko, 1975).

A classification scheme based on the textures of eclogites, that has subsequently been found to be fraught with complications, was first used by MacGregor and Carter (1970) in a pioneering study on RV eclogites. They recognised two basic textural groups—group I in which subhedral or rounded garnets are set in an interstitial matrix of clinopyroxene, with occasional large clinopyroxenes enclosing rounded garnets poikilitically; and group II, consisting of anhedral, sometimes-elongate, straight-edged garnets and pyroxenes that have an interlocking fabric (Figure 13.1b, c, d). They also noted that the pyroxenes in group I rocks are much more altered than those in group II eclogites, and that the garnets in group I are cloudy, contrasting with the clear group II garnets; a further distinction is the presence of oriented rutile needles in the garnets and pyroxenes in the group II rocks. MacGregor and Carter interpreted the group I rocks as igneous cumulates from a series of evolving liquids that eventually consolidated as the group II eclogites, thereby explaining the compositional continuum within and between the two texturally-different eclogite groups. This interpretation was held to be validated by a subsequent oxygen isotope study (Garlick *et al*, 1971).

This classification has been followed by some later workers, but more recent studies have shown that, even within the RV suite, the textures are more diverse and complex than initially believed. For example, Hatton and Gurney (1987) point out that all gradations exist between the two groups and that individual samples may grade from one textural group to the other over a distance of a few centimetres; they recognise fourteen textural types, though accepting that those with the texture of MacGregor and Carter's group I comprise 60% of the rocks; the remaining 40%, equivalent to MacGregor and Carter's original group II, are texturally much more diverse and thirteen new groups have been recognised from the colour (and, hence, composition) and shape of the minerals and the textures. The textures include lamellar exsolution textures and Smyth and Caporuscio (1984) also report that, in the Bobbejaan eclogite suite, all eclogites showing garnet or kyanite exsolution from pyroxene are type II eclogites. Even allowing for the textural diversity of the group II rocks, other textural features do not fit into the simple group I/group II classification with its explicitly igneous interpretation. For example, Lappin and Dawson (1975) describe a RV eclogite with an apparent group I texture, in which the subhedral garnets replace corroded cores of strained and kinked kyanite—a feature that results from interaction between the original phases in a previously-deformed kyanite-pyroxene rock, but incompatible with an igneous cumulate origin. Strained kyanite also occurs in a RV grospydite, in which the garnet is attributed to exsolution from an original high-Al clinopyroxene (Lappin, 1978), and replacement of kyanite by garnet (the so-called 'diablastic texture') is also recorded in grospydites from the Zagadochnaya kimberlite, USSR (Sobolev *et al*, 1968). In short, some rocks that are now eclogites, were originally garnet-free.

Another RV specimen which illustrates the pitfalls of a simple textural classification scheme is described by Harte and Gurney (1975); the specimen (JJG41) shows two main microstructural domains (i) a part with large clinopyroxene grains, up to 6 cm and containing multiple lamellae and equant grains of exsolved garnet; and (ii) a granular part with equant grains of garnet and pyroxene. The granular part is texturally a group I eclogite in the MacGregor and Carter classification, whereas the lamellar exsolution part would fall within the group II category of Hatton and Gurney (1987); moreover, oriented rutile needles (a criterion of MacGregor and Carter's group II eclogites) are present in globular garnets surrounding the large pyroxenes, but absent in small angular garnets within the same pyroxenes (Sautter and Harte, 1988). Thus, the classification and understanding of eclogite textures is still imperfect and genetic interpretations based on textural evidence must still be treated with circumspection.

Notwithstanding the drawbacks of the group I/group II classification, it appears that diamond is confined to eclogites with group I textural characteristics (Gurney *et al*, in press), although graphite occurs in both textural types (Robinson *et al*, 1984).

A further classification, linking textural and phase chemistry criteria, is based upon the Na_2O content of garnet and the K_2O content of the clinopyroxene. In a study of RV eclogites that have unambiguous group I or group II textures, McCandless and Gurney (1989) found that most (81%) of group I garnets contain >0.09 wt% Na_2O, whilst 89% of the group II garnets contain less than this value. For K_2O in clinopyroxenes, 94% of group I pyroxenes contain >0.08 wt%, and 76% group II pyroxenes contain less than these concentrations. The percentages of cases that satisfy these compositional cut-offs is improved when strongly-metasomatised (phlogopitised) group I eclogites, and kyanite eclogites that are texturally group II even though they contain high-Na garnets and high-K pyroxenes, are excluded from the statistics. Nonetheless, McCandless and Gurney do note that there are cases where high-Na group I garnets coexist with pyroxenes in which the K_2O content is <0.08 wt%.

Before the McCandless–Gurney proposed classification, it had been reported that RV diamondiferous eclogites contain high-Na garnets and high-K pyroxenes (Reid *et al*, 1976; Robinson *et al*, 1984); these earlier data, supplemented by new analyses on diamondiferous eclogites from RV, Orapa, Star, Jagersfontein, Excelsior, Mitchemanskraal and Sloan 2 kimberlites (McCandless and Gurney, 1989), show that, on the McCandless–Gurney compositional criteria, all diamondiferous eclogite garnets fall within the group I compositional range, and 67% of the pyroxenes fall above the 0.08 wt% K_2O group I cut-off. Although it remains to be seen how valid this classification will be in the case of other eclogite suites, it does not appear to be applicable to the Bobbejaan suite investigated by Smyth and Caporuscio (1984).

13.4 Mineral chemistry

There is now a considerable amount of data on the chemistry of minerals in these high-PT eclogites, in particular for the large, compositionally variable suite of eclogite xenoliths from the Roberts Victor Mine kimberlite. Consequently RV data figure prominently in this review, although reference is made to other eclogite xenolith suites where they provide important additional information. Comparative data are also included on mineral phases in eclogite parageneses included in diamonds.

13.4.1 *Garnet*

Eclogite garnets show a wide range in their Ca, Mg and Fe contents, and in a statistical analysis of xenolith garnets in general Dawson and Stephens (1975) identified six distinct compositional groups on their TiO_2, Cr_2O_3, FeO, MgO and CaO contents. There is a compositional continuum from deep-red high-pyrope varieties, through pyrope-almandines (usually red or red-brown in colour) to those with higher Ca contents and orange to yellow colour in the peraluminous eclogites and grospydites. The high-Mg garnets are generally low in Cr_2O_3, which distinguishes them from Mg- and Cr-rich garnets coexisting with Cr-diopside in 'eclogites' (containing small amounts of forsterite, enstatite and chromite) that are probably modal variants of garnet lherzolite. However, diamondiferous garnet pyroxenites from the Udachnaya pipe, Siberia, contain high-Mg phases (Mg/(Mg + Fe) is 0.72, 0.84 and 0.91 for garnet, diopside and enstatite respectively) similar to those in garnet lherzolites, but, unlike the lherzolite phases, they contain the very low Cr_2O_3 but high TiO_2 concentrations (0.08 and 1.13 wt% respectively in the garnet) found in the eclogite suite as a whole (Ponomarenko *et al*, 1980).

Garnets from alkremites range from Prp_{50-85}, Alm_{6-21} and Grs_{3-37}; they are typically low in Cr_2O_3—< 0.4 wt%. Although the more Fe- and Ca-rich ones can be matched with those in eclogites, two from Udachnaya (Prp_{85-86}) are the most magnesian found in eclogites *s.l.*

The garnets from many other eclogite suites fall on the trend shown on Figure 13.3. An exception is a high-Fe sample (Alm_{41}) in an eclogite from the Kao kimberlite, Lesotho, that equilibrated at about 850°C; Shervais *et al* (1988) point out the similarity of this phase chemistry to that in Coleman *et al*'s Group C eclogites, and it can also be matched in several lower-crustal eclogites from Lesotho (Griffin *et al*, 1979). Another exception is very high-Ca garnets (CaO 28–32 wt%) included in SE Australian alluvial diamonds (Sobolev, 1984).

In the RV suite, there are strong variations in garnet chemistry in specimens with visible banding or ovoid structures (Chinner and Cornell, 1974; Lappin and Dawson, 1975; Carswell *et al*, 1981); for example, the grossular molecule content can vary from 42% in a kyanite-bearing band to 17% in an adjacent kyanite-free band (Carswell *et al*, 1981) (Figure 13.2a). In addition, in the specimen JJG41 Harte and Gurney report compositional variations in the Ca, Mg and Fe contents in coarse, intermediate and fine exsolution lamellae of garnet. Compositional variations are not confined to the

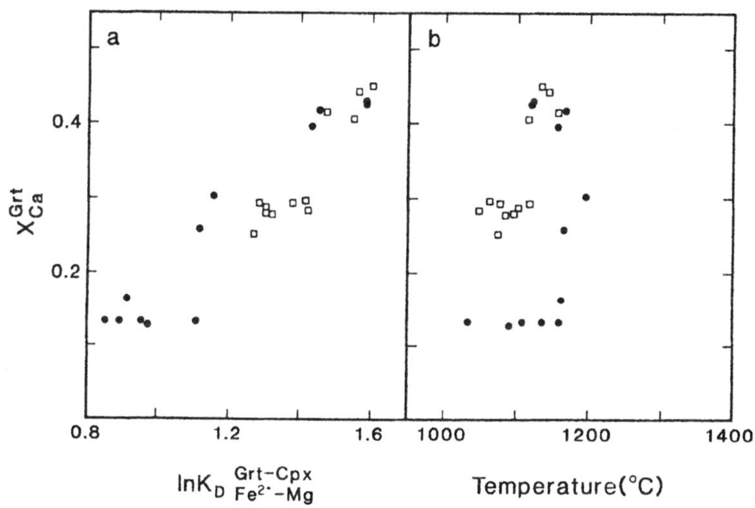

Figure 13.2 (a) Plot of X_{Ca}^{Grt} versus In $K_{D_{Fe-Mg}}^{Grt-Cpx}$ for multiple paired garnet-clinopyroxene analyses in samples 6913 (Figure 13.1a) (Lappin and Dawson, 1975) and RV9 (Carswell *et al*, 1981) from Roberts Victor Mine. Ornament: Open squares—6913; closed circles—RV9. (b) Calculated equilibrium temperatures based on the K_D formulation of Powell (1985b) against X_{Ca}^{Grt} for the same data sets as in (a).

eclogites. Bulanova *et al* (1988) have reported that garnet inclusions within the central part of a single diamond are richer in TiO_2 (0.93 against 0.42 wt%) and CaO (8.77 against 5.31 wt%), but poorer in FeO (16.95 against 17.74 wt%) and MgO (9.66 against 12.74 wt%) than garnet in the outer part of the diamond.

In most eclogites, individual grains tend to be homogeneous or, to be more correct, in most instances variable compositions have not been reported. However, Hatton and Gurney (1979) note garnets with cores enriched in Ca and Ti relative to Mg-rich rims in a RV diamond-graphite eclogite, Ater *et al* (1984) found zoning from Mg-rich cores to more calcic rims in peraluminous eclogites from N Colorado kimberlites, and Jagoutz (1988) found compositional differences between different generations of garnets in a Tanzanian eclogite. Within the peraluminous eclogites in the Orapa and N Colorado eclogite suites (Shee and Gurney, 1979; Ater *et al*, 1984), garnets in corundum eclogites tend to be less Fe-rich than in kyanite eclogites. Hatton and Gurney (1987) note that in RV eclogites with group I textures, the CaO content of the garnets correlates positively with Na_2O in the pyroxene; although a similar trend is found in the texturally more-diverse group II rocks, there is another trend (particularly shown in eclogites with secondary amphibole) in which these oxides correlate negatively.

Several studies (e.g. Reid *et al*, 1976; Shee and Gurney, 1979) show that the major-element compositions of garnets (and also the pyroxenes) in diamondiferous and graphite eclogites can be matched in eclogites that lack these carbon species.

For the minor elements (if the high-Mg varieties interpreted as lherzolite variants are excluded) most eclogite garnets contain only small amounts of Cr_2O_3—up to 0.49 wt% (Dawson 1980, table 30). Exceptions are rare kyanite eclogites and grospydites from the Zagadochnaya pipe, in which pale-violet or green Ca-rich garnets contain 1.5–7.2 wt% Cr_2O_3; also garnets in three Colorado peraluminous eclogites contain up to 1.3 wt% Cr_2O_3. High-precision electron microprobe analyses on garnets in Roberts Victor, Newlands and Bobbejaan eclogites for TiO_2 give up to 3000 ppm. Values for K_2O are <20 ppm, suggesting that earlier reports of up to 337 ppm in bulk analyses on garnet separates may be due to contamination. P_2O_5 contents are 160–940 ppm (average 530), and for Na_2O 100–1420 ppm (average 610) (Bishop *et al*, 1978). Small, but appreciable, amounts of Na_2O (up to 0.5 wt%) were first noted by Sobolev and Lavrentyev (1971) who suggested that, because sodium was especially high in the garnets in diamondiferous eclogites, its presence was possibly due to extreme pressures. Lower amounts of Na_2O (up to 0.2 wt%) are found in garnets in diamondiferous eclogites from southern Africa (Reid *et al*, 1976; Robinson *et al*, 1984); these concentrations can be matched in numerous examples of garnets in diamond-free eclogite, and McCandless and Gurney (1986) propose that values of >0.09 wt% characterise the garnets in group I eclogites. Na_2O concentrations in garnet inclusions in diamond are generally below 0.2 wt% (e.g. Sobolev, 1977; Gurney *et al*, 1984a,b; Gurney and Hatton, 1989), but up to 0.35 wt% is found in garnets in Koffiefontein diamonds (Rickard *et al*, 1989), and up to 0.52 wt% is reported in diamond inclusions from the Argyle lamproite, NW Australia (Hall and Smith, 1984). Exceptionally, Moore and Gurney (1985, 1989) report up to 1.08 wt% Na_2O in a group of forty-six garnets included in Monastery diamonds; these particular garnets, in which the Si cations range up to 3.429 and the Al cations as low as 0.962 (as opposed to expected stoichiometric values of 3 and 2 respectively), appear to reflect the effects of pyroxene solid solution in the garnets. By analogy with results from HP experimentation (Akaogi and Akimoto, 1979; Irifune *et al*, 1989), it is suggested that these diamonds have a particularly deep origin. Two garnets of similar composition have previously been reported, without comment, from the Jagersfontein kimberlite (Tsai *et al*, 1979a). It is impossible not to speculate if other similar garnets have been overlooked in the past due to the fairly common practice of rejecting garnet analyses with apparently spuriously-high SiO_2 values (e.g. Mottana, 1986a).

Bishop *et al* (1978) show that garnets from Roberts Victor and Bobbejaan eclogites, together with the diamondiferous eclogites of Reid *et al* (1978), lie close to a trend of atomic Ti = 1.5(Na–P), and propose the substitutions $Ca^{VIII}_{-1}Si^{IV}_{-1}Na^{VIII}P^{IV}$ and $Ca^{VIII}_{-1}Al^{VI}_{-1}Na^{VIII}Ti^{VI}$; these substitutions retain the normal valence and coordination numbers of Si, Ti and P, and preclude the need to invoke Si^{VI}, as proposed by Sobolev and Lavrentyev (1971) in the coupled substitution $Ca^{VIII}_{-1}Al^{VI}_{-1}Na^{VIII}Si^{VI}$. In the case of the garnets with pyroxene in solid solution, Moore and Gurney (1985) suggest substitution of two Al^{VI} ions by M^{2+} and Si^{4+}.

High precision data for other trace-elements in eclogite garnets are not abundant. Data up to 1980 were compiled by Dawson (1980) who notes the strong weighting by data from RV eclogites and that, owing to the widespread alteration of many of the specimens from this locality, trace element data obtained on mineral separates must be treated with some caution. With this caveat, the following data for garnets in eclogites from various localities, including RV, are (in ppm): Sc 40–129, V 13–80, Co 9–140, Ni 30–140, Rb 1.0–3.0, Sr 4.2–13.2, Zr 3–50 and Ba < 10–53. More recent data on thirty-two RV and two Bellsbank eclogites (MacGregor and Manton, 1986; Shervais *et al*, 1988) give values for Sc, Co and Ni falling within the range compiled by Dawson. Mass spectrometer analyses of garnet inclusions in diamond from Premier kimberlite and Argyle lamproite yield respectively: Rb 0.002–0.004 and 0.003–0.009 ppm; Sr 1.74–3.00 and 15.85–19.87; Sm 0.7463–0.8545 and 3.220–3.231; Nd 1.068–1.101 and 5.392–6.946 (Richardson, 1986). Proton microprobe analyses for inclusions from the Argyle and Ellendale lamproites, Australia (Griffin *et al*, 1988), give Ni < 18–64 ppm, Cu up to 21, Zn 91–173, Ga 9–23, Pb up to 9, Sr 4–88, Y 9–68 and Zr 23–184; Ni correlates roughly with Mg/(Mg + Fe), and there is positive correlation also between Zr and Y. Na, Y, and Zr are positively correlated with Ti and Fe, and negatively correlated with Mg and Ni, perhaps indicating an igneous differentiation trend. Sr is highest in the most Ca-rich garnets (88 ppm in garnet with Grs_{54} composition), shows no correlation with Na, Zr and Y, and negative correlation with P. The Sr values are higher than obtained for garnets in eclogites (< 5–14 ppm) (MacGregor and Manton, 1986).

In the context of coexisting phases, although kyanite and corundum only coexist with the more calcic members of the garnet compositional spectrum, both diamond and coesite coexist with garnets of widely varying compositions (Dawson, 1980; Shulze and Helmstaedt, 1988).

13.4.2 *Clinopyroxene*

Stephens and Dawson (1977) distinguished seven types of clinopyroxenes in eclogites and grospydites. They vary from diopside through jadeitic diopsides to Na- and Al-rich omphacites which contain a high proportion of the Ca-Tschermaks molecule ($CaAl_2SiO_6$). The main variations may be seen in terms of variable Jd:Di ratios, with the most jadeitic varieties occurring in the kyanite/corundum eclogites and grospydites. As in the case of garnets, large variations are found in the pyroxene compositions within single hand-specimens containing both kyanite-bearing and kyanite-free bands;

for example in one RV banded eclogite, the jadeite molecule content varies from 38% in pyroxene in the kyanite band to 14% in the kyanite-free area (Carswell *et al*, 1981). Pyroxenes in a zoned diamond from the Mir kimberlite are very variable in composition, with those in the central part of the diamond in general being lower in Di but higher in CaTs molecule than those in the periphery (Bulanova *et al*, 1988). In most pyroxenes high Na correlates with high Al and most contain <2 mol% CaTs. However, the pyroxenes in peraluminous eclogites and grospydites often contain excess Al^{VI} resulting from a substitution of Ca plus vacancy (probably in the M2 site) for 2Na ions in the jadeite molecule (Lappin, 1978; Smyth, 1980).

In very aluminous pyroxenes there is a deficiency of cations resulting from solid solution towards the non-stoichiometric end member Ca-Eskola ($Ca_{0.5}AlSi_2O_6$). This has been recognised in RV, Zagadochnaya, and Bobbejaan peraluminous eclogites and grospydites (Sobolev *et al*, 1968; Lappin, 1978; Smyth, 1980; Smyth *et al*, 1984). Some pyroxenes from RV contain up to 17 mol% CaEs, whereas those from Bobbejaan contain only up to 11% (Smyth *et al*, 1984); at Bobbejaan, group I eclogites are significantly lower in this end-member than group II rocks, and those that show garnet exsolution from pyroxene are particularly high in both CaEs and CaTs. In JJG41 from RV, the pyroxene exsolving garnet contains up to 25.6 mol% CaTs (Harte and Gurney, 1975). High Al in synthetic pyroxenes (O'Hara and Yoder, 1967) has been likened to kyanite in solid-solution, and lamellar kyanite exsolution from pyroxene has been observed in eclogites from the Dodoma Mine, Tanzania (O'Hara and Yoder, 1967) and, together with exsolved garnet, in kyanite and corundum eclogites from Bobbejaan (Carswell *et al*, 1981; Smyth *et al*, 1984; Smyth and Caporuscio, 1984). Fassaitic pyroxene, containing 40 mol% CaTs, in a diamondiferous eclogite from the Crown Mine, South Africa, is interpreted as secondary from the breakdown of omphacite to give fassaite plus albite; the albite expelled from the omphacite is now present as very small rounded blebs (Reid *et al*, 1976). Fassaite has also been reported as an alteration product of omphacite in a RV kyanite eclogite that has melted to give veinlets of Na-rich glass (Switzer and Melson, 1969). Very high Ca pyroxenes (CaO 19.0–23.5, Na_2O 1.7–2.3, Cr_2O_3 < 0.2 wt%) occur with coesite and high-Ca garnet inclusions in SE Australian alluvial diamonds (Sobolev, 1984).

Among the minor elements, P_2O_5 is 110–520 (average 300) ppm, and TiO_2 is up to 0.36 wt%, with the lowest values in kyanite eclogite and grospydite (Dawson, 1980). Cr_2O_3 is up to 0.3 wt% in eclogite, diamondiferous eclogite and diamond inclusions. Higher values of 0.89 wt% (RV—Bishop *et al*, 1978) and 2.46 wt% Cr_2O_3 (Bobbejaan—Smyth and Caporuscio, 1984) are found in pyroxenes coexisting with the Mg-, Cr-rich garnets in rocks that are probably modal variants of lherzolite. Up to 2.89 wt% Cr_2O_3 is found in the chrome-omphacites in the rare Ca- and Cr-rich grospydites from Zagadochnaya (Sobolev, 1977).

Perhaps the most intriguing minor oxide is K_2O which is commonly present in concentrations up 0.2 wt% in eclogite pyroxenes, and up to 0.3 wt% in pyroxenes from diamondiferous eclogites and grospydites, an exceptionally high value being 0.51 wt% in a specimen from Udachnaya (Safronov *et al*, 1980). Most values for clinopyroxene inclusions in diamond fall within the range for diamondiferous eclogites, but exceptionally high concentrations are reported in inclusions from West Africa (0.62–0.87 wt%—Prinz *et al*, 1975), the Sloan 2 kimberlite, Colorado (up to 0.78 wt%—Otter and Gurney, 1989), and from the Ellendale (0.77 wt%) and Argyle (1.3 wt%) lamproites,

Australia (Hall and Smith, 1984; Griffin *et al*, 1988). These high K_2O values set the high-PT eclogite clinopyroxenes apart from any other pyroxenes. Those containing > 0.1 wt% are believed to characterise the group I eclogites (McCandless and Gurney, 1989; Figure 13.1). Theoretically, K can be accommodated in the pyroxene structure under high pressures, and direct experimentation shows an increase in K with increasing pressure with 2200 ppm K_2O in pyroxenes synthesised at 100 kbar (Kushiro, 1971). An intriguing eclogite from the Mir kimberlite contains diamond that itself contains garnet and clinopyroxene inclusions (V.S. Sobolev *et al*, 1972); in most respects the compositions of the included phases match those of the phases of the host eclogite, but K_2O in the pyroxene inclusion is 0.30 wt% compared with 0.08 wt% in the eclogite pyroxene. This suggests the possibility of potassium diffusion from eclogite pyroxene as a result of pressure release unless the K-phase is encapsulated in diamond, with the inference that present values in the pyroxenes from many high-PT eclogites may represent values well below the original concentrations. Tiny (20 μm) blebs of a high-K, Al phase in the pyroxenes in some diamondiferous RV eclogites, tentatively identified as feldspar, may have arisen by exsolution from an originally high-K pyroxene (Reid *et al*, 1976). Glass blebs in kyanite eclogite, also from RV, believed to derive from omphacite melting as a result of pressure release, contain 1.9 wt% K_2O (Switzer and Melson, 1969).

For trace elements, the same caveats apply as for garnets. The pre-1980 compilation by Dawson (1980) gives Sc 10 ppm; V 450; Co 45; Ni 1000; Rb 0.56–3.4; Sr 58–249; Zr 25; Ba 28–178; and Pb < 3. More recent data for eclogite clinopyroxenes (MacGregor and Manton, 1986; Shervais *et al*, 1988) are Sc 27–29, Co 34–36; Ni 60–625; Rb 0.0012–0.0038; and Sr 16–23 ppm. Data on inclusions in diamonds from Argyle and Premier are respectively: Rb 0.381 and 0.167, Sr 427.8 and 103.6, Sm 1.084 and 0.572, and Nd 5.357 and 2.044 ppm (Richardson, 1986). Proton microprobe analyses for inclusions in diamonds from the Argyle and Ellendale lamproites give: Ni 44–222, Cu 20–62, Ga 17–35, Pb < 7–50, U 6–17, Rb 5–18, Sr 313–754, and Zr 3–70 (Griffin *et al*, 1988). Sr and Zr show good correlation, and Sr is generally higher than in eclogite xenolith pyroxenes, although MagGregor and Manton observe very high concentrations (up to 1880 ppm) in some group I RV eclogites. Ga correlates with Na, which perhaps indicates that Ga may be involved in the Na–Ca substitution. The Rb contents are higher than those reported by Richardson (1986) for a single pyroxene inclusion in an Argyle diamond, and are higher than others in eclogite pyroxenes. K/Rb increases with increasing K, reaching 1500; this contrasts with the usual pattern seen in rocks in which K/Rb decreases with increasing K, and may be due to crystal chemical effects at high pressure. These pyroxenes also have unusually high U and Pb concentrations that correlate positively with Rb, and K and U are also strongly correlated; the values are higher than those found in RV eclogite pyroxenes (Kramers, 1979), although similar high Pb levels are reported in pyroxenes from some group I RV eclogites (MacGregor and Manton, 1986). Griffin *et al* (1988) suggest their data may reflect bulk composition control; however, the likelihood of K diffusion from eclogite pyroxenes unarmoured by diamond has been discussed, and it may be that Rb, U and Pb diffusion may also be promoted by recrystallisation, metasomatism and pressure release. If this is the case, it could have profound implications for the Rb/Sr systematics of these rocks, and would suggest that the diamond inclusion data are more reliable.

13.4.3 *Orthopyroxene*

It has been mentioned that Hatton and Gurney (1987) infer the former presence of orthopyroxene, now represented by metasomatic phlogopite, in numerous group I RV mica-eclogites (Section 13.3). Inferential support comes from the presence of two high-Al (4.1 and 11.8 wt% Al_2O_3) bronzites in the eclogitic inclusion suite from Monastery diamonds (Moore and Gurney, 1989). Such highly-aluminous orthopyroxenes would, in the absence of armouring by diamond, be highly susceptible to metasomatic alteration, in a manner similar to those in the kelyphite reaction rims around lherzolite garnets (Reid and Dawson, 1972). Orthopyroxene (Mg/(Mg + Fe) 0.92, Al_2O_3 0.85 wt%, Cr_2O_3 0.07%) is present in garnet websterite from Orapa, occurring with high-Mg, low-Cr garnet and diopside (Shee and Gurney, 1979).

13.4.4 *Amphibole*

Amphiboles are generally secondary, although Meyer and Brookins (1971) report an apparently primary kaersutite from a Stockdale eclogite. Secondary hornblende, pargasite and sub-calcic edenite occur in RV and Orapa eclogites (Switzer and Melson, 1969; Bishop *et al*, 1978; Shee and Gurney, 1979).

13.4.5 *Mica*

Micas are secondary, replacing garnet, clinopyroxene and, in the inferred garnet websterites, former orthopyroxene. Analyses from Roberts Victor eclogites (Kushiro and Aoki, 1968; Bishop *et al*, 1978; Windom and Boettcher, 1980) show them to have widely varying TiO_2, Al_2O_3, MgO and FeO (and hence Mg/(Mg + Fe)), and K/Na. Unlike the secondary micas in peridotites, they are very low in Cr_2O_3.

13.4.6 *Kyanite*

Most kyanites are pure, containing only a little (<0.5 wt%) Fe_2O_3 in solid solution (Gurney *et al*, 1971; Bishop *et al*, 1978; Smyth *et al*, 1984). The same is true for inclusions in diamonds from West Africa (Prinz *et al*, 1975) and West Australia (Hall and Smith, 1984; Griffin *et al*, 1988). Exceptions are high-Cr kyanites, which contain up to 16.8 wt% Cr_2O_3, and occur with Cr-rich garnets, pyroxenes and corundum in some rare eclogites and grospydites from the Zagadochnaya kimberlite (Sobolev *et al*, 1968).

13.4.7 *Mullite*

Mullite in a RV kyanite eclogite (Switzer and Melson, 1969) is apparently pure Al silicate.

13.4.8 *Feldspar*

Sanidine (Or_{98-99} Ab_{1-2}) occurs in a RV coesite-bearing grospydite (Smyth and Hatton, 1977) and in a kyanite eclogite from Udachnaya (Spetsius *et al*, 1984), and a more sodic type ($Or_{95}Ab_5$) is reported from Colorado kyanite eclogite (Ater *et al*, 1984); all lack CaO. Sanidine has also been identified (although no analyses are given) in coesite-bearing grospydites, eclogites and kyanite eclogites from Udachnaya (Ponomarenko and Spetsius, 1978), in coesite eclogite from RV, in coesite-kyanite eclogite from the Lace kimberlite, and in coesite grospydite from Zagadochnaya (Schulze and Helmstaedt, 1988). Hence, it appears that sanidine appears to occur most frequently in kyanite eclogites and grospydites in which coesite is a common associate; however, Schulze and Helmstaedt (1988) report an exceptional example of sanidine in a RV eclogite containing pyrope-almandine garnet (as opposed to high-Ca garnet) and high-diopside pyroxene. Sanidine containing 0.13 wt% Na_2O is included in a West African diamond (Prinz *et al*, 1975), and Na- and Ca-free sanidine, assigned to the eclogite paragenesis, occurs in a Colorado diamond (Meyer, 1987).

13.4.9 *Yagiite*

A phase resembling yagiite $(Na, K)_3Mg_4(Al, Mg)_6(Si, Al)_{24}O_{60}$ occurs in an eclogite xenolith from the East Udachnaya kimberlite (Govorov *et al*, 1986); compared with the type yagiite from the Kolomera meteorite, the eclogite material is richer in K, Ca, Mg and Fe, but lower in Si, Al and Na

13.4.10 *Unidentified K-rich silicates*

Two high-K phases have been identified that form polyphase inclusions with rutile, pyroxene and an Fe-Ti silicate in a diamond from the Mir kimberlite (Bulanova *et al*, 1988). One of the phases contains mainly SiO_2 (67 wt%), Al_2O_3 (19–22 wt%) and K_2O (8 wt%), with small but significant amounts of TiO_2, FeO, CaO, Na_2O, Cl, F and S. The other is chemically more complex, with SiO_2 56.7–59.7 wt%, TiO_2 4.0–7.1, Al_2O_3 10.0–10.3, FeO 6.8–7.4, MgO 4.1–4.9, CaO 8.9–9.1, Na_2O 1.5–1.6, K_2O 2.7–2.8 wt%.

13.4.11 *Sulphides*

Four sulphides are reported from polymineralic aggregates in an eclogite from the RV kimberlite. The dominant sulphide has the formula $(Fe, Ni)_9S_{11}$ and contains 2.4–3.2 wt% Ni; the other phases are stoichiometric chalcopyrite, a violarite-like phase $(Fe, Ni)_9S_{11}$ with 21.5–30.0 wt% Ni, and pyrite, which contains 0.6–4.8 wt% Ni. The aggregates are considered to have developed upon cooling from a homogeneous monosulphide (Desborough and Czamanske, 1973). Ni-bearing pyrrhotite, chalcopyrite, pentlandite and monosulphide occur, also in aggregates, in eclogite from the Stockdale kimberlite, USA (Meyer and Boctor, 1975). Similar phases are present in eclogites from RV, Jagersfontein and Obnazhennaya, and a K, Fe, Cu sulphide, believed to be the result of interaction between a K-rich liquid and original Fe, Ni sulphide, also occurs in the Obnazhennaya sample (Tsai *et al*, 1979b). Djerfisherite, containing up to 11.6 wt% Cu, occurs as a rim on pyrrhotite-chalcopyrite globules within omphacite grains in eclogites from the East Udachnaya kimberlite (Govorov *et al*, 1986); there is an increase in K, from 8.05 to 10.15 wt%, towards the host omphacite, which suggests that the djerfisherite is the result of reaction between the original sulphide blebs and K derived from the surrounding omphacite.

Although sulphides are abundant as inclusions in diamond, there are very few data that relate them to either the eclogite or ultrabasic parageneses. One exception is the study by Chaussidon *et al* (1987) who report major element data and sulphur isotope ratios for four eclogite-paragenesis pyrrhotite inclusions from Orapa diamonds, and a pyrrhotite and a pentlandite of the peridotite-paragenesis from Premier diamonds. The range and values of the sulphur isotope ratios of the eclogitic sulphides ($d^{34}S + 4.2 - 9.5‰$). is considerably greater than those in the peridotitic sulphides ($+1.1 - +1.7‰$). Another study (Otter and Gurney, 1989) reports that pyrrhotite, usually with minor pentlandite exsolution blebs, is the dominant sulphide inclusion in Sloan, Colorado, diamonds.

13.4.12 *Rutile*

Smith and Dawson (1975) reviewed the chemistry of eclogite rutiles and showed that Fe_2O_3 (up to 2.3 wt%) is the only significant oxide additional to TiO_2; the specimens reviewed include RV eclogites, Zagadochnaya kyanite eclogite, and diamondiferous eclogites from Mir and Newlands. Cr (a significant element in peridotite rutiles), Al, Mn, Mg and Ca are present in only trivial amounts in most eclogite rutiles. Exceptions are rutile from a Cr-rich Zagadochnaya kyanite eclogite that contains 9.8 wt% Cr_2O_3 (Sobolev *et al*, 1972), and an inclusion in a diamond from the Ellendale 9 lamproite that has 0.65 wt% (Hall and Smith, 1984). Electron- and proton-microprobe analyses of three rutile inclusions in Argyle lamproite diamonds give: Cr 340 ppm (one sample, others below detection limit); Ni 11–24, Zn 10–112, Zr 1300–6800, Nb 900–1600, Ta 58–104, and Hf 88–207 (Griffin *et al*, 1988). The high Nb and Zr values are similar to those found in rutiles in metasomatised peridotites (Haggerty, 1987) but differ in having much lower Cr.

13.4.13 *Ilmenite*

Ilmenite is rare in eclogite; a magnesian variety (MgO 10 wt%) is reported in a Kansas eclogite (Meyer and Boctor, 1975). There are several reports of ilmenite inclusions in diamond (Meyer, 1987); some with low Cr may be assignable to the eclogitic paragenesis but other high-Cr examples resemble those from peridotite.

13.4.14 *Corundum*

Few analyses have been made. Corundums from Bobbejaan peraluminous eclogites contain up to 0.16 wt% Fe_2O_3 and 0.07 wt% Cr_2O_3 (Bishop *et al*, 1978; Smyth *et al*, 1984), and Carswell *et al*, (1981) report higher Cr_2O_3 (0.13 against 0.06 wt%) and Fe_2O_3 (0.46 against 0.14 wt%) in secondary corundums compared with primary crystals. The corundums in alkremites contain significant TiO_2 (0.08–0.21 wt%), Cr_2O_3 (0.17–0.46 wt%) and Fe_2O_3 (0.5–1.2 wt%) (Exley *et al*, 1983). A corundum inclusion in diamond from Monastery contains 3.05 wt% TiO_2 (Moore and Gurney, 1989) and two corundum inclusions in Sloan diamonds contain 1.1 and 2.0 wt% TiO_2 as the only additional oxide (Otter and Gurney, 1988).

Sobolev (1977) reports two phases intermediate between Al_2O_3 and Cr_2O_3(eskolaite), containing 34.9 and 64.4 wt% Cr_2O_3 in Zagadochnaya kyanite eclogites.

13.4.15 *Spinel*

Spinel in eclogites is confined to partial-melt glasses and no chemical data are available. Spinels in alkremites from Bellsbank and Jagersfontein have Mg/(Mg + Fe) of

0.7–0.8, Cr_2O_3 up to 4.92 wt% (highest in the more Mg-rich types), and up to 0.28 wt% ZnO (Nixon *et al*, 1978; Exley *et al*, 1983).

13.5 Eclogite bulk-rock chemistry

13.5.1 *Major and trace elements*

The major element chemistry of eclogites has been reviewed by Dawson (1980). Analyses of a compositional range of eclogites, peraluminous eclogites and grospydites (mainly from the Roberts Victor kimberlite) are given in Table 13.1. They vary from high magnesian varieties, through intermediate types with increasing iron to the highly-aluminous kyanite and corundum eclogites and grospydites. Corundum-bearing examples are low in SiO_2 compared with the others, particularly with the coesite-sanidine grospydite, and also are very low in iron. The CaO contents of kyanite eclogites and grospydites can be matched with those of other eclogites; it is their high Al_2O_3 contents, combined with low SiO_2 that are distinctive. TiO_2 is usually < 0.5 wt%, and Cr_2O_3 is mainly < 0.25 wt%. The high values for K_2O in some samples are undoubtedly due to the presence of secondary phlogopite and/or amphibole; note the relatively small amounts in analyses 1, 2 and 13 that are calculated from modal and phase chemistry data. From the aspect of normative mineralogy, RV eclogites range from hypersthene-normative to nepheline-normative; they range from types that are essentially pyroxene-free olivine-plagioclase norites, to gabbroic varieties with up to 50% normative clinopyroxene. However, there is no distinction between group I and group II eclogites in terms of their normative mineralogy. Smyth *et al* (1984) find that both group I and group II eclogites from Bobbejaan are mainly nepheline normative.

For the trace elements, many values for bulk rocks are spuriously high for the LILE due to metasomatic effects (Dawson, 1980). Using the individual phase data from Bishop *et al* (1978), Dawson (1980), MacGregor and Manton (1986), Shervais *et al* (1988) and Jagoutz (1988), and assuming an arbitrary 50:50 garnet:pyroxene ratio, the following are trace element estimates; Sc 45–60, V 250, Co 35–90, Ni 200–570, Ga 10–15, Sr 40–960, Ba 20–110, all values in ppm. Perhaps the most significant of these values is that the Ni concentrations of some eclogites (e.g. that for the high-Mg eclogite—analysis 1 of Table 13.1) are equal to, or higher than, 300 ppm. When combined with high *mg* ($mg = Mg/(Mg + 0.85\,Fe^T)$) values of > 0.68, these particular eclogites satisfy the two diagnostic chemical criteria for classifying basic liquids as primitive (i.e. unfractionated) upper mantle liquids (Frey *et al*, 1978).

For the rare earth elements, three samples from Roberts Victor (Haskin *et al*, 1966; Philpotts *et al*, 1972) have flat patterns with values of 10 × Cl—very similar to the values and patterns for MORB; another from Dutoitspan has a flat pattern, with the exception of slightly-higher LREE, and values approximately 20 × C1, and a further sample from Jagersfontein has LREE values similar to those in the RV samples (and MORB) but is HREE-depleted (1–2 × C1). Jagoutz *et al* (1984) found noritic group I RV eclogites are LREE-enriched but HREE-depleted (2 × C1), and have a positive Eu anomaly; by contrast, group II eclogites (one from Tanzania) are strongly LREE-depleted, although the HREE are 10 × C1. Two eclogites from Bellsbank have U-shaped REE patterns, interpreted as representing either cumulates or refractory residues but not magmatic compositions (Shervais *et al*, 1988).

Table 13.1 Analyses of eclogites, peraluminous eclogites and grospydites (From Roberts Victor Mine, unless stated otherwise).

	1	2	3	4	5	6	7
SiO_2	48.00	46.8	45.67	47.73	49.49	46.9	41.70
TiO_2	0.11	0.19	0.42	0.61	0.52	0.53	0.30
Al_2O_3	13.20	16.5	17.85	11.97	8.46	17.70	29.70
Cr_2O_3	0.32	0.04	0.07	0.12	0.24	0.03	0.03
Fe_2O_3	—	—	2.88	3.79	3.20	1.65	1.74
FeO	5.10[a]	11.9[a]	8.46	9.56	5.80	8.09	5.82
MnO	0.15	0.22	0.17	0.33	0.24	0.20	0.18
MgO	19.10	9.70	11.90	13.28	16.23	9.20	5.68
CaO	12.20	11.50	7.35	8.76	10.60	8.89	9.14
Na_2O	1.03	2.90	2.01	2.18	1.67	3.56	2.19
K_2O	0.01	0.0	0.39	0.40	1.12	0.86	0.83
H_2O^+	—	—	2.01	1.28	1.43	1.46	1.83
H_2O^-	—	—	1.07	0.33	0.64	0.55	0.49
CO_2	—	—	—	—	—	—	0.52
P_2O_5	—	—	0.04	0.01	0.64	0.14	0.18
Total	99.22	99.75	100.29	100.34	99.68	99.76	100.33

[a] Total iron as FeO
[b] Total iron as Fe_2O_3

1. High-Mg, Low-Cr eclogite 1698
2. High-Fe eclogite 1191

(1–2: Calculated from phase composition data on basis of 50% garnet, 50% pyroxene. Phase data from Bishop *et al* (1978).)

3–5. Eclogites (Kushiro and Aoki, 1968).
6. Eclogite band and 7. kyanite eclogite band in banded eclogite (Lappin and Dawson, 1975).
8. Kyanite eclogite, Zagadochnaya (Sobolev *et al*, 1968).
9. Kyanite eclogite, Rietfontein, S. Africa (Gurney *et al*, 1971).
10. Kyanite-corundum eclogite (Whitfield, 1971).
11. Corundum eclogite, Obnazhennaya, USSR (Sobolev and Kuznetsova, 1965).
12. Average grospydite, Zagadochnaya (Sobolev *et al*, 1968).
13. Coesite-sanidine grospydite (calculated from phase data) (Smyth and Hatton, 1977).

13.5.2 *Isotope chemistry*

13.5.2.1 *Carbon.* There have been several studies on the carbon isotope composition of diamond with inclusions of both the eclogite and peridotite parageneses. In a study of ninety-two inclusion-bearing diamonds from southern African, Yakutian, and SE Australian kimberlites and placers, Sobolev *et al* (1979) found that, compared with diamonds with peridotite inclusions ($d^{13}C$ −2 to −9‰), those with eclogitic inclusions have a much wider spread, +5 to −34.4‰; of especial note are the SE Australian diamonds that have the heaviest carbon, +2 to +5‰. Eclogitic diamonds from the Premier kimberlite have $d^{13}C$ −2.53 to −12.35‰, with most in the range −3 to −7‰ (Deines *et al*, 1984). Diamonds from the Roberts Victor kimberlite containing eclogitic inclusions fall into two distinct groups, one with $d^{13}C$ around −15.5‰ (Group A), and another around −5.6‰ (Group B) (Deines *et al*, 1987); pyroxenes in Group A diamonds are enriched in Al_2O_3, FeO, MnO,

Table 13.1 (*Contd.*)

	8	9	10	11	12	13
SiO_2	43.46	42.42	42.27	40.26	43.73	52.69
TiO_2	0.35	0.18	0.46	0.18	tr	0.09
Al_2O_3	24.11	24.60	22.16	29.43	27.48	21.43
Cr_2O_3	0.02	—	0.03	0.04	0.01	0.04
Fe_2O_3	1.99	7.39[b]	1.18	0.23	1.67	—
FeO	4.82	—	6.63	2.50	2.77	3.76[a]
MnO	0.17	0.10	0.25	0.06	—	0.08
MgO	9.05	6.55	6.51	9.48	7.53	5.62
CaO	10.55	13.91	9.75	14.02	12.14	11.77
Na_2O	1.87	2.41	4.31	1.63	1.53	4.08
K_2O	0.53	0.04	0.70	0.18	0.66	0.24
H_2O^+	—	—	2.12	—		—
H_2O^-	2.65	2.29	0.77	1.55	2.28	—
CO_2	—	—	0.39	—		—
P_2O_5	0.11	0.11	0.01	0.05	0.08	0.01
	99.92	100.00	99.56	99.61	99.88	99.56

Na_2O, but lower in MgO, CaO and K_2O than pyroxene inclusions in Group B diamonds which are similar in composition to those in two RV graphite-diamond eclogites. The diamonds in these eclogites (−5.76‰) are isotopically similar to the group B diamonds, whereas the graphite in the same samples is slightly heavier (−4.87‰). Diamonds with eclogite paragenesis inclusions from the Argyle lamproite also show range of d^{13}-depleted values from −9 to −12‰ (Jaques *et al*, 1989).

13.5.2.2 *Oxygen.* The pioneering work of Garlick *et al* (1971) found a range in the $d^{18}O$ from 2 to 8‰ in RV eclogites with the higher values (6–8‰) in the textural group I ('cumulate') eclogites and the lower values in the group II eclogites; the variations were attributed to fractional crystallisation, with ^{18}O partitioned preferentially into the earlier cumulate phases. Subsequent studies on RV eclogites (Jagoutz *et al*, 1984; MacGregor and Manton, 1986; Ongley *et al*, 1987) confirm this spread in values, and also show that there is little difference between the $d^{18}O$ values for coexisting garnet and pyroxene, though Ongley *et al* (1987) report that some secondary phlogopites have higher values than the primary phases. Shervais *et al* (1988) report values of 3 and 3.4‰ in two Bellsbank eclogites.

13.5.2.3 *Strontium and neodymium.* Early investigations considered strontium alone but most recent studies look at these isotopes together. Bulk rock analyses of RV eclogites show a wide range in $^{87}Sr/^{86}Sr$ values up to 0.7112 (Allsopp *et al*, 1969; Manton and Tatsumoto, 1971; MacGregor and Manton, 1986). Allsopp *et al* (1969) compared these high values with lower values of 0.703 for hand-picked omphacites, and suggested that the high bulk values could be attributed to the presence of varying amounts of secondary phlogopite ($^{87}Sr/^{86}Sr$ 0.738). Low values for fresh omphacites have been confirmed by later studies (Kramers, 1979; Jagoutz *et al*, 1984); two group I eclogites have $^{87}Sr/^{86}Sr$ 0.70155 and 0.70335, and $^{143}Nd/^{144}Nd$ 0.51142 and 0.51544 respectively, and a group II specimen contains the least radiogenic strontium value for any terrestrial material ($^{87}Sr/^{86}Sr$ 0.7009) combined with a high neodymium isotope ratio (0.53853) (Jagoutz *et al*, 1984). By comparison, group I eclogites (some diamondiferous) from Orapa show considerable variability (0.7024–0.7066) and bulk earth or higher neodymium ratios, whereas group II samples have 0.705–0.709 and less radiogenic neodymium ($^{143}Nd/^{144}Nd$ 0.5122 to 0.5125) (Smith *et al*, 1986). Other samples from Excelsior, Newlands and RV show similar wide ranges (Smith *et al*, 1989). High Sr isotope values (0.709359 and 0.710022 are reported from two Bellsbank eclogites (Shervais *et al*, 1988). In the studies by MacGregor and Manton (1986) and Shervais *et al* (1988), high $^{87}Sr/^{86}Sr$ correlates with low values for $d^{18}O$; the same is found by Jagoutz *et al* (1984) but over a relatively small Sr isotope range. In the most sophisticated study to date on the Sr–Nd systematics of a single eclogite sample, Jagoutz (1988) reports Nd isotope heterogeneities in different generations of garnet, (0.52229–0.52317) and zonation of Nd and Sr isotope compositions (0.51395–0.51434 and 0.70109–70143 respectively) within the pyroxene in an eclogite from the Vissuri kimberlite, Tanzania.

Two studies have been made on the Sr and Nd isotopes of eclogitic phases included in diamonds. Richardson (1986) reports Nd isotope ratios of 0.5104 and 0.5115, and Sr 0.7058 and 0.702, in garnets and clinopyroxenes included within diamonds from Argyle and Premier respectively; the asthenospheric isotopic signature of the Premier diamond inclusions, contrasts with an enriched lithosphere Nd and Sr component in the Argyle inclusions, which is similar to that in the W Australian lamproites (McCulloch *et al*, 1983). Smith *et al* (1989) report $^{143}Nd/^{144}Nd$ 0.51817 for eclogite garnet inclusions in a diamond from the Finsch kimberlite.

13.5.2.4 *Lead.* Lead isotopes on separated omphacites from RV eclogites show a considerable range 206/204 14.78–19.08, 207/204 14.86–15.47 and 208/204 34.93–37.41 (Lovering and Tatsumoto, 1968; Kramers, 1979); the work of Kramers (1979) suggests that the U–Pb system was not closed. Whole rock data by Manton and Tatsumoto (1971) identifies two isotopically distinct types of lead, one of which is interpreted to have been added from the host kimberlite during emplacement.

13.6 The age of eclogites

On the basis of low $^{87}Sr/^{86}Sr$ ratios in omphacite from RV eclogites, and assuming primitive mantle Rb/Sr to be > 0.023, Allsopp *et al* (1969) suggested that the formation of eclogites was a Precambrian event. Combined with the post-Jurassic age of the RV kimberlite, this effectively put an end to the theory that eclogites are crystal cumulates from the fractionating host kimberlite (e.g. Beck, 1899; O'Hara and Yoder, 1967). Some Rb–Sr (125 Ma on a secondary mica—Allsopp and Barrett, 1975) and Sm–Nd ages (90, 120, 220 Ma—Jagoutz *et al*, 1984; 144 and 188 Ma—McCulloch, 1989) indicate re-equilibration at, or just before the time of emplacement of the host kimberlite. However, other data are consistent with Precambrian ages. Jagoutz *et al* (1984) found that calculated values for three RV eclogites lie on an 'isochron' with an age of 2.7 Ga with an initial $^{143}Nd/^{144}Nd$ of 0.5096 ± 3 ($\varepsilon_0 + 9 \pm 6$), indicating derivation from a MORB-type LREE-depleted source; estimates of model ages range from 3.06 to 4.05 Ga but, in view of the implications of these very old model ages, further study of these rocks is continuing at the Max Planck Institute in Mainz. Other ages derived from Sm–Nd isochrons are: Vissuri, Tanzania 1.75 Ga (Jagoutz, 1988), Bellsbank 2.1 Ga (Shervais *et al*, 1988), inclusions in diamonds from Argyle and Premier—1.58 and 1.15 Ga respectively (Richardson, 1986), garnet inclusions in a Finsch diamond—1.67 Ga (Smith *et al*, 1989), and two Yakutian eclogites—1.54 and 2.60 Ga (McCulloch, 1989).

Whole-rock U–Pb data for five RV eclogites give ages ranging from 2.2 to 2.7 Ga, and Th–Pb data give an age of 2.3 Ga (Manton and Tatsumoto, 1971); a Pb–Pb isochron, however, indicates an age of 3.2 Ga. Kramers (1979) gives an apparent Pb–Pb age of 2.47 Ga for RV hand-picked omphacites, and a model lead age of about 1.2 Ga for presumed sulphide inclusions in Premier diamonds; the Premier age is very close to the 1.15 Ga model age obtained by Richardson (1986).

Taking all the data together, the dates appear to identify eclogite forming 'events' at around 3.2, 2.5–2.7, 1.67 and 1.2 Ga in southern Africa. The oldest event took place at about the time (3.2 Ga) of the growth of diamond hosting peridotite-suite inclusions (Richardson *et al*, 1984), and also around the 3.05 Ga age for upper crustal granite of the Vredefort Dome (Hart *et al*, 1981). The 2.7 Ga age is bracketed between the dates of 2.5–2.8 Ga for one of the major metamorphic episodes of the Limpopo mobile belt (Barton and Ryan, 1977) and close to a major isotope resetting event at 2.8 Ga that affects mafic granulites in the Vredefort Dome (Hart *et al*, 1981). The 1.67 Ga date is similar to some of the older ages in crustal rocks in the off-craton crustal rocks in the Namaqualand and Bushmanland areas, and the 1.2 Ga date, although virtually the same as that of the host kimberlite, is also similar to that of 1.4 + 0.1 Ga for granulite formation in the circum-cratonic Namaqua–Orange River fold belt (Hawkesworth *et al*, 1983). Outside southern Africa, the age of 1.75 Ga for the Tanzanian eclogite (Jagoutz, 1988) corresponds with that of granitic intrusions at 1.6–1.8 Ga in the central part of the Tanzania craton, and to post-orogenic thermal events in the circum-cratonic Usagaran and Ruwenzori fold-belts (Gabert and Wendt, 1974). The Argyle eclogitic diamond inclusion age of 1.58 Ga is older than the 1.13 Ga age of the host lamproite intrusion, and closer to the 1.8 Ga age for the stabilisation of the Halls Creek mobile zone (Richardson, 1986).

It appears therefore that each of the eclogite forming events can be broadly correlated either with deformation in the fold belts surrounding the cratons into which the eclogite-transporting magmas were subsequently emplaced, or with thermal events on the cratons; the deformation episodes themselves are possibly related to collision tectonics and crustal thickening. In southern Africa, formation of mantle eclogites beneath the craton contrasts with formation of crustal eclogites beneath the circum-cratonic fold belt (Gurney *et al*, in press). However, in contrast to the formation of LT- and MT-eclogites in subducted crust (Chapters 7–10), the timing of stabilisation of MT–HT eclogites in the lowermost continental crust (Chapter 12) and underlying lithospheric upper mantle may correlate more directly with post-orogenic, and post-magmatic, cooling.

13.7 *P–T* conditions for eclogite formation

Within the large PT field for eclogite stability (Figure 1.1), it is difficult to establish the precise conditions of equilibration for a given eclogite. The temperatures can be estimated from Fe–Mg partitioning between garnet and pyroxene (Chapter 4) but K_D

Table 13.2 Element partitioning data and temperatures of equilibration of diamondiferous eclogites and eclogitic inclusions in diamonds

Locality	Sample Number	Reference	X_{Na}^{Grt}	X_{Na}^{Cpx}	X_{Ca}^{Grt}	$K_D=\frac{Fe/Mg\ Grt}{Fe/Mg\ Cpx}$	T°C* at 50 kbar
			Diamondiferous eclogites				
Mir	5	1		0.4516	0.3802	5.02	1066
Mir	M46A	2	0.0250	0.4783	0.2322	3.92	1034
Mir	M45	2	0.0184	0.3368	0.0805	2.52	1059
Mir	A42/2	2	0.0301	0.5085	0.1655	3.96	951
Mir	TM8	2	0.0185	0.6013	0.1608	2.63	1131
Mir	M180	2	0.0267	0.5009	0.1697	2.78	1113
Mir	M52	2	0.0226	0.4077	0.1882	3.58	1016
Mir	M53	2	0.0196	0.5169	0.2504	4.72	967
Mir	M50	2	0.0308	0.5241	0.2458	4.27	939
Mir	Mb-1	2	0.0146	0.4901	0.2611	4.72	977
Mir	M49	2	0.0160	0.4885	0.2636	4.32	1014
Mir	M51	2	0.0220	0.5259	0.3122	2.37	1372
Mir	M33	2	0.0244	0.4192	0.2925	3.59	1124
Mir	M59	2	0.0160	0.5094	0.3466	4.25	1103
Mir	M821	2	0.0190	0.6079	0.3608	5.37	1027
Mir	M54	2	0.0216	0.4865	0.3802	5.47	1033
Mir	M75	2	0.0243	0.7123	0.3427	3.01	1265
Mir	BM432	3	0.0177	0.5286	0.3111	4.53	1041
Mir	BM702	3	0.0176	0.4668	0.1201	3.26	984
Mir	BM858	3	0.0301	0.4293	0.1381	1.95	1275
Mir	BM841	3	0.0183	0.3487	0.1279	2.73	1075
Mir	BM238	3	0.0179	0.4651	0.2299	4.62	955
Udachnaya	U259†	4	0.0144	0.4945	0.3978	4.92	1091
Udachnaya	U815†	5	0.0129	0.5232	0.4275	5.00	1112
Udachnaya	U820†	4	0.0100	0.6553	0.4812	2.56	1521
Udachnaya	U816†	4	0.0100	0.6399	0.4787	2.48	1539
Udachnaya	U210‡	4	0.0187	0.5976	0.5365	6.34	1116
Udachnaya	BGU947	3	0.0188	0.3422	0.3403	4.51	1071
Udachnaya	BGU891	3	0.0213	0.3440	0.1290	2.64	1092
Udachnaya	BGU637	3	0.0248	0.2590	0.2528	3.47	1097

Roberts Victor	F5108	2	0.0159	0.3375	0.1334	2.78	1072
Roberts Victor	111060	2	0.0139	0.4737	0.3243	4.58	1050
Roberts Victor	1	5	0.0172	0.4425	0.2322	3.97	1017
Roberts Victor	2	5	0.0152	0.2534	0.0987	2.65	1057
Roberts Victor	3	5	0.0098	0.2944	0.2830	4.55	1074
Roberts Victor	4	5	0.0188	0.4042	0.1847	3.30	1049
Roberts Victor	5	5	0.0098	0.2175	0.0946	2.70	1042
Roberts Victor	6	5	0.0114	0.3746	0.3562	3.97	1143
Roberts Victor	7	5	0.0198	0.3519	0.2315	3.53	1067
Roberts Victor	8	5	0.0141	0.4608	0.2379	3.54	1073
Crown	10	5	0.0086	0.3789	0.3832	3.28	1264
Crown	11	5	0.0129	0.3647	0.3601	4.14	1128
Crown	12	5	0.0112	0.2250	0.0980	2.51	1082
Orapa	18	5	0.0217	0.5155	0.3422	3.41	1201
Orapa	XM4	6	0.0100	0.3059	0.1857	2.87	1115
Orapa	XM5	6	0.0270	0.3581	0.3840	2.88	1336
Orapa	XM6	6	0.0099	0.2946	0.2068	3.03	1113
Orapa	XM7	6	0.0170	0.3446	0.2399	3.32	1104
Orapa	XM9	6	0.0201	0.3368	0.3380	2.82	1297
Orapa	XM26	6	0.0304	0.3499	0.2718	3.01	1187
Orapa	XM28	6	0.0230	0.2856	0.1914	3.23	1066
Orapa	XM31	6	0.0144	0.2968	0.1802	3.06	1077
Orapa	XM32	6	0.0142	0.2629	0.3647	2.81	1329
Orapa	JJG892	6	0.0226	0.3134	0.1731	2.27	1227
Orapa	JJG895	6	0.0157	0.3833	0.3003	3.64	1125
Orapa	XM25[§]	6	0.0233	0.3685	0.2412	3.95	1028
Orapa	JJG889[§]	6	0.0187	0.4096	0.1905	3.50	1028
Orapa	JJG891	6	0.0183	0.3953	0.1321	1.66	1380
Orapa	AK1/9	7	0.0229	0.3772	0.2903	6.07	1095
Orapa	AK1/10	7	0.0168	0.2867	0.3291	5.86	1300
Orapa	AK1/25	7	0.0114	0.3327	0.3383	3.43	1309
Orapa	AK1/11	7	0.0227	0.3604	0.1719	6.30	1182
Orapa	AK1/204	7	0.0293	0.4491	0.3290	6.52	1158
Star	1248	8	0.0112	0.4335	0.2992	3.84	1100
South African, Unknown	17	5	0.0116	0.3429	0.0657	2.20	1111

(Contd.)

Table 13.2 (*Contd.*)

Locality	Sample Number	Reference	X_{Na}^{Grt}	X_{Na}^{Cpx}	X_{Ca}^{Grt}	$K_D = \frac{\text{Fe/Mg Grt}}{\text{Fe/Mg Cpx}}$	T°C* at 50 kbar
			Diamond Inclusions				
'South Africa'	15	9	0.0787	0.1641	0.3650	2.58	1382
Finsch	F89	10	0.0341	0.3258	0.2978	4.06	1074
Premier	12	11	0.0203	0.3043	0.1343	1.97	1264
Roberts Victor	RV69	12	0.0147	0.3226	0.1572	2.08	1262
Orapa	OR21	13	0.0	0.3624	0.1592	2.42	1175
'West Africa'	33	14	0.0339	0.3170	0.2504	3.46	1097
'West Africa'	36	14	0.0087	0.1538	0.3577	2.78	1328
'West Africa'	38	14	0.0320	0.3995	0.1548	3.48	994
'West Africa'	42	14	0.0586	0.2684	0.2509	3.60	1078
Ellendale, W. Australia	E9	15	0.0254	0.3419	0.3437	3.38	1206
Sloan, Colorado	30	16	0.0281	0.3850	0.2601	3.59	1089
Sloan, Colorado	31	16	0.0228	0.3559	0.1947	3.20	1074
Sloan, Colorado	32	16	0.0202	0.2319	0.1608	2.80	1099
Sloan, Colorado	33	16	0.0256	0.1819	0.0915	3.17	965
Mir	M46	2	0.0247	0.4702	0.2209	4.21	983
Mir	AV4	2	0.0293	0.4735	0.2885	2.50	1310
Urals	880	2	0.0246	0.2764	0.2829	2.72	1254
Premier	P3	17	0.0211	0.2055	0.1481	1.90	1305
Premier	P5	17	0.0263	0.2831	0.1768	2.22	1244
Premier	P18	17	0.0254	0.1918	0.1354	1.78	1332
Premier	P35	17	0.0159	0.1576	0.3578	2.80	1322
Premier	P37	17	0.0256	0.3804	0.1630	2.13	1254
Premier	P38	17	0.0239	0.3379	0.0823	1.80	1256
Premier	P43	17	0.0235	0.1503	0.3584	2.98	1289

Premier	P44	17	0.0256	0.2162	0.1129	1.91	1258
Premier	P45	17	0.0291	0.2777	0.2268	2.03	1365
Premier	P46	17	0.0175	0.1555	0.3751	3.03	1297
Premier	P47	17	0.0216	0.2689	0.2324	2.27	1301
Premier	P48	17	0.0235	0.2136	0.0992	1.82	1269
Premier	P65	17	0.0383	0.3566	0.2135	2.38	1248
Premier	P66	17	0.0220	0.3403	0.2203	2.65	1198
Premier	P103	17	0.0348	0.3800	0.1582	1.98	1293
Premier	P104	17		0.1647	0.3580	2.99	1286
Premier	P105	17	0.0366	0.4271	0.2265	2.41	1257
Premier	P106	17		0.4084	0.1662	2.27	1219
Premier	P109	17		0.2197	0.1680	2.40	1189
Premier	P111	17	0.0229	0.2629	0.0950	1.85	1253
Premier	P113	17	0.0176	0.1578	0.3654	2.80	1332
Premier	P115	17	0.0219	0.1601	0.3705	2.92	1314

Notations: *Method of Powell (1985b); †Kyanite eclogite, ‡Grospydite; §With graphite, Detrital

References:
1. Sobolev *et al* (1970)
2. Sobolev (1977) (which summarises most pre-1977 Russian data)
3. Safronov *et al* (1980)
4. Ponomarenko *et al* (1976)
5. Reid *et al* (1976)
6. Robinson *et al* (1984)
7. Shee and Gurney (1979)
8. Gurney and Hatton (1986)
9. Meyer and Boyd (1972)
10. Gurney *et al* (1979)
11. Tsai *et al* (1979a)
12. Gurney *et al* (1984b)
13. Gurney *et al* (1984a)
14. Prinz *et al* (1975)
15. Hall and Smith (1984)
16. Otter and Gurney (1989)
17. Gurney *et al* (1985)

partition coefficients are both P and T dependent. Unlike the case for garnet lherzolites, it is not possible to determine absolute pressures for eclogite formation, so any temperature estimate is subject to an assumption of the pressure involved. However, two phase transitions, quartz⇌coesite and graphite⇌diamond, give minimum pressure conditions in eclogites containing coesite or diamond (Figure 1.1). For this reason we have mainly confined our P–T assessment to diamond-bearing eclogites and inclusions pairs in diamond. Table 13.2 shows the calculated equilibration temperatures assuming P of 50 kbar, and using the K_D formulation of Powell (1985b). The K_D calibration of Ellis and Green (1979) yields slightly higher T estimates, the difference increasing to about 20°C in the lower part of the T range of interest here, and the calibration of Krogh (1988) yields notably lower temperatures at $T < 1100°C$ for samples with lower X_{Ca}^{Grt}. Preference is here given to temperatures calculated from the K_D calibration of Powell (1985b) since the resultant 165°C spread of the T estimates

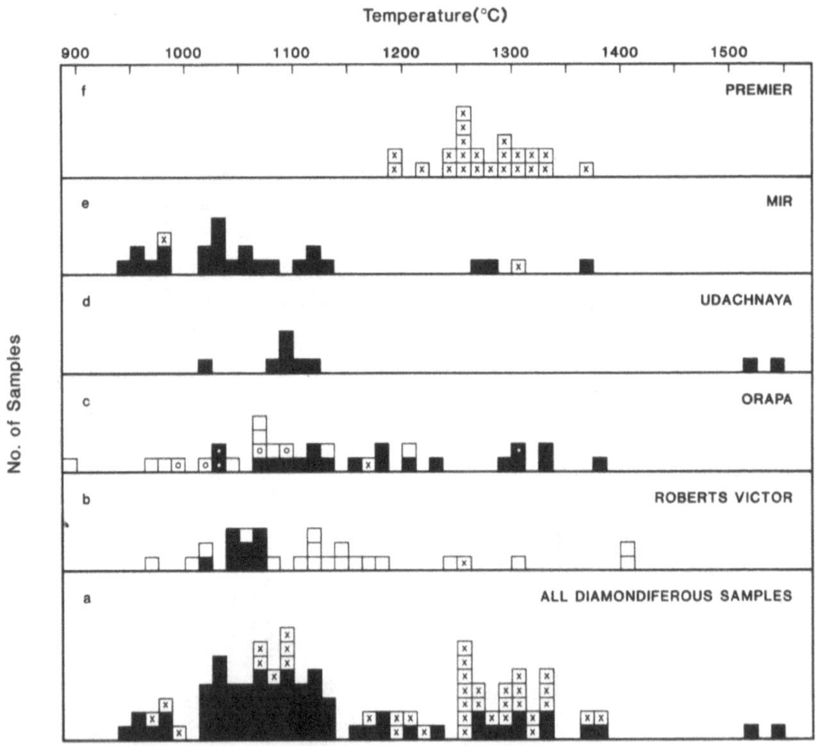

Figure 13.3 Histogram of calculated equilibrium temperatures (Table 13.2) based on the K_D calibration of Powell (1985) at an assumed $P = 50$ kbar for diamond eclogite (solid squares) and eclogite mineral inclusion pairs in diamond (open squares with cross). Also shown for comparison are equivalent calculated temperatures for diamond-free eclogites (open squares) from Roberts Victor and Orapa. Samples from Orapa containing both diamond and graphite are shown by solid squares with an inset open circle.

for the banded RV eclogite sample RV9 (see Figure 13.2b) is smaller than the 247°C spread from the calibration of Krogh (1988).

Figure 13.3 shows the wide *T* range of the samples listed in Table 13.2. For comparison, equivalent *T* estimates are also shown for diamond-free eclogites and grospydites from Orapa and Roberts Victor (data sources—Bishop *et al*, 1978; Carswell *et al*, 1981; Lappin, 1978; Robinson *et al*, 1984; Shee and Gurney, 1979; Smyth and Hatton, 1977). Reid *et al* (1976) had found a bimodal distribution of equilibration *T* for thirteen southern African samples, but the larger data set shows no convincing bimodality. In addition, the following points may be noted.

(i) There is an overall *T* range of $>400°C$ (Figure 13.3a). In the absence of distinctive petrographic or mineral chemistry features to substantiate them, the atypically-high ($>1500°C$) temperatures for two Udachnaya diamond-bearing kyanite eclogites (Ponomarenko *et al*, 1976) may well be spurious (perhaps due to chemical disequilibrium between analysed Grt and Cpx in samples described as banded) and are thus excluded from further discussion.

(ii) Gurney *et al* (1984a) note that inclusion pairs in diamonds from Roberts Victor have apparently formed at higher *T* than diamondiferous eclogites from the same kimberlite. Nonetheless, equilibration temperatures for both these sample types lie within the wide *T* range ($>400°C$) for the overall eclogite xenolith suite from Roberts Victor (Figure 13.3b).

(iii) From Figure 13.3b it appears likely that the absence of diamond in most eclogite and grospydite Roberts Victor xenoliths is due to a lack of free carbon, rather than to different physical conditions of formation of the diamondiferous eclogites.

(iv) There is also a substantial overlap in the formation temperatures for diamondiferous and diamond-free eclogites from Orapa (Figure 13.3c). Temperatures for graphite eclogites are at the lower end of the range, and two eclogites with coexisting graphite and diamond appear to have equilibrated at *T* consistent with proximity to the experimental graphite-diamond univariant reaction curve. However, another graphite-diamond eclogite (AK1/25) has equilibrated at higher *T*, suggesting that the graphite may be secondary.

(v) Taken all together, eclogite paragenesis inclusion pairs in diamonds appear to have equilibrated over a wide temperature range similar to eclogite xenoliths (Figure 13.3a). This is contrary to suggestions by Gurney *et al* (1984b, 1985), based on more limited RV and Premier Mine data, of higher temperatures for inclusions pairs, considered to indicate sub-solidus re-equilibration of the xenolith assemblages to lower temperatures.

(vi) Additional data for other eclogite suites also indicate equilibration over wide *T* ranges: e.g. Mir 940–1370°C (Figure 13.3e), Bobbejaan 940–1170°C (Smyth *et al*, 1984), Colorado 740–1230°C (Ater *et al*, 1984) and Argyle inclusions in diamond 1085–1575°C (Griffin *et al*, 1988).

Two contrasting interpretations of these wide *T* ranges, for both eclogite xenoliths (whether diamond-free or diamondiferous) and eclogite paragenesis inclusion pairs in diamond, merit consideration. The first interpretation is that the *T* range reflects varying degrees of sub-solidus re-equilibration of original high *T*, probably magmatic, assemblages. The second is that the range reflects derivation of the samples from a considerable depth range within the mantle. Although these two alternatives will be considered separately, they are not, in fact, mutually exclusive. In deciding between them, the key question is whether or not eclogite xenoliths (and mineral inclusion pairs in diamonds) comprise assemblages that have equilibrated under the ambient *T* (and *P*) of their upper mantle sites before entrainment in the transporting kimberlite.

Chemical disequilibrium indicating incomplete sub-solidus re-equilibration should be evidenced by reaction or exsolution textures and/or compositional zoning in the minerals. A significant degree of chemical disequilibrium is apparent in those eclogite and grospydite xenoliths (such as RV sample JJG41) in which there is obvious exsolution of garnet from originally more-aluminous clinopyroxene (Harte and Gurney, 1975; Lappin and Dawson, 1975; Lappin, 1978) and in which diffusion-controlled compositional gradients (notably for Al) exist in clinopyroxene adjacent to exsolved garnet lamellae (Sautter and Harte, 1988). Sautter and Harte interpret the development of garnet exsolution lamellae (and adjacent compositional gradients) in RV sample JJG41 as due to slow cooling from about 1400°C to 1000°C, with the lowest temperatures being recorded by Fe/Mg partitioning at garnet–clinopyroxene interfaces. However, well-documented samples with such obvious disequilibrium are uncommon, and even in JJG41 Fe/Mg zoning is absent in garnets and limited to only about 50 μm in clinopyroxene adjacent to garnet.

In the context of zoning, there are now some good examples of differences in the compositions of both Grt and Cpx inclusions in the cores and peripheries of individual diamonds (Bulanova *et al*, 1988; Griffin *et al*, 1988; Moore and Gurney, 1989; Otter and Gurney, 1989). These compositional differences, including changes in Fe–Mg partitioning between the Grt and Cpx, have been interpreted as resulting from changing temperatures during multi-stage growth of the host diamond. On the other hand, substantial inter-grain compositional variations in both Grt and Cpx in certain compositionally-banded eclogites (e.g. Carswell *et al*, 1981) are believed to reflect localised differences in *X* rather than in *P–T* conditions.

In other examples of zoning, Reid *et al* (1976) and Hatton and Gurney (1979) have reported examples of Na and Al depletion in the outermost zones of some eclogite clinopyroxenes. However, this Jd loss is almost certainly a late-stage decompression phenomenon associated with eruption of the host kimberlite.

Therefore good examples of eclogite xenoliths responding to decreasing sub-solidus temperatures by changes in Fe–Mg partitioning between Grt and Cpx are uncommon and, where apparent, there are visible signs of disequilibrium. In other samples, zoning is open to alternative interpretations. Hence, particularly for the large numbers of eclogites that lack obvious textural and chemical disequilibrium, it is appropriate to consider the alternative interpretation, i.e. that the wide range of calculated temperatures for eclogites and eclogite paragenesis inclusions is due to derivation from a wide depth range within the upper mantle. If this was the case, *P–T* conditions may be further constrained by reference to probable geothermal gradients for the shield areas to which kimberlite provinces are confined. If all samples from a given province equilibrated on a similar cratonic geotherm, then a wide range of both *T* and *P* is indicated. When the likely variations in calculated K_D temperatures with *P* are taken into account, the overall *T* range increases to nearly 600°C.

Intersections in PT space of the iso-K_D lines for individual samples with a preferred geotherm are shown on Figure 13.4 for diamond eclogites. Two graphite-diamond inversion curves are shown but, for reasons derived from the tighter constraints of garnet lherzolite thermobarometry (Carswell and Gibb, 1987), curve A (Bundy, 1980) is preferred to curve B (Kennedy and Kennedy, 1976). The reference geotherm lies between the continental 40 mW m^{-2} and 50 mW m^{-2} surface heat flow geotherms of Pollack and Chapman (1977), and corresponds closely with the geotherm deduced from the thermobarometry of garnet lherzolite xenoliths from the Mothae and Jagersfontein kimberlites, southern Africa (Carswell and Gibb, 1987; Carswell, unpublished data). The choice of this particular reference geotherm is arbitrary, but iso-K_D lines for diamond inclusion pairs also intersect this geotherm over a large T range (ca. 550°C).

Figure 13.4 shows that the intersections of the P–T iso-K_D lines for some of the lower-temperature samples lie slightly outside the diamond stability field as bounded even by curve A. A possible explanation might be that, in response to upper mantle diapiric movements, Fe–Mg re-equilibration between Grt and Cpx was readily achieved, whereas, for kinetic reasons, the diamond in some samples may have persisted metastably into the graphite stability field.

Fe^{2+}–Mg^{2+} partitioning between garnet and clinopyroxene is known to be affected by the Ca content of the garnet and the Na content of the clinopyroxene (see Section 4.7). Because of the wide range in garnet CaO contents, particularly in the kyanite eclogites and grospydites, it is appropriate to consider whether the CaO content could affect the calculated temperatures. Data calculated from the references given in Table 13.2 show that, whereas some samples with the combination of high X_{Ca}^{Grt}

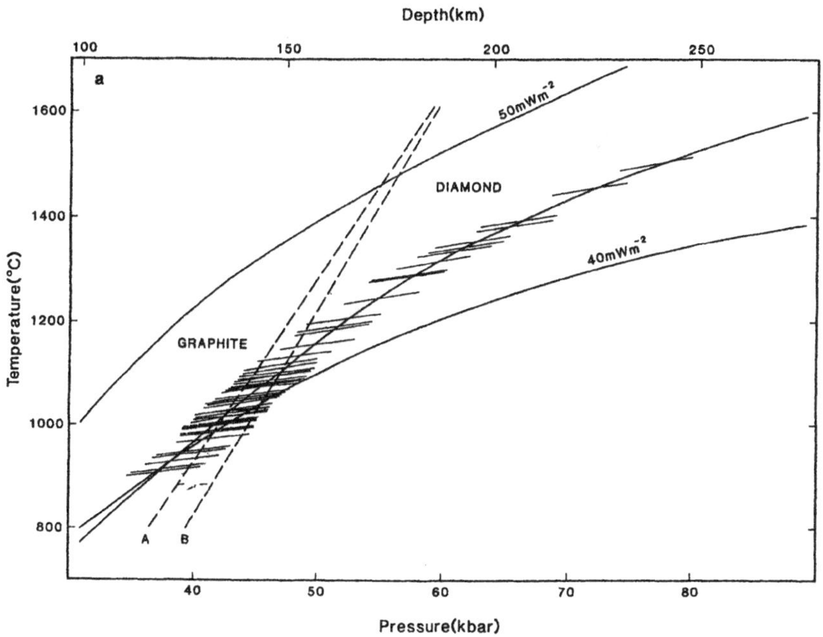

Figure 13.4 P–T plot for diamond eclogite samples showing the intersections of calculated iso-K_D lines, based on the K_D calibration of Powell (1985), with the possible upper mantle palaeogeotherm A and B represent alternative graphite⇌diamond inversion curves—see text.

and high X_{Na}^{Cpx} do have relatively-high equilibration temperatures, others do not. To test whether the Powell (1985b) calibration successfully takes into account the influence of X_{Ca}^{Grt} on K_D, we plotted these parameters on Figure 13.2a for two strongly-banded RV samples—6913 and RV9 that contain bands of both eclogite and kyanite eclogite (Lappin and Dawson, 1975; Carswell *et al*, 1981); for each specimen, the compositionally different bands are assumed to have equilibrated at the same temperatures. For both samples there is a positive correlation between ln K_D and X_{Ca}^{Grt}. This, together with no correlation between X_{Ca}^{Grt} and the calculated temperature (Figure 13.2b) gives confidence that the calibration of Powell takes adequate account of the influence of Ca concentrations in garnet.

Turning to the possible effects of X_{Na}^{Cpx}, Koons (1984) suggested that non-ideal solid-solution in clinopyroxenes with $X_{Na}^{Cpx} > 0.6$ affects thermometry based on Fe–Mg partitioning. However, the data for sample RV9 shows that variation on X_{Na}^{Cpx} from 0.28 to 0.70 does not greatly influence the calculated equilibration temperatures. After noting that garnets in diamond-bearing eclogites contain significant Na, Sobolev and Lavrentev (1971) suggested that Na/(Na + Ca) distribution between garnets and clinopyroxenes in eclogites might prove to be a useful pressure indicator. More recently, Bishop *et al* (1978) argued that there is a better correlation of the coefficient $K_{Na}(=X_{Na}^{Grt}/X_{Na}^{Cpx})$ with K_D and presumably, with P and/or T. However, on the basis of a larger data base, we find that X_{Na}^{Grt} in diamond eclogites varies considerably (0.010 to 0.031) and that there are no correlations between K_{Na} and K_D or calculated temperatures. Indeed, if (as seems likely) equilibration temperatures and pressures are positively correlated, there is no evidence of K_{Na} increasing with increasing pressure. Conversely, low K_{Na} values are typical of kyanite eclogites and grospydites which also have high X_{Ca}^{Grt} values. Overall, it appears that the combination of high X_{Ca}^{Grt} and X_{Na}^{Cpx} values, together with low K_{Na}, is largely controlled by bulk rock chemistry rather than P–T equilibration conditions.

13.8 Origin of high temperature kimberlite eclogites

Currently, there are two major hypotheses for the origin of HT kimberlite eclogites. In the first model, eclogites form from deep mantle melts that fractionate in upper mantle magma pools to give cumulates and frozen fractionated liquids (e.g. O'Hara and Yoder, 1967; MacGregor and Carter, 1970; Hatton, 1978; Kyser *et al*, 1982). A variant of the magmatic hypothesis, which suggests that eclogites are precipitates from the host kimberlites, is untenable on the grounds of gross age differences. A further magmatic variant, but on a global scale, postulates that eclogites are cumulates precipitated from an early magma ocean shortly after the accretion of the Earth (Anderson, 1982; McCulloch, 1982, 1989; Ohtani, 1985). The second major model is that eclogites are fragments of subducted oceanic crust (Ringwood and Green, 1966; Ater *et al*, 1984; Helmstaedt and Gurney, 1984; Jagoutz *et al*, 1984; MacGregor and Manton, 1986; Hatton and Gurney, 1987; Shervais *et al*, 1988).

13.8.1 *Model 1*

The magma-pool hypothesis arises from the assumption of broad similarities in chemical properties between mafic liquids and eclogite, the interpretation of group I textures as igneous cumulate textures, and experimental evidence for precipitation of garnet and pyroxene (some pyroxene containing kyanite in solid solution) from basic liquids. Mineral banding is interpreted as either reflecting original cumulate layering, or as sheeted intrusions of compositionally-different magmas. In their paper on the experimental conversion of basalt to eclogite, Yoder and Tilley (1962) proposed that 'for every major basalt-type there is a chemically equivalent eclogite'. Whereas this is broadly the case for many members of the broad chemical spectrum of basaltic rocks, and for certain of the major elements—Si, Fe, Mg, Ca and Na, there are sufficient exceptions to call in question an uncritical acceptance of this tenet. Even bearing in mind the fact that eclogites are metamorphic rocks, and it is thus perhaps simplistic to expect very close matches between eclogites and their protoliths, there are major discrepancies, particularly in the case of the peraluminous eclogites. For example, (i) TiO_2 is considerably lower than in most basaltic rocks (cf. Manson, 1967); (ii) with the exception of the low-K tholeiites, most basic rocks contain more K than occurs in unmetasomatised eclogites; (iii) the high-Mg eclogites are not matched by the high-Mg komatiites in which the high modal olivine and clinopyroxene contents lead to Al levels well below those of the eclogites; and (iv) no basaltic rocks, even the high-alumina Warner basalt (average of 3 specimens has 18.4 wt% Al_2O_3—Yoder and Tilley, 1962) match the >20% Al_2O_3 values found in the peraluminous eclogites. There are considerable difficulties in linking the samples in individual eclogite suites into groupings that are chemically meaningful. For example, Kramers (1979) and Smith *et al* (1989), in isotope studies on RV and Orapa eclogites respectively, conclude that the xenoliths do not comprise groups of related samples; the Orapa group II eclogites have a time-averaged enriched Rb/Sr and LREE character, suggestive of derivation from an enriched, lithospheric source, whereas the group I eclogites could derive from more primitive or depleted mantle with MORB characteristics. This evidence is not reconcilable with models envisaging a localised fractionating magma pool. Neither is the presence of coesite in eclogites of a wide compositional range since, in a fractionated sequence, it might be expected to occur only in the more fractionated Ca- and Fe-rich members (Shulze and Helmstaedt, 1988).

Turning to the textural evidence, we have pointed out in Section 13.3 the ambiguities of the original MacGregor and Carter (1970) classification, with its inherent igneous implications, and re-emphasise that some of the 'cumulate' textures are equally compatible with subsolidus recrystallisation. The development of some eclogites by exsolution of garnet (and sometimes kyanite) from original high-Al igneous clinopyroxene, and also formation of garnet by reaction between kyanite and clinopyroxene, has been noted. The former cases support the high-temperature 'magmatic' origin for at least some upper mantle eclogites, and may even indicate that, in some cases, garnet was not present as a cumulus phase in equilibrium with melt.

In the 'global eclogite precipitation' model, which has analogies with models for the early development of the Moon, there is a molten zone or 'magma ocean' with chondrite isotopic values in the outer layer of the accreting earth. Anderson (1982) envisages that an eclogite cumulate forms at the base of this magma zone, thereby setting up a mantle stratified in both its major element and isotopic characteristics; the 'eclogite' of the Anderson model would need to be garnetite (i.e. garnet with pyroxene in solid solution). This garnet layer could account for the long-lived LREE depleted character of eclogites with high εNd values (McCulloch, 1982, 1989). High pressure phase and liquid density modelling (Herzberg, 1984; Ohtani, 1985) supports the models for a stratified and heterogeneous mantle but some fractionation of perovskite- and spinel-structured phases may also be involved. Except for the work by McCulloch, these studies are more concerned with the development of isotopically-distinct reservoirs for generation of various magma types, rather than the generation of eclogite. However, subduction of LREE-depleted oceanic crust (Ringwood, 1982) is an alternative mechanism for producing a LREE-depleted (MORB reservoir) layer in the Earth.

13.8.2 *Model 2*

The recognition of the MORB-like characteristics of the predominant group I eclogites, together with the chemical similarities between many eclogites and fractionated basic rock units often found in ophiolite complexes, has given support to the second hypothesis (Ringwood and Green, 1966) that eclogite protoliths are ancient sea-floor rocks that have been deeply subducted (Helmstaedt and Carmichael, 1978; Helmstaedt and Gurney, 1984; Jagoutz *et al*, 1983, 1984; Ater *et al*, 1984; MacGregor and Manton, 1986; Hatton and Gurney, 1987; Shervais *et al*, 1988; Shulze and Helmstaedt, 1988). The mismatch between the chemistry of some eclogites, particularly the peraluminous eclogites, and basalts has been noted. For the peraluminous eclogites, closer matches can be found in fractionated basic rocks in the form of gabbroic anorthosite, and there is a close chemical similarity between kyanite eclogites and anorthositic rocks. Extreme examples of the basalt/eclogite mismatch are the corundum eclogites. For example, the corundum eclogite analysis shown in Table 13.1 (analysis 11) contains almost 30 wt% Al_2O_3 which is twice that found in most basalts.

Experimentally, kyanite eclogites have been synthesised using high-Al basalt and gabbroic anorthosite starting compositions (Green, 1967), and free silica (potentially coesite) has been shown experimentally to be stable in eclogites formed by high-PT metamorphism of tholeiitic compositions (Green and Ringwood, 1967a). An origin for the eclogite protoliths in ocean-floor layered complexes is compatible with other features: such as strong mineralogical and compositional banding; the low-K content of eclogites (gabbroic anorthosites or low-K MORB basalts); and the range in oxygen isotopes that can be matched by the values and ranges found in hydrothermally-altered MORB (Muehlenbachs and Clayton, 1976). Jagoutz *et al* (1984) interpret eclogites from Roberts Victor as noritic anorthosites (with positive Eu anomalies) and Ti-ferrogabbro, being complementary fractions of a basic magma differentiating at a shallow level in the crust.

Helmstaedt and Carmichael (1978) propose that grospydites are metarodingites—ocean floor high-Ca rocks that result from Ca-metasomatism, the Ca migrating from associated peridotites during serpentinisation. However, the problem with grospydites and kyanite/corundum eclogites is not so much high CaO values (which can often be matched in eclogites that lack peraluminous phases (Table 13.1)), but high Al_2O_3 combined with low SiO_2. This can, however, be matched in appropriate plagioclase-olivine combinations, and rodingitisation need not be invoked; in fact, development of metasomatic grossular and zoisite results in higher CaO values and much higher Ca/Al ratios than those found in the peraluminous eclogites.

There are two possible further attractions in the subduction hypothesis.

First, it can be argued that highly chloritised ocean-floor rocks could provide a suitable protolith for alkremites (Exley *et al*, 1983). The alternative is that alkremites represent the early precipitation of high-Mg garnets and spinel from an aluminous mantle melt before the onset of the additional pyroxene crystallisation to form eclogite (Nixon *et al*, 1978). This proposal, that alkremite is an early manifestation of an eclogite precipitation continuum, does not explain why the more Ca- and Fe-rich alkremite garnets (akin to eclogite garnets in composition) are also unaccompanied by pyroxene, nor the presence of corundum which, in the eclogite suite in general is associated with Ca-garnets in the peraluminous eclogites.

Second, subduction provides a means of transporting carbon and nitrogen into the upper mantle for diamond formation (Frank, 1969); and Ringwood (1977) suggests that diamonds originate in subducted oceanic spinel harzburgite cumulates. Subducted sedimentary material has also been suggested as the source of the SiO_2, CaO and Al_2O_3 found as coesite and calc-silicate inclusions in SE Australian diamonds (Sobolev, 1984). However, it would be necessary for the carbon in most diamonds to change its isotopic signature from the presumed biogenic character of ocean-floor carbon to the relatively high values ($d^{13}C$ -2 to -5‰) believed to be typical of the upper mantle. Nonetheless, the very large negative $d^{13}C$ values of some eclogite diamonds (Section 13.4.2) should be borne in mind; these apparently anomalous values could possibly be interpreted as resulting from incomplete re-equilibration of organic carbon. As an alternative to organic carbon, Schulze (1986) has suggested that diamond may form from graphite released during serpentinisation of lherzolite olivine, the serpentinisation also resulting in a low-Ca ultramafite; upon subduction into eclogite-facies conditions, the graphite-bearing low-Ca serpentinite is converted to harzburgite that contains diamond and low-Ca pyrope garnet. In suggesting olivine as the source of carbon, Schulze (1986) cited the observations of Pasteris (1981) who has found up to 9 wt% graphite in serpentinised olivines in kimberlites. Both Schulze (1986) and Pasteris (1981) are circumspect in attributing the graphite to release of carbon originally dissolved in olivine to the extent of several hundred ppm, as postulated by Freund *et al* (1980). More recent studies have found C at less than the analytical detection limits (30 ppm) in fresh upper mantle olivines (Mathez *et al*, 1987; Tsong *et al*, 1985), and it appears that experimentally-induced solubility of C in olivine is no more than 150 ppm at 30 kbar (Tingle and Green, 1987).

There are at least three problems with the subduction hypothesis. The first is that in all the chemical matching between eclogite and oceanic crust protolith, it must be assumed that the process of subduction, with its attendant recrystallisation and other metamorphic modifications, leaves the protolith bulk rock composition and isotopic characteristics virtually unchanged! The second is that of reconciliation of the range in isotopic characteristics found in individual eclogite suites, especially the 'enriched lithosphere' signature of some specimens. In some respects this is analogous to the case in upper mantle peridotites in which there is a range of isotope characteristics, and in some of which cryptic metasomatism results in enrichment in Rb and the LREE without visible changes in the mineralogy (Dawson, 1984). Perhaps an easier case may be made for metasomatic enrichment in the eclogites as many contain primary mica and amphibole as distinct from that developed during transport in kimberlite (Hatton and Gurney, 1987). The origin of this early metasomatism is obscure, but during subduction of hydrated oceanic crustal rocks, eclogitisation must have been accompanied by formation of a fluid into which incompatible elements would partition; such a fluid would have high metasomatic potential. The third problem is that, as pointed out by Ringwood (1982), the density of subducted and garnetised oceanic crust ensures that it is ultimately consigned to the lower mantle; the sampling problem is obvious, particularly in the case of entrainment of Archean eclogites in Phanerozoic transporting magmas, since some type of major upwelling (diapirism?) from depths of at least 600–700 km must be invoked. There is not such a problem for the sampling of oceanic crust that has been shallowly subducted relatively recently (e.g. that beneath the western USA that is sampled by Tertiary igneous activity on the Colorado Plateau—Helmstaedt and Doig, 1975).

If the formation of the deep-seated eclogites is in fact the result of subduction, it follows that subduction and major mantle convection are processes extending well back into the Archean.

13.9 Crustal exposures of ultra-high pressure eclogite facies lithologies

It was noted earlier that the presence of coesite and diamond indicates exceptionally high pressures in the formation of lithologies containing these minerals. Although indicative of ultra-high pressures, such crustal occurrences (in contrast to those in kimberlites) equilibrated within the MT rather than the HT part of the eclogite facies PT field.

Leaving aside meteorite craters where these phases are of impact metamorphic origin, there are now a number of examples of their occurrence in deeply-eroded metamorphic terrains. Diamond is an apparently primary phase in retrograded eclogite interbanded with biotite gneisses from the Precambrian Kokchetav massif of eastern Kazakhstan, USSR (Rozen *et al*, 1972; Sobolev and Shatskii, 1987, 1988). In another case, Nixon *et al* (1986) document graphite octahedra (interpreted as pseudomorphs after diamond) in pyroxenite and eclogite from the Beni Bousera complex of Morocco; $d^{13}C$ values of -15.5 to -22.95‰ (most in the range -19.1 to -22.95‰) for the Moroccan graphite (Slodkevich, 1982) suggest a biogenic origin.

Coesite occurs, together with almost pure pyrope garnet, talc and kyanite in the Dora Maira massif of the Western Alps (Chopin, 1984); the assemblage coesite + pyrope requires a minimum pressure of 30 kbar, equivalent to about 100 km depth, and is believed to derive from ultra-high P metamorphism of a Mg-rich (evaporitic clay?) sedimentary protolith as the result of crustal underthrusting and thickening during collision between the African and European plates (Schreyer, 1988b). Inclusions of coesite, or quartz pseudomorphing former coesite, have also been reported in garnets and pyroxenes in crustal eclogites from the Dora Maira massif (Chopin, 1987). Mosaics of quartz grains, interpreted as former coesite, are reported in eclogite from the Urals (Chesnokov and Popov, 1965) and Norway (Smith, 1984).

Although there is ambiguity as to whether some orogenic peridotites are of ophiolite or mantle origin, diamondiferous orogenic peridotites are known in Armenia (Pavlenko *et al*, 1974), in the Koryak mountains of eastern USSR (Shilo *et al*, 1978), and Tibet (Fang Chingsong and Bai Wenji, 1981). In addition to these authenticated peridotite sources for diamond, Dawson (1983) pointed out that there are many occurrences of alluvial diamonds (usually found during alluvial gold or platinum mining) in many orogenic belts (western USA—Alaska, Oregon, California; eastern USA—Carolina, Georgia, Alabama; Venezuela; eastern Australia—New South Wales, Victoria, Tasmania; and the east and west slopes of the Urals). In many of these fold belts there are lenses of peridotites that may be the possible source of the diamonds. Leaving aside these speculative examples, the authenticated diamondiferous orogenic peridotites provide firm evidence for the emplacement of deep upper mantle material to high crustal levels as the result of collision tectonics.

13.10 Concluding remarks

It is clear that no single simple model will satisfy the formation of all HT eclogites. Indeed any definitive model could be premature in view of the uncertainties in several fundamental aspects of eclogite xenolith petrology that have been recognised in this review. It is apparent that more research is required in certain fundamental areas.

1. A better eclogite xenolith classification must be established to distinguish clearly between disequilibrium igneous textures and those with well-equilibrated metamorphic textures. This, together with detailed microanalysis of zoned minerals, will permit a more conclusive interpretation of calculated K_D temperatures.
2. The implications of eclogites being open, rather than closed, chemical systems need to be explored. The effects of modal metasomatism are readily recognised, but there may also be the more subtle effects of pressure-release diffusion of LILE, as suggested from the inclusion data of V.S. Sobolev *et al* (1972); if so, what are the implications for isotope systematics?
3. The ranges in Nd and Sr isotope ratios within individual eclogite suites need to be re-evaluated, particularly in the light of the work of Jagoutz (1988) who has detected isotope heterogeneity within individual grains. Do the ranges reflect metasomatism, selective diffusion governed by crystallo-chemical effects or incomplete re-equilibration, or do they truly reflect events that affect the original eclogite protoliths?
4. Any model for eclogite genesis must be able to reconcile the juxtaposition within individual kimberlite pipes of eclogite xenoliths of differing texture, chemistry and radiogenic isotope values and, in some instances, containing diamonds with differing carbon isotope values.

In the context of point 4, an easy solution would be to propose a compromise model in which eclogites of different characteristics and origins (e.g. HP mantle cumulates or their recrystallised equivalents and/or subducted oceanic crust) are entrained from different upper mantle sites by the ascending kimberlite. An alternative would be to envisage the sampling of a mantle eclogite-facies equivalent of a crustal granite-intruded gneiss terrain in which, despite the overall similar bulk chemistries of the members, the fabrics, temperatures and isotope characteristics (and, hence, age) are highly variable.

References

Abraham, K., Hörmann, P.K. and Raith, M. (1974) Progressive metamorphism of basic rocks from the Hohe Tauern area, Tyrol (Austria), *Neues Jahrb. Mineral., Abh.* **122**, 1–35.

Abraham, K. and Schreyer, W. (1976) A talc–phengite assemblage in piemontite schist from Brezovica, Serbia, Yugoslavia. *J. Petrol.* **17**, 421–439.

Adams, G.E. and Bishop, F.C. (1982) Experimental investigation of Ca–Mg exchange between olivine, orthopyroxene and clinopyroxene: potential for geobarometry. *Earth Planet. Sci. Lett.* **57**, 241–250.

Adams, G.E. and Bishop, F.C. (1986) The olivine–clinopyroxene geobarometer: experimental results in the CaO–FeO–MgO–SiO_2 system. *Contrib. Mineral. Petrol.* **94**, 230–237.

Ahrens, T.J. and Schubert, G. (1975a) Rapid formation of eclogite in a slightly wet mantle. *Earth Planet. Sci. Lett.* **27**, 90–94.

Ahrens, T.J. and Schubert, G. (1975b) Gabbro–eclogite reaction rate and its geophysical significance. *Rev. Geophys. Space Phys.* **13**, No. 2, 383–400.

Akaogi, M. and Akimoto, S. (1979) High pressure phase equilibria in a garnet lherzolite, with special reference to Mg^{2+}–Fe^{2+} partitioning among constituent minerals. *Phys. Earth Planet. Int.* **19**, 31–51.

Akella, J. (1978) Garnet pyroxene equilibria in the system $CaSiO_3$–$MgSiO_3$–Al_2O_3 and in a natural mineral mixture. *Am. Mineral.* **61**, 589–598.

Allen, J.C. and Boettcher, A.L. (1978) Amphiboles in andesite and basalt: II. Stability as a function of P–T–f_{H_2O}–f_{O_2}. *Am. Mineral.* **63**, 1074–1087.

Allsopp, H.L. and Barrett, D.R. (1975) Rb–Sr age determinations on South African kimberlite pipes. *Phys. Chem. Earth* **9**, 605–617.

Allsopp, H.L, Nicolaysen, L.O. and Hahn-Weinheimer, P. (1969) Rb/K ratios and Sr-isotopic compositions of minerals in eclogitic and peridotitic rocks. *Earth Planet. Sci. Letts* **5**, 231–244.

Altherr, R., Schliestedt, M., Okrusch, M., Seidel, E., Kreuzer, H., Harre, W., Lenz, H., Wendt, I. and Wagner, G.A. (1979) Geochronology of high-pressure rocks on Sifnos (Cyclades, Greece). *Contrib. Mineral. Petrol.* **70**, 245–255.

Amundsen, H.E.F., Griffin, W.L. and O'Reilly, S.Y. (1987) The lower crust and upper mantle beneath northwestern Spitsbergen: evidence from xenoliths and geophysics. *Tectonophysics* **139**, 169–185.

Andersen, T., O'Reilly, S.Y. and Griffin, W.L. (1984) The trapped fluid phase in upper mantle xenoliths from Victoria, Australia: implications for upper mantle metasomatism. *Contrib. Mineral. Petrol.* **88**, 72–85.

Anderson, D.L. (1982) Isotopic evolution of the mantle: a model. *Earth Planet. Sci. Letts* **57**, 13–24.

Andreasson, P.-G. (1986) Seve terranes, Swedish Caledonides. *Geol. Fören. Stockholm Förh.* **108**, 261–263.

Andreasson, P.G., Gee D.G. and Sukotji, S. (1985) Seve eclogites in the Norbotten Caledonides. Sweden. In *The Caledonide Orogen—Scandinavia and Related Areas* (eds Gee, D.G. and Sturt, B.A.) pp. 887–901. John Wiley, Chichester.

Andreatta, C. (1935) La formazione gneisso-kinzigitica e le oliviniti di Val d'Ultimo (Alto Adige). *Mem. Mus. Storia. Natur. Ven. Trident.* **3**, 1–160.

Andreoli, M., Compagnoni, R. and Lombardo, B. (1976) Jadeite megablasts from Valchiusella (Sesia-Lanzo Zone, Western Alps). *Rend. Soc. Ital. Mineral. Petrol.* **32**, 681–698.

Andresen, A., Fareth, E., Berg, S., Kristensen, S.E. and Krogh, E. (1985) Review of Caledonian lithotectonic units in Troms, North Norway. In *The Caledonide Orogen—Scandinavia and Related Areas* (eds Gee, D.G. and Sturt, B.A.) pp. 569–578. John Wiley, Chichester.

Andriessen, P.A.M., Boelrijk, N.A.I.M., Hebeda, E.H., Priem, H.N.A., Verdurmen, E.A.Th. and Verschure, R.H. (1979) Dating the events of metamorphism and granitic magmatism in the Alpine orogen of Naxos (Cyclades, Greece). *Contrib. Mineral. Petrol.* **69**, 215–255.

Angel, F. (1957) Einige ausgewählte Probleme eklogitischer Gesteinsgruppen der österreichischer Ostalpen. *Neues Jahrb. Mineral. Abh.* **91**, 151–192.

Angel, F. and Schaider, F. (1950) Granat und Omphazit aus dem Eklogit des Gertrusk (Saualpe, Kärnten). *Carinthia* **II**, 58–60.

Ansilewski, J. (1966) Petrography of the gneiss area of the Bialskie Mts. (Polish with English summary). *Geol. Sudetica* **2**, 121–260.

Aranovich, L.Ya. and Podlesskii, K.K. (1989) Geothermometry of high-grade metapelites: simultaneously operating reactions. In: *Evolution of Metamorphic Belts* (eds. Daly, J.S., Cliff, R.A., and Yardley, B.W.D.). Geol. Soc. London Special Publ. **43**, 45–61.

Arculus, R.J., Ferguson, J., Chappell, B.W., Smith, D., McCulloch, M.T., Jackson, I., Hensel, H.D., Taylor, S.R., Knutson, J. and Gust, D.A. (1988) Trace element and isotopic characteristics of eclogites and other xenoliths derived from the lower continental crust of southeastern Australia and southwestern Colorado Plateau, U.S.A. In *Eclogites and Eclogite Facies Rocks*, (ed. Smith, D.C.) Elsevier, in press.

Arculus, R.J. and Smith, D. (1979) Eclogite, pyroxenite and amphibolite inclusions in the Sullivan Buttes latite, Chino Valley, Yavapai County, Arizona. In *The Mantle Sample: Inclusions in Kimberlites and other Volcanics*, (eds Boyd, F.R. and Meyer, H.O.A.) *Proc. 2nd Int. Kimberlite Conference* **2** American Geophysical Union, Washington, D.C., 59–86.

Arnold, A. and Scharbert, H.G. (1973) Rb–Sr-Altersbestimmungen on Granuliten der südlichen Bohmischen Masse in Österreich. *Schweiz. Mineral. Petrog. Mitt.* **53**, 61–78.

Arthaud, F. (1970) *Etude tectonique et microtectonique comparée de deux domaines hercyniens: les nappes de la Montagne Noire (France) et l'anticlinorium de l'Iglesiente (Sardaigne)*. Publ. de l'Université des Sciences et Techniques du Languedoc série Géologie Structurale, 175 pp.

Arthaud. F. and Matte, Ph. (1977) Late Paleozoic strike-slip faulting in southern Europe and Northern Africa: result of a right-lateral shear zone between the Appalachians and the Urals. *Geol. Soc. Am. Bull.* **88**, 1305–1320.

Ater, P.C., Eggler, D.H. and McCallum, M.E. (1984) Petrology and geochemistry of mantle eclogite xenoliths from Colorado–Wyoming kimberlites: recycled oceanic crust? *Kimberlites II: The Mantle and Crust-Mantle Relationships*. (ed. Kornprobst, J.) Devs. Petrology 11**B**, Elsevier, Amsterdam, 309–318.

Atlas, L. (1952) The polymorphism of $MgSiO_3$ and solid-state equilibria in the system $MgSiO_3$–$CaMgSi_2O_6$. *J. Geology* **60**, 125–147.

Aubouin, J. (1980) Introduction: the main structural complexes of Europe, pp. xiii–xxii. *Geology of the European Countries*. Comité National Francais de Geologie 1. Graham & Trotman, Ltd.

Audren, C. and Triboulet, C. (1984) Métamorphisme et déformation dans la ceinture de haute pression de l'Ile de Groix, Bretagne méridionale. *Bull. Soc. Géol. Mineral. Bretagne* **16**, 61–70.

Aurisicchio, C., Bocchio, R., Liborio, G. and Mottana, A. (1985). Petrogenesis of the eclogites from Soazza, Switzerland. *Chem. Geol.* **50**, 47–63.

Austrheim, H. (1978). *The metamorphic evolution of the granulite-facies rocks of Radøy, with special emphasis on the rocks of the Mangerite Complex*. Unpublished Cand. Real. thesis, University of Bergen.

Austrheim, H. (1987) Eclogitization of lower crustal granulites by fluid migration through shear zones. *Earth Planet. Sci. Lett.* **81**, 221–232.

Austrheim, H. and Griffin, W.L. (1985) Shear deformation and eclogite formation within granulite-facies anorthosites of the Bergen Arcs, western Norway. *Chem. Geol.* **50**, 267–281.

Autran, A. and Cogné, J. (1980) La zône interne de l'orogène varisque dans l'Ouest de la France et sa place dans le développement de la Chaîne hercynienne. In *Géologie de l'Europe du Précambrien aux Bassins Sédimentaires posthercyniennes*, (eds Cogné, J. and Slansky, M.) *Mém. B.R.G.M.* **108**, 90–111.

Bakun-Czubarow, N. (1983) Mineral chemistry of garnet lherzolites from the Sudetes, southwest Poland, *Yearb. Carnegie Inst. Wash.* **82**, 336–343.

Bakun-Czubarow, N. (1986) Mineral chemistry of ilmenite-bearing eclogite from Mountain Czernica in the Polish Central Sudetes. *XIVth IMA General Meeting*, Abstracts with Program, 49.

Baldelli, C., Dal Piaz, G.V. and Lombardo, B. (1985) Ophiolite eclogites from Verres, Val d'Aosta, Western Alps, Italy. *Chem. Geol.* **50**, 87–98.

Baldridge, W.S. (1979) Mafic and ultramafic inclusion suites from the Rio Grande Rift (New Mexico) and their bearing on the composition and thermal state of the lithosphere. *J. Volcanol. Geotherm. Res.* **6**, 319–351.

Balé, P. and Brun, J-P. (1986) Les complexes métamorphiques du Léon (NW Bretagne): un segment du domaine éo-hercynienne sud armoricain translaté au Dévonien. *Bull. Soc. geol. Fr.* **(8)** II. 471–477.

Ballard, S. and Pollack, H.N. (1988) Modern and ancient geotherms beneath southern Africa. *Earth Planet. Sci. Lett.* **88**, 132–142.

Ballèvre, M. (1988) Collision continentale et chemin P–T: l'unite pennique du Gran Paradis. *Mem. CAESS, Rennes*, **19**, 340 pp.

Ballèvre, M. Kienast, J.R. and Paquette, J.L. (1987) Le métamorphisme éclogitique dans la nappe hercynienne de Champtoceaux (Massif Armoricain), *Comptes-rendus de l'Académie des Sciences de Paris II, sé.* **305**, 127–131.

Banno, S. (1970) Classification of eclogites in terms of physical conditions of their origin. *Phys. Earth Planet. Int.* **3**, 405–421.

Banno, S. (1986) The high-pressure metamorphic belts of Japan: A review, In: *Blueschists and eclogites* (eds Evans, B.W. and Brown, E.H.) *Geol. Soc. Amer. Mem.* **164**, 365–374.

Barnicoat, A.C. (1988) Zoned high-pressure assemblages in pillow lavas of the Zermatt-Saas ophiolite zone, Switzerland, *Lithos* **21**, 227–236.

Barnicoat, A.C. and Fry, N. (1986) High-pressure metamorphism in the Zermatt–Saas Fee ophiolite zone, Switzerland. *J. Geol. Soc. London* **143**, 607–618.

Barton, J.M. and Ryan, B. (1977) A review of the geochronologic framework of the Limpopo mobile belt. *Bull. Geol. Surv. Botswana* **12**, 183–200.

Beard, B.L., Medaris, L.G., Jr., Valley, J.W., Brueckner, H.K. and Misař, Z. (1988) Mantle-derived Hercynian eclogites from the Bohemian Massif, Czechoslovakia. *Abstracts with Programs, Geol. Soc. Am.* **20**, A367.

Bearth, P. (1959) Über Eklogite, Glaukophanschiefer und metamorphe Pillowlaven, *Schweiz. Mineral. Petrogr. Mitt.* **39**, 267–286.

Bearth, P. (1967). Die Ophiolite der Zone von Zermatt–Saas Fee. Beiträge der Geologische Karte der Schweiz 130, 132 pp.

Bearth, P. (1973) Gesteins und Mineralparagenesen aus den Ophiolithen von Zermatt. *Schweiz. Mineral. Petrogr. Mitt.* **53**, 299–334.

Bearth, P. (1974) Zur Gliederung und Metamorphose der Ophiolithe der Westalpen. *Schweiz. Mineral. Petrogr. Mitt.* **54**, 385–397.

Bearth, P. and Schwander, H. (1981) The post-Triassic sediments of the ophiolite zone of Zermatt-Saas Fee and the associated manganese mineralisations. *Eclog. Geol. Helv.* **74**, 189–205.

Beccaluva, L., Dal Piaz, G.V. and Macciotta, G. (1984) Transitional to Normal MORB affinities in ophiolitic metabasites from the Zermatt–Saas, Combin and Antrona Units, Western Alps: implications for the palaeogeographic evolution of the Western Tethyan basin. *Geol. Mijnbouw* **63**, 165–177.

Beck, R. (1899) Neues von den Afrikanischen Diamantlagerstatten. *Zeit. für Prakt. Geologie* 417–419.

Beck-Mannagetta, P. (1961) Zur Deutung der Eklogite im Koralpenkristallin (Zentral-alpen). *Tschermaks Mineral. Petrogr. Mitt.* **7**, 437–450.

Behr, H-J., Engel, W. and Franke, W. (1980) *Münchberger Gneismasse und Bayerischer Wald.* Exc. Guide No. 2, Int. Conf. on the Effect of Deformation on Rocks, Gottingen, 100 pp.

Behr, H-J., Engel, W., Franke, W., Giese, P. and Weber, K. (1984) The Variscan belt in Central Europe: main structures, geodynamic implications, open questions. *Tectonophysics* **109**, 15–40.

Bellière, J. (1958) Contribution à l'étude pétrogénétique des schistes cristallins du massif des Aiguilles Rouge (Haute Savoie). *Ann. Soc. Géol. Belgique.* **81**, 1–198.

Bellière, J. and Streel, M. (1980) Roches d'âge viséen supérieure dans le massif des Aiguilles Rouges (Hte. Savoie). *C.R. Acad. Sci.* (quoted by Liégeois and Duchesne, 1981).

Berman, R. (1979) Thermal Properties. In *The Properties of Diamond*, (ed. Field, J.E.). Academic Press, London, 1–22.

Berman, R.G. (1988) Internally-consistent thermodynamic data for minerals in the system Na_2O–K_2O–CaO–MgO–FeO–Fe_2O_3–Al_2O_3–SiO_2–TiO_2–H_2O–CO_2. *J. Petrol.* **29**, 445–522.

Bernard-Griffiths, J., Cantagrel, J.M. and Duthou, J.L. (1977) Radiometric evidence for an Acadian tectonometamorphic event in Western Massif Central Français. *Contrib. Mineral. Petrol.* **61**, 199–212.

Bernard-Griffiths, J., Carpenter, M.S.N., Peucat, J.-J. and Jahn, B.M. (1986) Geochemical and isotopic characteristics of blueschist facies rocks from Île de Groix, Armorican Massif. *Lithos* **19**, 235–253.

Bernard-Griffiths, J. and Cornichet, J. (1985) Origin of eclogites from south Brittany, France: A Sm–Nd isotopic and REE study. In *Isotope Geochemistry and Geochronology of Eclogites* (eds Smith, D.C. and Vidal, Ph.) Chem. Geol. (Isot. Geosci. Sect.) 52, 185–201.

Bernard-Griffiths, J., Gebauer, D., Grünenfelder, M. and Piboule, M. (1985a) The tonalite belt of Limousin (French Central Massif): U–Pb zircon ages and geotectonic implications. *Bull. Soc. géol. Fr.* **(8) I**, 523–529.

Bernard-Griffiths, J. and Jahn, B.M. (1981) REE geochemistry of eclogites and associated rocks from Sauviat-sur-Vige. Massif Central. France. *Lithos* **14**, 263–274.

Bernard-Griffiths, J., Peucat, J-J., Cornichet, J., Iglesias Ponce de Léon, M. and Gil Ibarguchi, J.I. (1985b) U–Pb, Nd isotope and REE geochemistry in eclogites from the Cabo Ortegal Complex, Galicia, Spain: An example of REE immobility conserving MORB-like patterns during high-grade metamorphism. In *Isotope Geochemistry and Geochronology of Eclogites.* (eds Smith, D.C. and Vidal. Ph.) Chem. Geol. (Isot. Geosci. Sect.) **52**, 217–22.

Bertrand, P. and Mercier, J-C.C. (1985) The mutual solubility of coexisting ortho- and clinopyroxene; towards an absolute geothermometer for the natural system? *Earth Planet. Sci. Lett.* **76**, 109–122.

Bertrand, P., Sotin, C., Gaulier, J-M. and Mercier, J-C.C. (1987) La solubilité de l'aluminium dans l'orthopyroxène. Inversion globale des données expérimentales du système MgO–Al_2O_3–SiO_2. *Bull. Geol. Soc. Fr.* **8**, t. III, 821–832.

Bertrand, P., Sotin, C., Mercier, J-C.C. and Takahashi, E. (1986) From the simplest chemical system to the natural one: garnet peridotite barometry. *Contrib. Mineral. Petrol.* **93**, 168–178.

Best, N.F. and Graham, C.M. (1978) Redetermination of the reaction 2 zoisite + quartz + kyanite = 4 anorthite + H_2O. In: *Progress in Experimental Petrology*, (ed. McKenzie, W.S.) Natural Environment Research Council Dll, Manchester, 153–154.

Bickle, M.J. and Hawkesworth, C.J. (1978) Deformation phases and the tectonic history of the eastern Alps. *Bull. Geol. Soc. Am.* **89**, 293–306.

Binns, R.A. (1967) Barroisite-bearing eclogite from Naustdal, Sogn og Fjordane, Norway. *J. Petrol.* **8**, 349–371.

Bishop, F.C., Smith, J.V. and Dawson, J.B. (1978) Na, K, P and Ti in garnet, pyroxene and olivine from peridotite and eclogite xenoliths from African kimberlites. *Lithos* **11**, 155–173.

Black, P.M. (1974) Oxygen isotope study of metamorphic rocks from the Ouégoa district, New Caledonia. *Contrib. Mineral. Petrol.* **47**, 197–206.

Black, P.M. (1977) Regional high-pressure metamorphism in New Caledonia: phase equilibria in the Ouégoa district. *Tectonophysics* **43**, 89–107.

Blackburn, W.H. (1968) The spatial extent of chemical equilibrium in some high-grade metamorphic rocks from the Grenville of southeastern Ontario, *Contrib. Mineral. Petrol.* **19**, 72–92.

Blake, M.C., Brothers, R.N. and Lanphere, M.A. (1977) Radiometric ages of blueschists in New Caledonia. In: *International symposium on geodynamics in south-west Pacific, Nouméa 1976*, Editions Technip, 279–281.

Bloxam, T.W. and Allen, J.B. (1960) Glaucophane schist, eclogite, and associated rocks from Knocknormal in the Girvan-Ballantrae complex, south Ayrshire. *Trans. R. Soc. Edinburgh.* **64**, 1–27.

Blümel, P. (1983) The western margin of the Bohemian Massif in Bavaria. *Fortschr. Mineral.* **61**, 171–195.

Blümel, P. and Schreyer, W. (1976) Progressive regional low-pressure metamorphism in Moldanubian metapelites near Bodenmais, E Bavaria. *Krystalinikum* **12**, 7–30.

Bocchio, R. (1975). Le eclogiti retrometamorfiche di Gorduno (Canton Ticino, Svizzera). *Periodic. di Mineral.* **44**, 245–256.

Bocchio, R. (1977) Ferropargasite alluminifera in un' eclogite di Gorduno. *Rend. Soc. Ital. Mineral. Petrol.* **33**, 519–529.

Bocchio, R., Liborio, G. and Mottana, A. (1985) Petrology of the amphibolitised eclogites of Gorduno, Lepontine Alps, Switzerland. *Chem. Geol.* **50**, 65–86.

Bocchio, R., and Mottana, A. (1975) Trace element abundances of iron-rich eclogites, with implications on the geodynamical evolution of the Voltri Group (Pennidic Belt). *Chem. Geol.* **15**, 273–283.

Bocchio, R., Ungaretti, L. and Rossi, G. (1978) Crystal chemical study of eclogitic amphiboles from Alpe Arami, Lepontine Alps, southern Switzerland. *Rend. Soc. Ital. Mineral. Petrol.* **34**, 453–470.

Bocquet, J. (1974) *Études minéralogiques et pétrologiques sur les métamorphismes d'âges alpin dans les Alpes françaises.* Thèse Université de Grenoble, 489 pp.

Bocquet, J., Delaloye, M., Hunziker, J.C. and Krummenacher, D. (1974) K–Ar and Rb–Sr dating of blue amphiboles, micas and associated minerals from the Western Alps. *Contrib. Mineral. Petrol.* **47**, 7–26.

Bodewig, K. (1876) Über den Glaukophan von Zermatt. *Poggendorf Annals* **158**, 224.

Bodinier, J-L., Burg, J-P., Leyreloup, A. and Vidal, H. (1988). Reliques d'un bassin d'arrière-arc subducté, puis obducté dans la région de Marvejols (Massif Central), *Bull. Soc. géol. Fr.* **(8)** IV, 21–33.

Bodinier, J-L., Dupuy, C., Dostal, J. and Merlet, C. (1987) Distribution of trace transition elements in olivine and pyroxenes from ultramafic xenoliths: application of microprobe analysis. *Am. Mineral.* **72**, 902–913.

Bodinier, J.L., Giraud, D., Dupuy, C., Leyreloup, A. and Dostal, J. (1986) Caractérisation géochimique des métabasites associées à la suture méridionale hercynienne: Massif central français et Chamrousse (Alpes). *Bull. Soc. géol. Fr.* **(8)** II, 115–123.

Bodmer-Beder, A. (1902) Petrographische Untersuchungen von Steinwerkzeugen und ihrer Rohmaterialen aus schweizerischen Pfahlbaustatten. *Neues Jahrb. Mineral. Geol.* 16.

Boettcher, A.L. (1970) The system $CaO–Al_2O_3–SiO_2–H_2O$ at high pressures and temperatures. *J. Petrol.* **11**, 337–379.

Boettcher, A.L. and Wyllie, P.J. (1969) Phase relationships in the system $NaAlSiO_4–SiO_2–H_2O$ to 35 kilobars pressure. *Am. J. Sci.* **267**, 875–909.

Bohlen, S.R. (1987) Pressure–temperature–time paths and a tectonic model for the evolution of granulites. *J. Geol.* **95**, 617–632.

Bohlen, S.R. and Boettcher, A.L. (1982) The quartz-coesite transformation: a precise determination and the effects of other components. *J. Geophys. Res.* **87**, 7073–7078.

Bohlen, S.R., Boettcher, A.L. and Wall, V.J. (1982) The system albite-H_2O-CO_2: a model for melting and activities of water at high pressures. *Am. Mineral.* **67**, 451–462.

Bohlen, S.R. Wall, V.J. and Boettcher, A.L. (1983) Experimental investigations and geological applications of equilibria in the system $FeO–TiO_2–Al_2O_3–SiO_2–H_2O$. *Am. Mineral.* **68**, 1049–1058.

Bonatti, E. (1980) Metal deposits in the oceanic lithosphere. In *The Sea*, (ed. Emiliani, C.) John Wiley and Sons, 639–686.

Bonhomme, M.G. and Fluck, P. (1981) Nouvelles données isotopiques Rb–Sr obtenues sur les granulites des Vosges—Age protérozoique terminal de la série volcanique calco-alcaline et âge acadien du métamorphisme régional. *C.R. Acad. Sci. Paris.* **293**(D), 771–774.

Bonney, T.G. (1899) The parent rock of the diamond in South Africa. *Proc. R. Soc. London* **65**, 223–236.

Bonnot, H. and Piboule, M. (1980) Mise en évidence d'une dualité d'origine des ultrabasites à grenat du Limousin et recherche de la signification des péridotites d'origine mantellique dans le Massif Central français, *C.R. Acad. Sci. Paris* **291**, Ser. D, 129–132.

Borghi, A., Cadoppi, P., Porro, A. and Sacchi, R. (1985) Metamorphism in the north part of the Dora-Maira massif (Cottian Alps). *Boll. Museo Reg. Sci. Nat. Torino* **3**, 369–380.

Bouchardon, J.L. (1987) Évaluation des conditions de pression–température du 'top' métamorphique des éclogites du Lévezou (Rouergue, Massif central français). *C.R. Acad. Sci. Paris* **298**(II), 271–275.

Bouchardon, J.L. Briand, B., Gay, M., Piboule, P. and Santallier, D. (1987) Eclogites in the French Palaeozoic orogen. *Terra Cognita* **7**, 175–6.

Bougault, H. and Cande, S. (1983) Local versus regional mantle heterogeneties: IPOD Leg 82. *Terra Cognita* **3**, 121.

Boyd, F.R. (1970) Garnet peridotites and the system $CaSiO_3–MgSiO_3–Al_2O_3$, *Mineral. Soc. Am. Spec. Paper* **3**, 63–75.

Boyd, F.R. (1973) A pyroxene geotherm. *Geochim. Cosmochim. Acta* **37**, 2533–2546.

Boyd, F.R. and England, J.L. (1964) The system enstatite–pyrope. *Yearb. Carnegie Inst. Washington* **63**, 157–161.

Boyd, F.R. and Schairer, J.F. (1964) The system $MgSiO_3–CaMgSi_2O_6$. *J. Petrol.* **5**, 275–309.

Bradshaw, J.Y. (1978) *Mineralogy and petrology of interlayered eclogite and high-grade blueschist from California.* M.Sc. thesis, University of Calgary.

Brady, J.B. (1983) Intergranular diffusion in metamorphic rocks. *Am. J. Sci.* **283-A**, 181–200.

Brady, J.B. and McCallister, R.H. (1983) Diffusion data for clinopyroxenes from homogenization and self-diffusion experiments. *Am. Mineral.* **68**, 95–105.

Brastad, K. (1985) Relationships between peridotes, anorthosites and eclogites in Bjørkedalen, western Norway. In *The Caledonide Orogen—Scandinavia and Related Areas* (eds Gee, D.G. and Sturt, B.A.) pp. 859–872. John Wiley, Chichester.

Brey, G. and Huth, J. (1984) The enstatite–diopside solvus to 60 kbar. In *Kimberlites II: The Mantle and Crust Mantle Relationships,* (ed. Kornprobst, J.) Developments in Petrology **11B**, 257–264 Elsevier, Amsterdam.

Brey, G.P., Nickel, K.G. and Köhler, Th. (1989) Geothermobarometry in 4-phase peridotites. Part 2: New thermobarometers. (in preparation).

Briand, B., Piboule, M, and Bouchardon. J-L. (1988) Diversité géochimique des métabasites des groupes leptyno-amphiboliques du Rouergue et de Marvejols (Massif central). Origine et implications. *Bull. Soc. géol. Fr.* **(8)** IV, 489–498.

Brière, Y. (1920) Les éclogites françaises. *Bull. Soc. Mineral, Fr.* **42**, 72–222.

Briggs, R.M. (1978) Ferrocarpholite associated with low-grade blueschists, northern New Caledonia. *Mineral. Mag.* **42**, 147.

Bröcker, M. and Okrusch, M. (1987) Geochemistry of metabasites and associated metasediments from Tinos island, Cyclades. *Terra Cognita* **7**, 169.

Brodie, K.H. and Rutter, E.H. (1985) On the relationship between deformation and metamorphism, with special reference to the behavior of basic rocks. In *Metamorphic Reactions, Advances in Physical Geochemistry 4,* (eds Thompson, A.B. and Rubie, D.C.) Springer-Verlag, New York, 138–179.

Brothers, R.N. (1974) High-pressure schists in northern New Caledonia. *Contrib. Mineral. and Petrol.* **46**, 109–127.

Brothers, R.N. (1985) Regional mid-Tertiary blueschist–eclogite metamorphism in northern New Caledonia. *Géol. France* **1**; 37–44.

Brothers, R.N. and Yokoyama, K. (1982) Comparison of the high-pressure schist belts of New Caledonia and Sanbagawa, Japan. *Contrib. Mineral. Petrol.* **79**, 219–229.

Brown, E.H. and Bradshaw, J.Y. (1979) Phase relations of pyroxene and amphibole in greenstone, blueschist and eclogite of the Franciscan complex, California. *Contrib. Mineral. Petrol.* **71**. 67–83.

Brown, E.H. and Forbes, R.B. (1986) Phase petrology of eclogitic rocks in the Fairbanks district, Alaska. In: *Blueschists and eclogites.* (eds Evans, B.W. and Brown, E.H.) *Geol. Soc. Amer. Mem.* **164**, 155–167.

Brown, P., Essene, E.J. and Peacor D.R. (1978) The mineralogy and petrology of manganese-rich rocks from St. Marcel, Piedmont, Italy. *Contrib. Mineral. Petrol.* **67**, 227–232.

Brown, W.H. Fyfe, W.S. and Turner F.J. (1962) Aragonite in Californian glaucophane schists. *J. Petrol.* **3**, 566–587.

Brueckner, H.K. (1972) Interpretation of Rb–Sr ages from the Precambrian and Palaeozoic rocks of southern Norway. *Am. J. Sci.* **272**, 334–358.

Brueckner, H.K. (1974) 'Mantle' Rb/Sr and $^{87}Sr/^{86}Sr$ ratios for clinopyroxenes from Norwegian garnet peridotites and pyroxenites. *Earth Planet. Sci. Lett.* **24**, 26–32.

Brueckner, H.K. (1977a) A crustal origin for eclogites and a mantle origin for garnet peridotites: strontium isotopic evidence from clinopyroxenes. *Contrib. Mineral. Petrol.* **60**, 1–15.

Brueckner, H.K. (1977b) A structural, stratigraphic and petrologic study of anorthosites, eclogites, and ultramafic rocks and their country rocks, Tafjord area, western south Norway. *Nor. Geol. Unders.* **332**, 1–53.

Brueckner, H.K. (1979) Precambrian ages from the Geiranger-Tafjord-Grotli area of the Basal Gneiss Region, west Norway: *Nor. Geol. Tidsskr.* **59**, 141–153.

Bryhni, I. (1966) Reconnaissance studies of gneisses, ultrabasites, eclogites and anorthosites in outer Nordfjord, western Norway. *Nor. Geol. Unders.* **241**, 1–68.

Bryhni, I., Bockelie, J. and Nystuen, J.P. (1981) *Excursions in the Scandinavian Caledonides: The southern Norwegian Caledonides, Oslo–Sognefjord–Ålesund.* UCS Excursion No. A1, 99–111. Uppsala Caledonide Symposium.

Bryhni, I. and Griffin, W.L. (1971) Zoning in eclogite garnets from Nordjord, west Norway. *Contr. Mineral. Petrol.* **32**, 112–125.

Bryhni, I., Krogh, E.J. and Griffin, W.L. (1977) Crustal derivation of Norwegian eclogites: a review *Neues Jahrb. Mineral Abh.* **130**, 49–68.

Bryhni, I. and Sturt, B.A. (1985). Caledonides of southwestern Norway. In *The Caledonide Orogen—Scandinavia and Related Areas* (eds. Gee, D.G. and Sturt, B.A.) pp. 89–108. John Wiley, Chichester.

Bulanova, G.P., Novgorodov, P.G. and Pavlova, L.A. (1988) A first find of melt inclusion in diamond from the Mir pipe. Dokl. Akad. Nauk. SSSR Vol No: (5) 756–764 (In Russian).

Bundy, F.R. (1980) The P, T phase and reaction diagrams for elemental carbon, 1979, *J. Geophys. Res.* **85**, 6930–6936.

Bundy, F.R., Bovenkerk, H.P., Strong, H.M. and Wentorf, R.H. (1961) Diamond-graphite equilibrium line from growth and graphitisation of diamond. *J. Chem. Phys.* **35**, 383–391.

Burg, J.P., Leyreloup, A., Marchand, J. and Matte, Ph. (1984) Inverted metamorphic zonation and large-scale thrusting in the Variscan belt: an example in the French Massif Central. In *Variscan Tectonics of the North Atlantic Region* (eds Hutton, D.H.W. and Sanderson, D.J.) Spec. Pub. Geol. Soc. London 14, 47–61.

Burgath, K., Mohr, M., Drozak, J. and Klimainsky, M. (1987) Origin of metamorphic peridotites and amphibolites in the Central Schwarzwald. *Terra Cognita* 7, 168.

Busch, K. (1970) Die Eklogitvorkommen des Kristallinen Grundgebirges in NE-Bayern IV: Die Eklogit der Oberpfalz und ihr metamorpher Abbau. *Neues Jahrb. Mineral. Abh.* **113**, 138–178.

Cabanis, B. and Godard, G. (1987) Les éclogites du pays de Léon (Nord-Ouest du Massif armoricain): étude pétrologique et géochimique; implications géodynamiques. *Bull. Soc. géol. Fr.* **8** III, 1133–1142.

Caby, R., Kienast, J.R. and Saliot, O. (1978) Structure, métamorphisme et modèle d'évolution tectonique des Alpes Occidentales. *Rev. Geogr. Phys. Geol. Dyn.* **20**, 307–322.

Callegari, E., Compagnoni, R., Dal Piaz, G.V., Frisatto, V., Gosso, G. and Lombardo, B. (1976) Nuovi affioramenti di metagranitoidi nella zona Sesia–Lanzo. *Rend. Soc. Ital. Mineral. Petrol.* **32**, 97–111.

Cameron, M. and Papike, J.J. (1980) Crystal chemistry of silicate pyroxenes. In *Pyroxenes*, (ed. Prewitt, C.T.) *Mineral. Soc. Am. Reviews in Mineralogy*, **7**, 5–92.

Carlson, W.D. and Lindsley, D.H. (1988) Thermochemistry of pyroxenes on the join $Mg_2Si_2O_6$–$CaMgSi_2O_6$. *Am. Mineral.* **73**, 242–252.

Carlson, W.D. (1980) The calcite-aragonite equilibrium: effects of Sr substitution and anion orientational disordering. *Am. Mineral.* **65**, 1252–1262.

Carlson, W.D. and Rosenfeld, J.L. (1981) Optical determination of topotactic aragonite–calcite growth kinetics: metamorphic implications. *J. Geol.* **89**, 615–638.

Carman, J.H. and Gilbert, M.C. (1983) Experimental studies on glaucophane stability. *Am. J. Sci.* **283A**, 414–437.

Caron, J.-M. and Péquignot, G. (1986) The transition between blueschists and lawsonite-bearing eclogites based on observations from Corsican metabasalts. *Lithos* **19**, 205–218.

Carpena, J. (1985) Tectonic interpretation of an inverse gradient of zircon fission track ages with respect to altitude: alpine thermal history of the Gran Paradiso basement. *Contrib. Mineral. Petrol.* **90**, 74–82.

Carpenter, M.A. (1980) Mechanism of exsolution in sodic pyroxenes. *Contrib. Mineral. Petrol.* **71**, 289–300.

Carpenter, M.A. and Okay, A. (1978) Topotactic replacement of augite by omphacite in a blueschist rock from north-west Turkey. *Mineral. Mag.* **42**, 435–438.

Carswell, D.A. (1968a) Picritic magma–residual dunite relationships in garnet peridotite at Kalskaret near Tafjord, south Norway, *Contrib. Mineral. Petrol.* **19**, 97–124.

Carswell, D.A. (1968b) Possible primary upper mantle peridotite in Norwegian basal gneiss. *Lithos* **1**, 322–355.

Carswell, D.A. (1973) Garnet pyroxenite lens within Ugelvik layered garnet peridotite. *Earth Planet. Sci. Lett.* **20**, 347–352.

Carswell, D.A. (1974) Comparative equilibration temperatures and pressures of garnet lherzolites in Norwegian gneisses and in kimberlite. *Lithos* **7**, 113–121.

Carswell, D.A. (1975) Primary and secondary phlogopites and clinopyroxenes in garnet lherzolite xenoliths. *Phys. Chem. Earth.* **9**, 145–157.

Carswell, D.A. (1980) Mantle derived lherzolite nodules associated with kimberlite, carbonatite and basalt magmatism: a review. *Lithos* **13**, 121–138.

Carswell, D.A. (1981) Clarification of the petrology and occurrence of garnet lherzolites, garnet websterites and eclogite in the vicinity of Rodhaugen, Almklovdalen, west Norway, *Nor. Geol. Tidsskr.* **61**, 249–260.

Carswell, D.A. (1986) The metamorphic evolution of Mg–Cr type Norwegian garnet peridotites. *Lithos* **19**, 279–297.

Carswell, D.A. (1989) Hercynian high P–T metamorphism and uplift history in the Moldanubian Zone of the Bohemian Massif in Lower Austria. *28th Int. Geol. Congr. Abstract*, Washington. 1–248.

Carswell, D.A., Clarke, D.B. and Mitchell, R.H. (1979) The petrology and geochemistry of ultramafic nodules from Pipe 200, Northern Lesotho. In *The Mantle Sample: Inclusion in Kimberlite and Other Volcanis*, (eds Boyd, F.R. and Meyer, H.O.A.) Proceedings 2nd International Kimberlite Conference **2**, 127–144, American Geophysical Union, Washington, D.C.

Carswell, D.A. and Cuthbert, S.J. (1986) Eclogite facies metamorphism in the lower continental crust. In *Nature of the Lower Continental Crust*, (eds Dawson, J.B., Carswell, D.A., Hall, J. and Wedepohl, K.H., Geol. Soc. London Special Publ. 25, 193–209.

Carswell, D.A., Dawson, J.B. and Gibb, F.G.F. (1981) Equilibration conditions of upper-mantle eclogites: implications for kyanite-bearing and diamondiferous varieties. *Mineral. Mag.* **44**, 79–89.

Carswell, D.A. and Gibb, F.G.F. (1980a) The equilibration conditions and petrogenesis of European crustal garnet lherzolites. *Lithos* **13**, 19–29.

Carswell, D.A. and Gibb, F.G.F. (1980b) Geothermometry of garnet lherzolite nodules with special reference to those from the kimberlites of northern Lesotho. *Contrib. Mineral. Petrol.* **74**, 403–416.

Carswell, D.A. and Gibb, F.G.F. (1987a) Evaluation of mineral thermometers and barometers applicable to garnet lherzolite assemblages. *Contrib. Mineral. Petrol.* **95**, 499–511.

Carswell, D.A. and Gibb, F.G.F. (1987b). Garnet lherzolite xenoliths in the kimberlites of northern Lesotho: revised *P–T* equilibration conditions and upper mantle palaeogeotherm. *Contrib. Mineral. Petrol.* **97**, 473–487.

Carswell, D.A. and Griffin, W.L. (1981) Calculation of equilibration conditions for garnet granulite and websterite nodules in African kimberlite pipes. *Tschermaks Mineral Petrogr Mitt.* **28**, 229–244.
Carswell, D.A., Griffin, W.L. and Kresten, P. (1984). Peridotite nodules from the Ngopetsoeu and Lipelaneng kimberlites, Lesotho; A crustal or mantle origin. In *Kimberlites II: The Mantle and Crust–Mantle Relationships.* (ed. Kornprobst, J.); Developments in Petrology **11B**, 127–144. Elsevier, Amsterdam.
Carswell, D.A. and Harvey, M.A. (1985) The intrusive history and tectono-metamorphic evolution of the Basal Gneiss Complex in the Moldefjord area, west Norway. In *The Caledonide Orogen-Scandinavia and Related Areas* (eds Gee, D.G. and Sturt, B.A.) pp. 843–857, Wiley, Chichester.
Carswell, D.A., Harvey, M.A. and Al-Samman, A. (1983) The petrogenesis of contrasting Fe–Ti and Mg–Cr garnet peridotite types in the high grade gneiss complex of western Norway. *Bull. Mineral.* **106**, 727–750.
Carswell, D.A., Krogh, E.J. and Griffin, W.L. (1985) Norwegian orthopyroxene eclogites: calculated equilibrium conditions and petrogenetic implications. In *The Caledonide Orogen—Scandanavia and Related Areas*, (eds Gee, D.G. and Sturt, B.A.) John Wiley, London, 823–841.
Carter, N.L. and Tsenn, M.C. (1987). Flow properties of continental lithosphere. *Tectonophysics* **136**, 27–63.
Caruba, C. (1981) Les 'faciés éclogitiques' du Massif des Maures (Var, France). *C.R. Somm. Soc. Géol. Fr.* **1**, 18–21.
Castelli, D. (1987) *Il metamorfismo alpino delle rocce carbonatiche della zona Sesia Lanzo (Alpi Occidentali).* Thesis, Universita di Torino, 141 pp.
Cawthorn, R.G. and Collerson, K.D. (1974) The recalculation of pyroxene end-member parameters and the estimation of ferrous and ferric iron content from electron microprobe analyses. *Am. Mineral.* **59**, 1203–1208.
Cazes, M., Torreilles, G., Bois, C., Damotte, B., Galdéano, A., Hirn, A., Mascle, A., Matte, P., Van Ngoc, P. and Raoult, J-F. (1985) Structure de la croûte hercynienne du Nord de la France: premiers résultats du profil ECORS. *Bull. Soc. géol. Fr.* **8** I, 925–941.
Chatterjee, N.D. (1971) Phase equilibria in the alpine metamorphic rocks of the environs of the Dora–Maira–Massif, western Italian Alps. *Neues Jahrb. Mineral. Abh.* **114**, 181–210.
Chatterjee, N.D., Johannes, W. and Leistner, H. (1984) The system $CaO–Al_2O_3–SiO_2–H_2O$: new phase equilibria data, some calculated phase relations, and their petrological applications. *Contrib. Mineral. Petrol.* **88**, 1–13.
Chaussidon, M., Albarede, F. and Sheppard, S.M.F. (1987) Sulphur isotope heterogeneity in the mantle from ion microprobe measurements of sulphide inclusions in diamonds. *Nature* **330**, 242–244.
Chesnokov, B.V. and Popov, V.A. (1965) Increase in the volume of quartz grains in South Urals eclogite. *Dokl. Akad. Nauk SSSR* **162**, 176–178.
Chiesa, S., Liborio, G., Mottana, A. and Pasquarè, G. (1972) La paragonite nei calcescisti delle Alpi: Distribuzione e interpretazione geo-petrologica, *Memorie Scocietà Geologica Italiana* **11**, 1–30.
Chinner, G.A. and Cornell, D.H. (1974) Evidence of kimberlite–grospydite reaction. *Contrib. Mineral. Petrol.*, **45**, 153–160.
Chinner, G.A. and Dixon, J.E. (1973) Some high pressure parageneses of the Allalin gabbro, Valais, Switzerland, *J. Petrol.* **14**, 185–202.
Chopin, C. (1978) Les paragenèses réduites ou oxidées de concentrations manganésifères des 'schistes lustrés' de Haute-Maurienne (Alpes françaises), *Bull. Minéral.* **101**, 514–531.
Chopin, C. (1981) Talc-Phengite: a widespread assemblage in high-grade pelitic blue-schists of the Western Alps. *J. Petrol.* **22**, 628–650.
Chopin, C. (1984) Coesite and pure pyrope in high grade pelitic blueschists of the Western Alps: a first record and some consequences. *Contrib. Mineral. Petrol.* **86**, 107–118.
Chopin, C. (1985) Les relations de phase dans les métapelites de haute pression: approche expérimental et naturaliste, conséquences géodynamiques pour les Alpes occidentales. *Thèse d'état. Université de Paris* **6**, 80 pp.
Chopin, C. (1986) Phase relationships of ellenbergerite, a new high-pressure Ti–Al–Mg-silicate in pyrope-coesite-quartzite from the Western Alps., *Geol. Soc. Amer. Mem.* **164**, 31–42.
Chopin, C. (1987) Very-high pressure metamorphism in the Western Alps: implications for subduction of continental crust. *Phil. Trans. R. Soc. London* **A321**, 183–197.
Chopin, C., Klaska, R., Medenbach, O. and Dron, D. (1986) Ellenbergerite, a new high-pressure Mg–Al–(Ti, Zr)-silicate with a novel structure based on face-sharing octahedra. *Contrib. Mineral. Petrol.* **92**, 316–321.
Chopin, C. and Moniè, P. (1984) A unique magnesiochloritoid-bearing, high-pressure assemblage from the Monte Rosa: a petrologic and $^{40}Ar–^{39}Ar$ study. *Contrib. Mineral. Petrol.* **87**, 388–398.
Chopin, C. and Schreyer, W. (1983) Mangesiocarpholite and magnesiochloritoid: two index minerals of pelitic blueschists in the model system $MgO–Al_2O_3–SiO_2–H_2O$. *Am. J. Sci.* **283**(a), 72–96.
Chopin, C., Schreyer, W., and Baller, T. (1988) Synthesis and stability limits of ellenbergerite. *Terra Cognita* **8**, 59–60.
Church, W.R. (1964) Metamorphic eclogites from Co. Donegal, Eire, *Indian Mineral, Inst. Mineral. Ass. Sp. Nr.* 22–23.
Church, W.R. (1968) Eclogites. In *Basalts: The Poldervaart Treatise on Rocks of Basaltic Composition*, Volume 2 (eds H.H. Hess and A. Poldervaart), 755–798 Wiley, New York.
Cimmino, F., Messiga, B. and Piccardo, G. (1981) Le caratteristiche paragenetiche dell'evento eoalpino di alta pressione nei diversi sistemi (pelitici, femici, ultrafemici) delle ofioliti metamorfiche del Gruppo di Voltri (Liguria occidentale) *Rend. Soc. Ital. Mineral. Petrol.* **37**, 419–446.

Clague, D.A. and Straley P.F. (1977) Petrologic nature of the oceanic Moho, *Geology* **5**, 133–136.

Clark, S.P. and Ringwood, A.E. (1964) Density distribution and constitution of the mantle. *Rev. Geophys.* **2**, 35–88.

Cloos, M. (1986) Blueschists in the Franciscan Complex of California: Petrotectonic constraints on uplift mechanism. In: *Blueschists and eclogites.* (eds Evans, B.W. and Brown, E.H.) *Geol. Soc. Amer. Mem.* **164**, 77–94.

Coffrant, D. (1974) Les éclogites et les roches basique et ultrabasique associées du massif de Sauviat-sur-Vige, Massif Central Français. *Bull. Soc. Fr. Mineral. Cristallogr.* **97**, 70–78.

Cogné, J. (1966) La nappe cristallophyllienne de Champtoceaux en bordure méridionale du synclinal d'Ancenis (Bretagne–Anjou). *Bull. Serv. Carte géol. Alsace–Lorraine* **19**, 107–136.

Cogné, J. and Wright, A.E. (1980) L'Orogene cadomien. In *Géologie du l'Europe du Précambrien aux bassins sédimentaires posthercyniennes,* (eds Cogné, J. and Slansky, M.) *Mem. B.R.G.M.* **108**, 29–55.

Cohen, A.S., O'Nions, R.K., Siegentaler, R. and Griffin, W.L. (1988) Chronology of the pressure–temperature history recorded by a granulite terrain. *Contrib. Mineral. Petrol.* **98**, 303–311.

Coleman, R.G. and Lee, D.E. (1962) Metamorphic aragonite in the glaucophane schists of Cazadero, California. *Am. J. Sci.* **260**, 577–595.

Coleman, R.G. and Lee, D.E. (1963) Glaucophane-bearing metamorphic rocktypes of the Cazadero area, California. *J. Petrol.* **4**, 260–301.

Coleman, R.G., Lee, D.E., Beatty, L.B. and Brannock, W.W. (1965) Eclogites and eclogites: their differences and similarities. *Geol. Soc. Amer. Bull* **76**, 483–508.

Colombi, A. and Pfeifer, H.-R. (1986) Ferrogabbroic and basaltic metaeclogites from the Antrona mafic-ultramafic complex and the Centovalli–Locarno region (Italy and southern Switzerland)—first results. *Schweiz. Mineral. Petrogr. Mitt.* **66**, 99–110.

Compagnoni, R. (1977) The Sesia–Lanzo zone: high pressure—low temperature metamorphism in the Austroalpine continental margin, *Rend. Soc. Ital. Mineral. e Petrol.* **33**, 335–374.

Compagnoni, R., Dal Piaz, G.V., Fiora, L., Gosso, G., Lombardo, B., Maffeo, B. and Williams, P.F. (1977b) Excursion to the Sesia–Lanzo zone and Valtournanche metamorphic ophiolites. *Rend. Soc. Ital. Mineral. Petrol.* **33**, 473–491.

Compagnoni, R., Dal Piaz, G.V., Hunziker, J.C., Gosso, G., Lombardo, B. and Williams, P.F. (1977a) The Sesia–Lanzo zone, a slice of continental crust with apline high pressure–low temperature assemblages in the Western Italian Alps. *Rend. Soc. Ital. Mineral. Petrol.* **33**, 281–334.

Compagnoni, R., Elter, G., and Lombardo, B. (1974) Eterogeneita stratigrafica del complesso degli 'Gneiss minuti' nel Massiccio crisallino del Gran Paradiso. *Mem. Soc. Geol. Ital.* **13**, 227–239.

Compagnoni, R. and Lombardo, B. (1974) The Alpine age of the Gran Paradiso eclogites. *Rend. Soc. Ital. Mineral. Petrol.* **30**, 223–237.

Compagnoni, R. and Maffeo, B. (1973) Jadeite-bearing metagranites l.s. and related rocks in the Mount Mucrone area (Sesia–Lanzo Zone, Western Italian Alps), *Schweiz. Mineral. Petrogr. Mitt.* **53**, 355–378.

Compagnoni, R., Radicati di Brozolo, R. and Sandrone, R. (1984) Kaersutite-bearing mylonitic gabbros from the Lanzo perdotite (western Italian Alps), *Geologie en Mijnbouw* **63**, 189–196.

Compston, W., Kinny, P.D. Williams, I.S. and Foster, J.J. (1986) The age and Pb loss behaviour of zircons from the Isua supracrustal belt as determined by ion microprobe. *Earth Planet. Sci. Lett.* **80**, 71–81.

Compston, W., Williams, I.S., Campbell, I.H. and Gresham, J.J. (1985/86) Zircons xenocrysts from the Kambalda volcanics: age constraints and direct evidence for older continental crust below the Kambalda-Norseman greenstones. *Earth Planet. Sci. Lett.* **76**, 299–311.

Coney, P.J. and Harms, T.A. (1984) Cordilleran metamorphic core complexes. Geology. **12**, 550–554.

Conquere, F. and Kornprobst, J. (1972) Quelques précisions sur les caractères minéralogiques et chimiques des pyroxénolites à grenat du massif des Beni Bousera (Maroc) et sur leur signification pétrogénétique. *Bull. Soc. Fr. Mineral. Cristall.* **95**, 115–123.

Cordellier, F., Boudier, F. and Boullier, A.M. (1981) Structural study of the Almklovdalen massif (southern Norway). Tectonophysics **77**, 257–281.

Corstorphine, G.S. (1907) The occurrence in kimberlite of garnet–pyroxene nodules carrying diamonds. *Trans. Geol. Soc. S. Africa* **10**, 65–69.

Cortesogno, L. Ernst, W.G. Galli, M. Messiga, B. Pedemonte, G.M. and Piccardo, G.B. (1977) Chemical petrology of eclogitic lenses in serpentinite, Gruppo di Voltri, Ligurian Alps. *J. Geol.* **85**, 255–277.

Cortesogno, L. Lucchetti, G. and Massa, B. (1981) Rocce oficarbonatiche e marmi a silicati nel Massiccio di Voltri: origine e significato, chimismo dei minerali ed equilibri paragenetici. *Rend. Soc. Ital. Mineral. Petrol.* **37**, 481–507.

Cossa, A. (1879) Ricerche di chimica mineralogica su rocce del Piemonte. *Mem. R. Acc. Sci. Torino* **28**, 33 p.

Cotkin, S.J., Valley, J.W. and Essene, E.J. (1988) Petrology of margarite-bearing meta-anorthosite from Seljeneset, Nordfjord, Western Norway: Implications for the P–T history of the Western Gneiss Region during Caledonian uplift. *Lithos* **21**, 117–128.

Cressey, G. (1981) Entropies and enthalpies of aluminosilicate garnets. *Contrib. Mineral. Petrol.* **76**, 413–419.

Cull, J.P. (1982) An appraisal of Australian heat-flow data. *BMR J. Australian Geol. Geophysics* **7**, 11–21.

Cuthbert, S.J. (1985) *Petrology and tectonic settings of relatively low-temperature eclogites and related rocks in the Dalsfjord area, Sunnfjord, West Norway.* Unpublished PhD thesis, University of Sheffield.

Cuthbert, S.J., Harvey, M.A. and Carswell, D.A. (1983) A tectonic model for the metamorphic evolution of the Basal Gneiss Complex, Western South Norway. *J. Metamorphic Geol.* **1**, 63–90.
Cygan, C.T. and Lasaga, A.C. (1985) Self diffusion of magnesium in garnet at 750°C to 900°C. *Am. J. Sci.* **285**, 328–350.
Dachs, E. (1986) High-pressure mineral assemblages and their breakdown products in metasediments South of the Grossvenediger, Tauern Window, Austria. *Schweiz. Mineral. Petrogr. Mitt.* **66**, 145–161.
Dallmeyer, R.D. (1988) Polyphase tectonothermal evolution of the Scandinavian Caledonides. In *The Caledonian-Appalachian Orogen*, (eds Harris, A.L. and Fettes, D.J.) Geol. Soc. London Spec. Publ. 38, pp. 365–379.
Dallmeyer, R.D. and Gee, D.G. (1986) $^{40}Ar/^{39}Ar$ mineral ages from retrogressed eclogites within the Baltoscandinavian miogeocline: Implications for a polyphase Caledonian orogenic evolution. *Bull. Geol. Soc. Am.* **97**, 26–34.
Dallmeyer, R.D., Gee, D.G. and Beckholmen, N. (1983) $^{40}Ar/^{39}Ar$ mineral age record of early Caledonian tectonothermal activity in the Baltoscandian Miogeocline. *Am. J. Sci.* **285**, 532–568.
Dal Piaz, G.V. (1969) Filoni rodingitici e zone di reazione a bassa temperatura al contatto tettonico tra serpentine e rocce incassanti nelle Alpi occidentali italiane. *Rend. Soc. Ital. Mineral. Petrol.* 25, 262–315.
Dal Piaz, G.V. (1974) Le métamorphisme alpin de haute pression et basse température dans l'évolution structurale du bassin ophiolitique alpino-apenninique. *Schweiz. Mineral. Petrogc. Mitt.* **54**, 399–424.
Dal Piaz, G.V., Di Battistini, G., Kienast, J.R. and Venturelli, G. (1979) Manganiferous quartzitic schists of the Piemonte ophiolite nappe in the Valsesia-Valtournanche area (Italian Western Alps), *Memorie di Scienze Geologiche* **32**, 3–24.
Dal Piaz, G.V. and Ernst, W.G. (1978) Areal geology and petrology of eclogites and associated metabasites of the Piemonte ophiolite nappe, Breuil-St. Jacques area, Italian western Alps. *Tectonophysics* **51**, 99–126.
Dal Piaz, G.V., Hunziker, J.C. and Mattinotti, G. (1972) La zona Sesia-Lanzo e l'evoluzione tetonicometamorfica delle Alpi nord-occidentali interne. *Memorie della Societá Geologica Italiana* **11**, 433–460.
Dal Piaz, G.V. and Lombardo, B. (1986) Early alpine eclogite metamorphism in the Pennidic Monte Rosa—Gran Paradiso basement nappes of the north-western Alps. *Geol. Soc. Amer. Mem.* **164**, 249–265.
Dal Piaz, G.V., Lombardo, B. and Gosso, G. (1983) Metamorphic evolution of the Mt Emilius klippe, Dent Blanche nappe, western Alps, *Am. J. Sci.* **283-A**, 438–458.
Dal Piaz, G.V. and Omenetto, P. (1978) Brevi note su alcune mineralizzazioni della falda piemontese in Valle d'Aosta. *Ofioliti* **3**, 161–176.
Dal Piaz, G.V., Venturelli, G., and Spadea, P. Di Battistini, G. (1981) Geochemical features of metabasalts and metagabbros from the Piemonte ophiolite nappe, Italian western Alps, *Neues Jahrb. Mineral. Abh.* **142**, 248–269.
Dal Vesco, E. (1953) Genesi e metamorfosi delle rocce basiche e ultrabasiche nell'ambiante mesozonale dell'orogene Pennidico. *Schweiz, Mineral. Petrogr. Mitt.* **33**, 173–480.
Daly, J.S., Yardley, B.W.D. and Cliff, R.A. (eds) (1988) *Evolution of Metamorphic Belts.* Geol. Soc. London Spec Publ. **43**.
Damour, A. (1881) Nouvelles analyses sur la jadeite et sur quelques roches sodifères. *Bull. Soc. Mineral. Fr.* **4**, 157–164.
Danckwerth, P.A. and Newton, R.C. (1978) Experimental determination of the spinel peridotite to garnet peridotite reaction in the system $MgO–Al_2O_3–SiO_2$ in the range 900–1100°C and Al_2O_3 isopleths of enstatite in the spinel field. *Contrib. Mineral. Petrol.* **66**, 189–201.
Daniels, L.R.M. and Gurney J.J. (1989) The chemistry of garnets, chromites and diamond inclusions of the Dokolwayo kimberlite, Kingdom of Swaziland. In *Kimberlites and related rocks, Vol. 2. The mantle/crust setting, diamonds and diamond exploration.* Geol. Soc. Australia, Spec. Publn. 14, Blackwells, Carlton, 1012–1021.
Davidson, C.F. (1943) The Archaean rocks of the Rodil district, South Harris, Outer Hebrides. *Trans. R. Soc. Edinburgh* **61**, 71–112.
Davies, H.L. and Warren, R.G. (1988) Origin of eclogite-bearing, domed, layered metamorphic complexes ('core complexes') in the D'Entrecasteaux islands, Papua New Guinea. *Tectonics* **7**, 1–21.
Davis, B.T.C. and Boyd, F.R. (1966) $Mg_2Si_2O_6–CaMgSi_2O_6$ at 30 kilobars pressure and its application to pyroxenes from kimberlites. *J. Geophys. Res.* **71**, 3567–3576.
Davy, P. and Gillet, P. (1986) The stacking of thrust slices in collision zones and its thermal consequences *Tectonics* **5**, 913–929.
Dawson, J.B. (1980) *Kimberlites and their xenoliths.* Springer, Berlin.
Dawson, J.B. (1983) New developments in diamond geology. *Naturwissenschaften* **70**, 586–593.
Dawson, J.B. (1984) Contrasting types of upper mantle metasomatism? (ed. Kornprobst, J.) *Kimberlites II: The Mantle and Crust-Mantle Relationships*, Elsevier, Amsterdam, 289–284.
Dawson, J.B. and Smith, J.V. (1987) Reduced sapphirine granulite xenoliths from the Lace kimberlite, South Africa; implications for the deep structure of the Kaapvaal craton. *Contrib. Mineral Petrol.* **95**, 376–383.
Dawson, J.B. and Stephens, W.E. (1975) Statistical analysis of garnets from kimberlites and associated xenoliths. *J. Geol.* **83**, 589–607.
Davy, P. and Gillet, P. (1986) The stacking of thrust slices in collision zones and its thermal consequences. *Tectonics* **5**, 913–929.

Deer, W.A. Howie, R.A. and Zussman, J. (1978) *Rock Forming Minerals*, Vol. 2A, 2nd Edition *Single-Chain Silicates*, 668 pp. Longman, London.

Deer, W.A., Howie, R.A. and Zussman, J. (1982) *Rock Forming Minerals*, Vol. 1A, 2nd Edition *Orthosilicates*, 919 pp. Longman, London.

Deines, P., Gurney, J.J. and Harris, J.W. (1984) Associated chemical and carbon isotopic variations in diamonds from Finsch and Premier kimberlites, South Africa. *Geochim Cosmochim Acta* **48**, 325–342.

Deines, P., Harris, J.W. and Gurney, J.J. (1987) Carbon isotopic composition, nitrogen content and inclusion composition of diamonds from the Roberts Victor kimberlite, South Africa: evidence for ^{13}C depletion in the mantle. *Geochim. Cosmochim. Acta* **51**, 1227–1243.

Delaloye, M. and Desmons, J. (1976) K-Ar radiometric age determinations of white micas from the Piemont zone. French–Italian Western Alps. *Contrib. Mineral. Petrol.* **57**, 297–303.

Delaney, J.S., Smith, J.V., Carswell, D.A. and Dawson, J.B. (1980) Chemistry of micas from kimberlites and xenoliths. II. Primary- and secondary-textured micas from peridotite xenoliths. *Geochim. Cosmochim. Acta* **44**, 857–872.

Delaney, J.S., Smith, J.V., Dawson, J.B. and Nixon, P.H. (1979) Manganese thermometer for mantle peridotites. *Contrib. Mineral. Petrol.* **71**, 157–169.

Delor, C., Burg, J-P., Guiraud, M. and Leyreloup, A. (1987) Les métapelites à phengite-chloritoîde-grenat-staurotide-disthène de la klippe de Najac-Carmaux: nouveaux marqueurs d'un métamorphisme de haute pression varisque en Rouergue occidental. *C.R. Acad. Sci. Paris* **305** (II), 589–95.

Delor, C.P. and Leyreloup, A.F. (1986) Chromium rich kyanite in an eclogite from the Rouergue area. French Massif Central. *Mineral. Mag.* **50**, 535–7.

Delor, C., Leyreloup, A., Bodinier, J-L. and Burg, J-P. (1986) Découverte d'éclogites à glaucophane dans la klippe de Najac (Massif Central, France): nouveaux témoins océaniques d'un stade haute pression dans la chaîne de collision Varisque. *C.R. Acad. Sci. Paris.* **302**(II), 739–744.

Demange, M. (1985) The eclogite-facies rocks of the Montagne Noire, France. In *Chemistry and Petrology of Eclogites.* (eds. Smith, D.C., Franz, G. and Gebauer, D.) *Chem. Geol.* **50**, 173–188.

DePaolo, D.J. and Wasserburg, G.J. (1976a) Nd isotopic variations and petrogenetic models. *Geophys. Res. Lett*, **3**, 249–252.

DePaolo, D.J. and Wasserburg, G.J. (1976b) Inferences about magma sources and mantle structure from variations of 143Nd/144Nd. *Geophys. Res. Lett.* **4**, 743–746.

Desborough, G.A. and Czamanske, G.K. (1973) Sulfides in eclogite nodules from a kimberlite pipe, South Africa, with comments on violarite stoichiometry. *Am. Mineral*, **58**, 195–202.

Desmons J., and Ghent, E.D. (1977) Chemistry zonation and distribution coefficients of elements in eclogitic minerals from the eastern Sesia unit, Italian Western Alps. *Schweiz. Mineral. Petrol. Mitt.* **57**, 397–411.

Desmons, J. and O'Neil, J.R. (1978) Oxygen and hydrogen isotope compositions of eclogites and associated rocks from the Eastern Sesia Zone (Western Alps, Italy). *Contrib. Mineral. Petrol.* **67**, 79–85.

Dickey, J.S., Jr. (1970) Partial fusion products in alpine-type peridotites: Serrania de la Ronda and other examples. *Mineral. Soc. Am. Special Paper* **3**, 33–49.

Dietler, T.N., Koestler, A.G. and Milnes, A.G. (1985) A preliminary structural profile through the Western Gneiss Complex, Sognefjord, southwest Norway, *Nor. Geol. Tidsskr.* **65**, 233–235.

Dixon, J.E. (1969) *Petrology of the metamorphic rocks of Syros island (Greece).* Ph.D Thesis, University of Cambridge.

Dobretsov, N.L., Mísař, Z. and Popov, E.V. (1984) The PT conditions of equilibrium for some pyrope peridotites and country rocks in the Moldanubian area at Mohelno (Eastern Moravia, Czechoslovakia). *Mineral. Slov.* **16**, 87–94.

Dobretsov, N.L. and Sobolev, N.V. (1984) Glaucophane schists and eclogites in the folded systems of Northern Asia. *Ofioliti* **9**, 401–424.

Dodson, M.H. (1979) Theory of cooling ages. In *Lectures in Isotope Geology* (eds. Jager, E. and Hunziker, J.C.) pp. 194–202. Springer, Berlin.

Draper, G. and Bone, R. (1981) Denudation rates, thermal evolution, and preservation of blueschist terrains. *J. Geol.* **89**, 601–613.

Droop, G.T.R. (1983) Pre-Alpine eclogites in the Pennine Basement Complex of the Eastern Alps. *J. Metamorphic Geol.* **1**, 3–12.

Droop, G.T.R. (1985) Alpine metamorphism in the south-east Tauern Window, Austria: I. P–T variations in space and time. *J. Metamorphic Geol.* 3.

Droop, G.T.R. (1987) A general equation for estimating Fe^{3+} concentrations in ferromagnesian silicates and oxides from microprobe analyses, using stoichiometric criteria. *Mineral. Mag.* 51, 431–435.

Dubuisson, G., Hirn, A., Girardeau, J., Mercier, J-C.C. and Veinante, J-L. (1988) Multiple Variscan nappes in Limousin, western Massif Central, France: geophysical constraints to the geological model and geodynamic implications. *Tectonophysics* **147**, 19–31.

Ducrot, J., Lancelot, J.R. and Marchand, J. (1983) Datation U-Pb sur zircons de l'éclogite de la Borie (Haut Allier, France) et conséquences sur l'évolution anté-Hercynienne de l'Europe occidentale. *Earth Planet. Sci. Lett.* **62**, 385–394.

Dudek, A. and Fediuková, E. (1974) Eclogites of the Bohemian Moldanubicum. *Neues Jahrb. Mineral. Abh.* **121**, 127–159.

Dudek, A. and Kopecky, L. (1966) Ultrabasites in the Bohemian Massif. *Krystalinikum* **4**, 7–10.

Dudek, A., Matejovska, O. and Suk, M. (1974) Gföhl orthogneiss in the Moldanubicum of Bohemia and Moravia. *Krystalinikum* **10**, 67–78.

Dudek, A. and Misař, Z. (1985) *Second International Eclogite Conference Vienna. Post-conference excursion guide–Czechoslovak part.* 53 pp.

Dufour, E. (1985) Granulite facies metamorphism and retrogressive evolution of the Monts du Lyonnais metabasites (Massif Central. France). *Lithos* **18**, 97–113.

Dunn, S.R. and Medaris, L.G., Jr. (1989) Retrograded eclogites in the Western Gneiss Region, Norway, and thermal evolution of a portion of the Scandinavian Caledonides. *Lithos.* **22**, 229–245.

Dunning, G.R. and Pedersen, R.B. (1988) U/Pb ages of ophiolites and arc-related plutons of the Norwegian Caledonides: implications for the development of Iapetus. *Contrib. Mineral. Petrol.* **98**, 13–23.

Duparc, L. and Mrazec. L. (1898) Recherches géologiques et pétrographiques sur le Massif du Mont Blanc. *Mem. Soc. Phys. Hist. Nat. Génève* **33**, 1–227.

Dürr, S., Altherr, R., Keller, J., Okrusch, M. and Seidel, E. (1978) The Median Aegean crystalline belt: stratigraphy, structure, metamorphism. magmatism. In *Mediterranean orogens.* (eds. Closs, H., Roeder, D. and Schmidt, K.). *Inter-union Commission on Geodynamics. Scientific Report* 38, Schweizerbart (Stuttgart) 455–477.

Düsing, C. (1960) *Erläuterungen zur Geologischen Karte von Bayern 1:25,000 Blatt Nr 6540 Oberviechtach.* Bayerisches Geologisches Landesmt, München.

Edwards, A.B, Lovering, J.F. and Ferguson, J. (1979) High pressure basic inclusions from the Kayrunnera kimberlite diatreme in New South Wales, Australia. *Contrib. Mineral. Petrol.* **69**, 185–192.

Eggler, D.H., Dudas, F.O., Hearn, B.C., McCallum, M.E., McGee, E.S., Meyer, H.O.A. and Schulze, D.J. (1987) Lithosphere of the continental United States: xenoliths in kimberlites and othert alkaline magmas. In *Mantle Xenoliths* (ed. Nixon, P.H.) Wiley, New York, 41–58.

Ellenberger, F. and Tamain, A.L.G. (1980) Hercynian Europe, *Episodes* **1**, 22–27.

Ellis, D.J. and Green, D.H. (1979) An experimental study of the effect of Ca upon garnet–clinopyroxene Fe–Mg exchange equilibria. *Contrib. Mineral. Petrol.* **71**, 13–22.

Enami, M. and Banno, S. (1980) Zoisite-clinozoisite relations in low- to medium-grade high-pressure metamorphic rocks and their implications. *Mineral. Mag.* **43**, 1005–1013.

Enami, M. Wang, S., Zang, Q. and Hiraiwa, I. (1986) Eclogites from the Donghai district, Jiangsu province, east China. *Bull. Nagoyva Univ. Mus.* **2**, 55–70 (in Japanese).

Enami, M. and Zang, Q. (1988) Magnesian staurolite in garnet-corundum rocks and eclogite from Donghai district, Jiangsu province, east China, *Am. Mineral.* **73**, 48–56.

Engels, J.P. and Vogel, D.E. (1966) Garnet reaction-rims between plagioclase and hypersthene in a metanorite from Cabo Ortegal (NW Spain). *Neues Jahrb. Mineral. Monatsh.* **1**, 13–19.

England, P.C. and Holland, T.J.B. (1979) Archimedes and the Tauern eclogites: the role of buoyancy in the preservation of exotic eclogite blocks. *Earth Planet. Sci. Lett.* **44**, 287–294.

England, P.C. and Richardson, S.W. (1977) The influence of erosion upon the mineral facies of rocks from different metamorphic environments. *J. Geol. Soc. London.* **134**, 201–213.

England, P.C. and Thompson, A.B. (1984) Pressure–temperature–time paths of regional metamorphism I. Heat transfer during the evolution of regions of thickened continental crust. *J. Petrol.* **25**, 894–928.

Ernst, W.G. (1963a) Polymorphism in alkali amphiboles. *Am. Mineral.* **48**, 241–260.

Ernst, W.G. (1963b) Significance of phengitic micas from low-grade schists. *Am. Mineral.* **48**, 1357–1373.

Ernst, W.G. (1971) Metamorphic zonation on presumably subducted lithospheric plates from Japan, California and the Alps. *Contrib. Mineral. Petrol.* **34**, 43–59.

Ernst, E.G. (1973) Interpretative synthesis of metamorphism in the Alps. *Geol. Soc. Am. Bull.* **84**, 2053–2078.

Ernst, W.G. (1976) Mineral chemistry of eclogites and related rocks from the Voltri Group, Western Liguria, Italy. *Schweiz. Mineral. Petrogr. Mitt.* **56**, 293–343.

Ernst, W.G. (1977) Mineralogic study of eclogitic rocks from Alpe Arami, Lepontine Alps, southern Switzerland. *J. Petrol.* **18**, 371–398.

Ernst, W.G. (1978) Petrochemical study of lherzolitic rocks from the Western Alps. *J. Petrol.* **19**, 341–392.

Ernst, W.G. (1979) Coexisting sodic and calcic amphiboles from high-pressure metamorphic belts and the stability of barroisitic amphibole, *Mineral. Mag.* **43**, 269–278.

Ernst, W.G. (1981) Petrogenesis of eclogites and peridotites from the Western and Ligurian Alps. *Am. Mineral.* **66**, 443–472.

Ernst, W.G. (1984) Californian blueschist, subduction, and the significance of tectonostratigraphic terranes. *Geology*, **12**, 436–440.

Ernst, W.G. and Dal Piaz, G.V. (1978) Mineral parageneses of eclogite rocks and related mafic schists of the Piemonte ophiolite nappe, Breuil-St. Jacques area, Italian Western Alps, *Am. Mineral.* **63**, 621–640.

Ernst, W.G. and Piccardo, G.B. (1979) Petrogenesis of some Ligurian peridotites–I. Mineral and bulk-rock chemistry. *Geochim. Cosmochim. Acta* **43**, 219–237.

Ernst, W.G., Rambaldi, E. and Piccardo, G.B. (1983) Trade element geochemistry of iron + titanium-rich eclogitic rocks, Gruppo di Voltri, western Liguria, *J. Geol.* **91**, 413–425.

Ernst, W.G., Seki, Y., Onuki, H. and Gilbert, M.C. (1970) Comparative study of the low-grade metamorphism in the California coast ranges and outer metamorphic belt of Japan. *Geol. Soc. Amer. Mem.* **124**, 1–276.

Eskola, P. (1920) The mineral facies of rocks. *Nor. Geol. Tidsskr.* **6**, 143–194.
Eskola, P. (1921) On the eclogites of Norway. *Vidensk.-Selsk. Skr. I. Mat. Naturv. Kl.* **8**, 1–118.
Eskola, P. (1939) Die Metamorphen Gesteine. In *Die Entstehung der Gesteine* (eds Barth, T.F.W., Correns, C.W. and Eskola, P.) Julius Springer, Berlin, 263–407.
Esperanca, S. and Holloway, J.R. (1984) Lower crustal nodules from the Camp Creek latite, Carefree, Arizona. In *Kimberlites: II The Mantle and Crust–Mantle Relationships.* (ed. Kornprobst, J.) Developments in Petrology 11B, Elsevier, Amsterdam, 219–227.
Essene, E.J. (1982) Geologic thermometry and barometry. In *Characterization of Metamorphism through Mineral Equilibria.* (ed. Ferry, J.M.) *Mineral. Soc. Am. Reviews in Mineralogy* **10**, 153–206.
Essene, E. J. and Fyfe, W.S. (1967) Omphacite in Californian metamorphic rocks. Contrib. Mineral. Petrol. **15**, 1–23.
Essene, E.J., Hensen, B.J. and Green, D.H. (1970) Experimental study of amphibolite and eclogite stability. *Phys. Earth Planet. Inter.* **3**, 378–384.
Evans, B.W. (1986) Reactions among sodic, calcic, and ferromagnesian amphiboles, sodic pyroxene, and deerite in high-pressure metamorphosed ironstone, Siphnos, Greece, *Am. Mineral.* **71**, 1118–1125.
Evans, B. and Patrick, B. (1987) Phengite-3T in high-pressure granitic orthogneisses, Seward peninsula, Alaska. *Can. Mineral.* **25**, 141–158.
Evans, B.W. and Brown, E.H. (1987) Reply on blueschists and eclogites. *Geology* **15**, 773–775.
Evans, B.W. and Trommsdorff, V. (1978) Petrogenesis of garnet lherzolite, Cima di Gagnone, Lepontine Alps. *Earth Planet. Sci. Lett.* **40**, 333–348.
Evans, B.W. and Trommosdorff, V. (1983) Fluorine hydroxyl titanian clinohumite in alpine recrystallised garnet-peridotite: compositional controls and petrologic significance. *Am. J. Sci.* **283-A**, 355–369.
Evans, B.W. Trommsdorff, V. and Goles, G.G. (1981) Geochemistry of high-grade eclogites and metarodingites from the Central Alps. *Contrib. Mineral. Petrol.* **76**, 301–311.
Evans, B.W., Trommsdorff, V. and Richter, W. (1979) Petrology of an eclogite-metarodingite suite at Cima di Gagnone, Ticino, Switzerland. *Am. Mineral.* **64**, 15–31.
Exley, R.A., Smith, J.V. and Dawson, J.B. (1983) Alkremite, garnetite and eclogite xenoliths from Bellsbank and Jagersfontein, South Africa. *Am. Mineral.* **68**, 512–516.
Exner, C. and Wanderer, E. (1962) Zur Kenntnis des Eklogitamphibolites im Debanttal (Schobergruppe, Osttirol). *Karinthin* **45/46**, 228–234.
Fabriès, J. and Latouche, L. (1988) Granulite facies conditions in the Sainte Marie aux Mines 'Varied Group' (Central Vosges, France). *Terra Cognita* **8**, 249.
Fang, Chingson and Bai, Wenji (1981) The discovery of an Alpine-type diamond-bearing ultrabasic intrusion in Xizang (Tibet). *Geol. Review* **27**, 455–457 (in Chinese).
Faure-Muret, A. (1955) *Études géologiques sur le massif de l'Argentera-Mercantour et ses enveloppes sédimentaires.* Mém. Carte Géol. Fr. 336 pp.
Fediuková, E. and Dudek, A. (1977) Trace elements of the Moldanubian eclogites. *Neues Jahrb. Mineral. Abh.* **130**, 187–207.
Feininger, T. (1980) Eclogite and related high-pressure regional metamorphic rocks from the Andes of Ecuador, *J. Petrol.* **21**, 107–140.
Fiala, J. (1965) Pyrope of some garnet peridotites of the Czech Massif. *Kristalinikum* **3**, 55–74.
Fiala, J. (1966) The distribution of elements in mineral phases of some garnet peridotites from the Bohemian Massif. *Kristalinikum* **4**, 31–53.
Fiala, J. (1985) Becváry west of Kutná Hora. *Second Int. Eclogite Conf,* Post-Conference Excursion, Czechoslovak Part, 48–53.
Fiala, J., Matějovska, O. and Vaňková, V. (1987a) Moldanubian granulites: source material and petrogenetic considerations *Neues Johrb. Mineral. Abh.* **157**, 133–165.
Fiala, J., Matějovska, O. and Vanková, V. (1987b) Moldanubian granulites and related rocks: petrology, geochemistry, and radioactivity. *Rozpr. Česk. Akad. Ved.* **97**, 102 pp.
Fiala, J. and Paděra, K. (1977) The chemistry of the minerals of the pyrope dunite from borehole T-7 near Staré (Bohemia). *Tschermaks Mineral. Petrogr. Mitt.* **24**, 205–219.
Fiala, J. and Paděra, K. (1984) Eclogitic rocks from the basement of the České Středohoří Mts. *Krystalinikum* **17**, 33–45.
Finlayson, D.M. (1982) Geophysical differences in the lithosphere between Phanerozoic and Precambrian Australia. *Tectonophysics* **84**, 287–312.
Finnerty, A.A. (1989) Inflected mantle geotherm from xenoliths are real: evidence from olivine barometry. In: *Kimberlites and Related Rocks, Vol. 2: Mantle/Crust Setting, Diamonds and Diamond Exploration.* Geol. Soc. Aust. Spec. Publ., 14, 883–900.
Finnerty, A.A. and Boyd, F.R. (1978) Pressure-dependent solubility of calcium in forsterite coexisting with diopside and enstatite. *Yearb. Carnegie Inst. Washington* **77**, 713–717.
Finnerty, A.A. and Boyd, F.R. (1984). Evaluation of thermobarometers for garnet peridotites. *Geochim. Cosmochim. Acta* **48**, 15–27.
Finnerty, A.A. and Boyd, F.R. (1987) Thermobarometry for garnet–peridotite xenoliths: a basis from upper mantle stratigraphy. In *Mantle Xenoliths* (ed. Nixon, P.H.) pp. 381–402. J. Wiley and Sons, New York.

Finnerty, A. A. and Ridgen, S.M. (1981) Olivine barometry: application to pressure estimation for terrestrial and lunar rocks. *Lunar and Planet. Sci.* **XII**, 279–281.

Fischer, G. (1968) *Moldanubicum in Bavaria. Excursions to the Bavarian margin of the Bohemian Massif. III. Guide to excursion C35 (Fed. Repub. of Germany).* International Geological Congress, XXIII session, Prague.

Fisher, G.W. (1973) Nonequilibrium thermodynamics as a model for diffusion-controlled metamorphic processes. *Am. J. Sci.* **273**, 897–924.

Fisher, G.W. (1978) Rate laws in metamorphism. *Geochim. Cosmochim. Acta* **42**, 1035–1050.

Fluck, P., Maass, R. and Raumer, J.F. von (1980) The Variscan units east and west of the Rhine Graben. In *Géologie de l'Europe du Précambrien aux bassins sédimentaires posthercyniennes.* (eds. Cogné, J. and Slansky, M.) *Mém. B.R.G.M.* **108**, 112–131.

Forbes, R.B. (1965) The comparative chemical composition of eclogite and basalt. *J. Geophys. Res.* **70**, 1515–1521.

Forbes, R.B., Evans, B.W. and Thurston, S.P. (1984) Regional progressive high-pressure metamorphism, Seward peninsula, Alaska. *J. Metamorphic Geol.* **2**, 43–54.

Forestier, F-H., Lasnier, B., Leyreloup, A. and Marchand, J. (1973) Vues nouvelles sur la catazone dans le Massif Central français et le Massif Armoricain, de l'affleurement au Moho. *Bull. Soc. geol. Fr.* **47**, XV, 562–578.

Fountain, D.M. (1976) The Ivrea-Verbano and Strona-Ceneri zones, northern Italy: a cross section of the continental crust—new evidence from seismic velocities of rock samples. *Tectonophysics* **33**, 145–165.

Fountain, D.M. (1987) Geological and geophysical nature of the lower continental crust as revealed by exposed cross sections of the continental crust. *US Geol. Surv. Circular* **956**, 25–26.

Fountain, D.M. and Salisbury, M.H. (1981) Exposed cross-sections through the continental crust: implications for crustal structure, petrology and evolution. *Earth Planet. Sci. Lett.* **56**, 263–277.

Franchi, S. (1894) Relazione sui principali risultati dell rilevamento geologica nelle Alpe Maritime eseguito nelle campagne 1891–92–93. *Boll. R. Com. Geol.* **25**, 231–258.

Franchi, S. (1900) Sopra alcuni giacimenti di rocce giadeitiche nelle Alpi Occidentali e nell'Appennino ligure. *Boll. R. Com. Geol. Ital.* **31**, 119–158.

Frank, F.C. (1969) Diamonds and deep fluids in the upper mantle. In *The application of modern physics to the Earth and planetary interiors* (ed. Runcorn, S.K.) Wiley, New York, 247–250.

Frank, W., Esterlus, M., Frey, I., Jung, G., Krohe, A. and Weber, J. (1983) Die Entwicklungsgeschichte von Stub- and Koralpenkristallin und die Beziehung zum Grazer Paläzoikum. *Jahresbericht 1982 Hochschulschwerpunkt S15*, Graz, 263–293.

Frank, W., Höck, V. and Miller, C. (1987a) Metamorphic and tectonic history of the central Tauern Window. pp. 34–54 in *Geodynamics of the Eastern Alps* (eds Flügel, H.W. and Faupl, P.) Deuticke, Vienna.

Frank, W., Hoinkes, G., Purtscheller, F. and Thöni, M. (1987b) The Austroalpine unit west of the Hohe Tauern: The Ötztal-Stubai Complex as an example for the Eo-Alpine metamorphic evolution. pp. 179–225 in *Geodynamics of the Eastern Alps* (eds Flügel, H.W. and Faupl, P.) Deuticke, Vienna.

Frank, W., Kralik, M., Scharbert, S. and Thöni, M. (1987c) Geochronological data from the Eastern Alps. pp. 272–281 in *Geodynamics of the Eastern Alps* (eds Flügel, H.W. and Faupl, P.) Deuticke, Vienna.

Franz, G., and Althaus, E. (1977) The stability relations of the paragenesis paragonite–zoisite–quartz. *Neues Jahrb. Mineral. Abh.* **130**, 159–167.

Franz, G. and Spear, F.S. (1983) High pressure metamorphism in siliceous dolomites from the central Tauern Window, Austria, *Am. J. Sci.* **283** (A), 396–413.

Franz, G. and Spear, F.S. (1985) Aluminous titanite (sphene) from the eclogite zone, south-central Tauern window, Austria. *Chem. Geol.* **50**, 33–46.

Franz, G., Thomas, S. and Smith, D.C. (1986) High pressure phengite decomposition in the Weissenstein eclogite, Münchberg Gneiss Massif, Germany. *Contrib. Mineral. Petrol.* **92**, 71–85.

Fraser, A.R. and Pettifer, G.R. (1980) Reconnaissance gravity surveys in W.A. and S.A. 1969–1972. *Bureau of Mineral Resources Bull.* **196**, 60 pp.

Fraser, D. and Lawless, P. (1978) Paleogeotherms: implications of disequilibrium in garnet lherzolite xenoliths. *Nature* **273**, 220–222.

Freer, R., Carpenter, M.A., Long, J.V.P. and Reed, S.J.B. (1982) 'Null result' diffusion experiments with diopside: implications for pyroxene equilibria. *Earth Planet. Sci. Lett.* **58**, 285–292.

Freund, F., Kathrein, H., Wengeler, H. and Knobel, R. (1980) Carbon in solid solution in forsterite—a key to the untractable nature of reduced carbon in terrestrial and cosmogenic rocks. *Geochim. Cosmochim. Acta* **44**, 1319–1333.

Frey, F.A., Green, D.H. and Roy, S.D. (1978) Integrated models of basalt petrogenesis: a study of quartz tholeiites to olivine melilitites from South Eastern Australia utilizing geochemical and experimental petrological data. *J. Petrol.* **19**, 463–513.

Frey, F.A., Süen, C.J. and Stockman, H.W. (1985) The Ronda high temperature peridotite: geochemistry and petrogenesis. *Geochim. Cosmochim. Acta.* **49**, 2469–2491.

Frey, M., Bucher, K., Frank, E. and Mullis, J. (1980) Alpine metamorphism along the geotraverse Basel–Chiasso—a review, *Eclog. Geol. Helv.* **73**, 527–546.

Frey, M., Hunziker, J.C., Frank, W., Bocquet, J., Dal Piaz, G.V., Jäger, E. and Niggli, E. (1974) Alpine metamorphism of the Alps: a review. *Schweiz. Mineral. Petrogr. Mitt.* **54**, 247–290.

Frey, M., Jäger, E. and Niggli, E. (1976) Gesteinsmetamorphose in Bereich der Geotraverse Basel–Chiasso. *Schweiz. Mineral. Petrogr. Mitt.* **56**, 649–659.

Friedrich, M., Marignac, C. and Floc'h, J-P. (1988) Sur l'existence de trois chevauchements ductiles 'himalayens' successifs a vergence NW en Limousin. *C.R. Acad. Sci. Paris* **306** (II), 663–9.

Frisch, W., Neubauer, F. and Satir, M. (1984) Concepts of the evolution of the Austroalpine Basement Complex (Eastern Alps) during the Caledonian–Variscan cycle. *Geol. Rdsch.* **73**, 47–68.

Frueh-Green, G. (1985) Stable isotope indications of fluid involvement in ductile shear zones during eclogite facies metamorphism. *Trans. Am. Geophys. Union* **66**, 1126–1127.

Fry, N. and Barnicoat, A.C. (1987) The tectonic implications of high-pressure metamorphism in the western Alps. *J. Geol. Soc. London* **144**, 653–659.

Fry, N. and Fyfe, W.S. (1969) Eclogites and water pressure. *Contrib. Mineral. Petrol.* **24**, 1–6.

Fry, N. and Fyfe, W.S. (1971) On the significance of the eclogite facies in Alpine metamorphism. *Verh. Geol. B.-A.* **2**, 257–265.

Fuchs, G. (1971) Zur tektonik des Östlichen Waldviertels (N.Ö.) *Verh. Geol. B.-A.* **3**, 424–440.

Fuchs, G. and Scharbert, H.G. (1979) Kleinere Granulitvorkommen in Nieder-Österreichischen und ihre Bedeutung für Granulitgenese. *Verh. Geol. B.-A.* **2**, 29–49.

Fujii, T. (1977) Pyroxene equilibria in spinel lherzolite. *Yearb. Carnegie Inst. Washington* **76**, 569–572.

Fumasoli, M.W. (1974) *Geologie des Gebietes nördlich und südlich der Jorio*-Tonale-Linie im Westen von Gravedona Como (Italia). Dissertation, Universität Zürich, 230 pp.

Fyfe, W.S., Price, N.J. and Thompson, A.B. (1978) *Fluids in the Earth's Crust.* Elsevier, Amsterdam.

Gabert, G. and Wendt, I. (1974) Datierung von granitischen Gesteinen im Dodoman- und Usagaran-System und in der Ndembera-Serie (Tanzania). *Geol. Jb.* **11**, 3–35.

Ganguly, J. (1979) Garnet and clinopyroxene solid solutions, and geothermometry based on Fe–Mg distribution coefficients. *Geochim. Cosmochim. Acta* **43**, 1021–1029.

Ganguly, J. and Ghose, S. (1979) Aluminous orthopyroxene: order–disorder, thermodynamic properties and petrological implications. *Contrib. Mineral. Petrol.* **69**, 375–385.

Garlick, G.D., MacGregor, I.D. and Vogel, D.E. (1971) Oxygen isotope ratios in eclogites from kimberlites. *Science* **172**, 1025–1027.

Gasparik, T. (1984a) Two-pyroxene thermobarometry with new experimental data in the system CaO–MgO–Al_2O_3–SiO_2. *Contrib. Mineral. Petrol.* **87**, 87–97.

Gasparik, T. (1984b) Experimental study of subsolidus phase relations and mixing properties of pyroxene in the system CaO–Al_2O_3–SiO_2. *Geochim. Cosmochim. Acta* **48**, 2537–2545.

Gasparik, T. (1984c) Experimental determined stability of clinopyroxene + garnet + corundum in the system CaO–MgO–Al_2O_3–SiO_2. *Am. Mineral.* **69**, 1025–1035.

Gasparik, T. (1985a) Experimentally determined compositions of diopside–jadeite pyroxene in equilibrium with albite + quartz at 1200–1350°C and 15–34 kbar. *Geochim. Cosmochim. Acta.* **49**, 865–870.

Gasparik, T. (1985b) Experimental study of subsolidus phase relations and mixing properties of pyroxene and plagioclase in the system Na_2O–CaO–Al_2O_3–SiO_2. *Contrib. Mineral. Petrol.* **89**, 346–357.

Gasparik, T. (1987) Orthopyroxene thermobarometry in simple and complex systems. *Contrib. Mineral. Petrol.* **96**, 357–370.

Gasparik, T. and Lindsley, D.H. (1980) Phase equilibria at high pressure of pyroxenes containing monovalent and trivalent ions. In *Pyroxenes* (ed. Prewitt, C.T.) *Mineral. Soc. Am. Reviews in Mineralogy*, **7**, 309–339.

Gasparik, T. and Newton, R.C. (1984) The reversed alumina contents of orthopyroxene in equilibrium with spinel and forsterite in the system MgO–Al_2O_3–SiO_2. *Contrib. Mineral. Petrol.* **85**, 186–196.

Gebauer, D. (1983) FIEC Feedback, sections on Hercynides and Isotopes. *Terra Cognita* **3**, 320–323.

Gabauer, D. (1984) Ein Modell der Entwicklungsgeschichte des ostbayerischen Grundgebirges auf der Basis plattentektonisher Vorstellung und radiometrischer Datierungen. In *Erlauterungen zur Geologischen Karte von Bauern* (*Blatt Nr. 7446 Passau*) Bayerisches Geologisches Landesamt, München.

Gebauer, D. (1986) The development of the continental crust of the European Hercynides since the Archean based on radiometric data. In *Proceedings of the Third Workshop on the European Geotraverse (EGT) Project. The Central Segment.* (eds. Freeman, R., Mueller, St. and Giese, P.) ESF/ESRC, Strasbourg, 15–23.

Gebauer, D., Bernard-Griffiths, J. and Grünenfelder, M. (1981) U-Pb zircon and monazite dating of a mafic-ultramafic complex and its country rocks. Example: Sauviat-sur-Vige, French Massif Central. *Contrib. Mineral. Petrol.* **76**, 292–300.

Gebauer, D., Bernard-Griffiths, J., Krebs, O. and Grünenfelder, M. (1978) U-Pb systematics of zircons and monazites from a mafic to ultramafic complex and its country rocks (Sauviat-sur-Vige, French Central Massif). *Short papers of the Fourth International Conference, Geochronology, Cosmochronology, Isotope Geology, Geol. Survey Open File Report*, 78–701, 131–133.

Gebauer, D. and Grünenfelder, M. (1977) U-Pb Zirkondatierungen mafischer und ultramafischer Gesteine. *Arbeitstagung Österr. Mineral. Ges. gemeinsam mit SMPG* (abstract volume), 13.

Gebauer, D. and Grünenfelder, M. (1978) U-Pb dating of eclogites: Example, Münchberg Gneiss Massif (Northeastern Bavaria). *Short papers of the Fourth International Conference, Geochronology, Cosmochronology, Isotope Geology, Geol. Survey Open File Report*, 78–701, 137–138.

Gebauer, D. and Grünenfelder, M. (1979a) U-Pb zircon and Rb-Sr mineral dating of eclogites and their country rocks. Example: Münchberg Gneiss Massif, northeast Bavaria. *Earth Planet. Sci. Lett.* **42**, 35–44.

Gebauer, D. and Grünenfelder, M. (1979b) U–Th–Pb dating of minerals. In *Lectures in Isotope Geology*. (ed. Jäger E. and Hunziker, J.C.) Springer Verlag, 105–131.

Gebauer, D. and Grünenfelder, M. (1979c) U-Pb zircon dating of mafic and ultramafic rocks. *Proceedings of the International Ophiolite Symposium, Cyprus*, 215–218.

Gebauer, D. and Grünenfelder, M. (1982) Geological development of the Hercynian belt of Europe based on age and origin of high-grade and high-pressure mafic and ultramafic rocks. *Proc. 5th Int. Conf. Geochron. Cosmochron. Isotope Geol. Japan*, 111–112.

Gebauer, D., Lappin, M., Grünenfelder, M., Koestler, A. and Wyttenbach, A. (1982) Age and origin of some Norwegian eclogites: A U-Pb zircon- and REE-study. *Terra Cognita* **2**, 323.

Gebauer, D., Lappin, M., Grünenfelder, M. and Wyttenbach, A. (1985) The age and origin of some Norwegian eclogites: A U-Pb zircon- and REE-study. *Chem. Geol.* (*Isotope Geosc. Section*) **52**, 227–247.

Gebauer, D., Quadt, A., Williams, I.S. and Composton, W. (1987) Archean zircons in a retrograded eclogite of the Gotthard Massif. 3. Rundgespräch *Geodynamik des Europäischen Variszikums. Kaledonisch-Variszische Strukturen in den Alpen.* Fribourg/Schweiz. Abstracts, 7.

Gebauer, D., Quadt, A., Williams, I.S., Compston, W. and Grünenfelder, M. (1988) in Archean zircons in a retrograded, Caledonian eclogite of the Gotthard Massif (Central Alps, Switzerland). *Schweiz. Mineral. Petrogr. Mitt.* **68**, 485–490.

Gebauer, D., Williams, I.S., Compston, W. and Grünenfelder, M. (1989) The development of the Central European continental crust since the Early Archean based on conventional and ion-microprobe dating of up to 3.84 b.y. detrital zircons. In *The Evolution of the European Continental Crust: Deep Drilling, Geophysics, Geology and Geochemistry.* (eds Meissner, R. and Gebauer, D.) *Tectonophysics* **157**, 81–96.

Gee, D.G. (1975) A tectonic model for the central part of the Scandinavian Caledonides. *Am. J. Sci.* **275A**, 468–515.

Gee, D.G. (1987) Baltica the indentor, *Geol. Fören. Stockholm Förh.* **109**, 346–347.

Gee, D.G., Kumpulainen, R., Roberts, D., Stephens, M.B., Thon, A. and Zachrisson, E. (compilers) (1985) *Scandinavian Caledonides Tectonostratigraphic Map.* Sver. Geol. Unders. Ser. Ba NR 35.

Gee, D.G. and Sturt, B.A. (eds) (1985) *The Caledonide Orogen—Scandinavia and Related Areas.* John Wiley, Chichester.

Ghent, E.D., Black, P.M., Brothers, R.N. and Stout, M.Z. (1987) Eclogites and associated albite–epidote–garnet paragneisses between Yambe and Cape Colnett, New Caledonia. *J. Petrol.* **28**, 627–643.

Ghent, E.D. Stout, M.Z. Black, P.M. and Brothers, R.N. (1987) Chloritoid-bearing rocks associated with blueschists and eclogites, northern New Caledonia, *J. Metamorph. Geol.* **5**, 239–254.

Gilbert, M.C. and Troll, G. (1974) A comparison of the stabilities of OH- and F-tremolite. *International Mineral. Assoc., 9th General Meeting, Berlin and Regensburg*, Collected Abstracts.

Gil Ibarguchi, J.I., Jamaa, N.B., Girardeau, J., Mercier, J-C.C. and Agrinier, P. (1987) The ophiolitic sequence of the Cabo Ortegal complex (NW Spain): MORB-type mafic rocks and websterite-rich peridotites. *Terra Cognita* **7**, 176.

Gil Ibarguchi, J.I. and Ortega Gironés, E. (1985) Petrology, structure and geotectonic implications of glaucophane-bearing eclogites and related rocks from the Malpica-Tuy (MT) Unit, Galicia, Northwest Spain. In *Chemistry and Petrology of Eclogites.* (eds. Smith, D.C., Franz, G. and Gebauer, D.) *Chem. Geol.* **50**, 145–162.

Gillet, P., Choukroune, P., Ballèvre, M. and Davy, P. (1986) Thickening history of the Western Alps. *Earth Planet. Sci. Lett.* **78**, 44–52.

Gillet, Ph., Davy, Ph., Ballèvre, M. and Choukroune, P. (1985) Thermomechanical evolution of a collision zone: The example of the Western Alps. *Terra Cognita* **5**, 399–404.

Gillet, P. and Goffé, B. (1988) On the significance of aragonite occurrences in the Western Alps. *Contrib. Mineral. Petrol.* **99**, 70–81.

Giraud, A., Marchand, J., Dupuy, C. and Dostal, J. (1984) Geochemistry of leptyno-amphibolite complex from Haut Allier (French Massif Central). *Lithos* **17**, 203–214.

Giraud, M., Burg, J-P and Powell, R. (1987) Evidence for a Variscan suture zone in the Vendée, France: a petrological study of blueschist facies rocks from Bois de Cené. *J. Metamorphic Geol.* **5**, 225–238.

Gjelsvik, T. (1951) Oversikt over bergartene i Sunnmøre og til grensende deler av Nordfjord. *Nor. Geol. Unders.* **179**, 1–45.

Gjelsvik, T. (1952) Metamorphosed dolerites in the gneiss area of Sunnmøre on the west coast of Southern Norway. *Nor. Geol. Unders.* **30**, 33–134.

Godard, D. (1981) *Lambeaux probables d'une croûte océanique subductée: Les éclogites de Vendée (Massif Amoricain, France).* Thèse 3ème cycle, Université de Nantes, 153 pp.

Godard, G., Kienast, J-R. and Lasnier, B. (1981) Retrogressive development of glaucophane in some eclogites from 'massif Armoricain' (East of Nantes, France). *Contrib. Mineral. Petrol.* **78**, 126–135.

Goffé, B. and Chopin, C. (1986) High-pressure metamorphism in the western Alps: zoneography of metapelites, chronology and consequences. Schweiz, *Mineral. Petrogr. Mitt.* **66**, 41–52.

Goffé, B., Michard, A., Kienast, J-R. and Le Mer, O. (1988) A case of obduction-related high-pressure, low-temperature metamorphism in upper crustal nappes, Arabian continental margin, Oman: *P–T* paths and kinematic interpretation, *Tectonophysics* **151**, 363–386.

Goldsmith, J.R. (1980) The melting and breakdown reactions of anorthite at high pressures and temperatures. *Am. Mineral.* **65**, 272–284.

Goldsmith, J.R. (1981) The join $CaAl_2Si_2O_8$–H_2O (anorthite–water) at elevated temperatures and pressures. *Am. Mineral.* **66**, 1183–1188.

Goldsmith, J.R. (1982) Plagioclase stability at elevated temperatures and water pressures. *Am. Mineral.* **67**, 653–675.

Golovnya, S.V., Sidorenko, S.A., Gerasimova, N.A. and Zamysheva, Z.M. (1973) Moissanite in rocks of the granulite complex of Sal'nyye Tundry, Kola Peninsula. *Dokl. Akad. Nauk SSSR* **213**, 127–129 (English translation).

Gosso, G., Dal Piaz, G.V., Piovano, V. and Polino, R. (1979) *High pressure emplacement of early-Alpine nappes, postnappe deformations and structural levels.* Mem. Inst. Geol. Mineral. Univ. Padova 32, 15 pp.

Govorov, I.N., Blagodareva, N.S., Kiryukhina, N.I., Kharkiv, A.D. and Shcheglov, A.D. (1986) Primary potassium minerals in plutonic eclogite xenoliths from Yakutia. *Dokl. Akad. Nauk SSSR. Earth Sci. Sect.* **276**, 123–126.

Graham, C.M. and Powell, R. (1984) A garnet–hornblende geothermometer. Calibration, testing and application to the Pelone Schist, Southern California. *J. Metamorphic Geol.* **2**, 13–33.

Grant, S.M. (1988) Diffusion models for corona formation in metagabbros from the Western Grenville Province, Canada. *Contrib. Mineral. Petrol.* **98**, 49–63.

Grauert, B., Hanny, R. and Soptrajanova, G. (1974) Geochronology of a polymetamorphic and anatectic gneiss region: the Moldanubicum of the area Lam-Deggendorf, eastern Bavaria, Germany. *Contrib. Mineral. Petrol.* **45**, 37–63.

Green, D.H. (1973) Experimental melting studies of a model upper mantle composition at high pressure under water-saturated and water undersaturated conditions. *Earth Planet. Sci. Lett.* **19**, 37–53.

Green, D.H. and Hellman, P.L. (1982) Fe–Mg partitioning between coexisting garnet and phengite at high pressures, and comments on a garnet–phengite geothermometer. *Lithos* **15**, 253–266.

Green, D.H. and Lambert, I.B. (1965) Experimental crystallisation of anhydrous granite at high pressures and temperatures. *J. Geophys. Res.* **70**, 5259–5268.

Green, D.H., Lockwood, J.P. and Kiss, E. (1968) Eclogite and almandine–jadeite–quartz rock from the Guajira peninsula, Colombia, South America. *Am. Mineral.* **53**, 1320–1335.

Green, D.H. and Mysen, B.O. (1972) Genetic relationship between eclogite and hornblende and plagioclase pegmatite in western Norway. *Lithos.* **5**, 147–161.

Green, D.H. and Ringwood, A.E. (1967a) An experimental investigation of the gabbro to eclogite transformation and its petrological applications. *Geochim. Cosmochim. Acta* **31**, 767–833.

Green, D.H. and Ringwood, A.E. (1967b) The stability fields of aluminous pyroxene peridotite and garnet peridotite and their relevance in upper mantle structure. *Earth Planet. Sci. Lett.* **3**, 151–160.

Green, D.H. and Ringwood, A.E. (1972) A comparison of recent experimental data on the gabbro-garnet granulite-eclogite transition, *J. Geol.* **80**, 277–288.

Green, T.H. (1967) An experimental investigation of sub-solidus assemblages formed at high pressure in high-alumina basalt, kyanite eclogite and grosspydite compositions. *Contrib. Mineral. Petrol.* **16**, 84–114.

Green, T.H. (1969) The diopside-kyanite join at high pressures and temperatures. *Lithos* **2**, 334–341.

Grew, E.S. and Sandiford, M. (1984) A staurolite-talc assemblage in tourmaline-phlogopite-chlorite schist from northern Victoria Land, Antarctica, and its petrological significance, *Contrib. Mineral. Petrol.* **87**, 337–350.

Griffin, W.L. (1972) Formation of eclogites and the coronas in anorthosite, Bergen Arcs, Norway. *Mem. Geol. Soc. Am.* **135**, 37–63.

Griffin, W.L. (1987) 'On the eclogites of Norway'–65 years later. *Mineral. Mag.* **51**, 333–343.

Griffin, W.L., Austrheim, H., Brastad, K., Bryhni, I., Krill, A.G., Krogh, E.J., Mørk, M.B.E., Qvale, H. and Torudbakken, B. (1985) High-pressure metamorphism in the Scandinavian Caledonides. In *The Caledonide Orogen–Scandinavian and Related Areas* (eds Gee, D.G. and Sturt, B.A.) pp. 783–801, Wiley, Chichester.

Griffin, W.L. and Brueckner, H.K. (1980) Caledonian Sm–Nd ages and a crustal origin for Norwegian eclogites. *Nature* **285**, 319–321.

Griffin, W.L. and Brueckner, H.K. (1985) REE, Rb–Sr and Sm–Nd studies of Norwegian eclogites. *Chem. Geol. (Isot. Geosci.)* **52**, 249–271.

Griffin, W.L. and Carswell, D.A. (1985) In situ metamorphism of Norwegian eclogites: an example. In *The Caledonide Orogen–Scandinavia and Related Areas* (eds Gee, D.G. and Sturt, B.A.) pp. 813–822. John Wiley, Chichester.

Griffin, W.L., Carswell, D.A. and Nixon, P.H. (1979) Lower crustal granulites and eclogites from Lesotho, southern Africa. In *The Mantle Sample: Inclusions in Kimberlites and Other Volcanics.* eds Boyd, F.R. and Meyer, H.O.A. Am. Geophys. Union, Washington, D.C., 59–86.

Griffin, W.L. and Heier, K.S. (1973) Petrological implications of some corona structures. *Lithos* **6**, 315–335.

Griffin, W.L., Jaques, A.L., Sie, S.H., Ryan, C.G., Cousens, D.R. and Suter, G.F. (1988) Conditions of diamond growth: a proton microprobe study of inclusions in West Australian diamonds. *Contrib. Mineral. Petrol.* **99**, 143–158.

Griffin, W.L. and Mørk, M.B.E. (1981) *Eclogites and basal gneisses in west Norway.* Uppsala Caledonide Symposium, Excursion No. B1.

Griffin, W.L. and Mottana, A. (1982) Crystal chemistry of clinopyroxenes from the St Marcel manganese deposit. Val d'Aosta. Italy. *Am. Mineral.* **87**, 568–586.
Griffin, W.L. and O'Reilly, S.Y. (1987a) The composition of the lower crust and the nature of the continental Moho—xenolith evidence. In *Mantle Xenoliths.* (ed. Nixon, P.H.) Wiley, 413–430.
Griffin, W.L. and O'Reilly, S.Y. (1987b) Is the continental Moho the crust–mantle boundary? *Geology* **15**, 241–244.
Griffin, W.L. and Qvale, H. (1985) Superferrian eclogites and the crustal origin of garnet peridotites, Almklovdalen, Norway. In *The Caledonide Orogen—Scandinavia and Related Areas* (eds Gee, D.G. and Sturt, B.A.) pp. 803–812, Wiley, Chichester.
Griffin, W.L. and Råheim, A. (1973) Convergent metamorphism of eclogites and dolerites, Kristiansund area, Norway. *Lithos* **6**, 21–40.
Griffin, W.L., Wass, S.Y. and Hollis, J.D. (1984) Ultramafic xenoliths from Bullenmerri and Gnotuk maars, Victoria, Australia: petrology of a subcontinental crust–mantle transition. *J. Petrol.* **25**, 53–87.
Griffin, W.L. and Råheim, A. (1973) Covergent metamorphism of eclogites and dolerites, Kristiansund area, Norway. *Lithos* **6**, 21–40.
Grocholski, W. (1967) Structure of the Sowie Mts. (Polish with English summary). *Geol. Sudetica* **3**, 181–249.
Grubenmann, U. (1908) Der Granatolivinfels des Gordunotales and seine Begleitgesteine. *Vjschr. Naturforsch. Ges. Zürich* **53**, 129–156.
Gurney, J.J. and Boyd, F.R. (1982) Mineral intergrowths with polycrystalline diamonds from the Orapa Mine, Botswana. *Yearb. Carnegie Inst. Washington*, **81**, 267–283.
Gurney, J.J., Harris, J.W. and Rickard, R.S. (1979) Silicate and oxide inclusions in diamonds from the Finsch kimberlite pipe. In *Kimberlites, Diatremes and Diamonds* (eds Boyd, F.R. and Mayer, H.O.A.) Am. Geophys. Union, Washington, D.C., 1–15.
Gurney, J.J., Harris, J.W. and Rickard, R.S. (1984a) Silicate and oxide inclusions in diamonds from the Orapa Mine, Botswana. In *Kimberlites II: The Mantle and Crust–Mantle Relationship* (ed. Kornprobst, J.) Elsevier, Amsterdam, 3–9.
Gurney, J.J., Harris, J.W. and Rickard, R.S. (1984b) Minerals associated with diamonds from the Roberts Victor Mine. In *Kimberlites II: The Mantle and Crust–Mantle Relationships* (ed. Kornprobst, J.) Elsevier, Amsterdam, 25–32.
Gurney, J.J., Harris, J.W., Rickard, R.S. and Moore, R.O. (1985) Inclusions in Premier Mine diamonds. *Trans. Geol. Soc. Africa* **88**, 301–310.
Gurney, J.J. and Hatton, C.J. (1989) Diamondiferous minerals from the Star Mine, South Africa. In *Kimberlites and related rocks, Vol. 2. Their mantle/crust setting, diamonds and diamond exploration.* Geol. Soc. Australia, Spec. Publn 14, Blackwells, Carlton, 1022–1028.
Gurney, J.J., Mathias, M., Siebert, C. and Moseley, G. (1971) Kyanite eclogites from the Rietfontein kimberlite pipe, Mier Coloured Reserve, Gordonia, Cape Province, South Africa. *Contrib. Mineral. Petrol.* **30**, 46–52.
Gurney, J.J., Moore, R.O., Otter, M.L., Kirkley, M.B., Hops, J.J. and McCandless, T.E. (in press) Southern African kimberlites. In, *Phanerozoic magmatism and structural evolution of the African Plate* (eds Kampunzu, A.B. and Lubala, R.T.).
Haggerty, S.E. (1987) Metasomatic mineral titanates in upper mantle xenoliths. In *Mantle Xenoliths*, (ed. Nixon, P.H.), Wiley, Chichester, 671–690.
Halbach, H. and Chatterjee, N.D. (1984) An internally consistent set of thermodynamic data for twenty-one $CaO–Al_2O_3–SiO_2–H_2O$ phases by linear parametric programming. *Contrib. Mineral. Petrol.* **88**, 14–23.
Halbase, T. and Kleppa, O.J. (1968) The thermochemistry of Jadeite. *Am. Mineral.* **53**, 1281–1292.
Hales, A.L. and Sacks, I.S. (1959) Evidence for an intermediate layer from crustal seismic studies in the Eastern Transvaal. *Geophys. J. R. Astr. Soc.* **2**, 15–33.
Hall, A.E. and Smith, C.B. (1984) Lamproite diamonds—are they different? In *Kimberlite occurrence and origin: a basis for conceptual models in exploration* (eds Glover, J.E. and Harris, P.G.) Univ. W. Australia, Perth, 167–212.
Hameurt, J. (1967) *Les terrains cristallins et cristallophylliens du versant occidental des Vosges moyennes.* Mém. Serv. Carte geol. Alsace-Lorraine 26, 402 pp.
Hammer, W. (1899) Olivingesteine aus dem Nonsberg, Sulzberg und Ultental. *Zeitschr. Naturwiss.* **72**, 1–48.
Hammer, W. (1926) Eklogit und Peridotit in den mittleren Ötztaler Alpen. *Jahrb. Geol. Bundesanstalt, Wien* **76**, 97–123.
Hammer, W. (1933) Über die Peridotite am Loibiskogel im Ötztal. *Verh. Geol. Bundesanstalt, Wien*, 59–62.
Hänny, R., Grauert, B. and Soptrajanova, G. (1975) Palaeozoic migmatites affected by high-grade Tertiary metamorphism in the Central Alps. *Contrib. Mineral. Petrol.* **51**, 173–196.
Hanselmeyer, J. (1977) Eín interessantes Gesteinsvorkommen der Schobergruppe (Eklogitamphibolit mit Hornblendekelyphit). *Carinthia* II **167**, 53–58.
Hansen, B. (1981) The transition from pyroxene granulite facies to garnet clinopyroxene granulite facies. Experiments in the system $CaO–MgO–Al_2O_3–SiO_2$. *Contrib. Mineral. Petrol.* **76**, 234–242.
Harland, W.B., Cox, A.V., Llewellyn, P.G., Picton, C.A.G., Smith, A.G. and Walters, R. (1982) *A Geological Time Scale.* Cambridge University Press, Cambridge, England.
Harley, S.L. (1984a) An experimental study of the partitioning of Fe and Mg between garnet and orthopyroxene. *Contrib. Mineral. Petrol.* **86**, 359–373.

Harley, S.L. (1984b) The solubility of alumina in orthopyroxene coexisting with garnet in $FeO-MgO-Al_2O_3-SiO_2$ and $CaO-FeO-MgO-Al_2O_3-SiO_2$. *J. Petrol.* **25**, 665–696.

Harley, S.L. (1984c) Comparison of the garnet–orthopyroxene geobarometer with recent experimental studies, and applications to natural assemblages. *J. Petrol.* **25**, 697–712.

Harley, S.L. Green, D.H. (1981) Petrogenesis of eclogite inclusions in the Moses Rock Dyke, Utah, U.S.A. *Tschermaks Mineral. Petrogr. Mitt.* **28**, 131–155.

Harley, S.L. and Green, D.H. (1982) Garnet–orthopyroxene barometry for granulites and peridotites. *Nature* **300**, 697–701.

Harrison, T.M. (1981) Diffusion of ^{40}Ar in hornblende. *Contrib. Mineral. Petrol.* **78**, 324–331.

Hart, R.J., Welke, H.J. and Nicolaysen, L.O. (1971) Geochronology of the deep profile through Archaean basement at Vredefort, with implications for early crustal evolution. *J. Geophys. Res.* **86** (B11), 10663–10680.

Harte, B. and Gurney, J.J. (1975) Evolution of clinopyroxene and garnet in an eclogite nodule from the Roberts Victor kimberlite pipe, South Africa. *Phys. Chem. Earth* **9**, 367–387.

Harte B. and Hudson N.F.C. (1979) Pelite facies and pressures of Dalradian metamorphism in E-Scotland. In *The Caledonides of the British Isles–reviewed.* Geol. Soc. Lond. Spec. Pub. **8**, 323–337.

Harvey, M.A. (1983) *A geochemical and structual study of the gneisses and eclogites on the Molde peninsula, west Norway.* Unpublished PhD thesis, University of Sheffield.

Harvey, M.A. (1983) A geochemical and Rb–Sr study of the Proterozoic augen orthogneisses on the Molde Peninsula west Norway. *Lithos* **16**, 325–338.

Haselton, H.T. and Newton, R.C. (1980) Thermodynamics of pyrope-grossular garnets and their stabilities at high temperatures and high pressures. *J. Geophys. Res.* **85**, B12, 6973–6982.

Haskin, L.A., Frey, F.A. Schmitt, R.A. and Smith, R.H. (1966) Meteoritic, solar and terrestrial rare-earth distribution. *Phys. Chem. Earth* **7**, 167–321.

Hatton, C.J. (1978) *Geochemistry and origin of eclogite xenoliths from the Roberts Victor Mine.* Unpublished. Ph.D. thesis, University of Cape Town.

Hatton, C.J. and Gurney, J.J. (1979) A diamond-graphite eclogite from the Roberts Victor Mine. In *The Mantle Sample: Inclusions in Kimberlites and Other Volcanics.* (eds Boyd, F.R. and Meyer, H.O.A.) Am. Geophys. Union, Washington, D.C., 29–36.

Hatton, C.J. and Gurney, J.J. (1987) Roberts Victor eclogites and their relation to the mantle. In *Mantle Xenoliths*, (ed. Nixon, P.H.) Wiley, Chichester, 453–463.

Hawkesworth, C.J., Erlank, A.J., Marsh, J.S. Menzies, M.A. and van Calsteren, P. (1983) Evolution of the continental lithosphere: evidence from volcanics and xenoliths in southern Africa. In *Continental Basalts and Mantle Xenoliths.* (eds Hawkesworth, C.J. and Norry, M.J.) Shiva, Nantwich, 111–138.

Hays, J.F. (1966) Lime–alumina–silica. *Yearb. Carnegie Inst. Washington* **65**, 234–239.

Haüy, R.J. (1822) *Traité de Mineralogie.* 2nd Edition. Bachelier, Paris Delance, IV, 584.

Heinrich, C.A. (1982) Kyanite-eclogite to amphibolite facies evolution of hydrous mafic and pelitic rocks, Adula Nappe, central Alps. *Contrib. Mineral. Petrol.* **81**, 30–38.

Heinrich, C.A. (1986) Eclogite facies regional metamorphism of hydrous mafic rocks in the central alpine Adula nappe. *J. Petrol.* **27**, 123–154.

Heinrich, W. and Althaus, E. (1980) Die obere Stabilitätsgrenze von Lawsonit plus Albit bzw. Jadeit. *Fortschr. Mineral.* **58**, 49–50.

Heinrich, W. and Althaus, E. (1980) Die obere Stabilitätsgrenze von Lawsonit plus Albit bzw. Jadeit. *Fortschr. Mineral* **58**, 509–528.

Heinrich, W. and Althaus, E. (1988) Experimental determination of the reactions 4 lawsonite + 1 albite = 1 paragonite + 2 zoisite + 2 quartz + $6H_2O$ and 4 lawsonite + 1 jadeite = 1 paragonite + 2 zoisite + 1 quartz + $6H_2O$. *Neues. Jahrb. Mineral. Monatsh.* **11**, 516–528.

Helmstaedt, H. and Carmichael, D.M. (1978) Are grospydites metarodingites? Implications for upper mantle models based on xenoliths from some diamondiferous kimberlites. *XI Gen. Meeting Int. Mineral. Ass. Novosibirsk Abst.* **2**, 40–41.

Helmstaedt, H., Carmichael, D.M. and Percival, J.A. (1979) Grospydite xenoliths from the Zagadochnaya kimberlite pipe, Yakutia-high grade metamorphic rodingites, *Geol. Soc. Amer. Abstr. Progr.* **11**, 442.

Helmstaedt, H. and Doig, R. (1975) Eclogite nodules from kimberlite pipes of the Colorado Plateau—samples of subducted Franciscan-type oceanic lithosphere. *Phys. Chem. Earth* **9**, 95–112.

Helmstaedt, H. and Gurney, J.J. (1984) Kimberlites of southern Africa—Are they related to subduction processes? In *Kimberlites I: Kimberlites and Related Rocks* (ed. Kornprobst, J.) Elsevier, Amsterdam, 425–434.

Helmstaedt, H. and Schulze, D.J. (1979) Garnet pyroxenite–chlorite eclogite transition in a xenolith from Moses Rock: further evidence for metamorphosed ophiolites under the Colorado Plateau. In *The Mantle Sample: Inclusions in Kimberlites and other Volcanics.* (eds Boyd, F.R. and Meyer H.O.A.) *Proc. 2nd Int. Kimberlite Conference* **2**, American Geophysical Union, Washington D.C., 357–365.

Helz, R.T. (1982) Phase relations and compositions of amphiboles produced in studies of the melting behaviour of rocks. In *Amphiboles: Petrology and Experimental Phase Relations*, eds. Veblen, D.R. and Ribbe, P.H., *Mineral Soc. Am.* Washington. Reviews in *Mineralogy*, **9B**, 279–347.

Henry, C., Chopin, C. and Michard, A. (1987) Tectonic setting of the very-high-pressure rocks within the Dora-Maira massif, Western Alps. *Terra Cognita* **7** (2–3), 94.

Hensen, B.J. (1976) The stability of pyrope-grossular garnet with excess silica. *Contrib. Mineral. Petrol.* **55**, 279–292.

Hensen, B.J. and Essene, E.J. (1971) Stability of pyrope-quartz in the system $MgO–Al_2O_3–SiO_2$. *Contrib. Mineral. Petrol.* **83**, 72–83.

Hensen, B.J., Schmid, R. and Wood, B.J. (1975) Activity-composition relationships for pyrope-grossular garnet. *Contrib. Mineral. Petrol.* **51**, 161–166.

Heritsch, H. (1966) Bericht über Untersuchungen an eklogitischen Gesteinen der Koralpe. *Öster. Akad. Naturwiss. Math.-Nat. Kl.* **14**, 268–271.

Heritisch, H. (1973) Die Bildungsbedingungen von alpinotypem Eklogitamphibolit und Metagabbro, erläutert an Gesteinen der Koralpe, Steiermark. *Tschermaks Mineral. Petrogr. Mitt.* **19**, 213–271.

Heritsch, H. and Bossert, F. (1969) Chemische Analysen gabbroider und eklogitischer Gesteine und ihrer Mineralien vom Fundpunk Lenzbauer in Gressenburg bei Schwanberg, Koralpe, Steiermark. *Mitt. Naturwiss. Ver. Steiermark* **99**, 5–17.

Heritsch, H. and Wittman, R. (1969) Chemische Analysen eklogitischer Gesteine und ihrer Mineralien vom Fundpunkt Hohl bei Wies, Koralpe, Steiermark. *Mitt. Naturwiss. Ver. Steiermark* **99**, 18–29.

Hervig, R.L., Smith, J.V. and Dawson, J.B. (1986) Lherzolite xenoliths in kimberlites and basalts: petrogenetic and crystallographic signficance of some minor and trace elements in olivine, pyroxenes, garnet and spinel. *Trans, R. Soc. Edinburgh, Earth Sci.* **77**, 181–201.

Herzberg, C.T. (1978) Pyroxene geothermometry and geobarometry: experimental and thermodynamic evaluation of some subsolidus phase relations involving pyroxenes in the system $CaO–MgO–Al_2O_3–SiO_2$. *Geochim. Cosmochim. Acta* **42**, 9454–9457.

Herzberg, C.T. (1984) Chemical stratification in the silicate Earth. *Earth Planet. Sci. Letts* **67**, 249–260.

Hezner, L. (1903) Ein Beitrage zur Kenntnis der Eklogite und Amphibolite mit besonderer Berücksichtigung der Vorkommen des mittleren Ötztales. *Tschermaks Mineral. Petrogr. Mitt.* **22**, 437–471.

Hill, R.E.T. and Boettcher, A.L. (1970) Water in the earth's mantle: melting curves of basalt–water and basalt–water–carbon dioxide. *Science* **167**, 980–982.

Hinterlechner-Ravnik, A. (1971) The metamorphic rocks of Pohorje (English summary). *Geologija Razprave Porocila* **14**, 187–226.

Hirajima, T. (1983) Jadeite + quartz rock from Kanto Mountains, *J. Jap. Ass. Mineralogists Petrologists Economic Geologists* **78**, 77–83.

Hirajima, T., Banno, S., Hiroi Y. and Ohta, Y. (1988) Phase petrology of eclogites and related rocks from the Motalafjella high-pressure metamorphic complex in Spitsbergen (Arctic Ocean) and its significance. *Lithos* **22**, 75–97.

Höck, V. and Miller, C. (1987) Mesozoic ophiolitic sequences and non-ophiolitic metabasites in the Hohe Tauern. pp. 16–23. In *Geodynamics of the Earth Alps* (eds Flügel, H.W. and Faupl, P.) Deuticke, Vienna.

Hodges, K.V., Bartley, J.M. and Burchfiel, B.C. (1982) Structural evolution of an A-type subduction zone, Lofoten-Rombak area, northern Scandinavian Caledonides. *Tectonics*, **1**, 441–462.

Hodges, K.V and McKenna, L.W. (1987) Realistic propagation of uncertainties in geologic thermobarometry. *Am. Mineral.* **72**, 671–680

Hoke, L. (1987) *The geology of the Altkristallin Complex, northern Kreuzeck region*, Austria. Unpublished Ph.D thesis, University of Cambridge.

Holdaway, M.J. (1971) Stability of andalusite and the aluminium-silicate phase diagram. *Am. J. Sci.* **271**, 97–131.

Holland, T.J.B. (1979a) High-water activities in the generation of high-pressure kyanite eclogites of the Tauern Window, Austria, *J. Geol.* **87**, 1–27.

Holland, T.J.B. (1979b) Experimental determination of the reaction paragonite = jadeite + kyanite + water, and internally consistent thermodynamic data for part of the system $Na_2O–Al_2O_3–SiO_2–H_2O$, with application to eclogites and blueschists. *Contrib. Mineral. Petrol.* **68**, 293–301.

Holland, T.J.B. (1980) The reaction albite = jadeite + quartz determined experimentally in the range 600–1200°C. *Am Mineral* **65**, 129–134.

Holland, T.J.B., (1983) The experimental determination of activities in disordered and short-range ordered jadeitic pyroxenes. *Contrib. Mineral. Petrol.* **82**, 214–220.

Holland, T.B.J. (1988) Preliminary phase relations involving glaucophane and applications to high pressure petrology: new heat capacity and thermodynamic data. *Contrib. Mineral. Petrology* **99**, 134–142.

Holland, T.J.B., Navrotsky, A. and Newton, R.C. (1979) Thermodynamic parameters of $CaMgSi_2O_6$–$Mg_2Si_2O_6$ pyroxenes based on regular solution and cooperative disordering models. *Contrib. Mineral. Petrol.* **69**, 337–344.

Holland, T.J.B. and Powell, R. (1985) An internally consistent thermodynamic dataset with uncertainties and correlations: 2. Data and results. *J. Metamorphic Geol.* **3**, 343–370.

Holland, T.J.B. and Ray, N.J. (1985) Glaucophane and pyroxene breakdown reactions in the Pennine units of the Eastern Alps. *J. Metamorphic Geol.* **3**, 417–438.

Holland, T.J.B. and Richardson, S.W. (1979) Amphibole zonation in metabasites as a guide to evolution of metamorphic conditions. *Contrib. Mineral. Petrol.* **70**, 143–148.

Hollister, L.S., Burruss, R.C., Henry, D.L. and Hendel, E.M. (1979) Physical conditions during uplift of metamorphic terranes as recorded by fluid inclusions, *Bull. Mineral.* **102**, 555–561.

Holm, J.L., Kleppa, O.J. and Westrum, E.F. (1967) Thermodynamics of polymorphic transformation in silica:

thermal properties from 5 to 1070 K and pressure–temperature stability fields for coesite and stishovite. *Geochim. Cosmochim. Acta* **31**, 2289–2307.

Holub, F.V. and Munshi, R.L. (1984) Subsolidus reaction rims between olivine and calcic plagioclase in the Svatý Kříž massif, Eastern Bohemia. *Krystalinikum* **17**, 47–58.

Hopson, C.A., Coleman, R.G., Gregory, R.T., Henry, D.L. and Hendel, E.M. (1981) Geologic section through the Semail ophiolite and associated rocks along a Muscat-Ibra transect, southeastern Oman Mountains, *J. Geophys. Res.* **86**, 2527–2544.

Hossack, J.R. (1984) The geometry of listric growth faults in the Devonian basins of Sunnfjord, West Norway. *J. Geol. Soc.* London **141**, 629–637.

Hossack, J.R. and Cooper, M.A. (1986) Collision tectonics in the Scandinavian Caledonides. In *Collision Tectonics* (eds Coward, M.P. and Ries, A.C.) *Geol. Soc. London* Spec. Publ. No. 19, pp. 287–304.

Hossack, J.R., Garton, M.R. and Nickelsen, R.P. (1985) The geological section from the foreland up to the Jotun thrust sheet in the Valdres area, south Norway. In *The Caledonide Orogen—Scandinavia and Related Areas* (eds Gee, D.G. and Sturt, B.A.) pp. 443–456, John Wiley, Chichester.

Howells, S. and O'Hara, M.J. (1978) Low solubility of alumina in enstatite and uncertainties in estimated palaeogeotherms. *Phil. Trans. R. Soc. London A* **288**, 471–486.

Huang, W.L. and Wyllie, P.J. (1973) Melting relations of muscovite granite to 35 kbar as a model for fusion of metamorphosed subducted oceanic sediments. *Contrib. Mineral. Petrol.* **42**, 1–14.

Humphries, F.J. and Cliff, R.A. (1982) Sm–Nd dating and cooling history of Scourian granulites, Sutherland. *Nature* 295, 515–517.

Hunziker, J.C. (1974) Rb–Sr and K–Ar age determinations and the Alpine tectonic history of the Western Alps. Università di Padova. *Mem. Inst. Geol. Mineral.* **31**, 54 pp.

Indares, A. and Martignole, J. (1985) Biotite–garnet geothermometry in the granulite facies: the influence of Ti and Al in biotite. *Am. Mineral.* **70**, 272–278.

Irifune, T., Hibberson, W.O. and Ringwood, A.E. (1989) Eclogite–garnetite transformation at high pressure and its bearing on the occurrence of garnet inclusions in diamond. In *Kimberlites and related rocks, Vol. 2. Their mantle/crust setting, diamonds and diamond exploration.* Geol. Soc. Australia Spec. Publn. **14**, Blackwells, Carlton, 877–882.

Irifune, T., Sekine, T., Ringwood, A.E. and Hibberson, W.O. (1986) The eclogite-garnetite transformation at high pressure and some geophysical implications. *Earth Planet. Sci. Lett.* 77, 245–256.

Itaya, T., Brothers, R.N. and Black, P.M. (1985) Sulfides, oxides and sphene in high-pressure schists from New Caledonia, *Contrib. Mineral. Petrol.* **91**, 151–162.

Ito, K. and Kennedy, G.C. (1971) An experimental study of the basalt-garnet granulite-eclogite transition. In: *The Structure and Physical Properties of the Earth's Crust, Geophys. Monograph* 14, 303–314. Am. Geophys. Union, Washington.

Jackson, P.M. and Harte, B. (1977) The nature and conditons of formation of granulite facies xenoliths from the Matsoku kimberlite pipe, Lesotho. Extended abstr. *2nd Int. Kimberlite Conference*, unpaginated.

Jacobsen, S.B. and Wasserburg, G.J. (1980) Nd and Sr isotopes of the Norwegian garnet peridotites and eclogites. *EOS* (*Trans. Am. Geophys. Union*) **61/17**, 389.

Jagoutz, E. (1988) Nd and Sr systematics in an eclogite xenolith from Tanzania: Evidence for frozen mineral equilibria in the continental lithosphere. *Geochim. Cosmochim. Acta* **52**, 1285–1293.

Jagoutz, E., Dawson, J.B., Spettel, B. and Wanke, H. (1983) Identification of early differentiation processes on the Earth. *46th Ann. Mtg. Meteor. Soc. Mainz, Abstract* (unpaged).

Jagoutz, E., Dawson, J.B., Hoernes, S., Spettel, B. and Wanke, H. (1948) Anorthositic oceanic crust in the Archean Earth. *Lunar Planet. Sci.* **XV**, 395–396.

Jakeš, P. (1969a) Distribution of the granulites within the Bohemian Massif and their petrological features. *Geol. Soc. Australia. Special Publication* **2**, 269–277.

Jakeš, P. (1969b) Retrogressive changes of granulite-facies rocks—an example from the Bohemian Massif. *Geol. Soc. Australia, Special Publication* **2**, 367–374.

Jamtveit, B. (1984) High-P metamorphism and deformation of the Gurskebotn, garnet peridotite, Sunnmøre, western Norway. *Nor. Geol. Tidsskr.* **64**, 97–110.

Jamtveit, B. (1987a) Metamorphic evolution of the Eiksunddal eclogite complex, western Norway, and some tectonic implications. *Contrib. Mineral. Petrol.* **95**, 82–99.

Jamtveit, B. (1987b) Magmatic and metamorphic controls on chemical variations within the Eiksunddal eclogite complex, Sunnmøre, western Norway. *Lithos* **20**, 369–389.

Jamtveit, B., Carswell, D.A. and Mearns, E.W. (1989). Chronology of the high-P, T metamorphism of Norwegian garnet peridotites. *J. metamorphic Geol.* (submitted).

Jaques, A.L., Hall, A.E., Sheraton, J.W., Smith, C.B., Sun, S.S., Drew, R.M., Foudoulis, C. and Ellingsen, K. (1989) Composition of crystalline inclusions and C-isotopic compositions of Argyle and Ellendale diamonds. In *Kimberlites and related rocks, Vol. 2. Their mantle/crust setting, diamonds and diamond exploration.* Geol. Soc. Australia Spec. Publn. **14**, Blackwells, Carlton, 966–989.

Jegouzo, P., Peucat, J.-J. and Audren, C. (1986) Caractérisation et signification géodynamique des orthogneiss calco-alcalins d'ăge ordovicien de Bretagne méridionale. *Bull. Soc. Geol. Fr.* (8) II, 839–848.

Jelínek, E., Pačesová, M., Mísař, P., Martinec, P. and Weiss, Z. (1984) Geochemistry of a dismembered metaophiolite complex, Letovice, Czechoslovakia. *Trans. R. Soc. Edinburgh Earth Sci.* **73**, 89–108.

Jenkins, D.M. (1981) Experimental phase relations of hydrous peridotites modelled in the system H_2O–CaO–MgO–Al_2O_3–SiO_2. *Contrib. Mineral. Petrol.* **77**, 166–176.

Jenkins, D.M. (1984) Upper-pressure stability of synthetic margarite plus quartz. *Contrib. Mineral. Petrol.* **88**, 332–339.

Jenkins, D.M. and Newton, R.C. (1979) Experimental determination of the spinel peridotite to garnet peridotite inversion at 900°C and 1100°C in the system CaO–MgO–Al_2O_3–SiO_2, and at 900°C with natural garnet and olivine. *Contrib. Mineral. Petrol.* **68**, 407–419.

Joesten, R. (1977) Evolution of mineral assemblage zoning in diffusion metasomatism. *Geochim. Cosmochim. Acta* **41**, 649–670.

Joesten, R. (1986) The role of magmatic reaction, diffusion and annealing in the evolution of coronitic microstructure in troctolitic gabbro from Risør, Norway. *Mineral. Mag.* **50**, 441–467.

Johannes, W. (1980) Melting and subsolidus reactions in the system K_2O–CaO–Al_2O_3–SiO_2-H_2O. *Contrib. Mineral. Petrol.* **74**, 29–34.

Johannes, W. (1984) Beginning of melting in the granite system Qz–Or–Ab–An– H_2O. *Contrib. Mineral. Petrol.* **86**, 264–273.

Julivert, M., Martinez, F.J. and Ribeiro, A. (1980) The Iberian segment of the European Hercynian foldbelt. In *Géologie de l'Europe du Précambrien aux bassins sédimentaires posthercyniennes.* (eds Cogné, J. and Slansky, M.) *Mém. B.R.G.M.* **108**, 132–158.

Kappel, F. (1967) Die Eklogite Meidling im Tal und Mitterbachgraben im niederösterreichischen Moldanubikum südlich der Donau: *Neues Jahrb. Mineral. Abh.* **107**, 266–298.

Katchalsky, A. and Curran, P.F. (1967) *Nonequilibrium Thermodynamics in Biophysics.* Harvard University Press, Cambridge.

Kawasaki, T. (1979) Thermodynamic analysis on the Mg–Fe exchange equilibrium between olivine and garnet: an application to the estimation of P–T relations of ultramafic rocks. *J. Japan Ass. Mineral. Petrol. Econ. Geol.* **74**, 395–405.

Kawasaki, T. and Matsui, Y. (1983) Thermodynamic analysis of equilibria involving olivine, orthopyroxene and garnet. *Geochim. Cosmochim. Acta* **47**, 1661–1679.

Kennedy, C.S. and Kennedy, G.C. (1976) The equilibrium boundary between graphite and diamond. *J. Geophys. Res.* **81**, 2467–2470.

Kepezhinkas, V.V. (1979) Cenozoic alkaline basaltoids of Mongolia and related deep inclusions. *Trans. Joint Soviet-Mongolian Scient. Res. Geol. Exp.* **25**, 259 pp; also Nauk press, 311 pp. Moscow (in Russian).

Kienast, J.R. (1983) *Le métamorphisme de haute pression at basse température (éclogite et schistes bleus): données nouvelles sur la pétrologie des roches de la croûte océanique subductée et des sédiments associés.* Thesis Université de Paris VI, 474 pp.

Kienast, J.R. and Lombardo, B. (1987) Eclogites and polymetamorphic gneisses from the southern Dora-Maira (Western Alps). *Terra Cognita* **7** (2–3), 94.

Kienast, J.R. and Messiga, B. (1987) Cr-rich Mg-chloritoid, a first record in high pressure metagabbros from Monviso (Cottian Alps, Italy). *Mineral. Mag.* **51**, 681–687.

Kienast, J.R. and Pognante, U. (1988) Chloritoid-bearing assemblages in eclogitised metagabbros of the Lanzo peridotite body (Western Italian Alps). *Lithos* **21**, 1–11.

Kienast, J.-R. and Triboulet, C. (1972) Le chloritoïde dans les paragenèses à glaucophane, albite ou paragonite, *Bull. Soc. Fr. Minéral. Cristallogr.* **95**, 565–573.

Kieslinger, A. (1928) Geologie und Petrographie der Koralpe: VII Eklogite und Amphibolite. *Akad. Wiss. Math.-Naturwiss. Kl. I Sitzungsüber* **147**, 401–454.

Kildal, E.S. (1971) *Geologisk Karte over Norge, berggrunsskart, Måløy 1:250,000.* Nor. Geol. Unders.

Kinny, P.D., Compston W., Bristow, J.W. and Williams, I.S. (1986) Origin of zircon and diamond in kimberlites. ANU Research School of Earth Sciences, Annual report 1986, 64–65.

Kirby, S.H. (1985) Rock mechanic observations pertinent to the rheology of the continental lithosphere and the localization of strain along shear zones. *Tectonophysics* **119**, 1–27.

Klapova, H. (1985) Petrology and geochemistry of eclogites from the Krusné Hory area, Erzgebirge Mts, NW Bohemia. *Terra Cognita* **5**, 422–3.

Klein, H. and Wimmenauer, W. (1984) Eclogites and their retrograde transformation in the Schwarzwald (Fed. Rep. Germany). *Neues Jahrb. Mineral. Monatsh.* **1**, 25–38.

Klinkhammer, B. and Rost, F. (1975) Die Serpentinite des Oberpfälzer Waldes. *Der Aufschluss Heidelberg* **26**, 39–64.

Knauer, E. (1975) Die Eklogitvorkommen des Kristallinen Grundgebirges in NE-Bayern VIII. Uber den Zonarbau des Granats in den Eklogiten des Münchberger Gneisgebietes. *Neues Jahrb. Mineral. Abh.* **125**, 27–62.

Knauer, E. and Matthes, S. (1970) Die Eklogitvorkommen des kristallinen Grundgebirges in NE-Bayern, V. Die Opakminerale und Rutil der Eklogite und Eklogit-amphibolite des Münchberger Gneisgebietes. *Neues Jahrb. Mineral. Abh.* **114**, 1–17.

Kohlmann, A. (1978) *Die Überprägung hochdruckmetamorpher Serien auf der Insel Tinos, Kykladen (Griechenland).* Diplom Thesis, Technische Universität Braunschweig.

Kolčeva, K., Želgažkova-Panajotova, M., Dobrecov, N.L. and Stojanova, V. (1986) Eclogites in Central Rhodope metamorphic group and their retrograde metamorphism (Bulgarian with English abstract). *Geohimija Mineralogija Petrologija* **20–21**, 130–144.

Koons, P.O. (1982) An experimental investigation of the behaviour of amphibole in the system Na_2O–MgO–Al_2O_3–SiO_2-H_2O at high pressures. *Contrib. Mineral. Petrol.* **79**, 258–267.

Koons, P.O (1984) Implications to garnet–clinopyroxene geothermometry of non-ideal solid solution in jadeitic pyroxenes. *Contrib. Mineral. Petrol.* **88**, 340–347.

Koons, P. (1986) Relative geobarometry from high pressure rocks of quartzofeldspathic composition from the Sesia Zone, Western Alps, Italy. *Contrib. Mineral. Petrol.* **93**, 322–334.

Koons, P.O. (1987) Some thermal and mechanical consequences of rapid uplift: an example from the Southern Alps, New Zealand. *Earth Planet Sci. Lett.* **86**, 307–319.

Koons, P.O. Rubie, D.C. and Frueh-Green, G. (1987) The effects of disequilibrium and deformation on the mineralogical evolution of quartz-diorite during metamorphism in the eclogite facies. *J. Petrol.* **28**, 679–700.

Koons, P.O. and Thompson, A.B. (1985) Non-mafic rocks in the greenschist, blueschist and eclogite facies. *Chem. Geol.* **50**, 3–30.

Kopecky, L. and Sattran. V. (1966) Buried occurrences of pyrope-peridotite and the structure of the crystalline basement in the extreme SW of the Česke Středohoři Mountains. *Kristalinikum* **4**, 65–86.

Korikovskiy, S.P., Talitskiy, V.G., Boronikhin, V.A. and Ivanov V.P. (1983) Paragenezis talk-khloritoid v metapelitakh i yego petrologicheskoye znacheniye (na primere Makbalskogo antiklinoriya Tyan-Shania). *Doklady Akad. Nauk. S.S.S.R.*, **268**, 1454–1457.

Kornprobst, J. (1969) Le massif ultrabasique des Beni Bouchera (Rif Interne, Maroc): étude des péridotites de haute température et de haute pression, et des pyroxénolites, á grenat ou sans grenat, que leur sont associées. *Contrib. Mineral. Petrol.* **23**, 283–322.

Kornprobst, J. Kienast, J.-R. and Vilminot, J.-C. (1979) The high-pressure assemblages at Milos, Greece. A contribution to the petrological study of the basement of the Cyclades archipelago, *Contrib. Mineral. Petrol.* **69**, 49–63.

Kossmatt, F. (1927) Gliederung des varistischen Gebirgsbaues. *Abh Sächs. Geol. Landesamtes* **1**, 39pp.

Koziol, A.M. and Newton, R.C. (1988) Redetermination of the anorthite breakdown reaction and improvement of the plagioclase-garnet-Al_2SiO_5-quartz geobarometer. *Am. Mineral.* **73**, 216–223.

Kozlowski, (1958) On the eclogite-like rocks of Stary Gieraltow (East Sudeten). *Bull. Acad. Pol. Sc. Ser. Chim. Geol. Geogr.* **6**, 723–728.

Kramers, J.D. (1979) Lead, uranium, strontium, potassium and rubidium in inclusion-bearing diamonds and mantle-derived xenoliths from southern Africa. *Earth Planet. Sci. Letts* **42**, 58–70.

Krebs, W. and Wachendorf, H. (1973) Proterozoic–Paleozoic geosynclinal and orogenic evolution of Central Europe. *Bull. Geol. Soc. Am.* **84**, 2611–2629.

Krieger, M.H. (1965) *Geology of the Prescott and Paulden Quadrangles, Arizona.* US. Geol. Surv. Prof. Paper 467.

Krill, A.G. (1980) Tectonics of the Oppdral area, central Norway. *Geol. Foren. Stockholm Forh.* **102**, 523–530.

Krogh, E.J., (1977) Evidence for a Precambrian continent–continent collision in western Norway. *Nature* **267**, 17–19.

Krogh, E.J. (1977) Crustal and in-situ origin of Norwegian eclogites. Reply. *Nature* **269**, 730.

Krogh, E.J. (1980a) Geochemistry and petrology of glaucophane-bearing eclogites and associated rocks from Sunnfjord, western Norway, *Lithos.* **13**, 355–380.

Krogh, E.J. (1980b) Compatible P–T conditions for eclogite and surrounding gneisses in the Kristiansund area, western Norway, *Contrib. Mineral. Petrol.* **75**, 387–393.

Krogh, E.J. (1982) Metamorphic evolution deduced from mineral inclusions and compositional zoning in garnets from Norwegian country-rock eclogites. *Lithos* **15**, 305–321.

Krogh, E.J., (1988) The garnet–clinopyroxene Fe–Mg geothermometer—a reinterpretation of existing experimental data. *Contrib. Mineral. Petrol.* **99**, 44–48.

Krogh, E.J., Andresen, A., Bryhni, I. and Kristensen, S.-E. (in press). Tectonometamorphic evolution of the eclogite-bearing uppermost allochthon in Troms, northern Norwegian Caledonides. Submitted to *J. metamorphic Geol.*

Krogh, E.J. and Råheim, A. (1978) Temperature and pressure dependence of Fe–Mg partitioning between garnet and phengite with particular reference to eclogites. *Contrib. Mineral. Petrol.* **66**, 75–80.

Krogh, T.E., Mysen, B.O. and Davis, G.L. (1974) A Palaeozoic age for the primary minerals of a Norwegian eclogite. *Yearb. Carnegie Inst. Washington* **73**, 575–576.

Kroner, A., Wendt, I., Liew, T.C., Compston, W., Todt, W., Fiala, J., Vankova, V. and Vanek, J. (1988) U–Pb zircon and Sm–Nd model ages of high-grade Moldanubian metasediments, Bohemian Massif, Czechoslovakia. *Contrib. Mineral. Petrol.* **99**, 257–266.

KTB (1986) *Kontinentales Tiefbohrprogramm der Bundesrepublik Deutschland, Ergenbnisse der Vorerkundungsarbeiten Lokation Oberpfalz.* 2. KTB-Kolloquium Seeheim/Odenwald, 19.9.–21.9.1986, (eds Weber, K. and Vollbrecht, A.) 186pp.

Kuijper, R.P. (1979) U–Pb systematics and the petrogenetic evolution of infracrustal rocks in the Palaeozoic basement of western Galicia (N.W. Spain). *Z.W.O. (Neth. Org. Adv. Pure Res.) Lab. Isot. Geol.*, Amsterdam, Verhand. **5**, 1–100.

Kuijper, R.P., Vogel, D.E. and Den Tex, E. (1985) Eclogite–plagiopyrigarnite relations in the catazonal complexes of Northwest Spain. In *Chemistry and Petrology of Eclogites*. (eds Smith, D.C., Franz, G. and Gebauer, D.) *Chem. Geol.* 50, 163–171.

Kullerud, K. (1987) *Opprinnelse og tektonometamorf utvikling for eklogittene i Tsäkkoklinsen, (Sevedekene), Søndre Norrbotten Kaledonidene, Sverige*. Cand. Sci. Thesis, University of Oslo, 1–211.

Kullerud, L., Tørudbakken, B.O. and Ilebekk S. (1986) A compilation of radiometric age determinations from the Western Gneiss Region, South Norway. *Nor. Geol. Unders.* **406**, 17–42.

Kushiro, I. (1969) Clinopyroxene solid solutions formed by reactions between diopside and plagioclase at high pressures. *Mineral. Soc. Am. Spec.* Paper **2**, 179–191.

Kushiro, I. (1971) Potassium contents of synthetic clinopyroxenes at high pressures and temperatures. *Earth Planet. Sci. Letts* **11**, 374–380.

Kushiro, I. and Aoki, K. (1968) Origin of some eclogite inclusions in kimberlite. *Am. Mineral.* **53**, 1347–1367.

Kyser, T.K., O.'Neil, J.R. and Carmichael, I.S.E. (1982) Genetic relations among basic lavas and ultramafic nodules: Evidence from oxygen isotope composition. *Contrib. Mineral. Petrol.* **81**, 88–102.

Lane, D.L. and Ganguley, J. (1980) Al_2O_3 solubility in orthopyroxene in the system $MgO–Al_2O_3–SiO_2$: a reevaluation and mantle geotherm. *J. Geophys. Res.* **85**, 6963–6972.

Lappin, M.A. (1966) The field relationships of basic and ultrabasic masses in the basal gneiss complex of Stadlandet and Almklovdalen, Nordfjord, southwestern Norway. *Nor. Geol. Tidsskr.* **46**, 439–496.

Lappin, M.A. (1973) An unusual clinopyroxene with complex lamellar intergrowths from an eclogite in the Sunndal-Grubse ultramafic mass, Almklovdalen, Nordfjord, Norway. *Mineral. Mag.* **39**, 313–320.

Lappin, M.A. (1974) Eclogites from the Sunndal-Grubse ultramafic mass, Almklovdalen, Norway and the T–P history of the Almklovdalen masses. *J. Petrol.* **15** 567–601.

Lappin, M.A. (1977) Crustal and in-situ origin of Norwegian eclogites. *Nature* **289**, 730.

Lappin, M.A. (1978) The evolution of a grospydite from the Roberts Victor Mine, South Africa. *Contrib. Mineral. Petrol.* **66**, 229–241.

Lappin, M.A. and Dawson, J.B. (1975) Two Roberts Victor cumulate eclogites and their re-equilibration. *Phys. Chem. Earth* **9**, 351–366.

Lappin, M.A., Pidgeon, R.T. and van Breemen, O. (1979) Geochronology of basal gneisses and mangerite syenites of Stadlandet, west Norway. *Nor. Geol. Tidsskr.* **59**, 161–181.

Lappin, M.A. and Smith, D.C. (1978) Mantle-equilibrated eclogite pods from the Basal Gneisses in the Selje district, Western Norway. *J. Petrol.* **19**, 530–584.

Lappin, M.A. and Smith, D.C. (1981) Carbonate–silicate and fluid relationships in eclogites, Selje district and environs, S.W. Norway. *Trans. R. Soc. Edinburgh: Earth Sci.* **72**, 171–193.

Lardeaux, J.M., Caron, J.M., Nisio, P., Pequignot, G. and Boudelle, M. (1986) Microstructural criteria for reliable thermometry in low-temperature eclogites. *Lithos* **19**, 187–203.

Lardeaux, J.M., Gosso, G., Kienast, J.R. and Lombardo, B. (1982) Relations entre le métamorphisme et la déformation dans la zone Sesia-Lanzo (Alpes Occidentales) et le problème de l'éclogitisation de la croûte continentale. *Bull. Soc. Geol. Fr.* **24**, 793–800.

Lardeaux, J.-M. Gosso, G. Kienast, J.-R. and Lombardo, B. (1983) Chemical variations in phengitic micas of successive foliations within the eclogite micaschists complex, Sesia-Lanzo Zone (Italy, Western Alps). *Bull. Mineral.* **106**, 673–689.

Lardeaux, J.M., Lombardo, B., Gosso, G. and Kienast, J.R. (1983) Découverte de paragénèses à ferro-omphacites dans les orthogneiss de la zone Sesia-Lanzo septentrionale (Alpes Italiennes). *C.R. Acad. Sc. Paris* **296**, 453–456.

Lasaga, A.C. (1983) Geospeedometry: an extension of geothermometry. In *Advance in Physical Geochemistry 3. Kinetics and Equilibrium in Mineral Reactions*. (ed. Saxena, S.K.) pp. 81–114, Springer, New York.

Lasnier, B. (1970) *Le Métamorphisme régionale des Gabbros d'après la littérature internationale. Etude préliminaire des gabbros coronitiques du Massif armoricain et du massif des Maures (France)*. Thèse 3ème cycle, Université de Nantes, 297 pp.

Lasnier, B. (1971) Les péridotites et pyroxénolites à grenat des Bois des Feuilles (Monts du Lyonnais, France). *Contrib. Mineral. Petrol.* **34**, 29–42.

Lasnier B., Leyreloup, A. and Marchand, J. (1973) Découverte d'un granite 'Charnockitique' au sein de 'gneiss oeillés'. Perspectives nouvelles sur l'origine de certaines leptynites du Massif Armoricain Méridional (France). *Contrib. Mineral. Petrol.* **41**, 131–144.

Lattard D., Schreyer W. (1981) Experimental results bearing on the stability of the blueschist-facies minerals deerite, howieite, and zussmanite, and their petrological significance. *Bull.* Minéral. **104**, 431–440.

Leake, B.E. (1978a) Nomenclature of amphiboles. *Canadian Mineral.* **16**, 501–520.

Leake, B.E. (1978b) Nomenclature of amphiboles. *Mineral. Mag.* **42**, 533–563.

Lee, H.Y. and Ganguly, J. (1988) Equilibrium compositions of coexisting garnet and orthopyroxene: experimental determinations in the system $FeO–MgO–Al_2O_3–SiO_2$, and applications. *J. Petrol.* **29**, 93–113.

Leupolt, L. and Franz, G. (1986) Zink-staurolith aus Metasedimenten der Eklogitzone. Tauern/Österreich, *Fortschr. Mineral.* **64**, 96.

Liati, A. (1988) Amphibolitised eclogites in the Rhodope Crystalline Complex, near Xanthi (N. Greece). *Neues Jahrb. Mineral. Monatsh.* **1**, 1–8.

Liborio, G. and Mottana, A. (1973) I carbonati dei calcescisti in relazione all a distribuzione delle temperature metamorfiche alpine. *Rend. Soc. Ital. Mineral. Petrol.* **29**, 43–80.

Liborio, G. and Mottana, A. (1975) White micas with 3T polymorph from the Calcescisti of the Alps. *Neues Jahrb. Mineral. Monatsh.* 546–555.

Liegeois, J.P. and Duchesne, J.C. (1981) The Lac Cornu retrograded eclogites (Aiguilles Rouges massif, western Alps, France): evidence of crustal origin and metasomatic alteration. *Lithos* **14**, 35–48.

Lindsley, D.H. (1983) Pyroxene thermometry. *Am. Mineral.* **68**, 477–493.

Lindsley, D.H. and Anderson, D.J. (1983) A two-pyroxene thermometer. *J. Geophys. Res.* **88** (supplement), A887-A906.

Lindsley, D.H. and Dixon, S.A. (1976) Diopside–enstatite equilibria at 850 to 1400°C, 5 to 35 kb. *Am. J. Sci.* **276**, 1285–2301.

Lindsley, D.H., Grover, J.E. and Davidson, P.M. (1981) The thermodynamics of the $Mg_2Si_2O_6$–$CaMgSi_2O_6$ join: a review and an improved model. In *Thermodynamics of Minerals and Melts* (eds Newton, R.C., Navrotsky, A. and Wood, B.J.) Advances in Physical Geochemistry **1**, 149–175 Springer-Verlag, New York.

Lindsley, D.H. and Munoz, J.L. (1969) Subsolidus relations along the join hedenbergite–ferrosilite. *Am. J. Sci.* **276**, 1285–1301.

Lister, G.S., Banga, G. and Feenstra, A. (1984) Metamorphic core complexes of Cordilleran type in the Cyclades, Aegean Sea, Greece, *Geology* **12**, 221–225.

Livingstone, A. (1967) A garnet peridotite and a garnet-amphibolite pyroxenite from South Harris, Outer Hebrides and their bearing on the South Harris eclogite facies status. *Mineral. Mag.* **36**, 380–388.

Livingstone, A. (1976) The paragenesis of spinel- and garnet-amphibole lherzolite in the Rodel area, South Harris. *Scott. J. Geol.* **12**, 293–300.

Lombardo, B., Compagnoni, R., Fiora, L. and Facchinelli, A. (1977) Composition of some sodic pyroxenes from the eclogitic mica schists of lower Val d'Aosta (Sesia–Lanzo Zone, Western Alps). *Rend. Soc. Ital. Mineral. Petrol.* **33**, 375–387.

Lombardo, B., Nervo, R., Compagnoni, R., Messiga, B., Kienast, J.R., Mevel, C., Fiora, L., Piccardo, G. and Lanza, R. (1978) Osservazioni preliminari sulle ofioliti metamorfiche del Monviso (Alpi occidentali). *Rend. Soc. Ital. Mineral. Petrol.* **34**, 253–305.

Loomis, T.P. (1972a) Contact metamorphism of the Ronda high-temperature ultramafic intrusion, southern Spain. *Geol. Soc. Am. Bull.* **83**, 2449–2474.

Loomis, T.P. (1972b) Diapiric emplacement of the Ronda high-temperature ultramafic intrusion, southern Spain. *Geol. Soc. Am. Bull.* **83**, 2475–2496.

Loomis, T.P. (1975) Tertiary mantle diapirism, orogeny, and plate tectonics east of the Strait of Gibraltar. *Am. J. Sci.* **275**, 1–30.

Lovering, F.J., (1964) The eclogite-bearing basic igneous pipe at Ruby Hill near Bingera, N.S.W. *J. Proc. R. Soc. N.S.W.* **97**, 73–79.

Lovering, J.F. and Tatsumoto, M. (1968) Lead isotope and the origin of granulite and eclogite inclusions in deep-seated pipes. *Earth Planet. Sci. Letts.* **4**, 350–356.

Lovering, J.F. and White, A.J.R. (1969) Granulitic and eclogite inclusions from basic pipes at Delegate, Australia. *Contrib. Mineral. Petrol.* **21**, 9–52.

Löw, S. (1986) Ein tektono-metamorphes Entwicklungsmodell der nördlichen Adula Decke. *Schweiz. Mineral Petrogr. Mitt.* **66**, 129–134.

Luckscheiter, B. and Morteani, G. (1980) The fluid phase in eclogites, glaucophane-bearing rocks and amphibolites from the central Tauern Window as deduced from fluid inclusion studies. *Tschermaks Mineral. Petrogr. Mitt.* **27**, 99–111.

Lugmair, G.W. (1974) Sm–Nd ages; a new dating method. *Meteorites* **9**, 369.

Lugmair, G.W., Scheinin, N.B. and Marti, K. (1975) Sm–Nd age of Apollo 17 basalt 75075: evidence for early differentiation of the lunar exterior. *Proc. Lunar Planet. Sci. Conf. 6th*, 1419–1429.

Lundeen, M.T. (1978) Emplacement of the Ronda peridotite, Sierra Bermeja, Spain. *Geol. Soc. Am. Bull.* **89**, 172–180.

Lux, D.R. (1985) K/Ar ages from the Basal Gneiss Region, Stadlandet area, western Norway. *Norsk. Geol. Tidsskr.* **65**, 277–286.

McCandless, T.E. and Gurney, J.J. (1989) Sodium in garnet and potassium in clinopyroxene: criteria for classifying mantle eclogites. In *Kimberlites and related rocks, Vol. 2. Their mantle/crust setting diamonds and diamond exploration.* Geol. Soc. Australia Spec. Publin. 14, Blackwells, Carlton, 827–832.

McClay, K.R., Norton, M.G., Coney, P. and Davis, G.H. (1986) Collapse of the Caledonide orogen and the Old Red Sandstone. *Nature* **323**, 147–149.

McCulloch, M.T. (1982) Identification of Earth's first differentiates. *Abstracts 5th Int. Conf. Geochron. Cosmochron. Isotope Geol.*, Nikko, Japan, 244–246.

McCulloch, M.T., (1989) Sm–Nd systematics in eclogite and garnet peridotite nodules from kimberlites: implications for the early differentiation of the Earth. In *Kimberlites and related rocks, Vol. 2. Their mantle/crust setting, diamonds and diamond exploration.* Geol. Soc. Australia Spec. Publn. 14, Blackwells, Carlton, 864–876.

McCulloch, M.T., Arculus, R.J., Chappell, K.W. and Ferguson, J. (1982) Lower continental crust: inferences from isotopic and geochemical studies of nodules in kimberlite, *Nature* **300**, 166–169.

McCulloch, M.T., Jaques, A.L., Nelson, D.R. and Lewis, J.D. (1983) Nd and Sr isotopes in kimberlites and lamproites from Western Australia: an enriched mantle origin. *Nature* **302**, 400–403.

McDougall, I. and Green, D.H. (1964) Excess radiogenic argon in pyroxenes and isotopic ages on minerals from Norwegian eclogites. *Nor. Geol. Tidsskr.* **44**, 183–196.

McDowell, F.W., Lehman, D.H., Gucwa, P.R., Fritz, D. and Maxwell, J.C. (1984) Glaucophane schists and ophiolites of the northern Coast Ranges: Isotopic ages and their tectonic implications. *Geol. Soc. Amer. Bull.* **95**, 1373–1382.

McGee, E.S. and Hearn, B.C. (1984) The Lake Ellen kimberlite, Michigan, U.S.A. In *Kimberlites I: Kimberlites and Related Rocks.* (ed. Kornprobst, J.) Elsevier, Amsterdam, 143–154.

MacGregor, I.D. (1970) The effect of CaO, Cr_2O_3, Fe_2O_3 and Al_2O_3 on the stability of spinel and garnet peridotite. *Phys. Earth Planet. Inter.* **3**, 372–377.

MacGregor, I.D. (1974) The system $MgO–Al_2O_3–SiO_2$: solubility of Al_2O_3 in enstatite for spinel and garnet peridotite compositions. *Am. Mineral.* **59**, 110–119.

MacGregor, I.D. and Carter, J.L. (1970) The chemistry of clinopyroxenes and garnets of eclogite and peridotite xenoliths from the Roberts Victor Mine, South Africa. *Phys. Earth Planet. Int.* **3**, 391–397.

MacGregor, I.D. and Manton, W.I. (1986) Roberts Victor eclogites: ancient oceanic crust. *J. Geophys. Res.* **91**, 14063–14079.

MacGregor, I.D. and Ringwood, A.E. (1964) The natural system enstatite–pyrope. *Yearb. Carnegie Inst. Washington* **63**, 161–165.

McKenzie, D. and O'Nions, R.K. (1983) Mantle reservoirs and ocean island basalts. *Nature* **301**, 229–231.

Maaskant, P. (1970) Chemical petrology of polymetamorphic ultramafic rocks from Galicia, NW Spain. *Leidse Geol. Meded.* **45**, 237–325.

Machart, J. (1984) Ultramafic rocks in the Bohemian part of the Moldanubicum and Central Bohemian. Islet Zone (Bohemian Massif). *Kristalinikum* **17**, 13–32.

Machart, J. and Paděra, K. (1982) Websterite a griquait ze serpentinitu od Bezděčina u Tábora (jižni Čechy). *Cas. Mineral. Geol.* **27**, 95–102.

Machatschki, K. and Walitzi, E.M. (1962) Hornblenden aus Eklogiten und Amphiboliten der südkucgeb Koralpe. *Tschermaks Mineral. Petrogr. Mitt.* **8**, 140–151.

Maggetti, M., Galetti, G. and Stosch, H-G. (1987) Eclogites from the Silvretta Nappe, Switzerland: geochemical constraints on the nature and geotectonic setting of their protoliths. Chem. Geol. **64**, 319–334.

Malaroda, R. (1970) Carta geologica del massiccio dell'Argentera alla scala 1:50000. *Mem. Soc. Geol. Ital.* **9**, 557–663.

Manby, G.M. and Thiedig, F. (1988) Petrology of eclogites from the Saualpe, Austria. *Schweiz. Mineral. Petrogr. Mitt.* **68**, 441–466.

Manson, V. (1967) Geochemistry of basaltic rocks: major elements. In *Basalts: the Poldervaart Treatise on Rocks of Basaltic Composition*, (eds Hess, H.H. and Poldervaart A.) Interscience, New York, 215–269.

Manton, W.I. and Tatsumoto, M.(1971) Some Pb and Sr isotopic measurements in eclogites from the Roberts Victor Mine, South Africa. *Earth Planet. Sci. Letts* **10**, 217–226.

Marchand, J. (1982) Une véritable eclogite en Bretagne occidentale (Baie d'Audierne—France). *Terra Cognita* **2**, 312.

Maresch, W.V. (1972) Eclogite–amphibolitic rocks on Isla Margarita, Venezuela. *Geol. Soc. Amer. Mem.* **132**, 429–437.

Maresch, W.V. (1977) Experimental studies on glaucophane: an analysis of present knowledge. *Tectonophysics* **43**, 109–125.

Maresch, W.V. (1983) Die Bildung von Eklogit aus Amphibolit bei progressiver Hochdruckmetamorphose. *Fortschr. Mineral.* **60**, 137–138.

Maresch, W.V. and Abraham, K. (1981) Petrography, mineralogy, and metamorphic evolution of an eclogite from the island of Margarita, Venezuela. *J. Petrol.* **22**, 337–362.

Maresch, W.V. Medenbach, O. and Rudolph, A. (1982) Winchite and the actinolite—glaucophane miscibility gap. *Nature* **296**, 731–732.

Martin, S. and Kienast, J.R. (1987) The HP-LT manganiferous quartzites of Praborna, Piemonte ophiolite nappe, Italian Western Alps. *Schweiz. Mineral. Petrogr. Mitt.* **67**, 339–360.

Martin-Vernizzi, S. (1982) La mine de Praborna (Val d'Aoste, Italie): une série manganesifère metamorphisée dans la facies eclogite. *Thèse 3ème cycle Univ. Paris*, 215 pp.

Maruyama, S. and Liou, J.G. (1987) Clinopyroxene—a mineral telescoped through the presence of blueschist facies metamorphism. *J. Metamorphic Geol.* **5**, 529–552.

Maruyama, S. and Liou, J.G. (1988) Petrology of Franciscan metabasites along the jadeite-glaucophane facies series, Cazadero, California. *J. Petrol.* **29**, 1–37.

Massonne, H.-J. (1983) Experiments on melting to 50 kbar in the system $MgO–Al_2O_3–SiO_2–H_2O$ (MASH) with excess SiO_2 and H_2O. *EOS* **64**, 875.

Massonne, H.J. (1989) The upper thermal stability of chlorite + quartz: an experimental study in the system $MgO–Al_2O_3–SiO_2–H_2O$. *J. metamorph. Geol.* (in press).

Massonne, H.-J. Mirwald, P.W. and Schreyer, W. (1981) Experimentelle Uberprüfung der Reactionskurve Chlorit + Quarz = Talk + Disthen im system $MgO–Al_2O_3–SiO_2–H_2O$. *Fortschr. Mineral.* **59**, 122–123.

Massonne, H.J. and Schreyer, W. (1983) Stability of the talc-kyanite assemblage revisited (abstract). *Terra Cognita* **3**, 187.

Massonne, H.J. and Schreyer, W. (1985) Phengite barometry in assemblages with kyanite, Mg-rich silicates and a SiO_2 phase (abstract). *Terra Cognita* **5**, 432.

Massonne, H.J. and Schreyer, W. (1987) Phengite geobarometry based on the limiting assemblage with K-feldspar, phlogopite, and quartz, *Contrib. Mineral. and Petrol.* **96**, 212–224.

Massonne, H.J. and Schreyer, W. (1989) The stability of the high-pressure assemblage talc-phengite, and two new phengite geobarometers. *Europ. J. Mineral.* **1**, 391–410.

Mathe, G. (1969a) Die Metabasite des sachsischen Granulitgebirges. *Freiberg Forsch* **251**, 130 pp.

Mathé G. (1969b) Granat–klinopyroxen Paragenesen des sächsischen Grundgebirges und ihre petrographisch-mineralifazielle Einordnung. *Freiberg. Forsch.* **C266**, 49–64.

Mathez, E.A., Blacic, J.D., Beery, J., Hollander, M. and Maggiore, C. (1987) Carbon in olivine; results from nuclear reaction analysis. *J. Geophys. Res.* **B92**, 3500–3506.

Matte, P. (1986) Tectonics and plate tectonics model for the Variscan belt of Europe. *Tectonophysics* **126**, 329–374.

Matte, P. and Burg, J.P. (1981) Sutures, thrusts, and nappes in the Variscan arc of western Europe, plate tectonic implications. In *Thrust and Nappe Tectonics* (eds McClay, K.R. and Price, N.J.) Spec. Publ. Geol. Soc. London **9**, 353–358.

Matthes, S. (1978) The eclogites of Southern Germany. A summary. *Neues Jahrb. Mineral. Monatsh.* **3**, 93–109.

Matthes, S., Richter, P. and Schmidt, K. (1974) Die Eklogitvorkommen des kristallinen Grundgebirges NE-Bayerns, VII. Ergebnisse aus einer Kernbohrung durch den Eklogitkörper des Weissensteins. *Neues Jahrb. Mineral. Abh.* **120**, 270–314.

Matthes, S., Richter, P. and Schmidt, K. (1975) Die Eklogitvorkommen des kristallinen Grundgebirges NE-Bayerns, IX. Petrographie, Geochemie und Petrogenese der Eklogite des Münchberger Gneisgebietes. *Neues Jahrb. Mineral. Abh.* **126**, 45–86.

Matthes, S. and Schmidt, K. (1974) Excursion A2. Eclogites of the Münchberg Mass, NE Bavaria. *Fortschr. Mineral.* **52**, 33–57.

Matthes, S. and Seidel, E. (1977) Die Eklogitvorkommen des Kristallinen Grundgebirges in NE-Bayern X. Bestehen genetische beziehungen zwischen eklogit und meta-gabbro innerhalb des Münchberger Gneisgebietes? *Neues Jahrb. Mineral. Abh.* **129**, 269–291.

Matthews, A. (1988) Oxygen isotope thermometry in regional metamorphism. *Chem. Geol.* **70**, 185.

Matthews, A., Beckinsale, R.D. and Durham, J.J. (1979) Oxygen isotope fractionation between rutile and water and geothermometry of metamorphic eclogites. *Mineral. Mag.* **43**, 405–413.

Matthews, A., Goldsmith, J.R. and Clayton, R.N. (1983) Oxygen isotope fractionations involving pyroxenes: the calibration of mineral pair geothermometers. *Geochim. Cosmochim. Acta* **47**, 631–644.

Matthews, A. and Schliestedt, M. (1984) Evolution of the blueschist and greenschist facies rocks of Sifnos, Cyclades, Greece. A stable isotope study of subduction-related metamorphism. *Contrib. Mineral. Petrol.* **88**, 150–163.

Maynew, M.A. and Johnson, B.D. (1987) An equivalent layer magnetization model for Australia based on MAGSAT data. *Earth Planet. Sci. Lett.* **83**, 167–174.

Mazzone, P. and Haggerty, S.E. (1989) Corganites and corgspinites: two new types of aluminous assemblages from the Jagersfontein pipe. In *Kimberlites and related rocks. Vol. 2. Their mantle/crust setting, diamonds and diamond exploration.* Geol. Soc. Australia Spec. Publn. 14, Blackwells, Carlton, 795–808.

Meagher, E.P. (1982) Silicate garnets. In *Orthosilicates* (ed. Ribbe, P.H.) *Min. Soc. America-Reviews in Mineralogy* **5**, 25–66.

Mearns, E.W. (1986) Sm–Nd ages for Norwegian garnet peridotite. *Lithos* **19**, 269–278.

Mearns, E.W. and Lappin, M.A. (1982) A Sm–Nd isotopic study of 'internal' and 'external eclogites', garnet lherzolite and grey gneiss from Almklovdalen, western Norway. *Terra Cognita* **2**, 324–325.

Medaris, L.G. Jr. (1980a) Convergent metamorphism of eclogite and garnet-bearing ultramafic rocks at Lien, west Norway. *Nature* **283**, 470–472.

Medaris, L.G. Jr. (1980b) Petrogenesis of the Lien peridotite and associated eclogites, Almklovdalen, western Norway. *Lithos* **13**, 339–353.

Medaris, L.G. Jr. (1984) A geothermobarometric investigation of the garnet-peridotites in the western Gneiss Region of Norway. *Contrib. Mineral. Petrol.* **87**, 72–86.

Medaris, L.G. Jr. and Wang, H.F. (1986) A thermal–tectonic model for high-pressure rocks in the Basal Gneiss Complex of western Norway. *Lithos* **19**, 299–316.

Medaris, L.G., Jr., Wang, H.F. and Mísař, Z. (1988a) An unusual high-temperature spinel/garnet peridotite in the Bohemian Massif of Czechoslovakia. *Trans. Am. Geophys. Union* **69**, 509–510.

Medaris, L.G., Jr., Wang, H.F., Peterson, D.E., Mísař, Z. and Jelinek, E. (1988b) Contrasting thermal histories for two crustal garnet peridotites in the Bohemian Massif of Czechoslovakia. *Abstracts with Programs, Geol. Soc. Am.* **20**, A99.

Meissner, R. (1986) Twenty years of deep reflection profiling in Germany—a contribution to our knowledge of the nature of the lower Variscan crust. In *The Nature of the Lower Continental Crust.* (eds Dawson, J.B., Carswell, D.A., Hall, J. and Wedepohl, K.H.) Spec. Pub. Geol. Soc. London **24**, 1–10.

Ménot, R.-P., Peucat, J.-J., Scarenzi, D. and Piboule, M. (1988) 496 Ma age of plagiogranites in the Chamrousse ophiolite complex (external crystalline massifs in the French Alps): evidence of a lower Paleozoic oceanization. *Earth Planet. Sci. Lett.* **88**, 82–92.

Menot, R.P. and Seddoh, K.F. (1985) The eclogites of the Lato hills, south Togo, West Africa: relics from the early tectonometamorphic evolution of the Pan-African orogeny. *Chem. Geol.* **50**, 313–330.

Mercier, J-C.C., Pozzo Di Borgo, M., Frison, J.Y. and Girardeau, J. (1982) Les associations basiques et ultrabasiques du Bas-Limousin, restes d'un complexe ophiolitique démembré d'une fraicheur remarquable. *9ème Réun. Ann. Sci. Terre Paris*, 430.

Mercy, E.L.P. and O'Hara, M.J. (1965a) Chemistry of some garnet-bearing rocks from south Norwegian peridotites. *Nor. Geol. Tidsskr.* **45**, 323–332.

Mercy, E.L.P. and O'Hara, M.J. (1965b) Olivines and orthopyroxenes from garnetiferous peridotites and related rocks. *Nor. Geol. Tidsskr.* **45**, 457–461.

Merrill, R.B. and Wyllie, P.J. (1975) Kaersutite and kaersutite eclogite from Kakanui, New Zealand—water excess and water-deficient melting to 30 kilobars. *Bull. Geol. Soc. Am.* **86**, 555–570.

Messiga, B. (1981) Evidenze strutturale e paragenetiche dell'evoluzione polifasica pre-Alpina del Massiccio Cristallino di Savona. *Rend. Soc. Ital. Mineral. Petrol.* **37**, 739–754.

Messiga, B. (1987) Alpine metamorphic evolution of Ligurian Alps (North-West Italy): chemographic and petrological constraints inferred from metamorphic climax assemblages. *Contri. Mineral. Petrol.* **95**, 269–277.

Messiga, B., Oxilia, M., Piccardo, G.B. and Vanossi, M. (1983a) Fasi metamorfiche e deformazioni alpine nel Brianzonese e nel Pre-Piemontese–Piemontese esterno delle Alpi Liguri: un possibile modello evolutivo. *Rend. Soc. Ital. Mineral. Petrol.* **38**, 261–280.

Messiga, B., Piccardo, G.B. and Ernst, W.G. (1983b) High-Pressure Eo-Alpine Parageneses Developed in Magnesian Metagabbros, Gruppo di Voltri, Western Ligurian Italy. *Contrib. Mineral. Petrol.* **83**, 1–15.

Messiga, B. Piccardo, G.B. and Mazzucotelli, A. (1979) Le prasiniti del Gruppo di Voltri, Le caratteristiche petrografiche e chimiche. *Rend. Soc. Ital. Mineral. Petrol.* **33**, 549–581.

Messiga, B. and Scambelluri, M. (1988) Comparison between two types of coronitic eclogites from the Western Alps: Implications for pre-eclogite evolution. *Schweiz. Mineral. Petrogr. Mitt.* **68**, 225–235.

Meyer, H.O.A. (1987) Inclusions in diamond. In *Mantle Xenoliths.* (ed. Nixon, P.H.) Wiley, Chichester, 501–522.

Meyer, H.O.A. and Boctor, N.Z. (1975) Sulfide-oxide minerals in eclogite from Stockdale kimberlite, Kansas. *Contrib. Mineral. Petrol.* **52**, 57–68.

Meyer, H.O.A. and Boyd, F.R. (1972) Composition and origin of crystalline inclusions in natural diamonds. *Geochim. Cosmochim. Acta* **36**, 1255–1272.

Meyer, H.O.A., and Brookins, D.G. (1971) Eclogite xenoliths from the Stockdale kimberlite, Kansas, *Contrib. Mineral. Petrol.* **34**, 60–72.

Meyer, J. (1983a) The development of the high pressure metamorphism in the Allalin metagabbro (Switzerland). *Terra Cognita* **3**, 187.

Meyer, J. (1983b) Mineralogie und Petrologie des Allalin gabbros. Ph.D. Thesis, University of Basel.

Mielke, H., Blümel, P. and Langer. K. (1979) Regional low-pressure metamorphism of low and medium grade in metapelites and -psammites of the Fichtelgebirge area, NE-Bavaria. *Neues Jahrb. Mineral. Abh. 137*, 83–112.

Mikhailov, N.P. and Rovsha, V.S. (1966) Pyrope-bearing peridotites of the Bohemian Massif and their genesis. *Kristalinikum* **4**, 87–107.

Miller, C. (1969) Zur Metamorphose der Eklogite und Metagabbros im Ötztaler Altkristallin. *Fortschr. Mineral.* **47**, 43–44.

Miller, C. (1970) Petrology of some eclogites and metagabbros of the Oetztal Alps, Tirol. Austria. *Contrib. Mineral. Petrol.* **28**, 42–56.

Miller, C. (1974a) Reaction rims between olivine and plagioclase in metaperidotites, Ötztal Alps, Austria. *Contrib. Mineral. Petrol.* **43**, 333–342.

Miller, C. (1974b) On the metamorphism of eclogites and high grade blueschists from the Penninic terrain of the Tauern Window, Austria. *Schweiz. Mineral. Petrogr. Mitt.* **54**, 371–384.

Miller, C. (1976) Zur Metamorphose der eklogitischen Gesteine der Hohen Tauern. *Fortschr. Mineral.* **54**, 60.

Miller, C. (1977a) Mineral parageneses recording the P–T history of Alpine eclogites in the Tauern Window, Austria. *Neues Jahrb. Mineral., Abh.* **130**, 69–77.

Miller, C. (1977b) Chemismus und phasenpetrologische Untersuchungen der Gesteine aus der Eklogitzone des Tauernfensters, Österreich. *Tschermaks Mineral. Petrogr. Mitt.* **24**, 221–277.

Miller, C. (1985) *Metagabbros and eclogites in the Kor- and Saualpe, Austria. Second International Eclogite Conference*, Guide Book, 48 pp.

Miller, C. (1986) Alpine high-pressure metamorphism in the Eastern Alps. *Schweiz. Mineral. Petrogr. Mitt.* **66**, 139–144.

Miller, C. and Frank, W. (1983) Das Alter der Metamorphose von Metabasiten und Eklogiten in Kor- und Saualpe. *Jahresbericht 1982. Hochschulschwerpunkt S15*, Graz, 229–236.

Miller, C., Sassi, F.P. and Armari, G. (1976) On the occurrence of altered eclogitic rocks in northeastern Sardinia and their implications. *Neues Jahrb. Geol. Paläont. Monatsh.* **11**, 683–689.

Miller, C., Satir, M. and Frank, W. (1980) High-pressure metamorphism in the Tauern Window. *Mitt. Österr. Geol. Ges.* **71/72**, 89–97.

Miller, Ch., Stosch, H-G. and Hoernes, St. (1988) Geochemistry and origin of eclogites from the type locality Koralpe and Saualpe, Eastern Alps, Austria. *Chem. Geol.* **67**, 103–118.

Milnes, A.G. and Koestler, A.G. (1985) Geological structure of Jotunheimen, southern Norway (Sognefjell–Valdres cross-section). In *The Caledonide Orogen—Scandinavia and Related Areas* (eds Gee, D.G. and Sturt, B.A.) pp. 457–474. John Wiley, Chichester.

Mirwald, P.W. and Massonne, H.-J. (1980) The low-high quartz and quartz-coesite transition to 40 kbar between 600° and 1600°C and some reconnaissance data on the effect of $NaAlO_2$ component on the low quartz–coesite transition. *J. Geophys. Res.* **85**, B12, 6983–6990.

Misař, Z. and Jelínek, E. (1981) Inclusions of peridotites, pyroxenites, eclogites and opals in the leptynites of the Moldanubicum at the locality Bory (SW Moravia) (in Czech) *Věsn. Ustřed. Ustava Geol CAV* **56**, 13–20.

Misař, Z., Jelínek, E. and Jakeš, P. (1984) Inclusions of peridotite, pyroxenite and eclogite in granulite rocks of pre-Hercynian upper mantle and lower crust in the eastern Bohemian Massif (Czechoslovakia). *Ann. Sci. Univ. Clermont-Fd* **74**, 85–95.

Miyashiro, A. (1961) Evolution of metamorphic belts. *J. Petrol.* **2**, 277–311.

Miyashiro, A. (1973) *Metamorphism and Metamorphic Belts.* John Wiley, New York 492 pp.

Miyashiro, A. (1974) Volcanic rock series in island arcs and active continental margins. *Am. J. Sci.* **274**, 321–355.

Möckel, J.R. (1969) Structural geology of the garnet peridotite of Alpe Arami (Ticino, Switzerland). *Leidse Geol. Meded.* **42**, 61–130.

Mongkoltip, P. and Ashworth, J.R. (1983) Quantitative estimation of an open-system symplectite-forming reaction: restricted diffusion of Al and Si in coronas around olivine. *J. Petrol.* **24**, 635–661.

Moore, D.E. and Liou, J.G. (1979) Mineral chemistry of some Franciscan blueschist facies metasedimentary rocks from the Diablo Range, California. Part II. *Bull. Geol. Soc. Am.* **90**, 1737–1781.

Moore, R.O. and Gurney, J.J. (1985) Pyroxene solid solution in garnet included in diamond. *Nature* 318, 553–555.

Moore, R.O. and Gurney, J.J. (1989) Mineral inclusions in diamonds from the Monastery kimberlite, South Africa. In *Kimberlites and related rocks, Vol. 2. Their mantle/crust setting, diamonds and diamond exploration.* Geol. Soc. Australia Spec. Publn 14, Blackwells, Carlton, 1029–1041.

Moore, R.O., Otter, M.L., Rickard, R.S., Harris, J.W. and Gurney, J.J. (1986) The occurrence of moissanite and ferri-periclase as inclusions in diamond. *Fourth Int. Kimberlite Conf., Geol Soc. Australia, Abstracts* **16**, 409–411.

Mori, T. (1978) Experimental study of pyroxene equilibria in the system. $CaO–MgO–FeO–SiO_2$. *J. Petrol.* **19**, 45–65.

Mori, T. and Green, D.H. (1975) Pyroxenes in the system $Mg_2SiO_2O_6$–$CaMgSi_2O_6$ at high pressure. *Earth Planet. Sci. Lett.* **26**, 277–286.

Mori, T. and Green, D.H. (1978) Laboratory duplication of phase equilibria observed in natural garnet lherzolites. *J. Geol.* **86**, 83–97.

Morimoto, N., Fabries, J., Ferguson, A.K., Ginzburg, I.V., Ross, M., Seifert, F.A., Zussman, J., Aoki, K. and Gottardi, G. (1988) Nomenclature of pyroxenes. *Mineral. Mag.* **52**, 535–550.

Mørk, M.B.E. (1985a) A gabbro to eclogite transition on Flemsøy, Sunnmøre, western Norway. *Chem. Geol.* **50**, 283–310.

Mørk, M.B.E. (1985b) Incomplete high *P–T* metamorphic transitions within the Kvamsøy pyroxenite complex, west Norway: a case study of disequilibrium. *J. metamorphic Geol.* **3**, 245–264.

Mørk, M.B.E. (1986) Coronite and eclogite formation in olivine gabbro (Western Norway): reaction paths and garnet zoning. *Mineral. Mag.* **50**, 417–426.

Mørk, M.B.E., Kullerud. K. and Stabel, A. (1988) Sm–Nd dating of Seve eclogites. Norbotten, Sweden—evidence for early Caledonian (505 Ma) subduction. *Contrib. Mineral. Petrol.* **99**, 344–351.

Mørk, M.B.E. and Mearns, E.W. (1986) Sm–Nd isotopic systematics of a gabbro—eclogite transition. *Lithos* **19**, 255–267.

Morten, L., Bargossi, G.M., Martinez, J.M., Puga, E. and Diaz de Federico, A. (1987) Metagabbro and associated eclogites in the Lubrin area, Nevado–Filabride complex, Spain. *J. metamorphic Geol.* **5**, 155–174.

Morten, L. Bocchio, R. and Mottana, A. (1985) Transformation of a superferrian eclogite within serpentinite: a case study from the Voltri Group, Italy, *Chem. Geol.* **50**, 111–127.

Morten, L. and Bondi, M. (1984) Disequilibrium in co-existing clinopyroxene and garnet from blueschist-facies Fe-rich eclogites, Voltri Group, Italy. *J. metamorphic Geol.* **2**, 205–218.

Morten, L., Brunfelt, A.O. and Mottana, A. (1979) Rare earth abundances in superferrian eclogites from the Voltri Group (Pennidic Belt, Italy). *Lithos* **12**, 25–32.

Morten, L. and Obata, M. (1983) Possible high-temperature origin of pyroxenite lenses within garnet peridotite, northern Italy. *Bull. Mineral.* **106**, 775–780.

Mottana, A. (1972) Eclogite occurrences in gneiss as indicators of a Caledonian high-pressure metamorphism in the Western Alps. *Proc. 24th Int. Geol. Congress* **2**, 35–41.

Mottana, A. (1977) Petrochemical characteristics of the Western Alps eclogites. *Neues Jahrb. Mineral. Abh.* **130**, 78–88.

Mottana, A. (1986a) Crystal-chemical evaluation of garnet and omphacite microprobe analyses: Its bearing on the classification of eclogites. *Lithos* **19**, 171–186.

Mottana, A. (1986b) Blueschist-facies metamorphism of manganiferous cherts: A review of the alpine occurrences. *Geol. Soc. Amer. Memoir* **164**, 267–299.

Mottana, A. and Bocchio, R. (1975) Superferric eclogites of the Voltri Group (Pennidic Belt, Apennines). *Contrib. Mineral. Petrol.* **49**, 201–210.

Mottana, A., Bocchio, R., Liborio, G., Morten, L. and Maresch, W.V. (1985) The eclogite-bearing metabasaltic sequence of Isla Margarita, Venezuela: a geochemical study, *Chem. Geol.* **50**, 351–368.

Mottana, A., Church, W.R. and Edgar, A.D. (1968) Chemistry, mineralogy and petrology of an eclogite from the type locality (Saualpe, Austria). *Contrib. Mineral. Petrol.* **18**, 388–346.

Mottana, A. and Griffin, W.L. (1986) Crystal chemistry of two coexisting K-richterites from St. Marcel (Val d'Aosta, Italy). *Am. Mineral.* **71**, 1426–1433.

Muehlenbachs, K. and Clayton, R.N. (1976) Oxygen isotope composition of the oceanic crust and its bearing on seawater. *J. Geophys. Res.* **81**, 4365–4369.

Muir Wood, R. (1982) The Laytonville Quarry (Mendocino County California) exotic block: iron-rich blueschist-facies subduction-zone metamorphism. *Mineral. Mag.* **45**, 87–99.

Munhá, J. and Marques, F. (1987) A model for the Lower Continental Crust (Morais and Bragança Massifs, NE Portugal). II. Petrology. *Terra Cognita* **7**, 161.

Munhá, J., Ribeiro, A. and Ribeiro, M.L. (1984) Blueschists in the Iberian Variscan Chain (Trás-os-Montes, NE Portugal). *Comun. Serv. Geol. Portugal* **70**, 31–53.

Mykkelveit, S., Husebye, E.S. and Oftedahl, C. (1980) Subduction of the Iapetus Ocean beneath the Møre Gneiss Region, Southern Norway. *Nature* **288**, 473–5.

Mysen, B.O. (1972) Five clinopyroxenes in the Hareidlandet area, western Norway. *Contrib. Mineral. Petrol.* **34**, 315–325.

Mysen, B.O. (1976) Experimental determination of some geochemical parameters relating to conditions of equilibration of peridotite in the upper mantle. *Am. Mineral.* **61**, 677–683.

Mysen, B.O. and Griffin, W.L. (1973) Pyroxene stoichiometry and the breakdown of omphacite. *Am. Mineral.* **58**, 60–63.

Mysen, B.O. and Heier, K.S. (1972) Petrogenesis of eclogites in high grade metamorphic gneisses exemplified by the Hareidland eclogite, west Norway. *Contrib. Mineral. Petrol.* **36**, 73–94.

Neumann, E.R. (1976) Two refinements for the calculation of structural formulae for pyroxenes and amphiboles. *Nor. Geol. Tidsskr.* **56**, 1–6.

Newton, R.C. (1966) Some calc–silicate equilibrium reactions. *Am. J. Sci.* **264**, 204–222.

Newton, R.C. (1986) Metamorphic temperatures and pressures of Group B and C eclogites. In *Blueschists and eclogites.* (eds Evans, B.W. and Brown, E.H.) *Geol. Soc. Amer. Mem.* **164**, 17–30.

Newton, R.C., Charlu, T.V. and Kleppa, O.J. (1980) Thermochemistry of high structural state plagioclases. *Geochim. Cosmochim. Acta* **44**, 933–941.

Newton, R.C. and Fyfe, W.S. (1976) High pressure metamorphism. In: *The Evolution of the Crystalline Rocks.* (eds Bailey, D.K. and Macdonald, R.) 101–186. Academic Press, London.

Newton, R.C. and Haselton, H.T. (1981) Thermodynamics of the garnet-plagioclase-Al_2SiO_5-quartz geobarometer. In *Advances in Physical Geochemistry* Vol. 1, 131–147 (eds Newton, R.C., Navrotski, A. and Wood, B.J.) Springer-Verlag, New York.

Newton, R.C. and Kennedy G.C. (1963) Some equilibrium reactions in the join $CaAl_2Si_2O_8$–H_2O *J. Geophys. Res.* **68**, 2967–2983.

Newton, R.C. and Perkins, D. (1982) Thermodynamic calibration of geobarometers based on the assemblages garnet-plagioclase-orthopyroxene-(clinopyroxene)-quartz. *Am. Mineral.* **67**, 203–222.

Newton, R.C. and Smith, J.V. (1967) Investigations concerning the breakdown of albite at depth in the earth. *J. Geol.* **75**, 268–286.

Nickel, K.G. (1986) Phase equilibria in the system SiO_2–MgO–Al_2O_3–CaO–Cr_2O_3(SMACCR) and their bearing on spinel/garnet lherzolite relationships. *Neues Jahrb. Mineral Abh.* **155**, 259–287.

Nickel, K.G. and Brey, G. (1984) Subsolidus orthopyroxene–clinopyroxene systematics in the system CaO–MgO–SiO_2 to 60 kb; a re-evaluation of the regular solution model. *Contrib. Mineral. Petrol.* **87**, 35–42.

Nickel, K.G., Brey, G.P. and Kogarko, L. (1985) Orthopyroxene–clinopyroxene equilibria in the system CaO–MgO–Al_2O_3–SiO_2 (CMAS); new experimental results and implications for two-pyroxene thermometry. *Contrib. Mineral. Petrol.* **91**, 44–53.

Nickel, K.G. and Green, D.H. (1985) Empirical geothermobarometry for garnet peridotites and implications for the nature of the lithosphere, kimberlites and diamonds. *Earth Planet. Sci. Lett.* **73**, 158–170.

Nicolas, A. and Poirier, J.P. (1976) *Crystalline Plasticity and Solid State Flow in Metamorphic Rocks.* John Wiley, London.

Nicollet, C. (1977) Une nouvelle éelogite à disthène et corindon primaires dans les complexes leptyno-amphiboliques du Massif Central francais (Lévezou, Rouergue), *Bull. Soc. Fr. Minéral. Cristallogr.* **100**, 334–337.

Nicollet, G., Leyreloup, A. and Dupuy, C. (1979) Petrogenesis of high pressure trondhjemitic layers in eclogites and amphibolites from Southern Massif Central, France. In *Trondhjemites, Dacites and Related Rocks.* (ed. Barker, F.) Elsevier, Amsterdam, 434–63.

Nishiyama, T. (1983) Steady diffusion model for olivine–plagioclase corona growth. *Geochim. Cosmochim. Acta* **47**, 283–294.

Nisio, P. (1985) *Les domaines d'antiphase des omphacites et la petrologie des eclogites: contribution à l'étude de l'évolution tectonométamorphique du Monviso (Alpes italiennes occidentales).* Thesis Université Claude Bernard, Lyon, 117 pp.

Nisio, P. and Lardeaux, J.-M. (1987) Retromorphic Fe-rich talc in low-temperature eclogites: example from Monviso (Italian Western Alps), *Bull. Minéral.* **110**, 427–437.

Nisio, P. Lardeaux, J.-M. and Bertrand, H. (1986) Témoins chimiques et minéralogiques du métamorphisme océanique dans la croute océanique subductée: example des éclogites du Monviso (Alpes Italiennes). *Comptes Rendus de l'Académie des Sciences de Paris. II, sé.* **303**, 1111–1116.

Nitsch, K.-H. (1968) Die Stabilität von Lawsonite. *Naturwissenschaften* **55**, 388.

Nitsch, K.H. (1971) Stabilitätsbeziehungen von Prehnit im Pumpellyit-haltigen Paragenesen. *Contrib. Mineral. Petrol.* **30**, 240–260.

Nixon, P.H. (1987) *Mantle Xenoliths*, John Wiley and Sons, Chichester 844 pp.

Nixon, P.H. and Boyd, F.R. (1973) Petrogenesis of the granular and sheared ultrabasic nodule suite in kimberlites. In *Lesotho Kimberlites* (ed. Nixon, P.H.) pp. 48–56. Lesotho Nat. Dev. Corpn. Maseru.

Nixon, P.H. and Boyd, F.R. (1973) The discrete nodule (megacryst) association in kimberlites from northern Lesotho. In *Lesotho Kimberlites* (ed. Nixon, P.H.) Lesotho Nat. Dev. Corpn. Maseru, 67–75.

Nixon, P.H., Chapman, N.A. and Gurney, J.J. (1978) Pyrope-spinel (alkremite) xenoliths from kimberlite. *Contrib. Mineral. Petrol.* **65**, 341–346.

Nixon, P.H., Davies, G.R. Slodkevich, V.V. and Bergman, S.C. (1986) Graphite pseudomorphs after diamond in the eclogite-peridotite massif of Beni-Bousera, Morocco, and a review of anomalous diamond occurrences. *Fourth Int. Kimberlite Conf., Geol. Soc. Australia Abstracts* **16**, 412.

Norris, R.J. and Henley, R.W. (1976) Dewatering of a metamorphic pile. *Geology* **4**, 333–336.

Norton, M.G. (1986) Late Caledonide extension in western Norway a response to extreme crustal thickening. *Tectonics* **5**, 195–204.

Norton, M.G. (1987) The Nordfjord Sogn Detachment, W. Norway. *Nor. Geol. Tidsskr.* **67**, 93–106.

Notsu, K., Mabuchi, H., Yoshioka, O. and Ozima, M. (1973) 147Sm–143Nd dating of Robert Victors eclogite. *Geochim. J.* **7**, 51–53.

Obata, M. (1980) The Ronda peridotite: garnet-, spinel-, and plagioclase-lherzolite facies and the P–T trajectories of a high-temperature mantle intrusion. *J. Petrol.* **21**, 533–572.

Obata, M. and Morten, L. (1982) Remnants of some high-temperature peridotite and the transformation from spinel-lherzolite to garnet-peridotite observed in the ultramafic bodies in the Austridic metamorphic terrain, between Non and Ultimo Valleys, northern Italy. *Terra Cognita* **2**, 239.

Obata, M. and Morten, L. (1987) Transformation of spinel lherzolite to garnet lherzolite in ultramafic lenses of the Austridic Crystalline Complex, Northern Italy. *J. Petrol.* **28**, 599–623.

Obata, M. Suen, C.J. and Dickey, J.S. Jr. (1980) The origin of mafic layers in the Ronda high temperature peridotite intrusion, S. Spain: an evidence of partial fusion and fractional crystallization in the upper mantle. *Colloques Internationaux du CNRS–Les Orogénes* **272**, 257–268.

Obata, M. and Thompson, A.B. (1981) Amphibole and chlorite in mafic and ultramafic rocks in the lower crust and upper mantle. *Contrib. Mineral. Petrol.* **77**, 74–81.

Oberhänsli, R. (1980) P–T Bestimmungen anhand von Mineralanalysen in Eklogiten und Glaucophaniten der Ophiolite von Zermatt. *Schweiz. Mineral. Petrogr. Mitt.* **60**, 215–235.

Oberhänsli, R., Hunziker, J.C., Martinotti, G. and Stern, W.B. (1982) Mucronites: an example of Eo-Alpine eclogitization of Permian granitoids, Italy (abstr.) *Terra Cognita* **2**(3), 325.

Oberhänsli, R., Hunziker, J.C., Martinotti, G. and Stern, W.B. (1985) Geochemistry, geochronology and petrology of Monte Mucrone: an example of Eo-Alpine eclogitization of Permian granitoids in the Sesia-Lanzo zone, Western Alps, Italy. *Chem. Geol.* **52**, 165–184.

O'Brien, P.J. (1989a) The petrology of retrograded eclogites of the Oberpfalz Forest, NE Bavaria, West Germany. In *The Evolution of the European Continental Crust: Deep Drilling, Geophysics, Geology and Geochemistry*, (eds Meissner, R. and Gebauer, D.) *Tectonophysics* **157**, 195–212.

O'Brien, P.J. (1989b) A study of retrogression in eclogites of the Oberpfalz Forest, NE Bavaria, West Germany, and their significance in the tectonic evolution of the Bohemian Massif. In *The Evolution of Metamorphic Belts.* (eds Daly, J.S., Cliff, R.A. and Yardley, B.W.D.) Spec. Pub. Geol. Soc. London **43**, 507–512.

O'Brien, P.J. (1989c) *The petrology of eclogites and related rocks from the Bavarian margin of the Bohemian Massif., West Germany.* Unpubl. PhD Thesis, University of Sheffield, UK.

O'Hara, M.J. (1967a) Mineral facies in ultrabasic rocks. In *Ultramafic and Related Rocks*, (ed. Wyllie, P.J.) John Wiley and Sons, New York, 7–18.

O'Hara, M.J. (1967b) Mineral parageneses in ultrabasic rocks. In *Ultramafic and Related Rocks*, (ed. Wyllie, P.J.) pp. 393–403, John Wiley and Sons, New York.

O'Hara, M.J. (1975) Pressure required to stabilise garnet-peridotite and eclogite at low temperatures. *Abstr. Int. Conf. Geotherm. Geobarom. Penn.* State University, U.S.A.

O'Hara, M.J. (1976) Origin of the Norwegian eclogites. *Progr. Expt. Petrol. NERC* **3**, 252.

O'Hara, M.J. and Mercy, E.L.P. (1963) Petrology and petrogenesis of some garnetiferous peridotites. *Trans. R. Soc. Edinburgh.* **65**, 251–314.

O'Hara, M.J. and Mercy, E.L.P. (1966) Garnet-peridotite and eclogite from Bellinzona, Switzerland. *Earth Planet. Sci. Lett* **1**, 295–300.

O'Hara, M.J., Richardson, S.W. and Wilson, G. (1971) Garnet peridotite stability and occurrence in crust and mantle. *Contrib. Mineral. Petrol.* **32**, 48–68.

O'Hara, M.J. and Yoder, H.S. (1967) Formation and fractionation of basic magmas at high pressure. *Scott. J. Geol.* **3**, 67–117.

Ohta, Y. (1979) Blue schists from Motalafjella, Western Spitsbergen. *Norsk Polarinstitut Skriften* **167**, 171–215.
Ohta, Y., Hirajima, T. and Hiroi, Y. (1986) Caledonian high-pressure metamorphism in central western Spitsbergen. In *Blueschists and Eclogites* (eds Evans, B.W. and Brown, E.H.) *Geol. Soc. Amer. Mem.* **164**, 205–216.
Ohtani, E. (1985) The primordial terrestrial magma ocean and its implication for stratification of the mantle. *Phys. Earth. Planet. Int.* **38**, 70–80.
Okay, A.J. (1986) High-pressure/low-temperature metamorphic rocks of Turkey. In *Blueschists and eclogites.* (ed Evans, B.W. and Brown, E.H.) *Geol. Soc. Amer. Mem.* **164**, 333–347.
Okrusch, M. (1981) Chloritoid-führende Parageneses in Hochdruckgesteinen von Samos. *Fortschr. Mineral.* **59**, 145–146.
Okrusch, M., Richter, P. and Katsikatsos, G. (1984) High-pressure rocks of Samos, Greece. In *The Geological Evolution of the Eastern Mediterranean* (eds Dixon, J.E. and Robertson, A.H.F.) *Geol. Soc. London Spec. Publ.* **17**, 529–536.
Okrusch, M., Seidel, E. and Davis, E.N. (1978) The assemblage jadeite–quartz in the glaucophane rocks of Sifnos (Cycladic Archipelago, Greece). *Neues Jahrb. Mineral. Abh.* **132**, 284–308.
O'Neill, H. StC. 1981) The transition between spinel lherzolite and garnet lherzolite, and its use as a geobarometer. *Contrib. Mineral. Petrol.* **77**, 185–194.
O'Neill, H. StC. and Wood, B.J. (1979) An experimental study of Fe–Mg partitioning between garnet and olivine and its calibration as a geothermometer. *Contrib. Mineral. Petrol.* **70**, 59–70.
O'Neill H. StC. and Wood, B.J. (1980) An experimental study of Fe–Mg partitioning between garnet and olivine and its calibration as a geothermometer: corrections. *Contrib. Mineral. Petrol.* **72**, 337.
Ongley, J.S., Basu, A.R. and Kyser, T.K. (1987) Oxygen isotopes in coexisting garnets, clinopyroxenes and phlogopites of Roberts Victor eclogites: implications for petrogenesis and mantle metasomatism. *Earth Planet. Sci. Lett.* **83**, 80–84.
O'Nions, R.K., Evensen, N.M., Hamilton, P.J. and Carter, S.R. (1978) Melting of the mantle past and present: isotope and trace element evidence. *Phil. Trans. Roy. Soc. Lond. (A)* **288**, 547–559.
O'Reilly, S.Y. and Griffin, W.L. (1985) A xenolith-derived geotherm for southeastern Australia and its geophysical implications. *Tectonophysics* **111**, 41–64.
O'Reilly, S.Y. and Griffin, W.L. (1987) Eastern Australia: 4,000 km of mantle samples. In *Mantle xenoliths.* (ed. Nixon, P.H.) Wiley, London, 267–280.
O'Reilly, S.Y., Griffin, W.L. and Pearson, N.J. (in press) Xenoliths and crustal origin. In *East Australian Volcanism* (eds. Johnson, R.W. and Taylor, S.R.) Cambridge Univ. Press.
O'Reilly, S.Y., Griffin, W.L. and Stabel, A. (1988) Evolution of Phanerozoic eastern Australian lithosphere: isotopic evidence for magmatic and tectonic underplating. *J. Petrol.* in press.
Otten, M. and Brouwer, H. (1974) High-pressure metamorphism in a ferrogabbro from Fenestrelle (Western Italian Alps). *Schweiz. Mineral. Petrogr. Mitt.* **59**, 281–297.
Otten, M.T. and Buseck, P.R. (1987) TEM study of the transformation of augite to sodic pyroxene in eclogitized ferrogabbro. *Contrib. Mineral. Petrol.* **96**, 529–538.
Otter, M.L. and Gurney, J.J. (1989) Mineral inclusions in diamonds from the Sloan diatremes, Colorado–Wyoming State Line kimberlite district, North America. In *Kimberlites and related rocks, Vol. 2. Their mantle/crust setting, diamonds and diamond exploration.* Geol. Soc. Australia Spec. Publn 14, Blackwells, Carlton, 1042–1053.
Oxburgh, E.R. and Turcotte, D.L. (1974) Themal gradients and regional metamorphism in overthrust terrains with special reference to the eastern Alps. *Schweiz. Mineral. Petrogr. Mitt.* **54**, 641–662.
Paděra, K. and Pouba, Z. (1984) White's clinopyroxene diagram applied to the differentiation of the eclogite and granulite facies. *Mineralia slov.* **16**, 39–42.
Paděra, K. and Procházka, J. (1970) Eclogite (Griquaite) with orthorhombic pyroxene from Nihov, Western Moravia. *Acta. Univ. Carolinae–Geol.* **3**, 171–186.
Paděra, K., Procházka, J. and Malec, J. (1971) Griquait mit rhombischem Pyroxen (Granatwebsterit) von Borek bei Stupná bei Křemže (Krems) in Böhmen. *Tschermaks Mineral. Petrogr. Mitt.* **16**, 59–78.
Padovani, E.R. and Carter, J.L. (1977) Aspects of deep crustal evolution beneath south central New Mexico. In *The Earth's Crust, its Nature and Physical Properties.* (ed. Heacock, J.G.) *Geophys. Monogr.* **20**, American Geophysical Union, 19–55.
Padovani, E.R. and Tracy, R.J. (1981) A pyrope-spinel (alkremite) xenolith from Moses rock Dike: first known North American occurrence. *Am. Mineral.* **66**, 741–745.
Papanikolaou, D.J. (1987) Tectonic evolution of the Cycladic blueschist belt (Aegean Sea, Greece). In *Chemical transport in metasomatic processes.* (ed. Helgeson, H.C.), Reidel (Dordrecht) 429–450.
Paquette, J.L. (1987) *Comportement des systèmes isotopiques U-Pb et Sm–Nd dans le métamorphisme éclogitic. Chaîne Hercynienne et chaîne Alpine.* Thesis, University of Rennes, Mém. et Doc. Centre Armo. Etude Struct. Socles, 14, 190 pp.
Paquette, J.L., Balé, P., Ballèvre, M. and Georget, Y. (1987) Géochronologie et géochime des éclogites du Léon: nouvelles contraintes sur évolution géodynamique du Nord-Ouest du Massif Armoricain. *Bull. Minéral.* **110**, 683–696.
Paquette, J.L., Peucat, J., Bernard-Griffiths, J. and Marchand, J. (1985) Evidence for old Precambrian relics shown by U-Pb zircon dating of eclogites and associated rocks in the Hercynian belt of South Brittany, France. *Chem. Geol. (Isot. Geosci. Sect.)* **52**, 203–216.

Passchier, C.W. (1986) Mylonites in the continental crust and their role as seimic reflectors. *Geol. Mijnbouw* **65**, 167–176.

Passchier, C.W., Urai, J.L., Van Loon, J. and Williams, P.F. (1981) Structural geology of the central Sesia Lanzo zone. *Geol. Mijnbouw* **60**, 497–507.

Pasteris, J.D. (1981) Occurrence of graphite in serpentinized olivines in kimberlite. *Geology* **9**, 356–359.

Pattison, D.R.M. and Newton, R.C. (1987) Garnet–clinopyroxene K_D(Fe–Mg): Variations due to X_{Ca}^{Grt} and pressure (abstract). *EOS Trans. AGU* **68**, 461.

Pavelescu, L. (1957) Contributions to the study of some eclogites from the Sebesului Mountains. *Bulletin Acad. Repub. Rom. Sec. Geol. Geogr.* **2**, 187–199.

Pavlenko, A.S., Gevorkin, R.G., Aslanyan Gulyan, E.K. and Yegorov, O.S. (1974) On the diamonds in the ultramafic belts of Armenia. *Geochem. Int.* **11**, 282–294.

Peacock, S.M. (1987) Thermal effects of metamorphic fluids in subduction zones. *Geology* **15**, 1057–1060.

Pearce, J.A. and Cann, J.R. (1973) Tectonic setting of basic volcanic rocks determined using trace element analysis. *Earth. Planet. Sci. Lett.* **19**, 290–300.

Péquignot, G., Lardeaux, J.M. and Caron, J.M. (1984) Recristallisation d'éclogites de basse température dans les métabasalts corses. *C.R. Académie des Sciences Paris.* **299**, 871–874.

Perkins, D. (1983) The stability of Mg-rich garnet in the system $CaO–MgO–Al_2O_3–SiO_2$ at 1000–1300°C and high pressures. *Am. Mineral.* **68**, 355–364.

Perkins, D., Holland, H.J.B. and Newton, R.C. (1981) The Al_2O_3 contents of enstatite in equilibrium with garnet in the system $MgO–Al_2O_3–SiO_2$ at 15–40 kbar and 900–1600°C. *Contrib. Mineral. Petrol.* **78**, 99–109.

Perkins, D. and Newton, R.C. (1980) The compositions of coexisting pyroxenes and garnet in the system $CaO–MgO–Al_2O_3–SiO_2$ at 900°–1100°C and high pressures. *Contrib. Mineral. Petrol.* **75**, 291–300.

Perkins, D. and Newton, R.C. (1981). Charnockite geobarometers based on coexisting garnet–pyroxene–plagioclase–quartz. *Nature* **292**, 144–146.

Perseil, E.A. (1985) Quelques caractèristiques des faciès à oxydes de manganèse dans le gisement de St. Marcel-Praborna-V. Aoste, Italie, *Mineralium Deposita* **20**, 271–276.

Peucat, J.-J. (1986). Rb-Sr and U-Pb dating of the blueschists of the Ile de Groix (France) In *Blueschist and eclogites.* (eds Evans, B.W. and Brown, E.H.). *Geol. Soc. Amer. Mem.* **164**, 229–238.

Peucat, J.J. and Cogné, J. (1974) Les schistes cristallins de la Baie d'Audierne (Sud-Finistère), un jalon intermédiare dans le socle antécambrien entre la Meseta ibérique et les régions sud-armoricaines. *C.R. Acad. Sci. Paris.* **278(D)**, 1809–1812.

Peucat, J.J. and Cogné, J. (1977) Geochronology of some blueschists from Île de Groix, France, *Nature* **268**, 131–2.

Peucat, J.J., Vidal, Ph., Godard, G. and Postaire, B. (1982) Precambrian U–Pb zircon ages in eclogites and garnet-pyroxenites from south Brittany (France): an old oceanic crust in the West European Hercynian belt? *Earth Planet. Sci. Lett.* **60**, 70–78.

Philpotts, J.A., Schnetzler, G.C. and Thomas, H.M. (1972) Petrogenetic implications of some new geochemical data on eclogitic and ultrabasic inclusions. *Geochim. Cosmochim. Acta* **36**, 1131–1166.

Piboule, M. and Coffrant, D. (1970) Métagabbros coronitiques dans le Rouergue et le Limousin (Massif Central français), une stade banal de l'éclogitisation. *C.R. Acad. Sci. Paris* **270(D)**, 22–24.

Piccardo, G.B., Cimmino, F. and Messiga, B. (1980) Antigoritic serpentinites and rodingites of the Voltri Massif: some petrological evidences for their evolutive history. *Ofioliti* **5**, 111–114.

Piccardo, G.B., Messiga, B. and Mazzucotelli, A. (1979) Chemical petrology and geodynamic evolution of the ophiolitic metavolcanites (prasinites) from the Voltri Massif, Piemontese ophiolite nappe (western Liguria, Italy), *Ofioliti* **4**, 373–402.

Pidgeon, R.T. and Råheim, A. (1972) Geochronological investigation of the gneisses and minor intrusive rocks from Kristiansund, West Norway, *Nor. Geol. Tidsskr.* **52**, 241–256.

Pin, C. (1981) Old inherited zircons in two synkinematic variscan granitoids, the 'granite du Pinet' and the 'Orthogneiss de Marvejols' (southern French Massif Central). *Neues Jahrb. Mineral. Abh.* **142**, 27–48.

Pin, C. and Carme, F. (1987) A Sm–Nd isotopic study of 500 Ma old crust in the Variscan belt of Western Europe: the Chamrousse ophiolite complex, Western Alps (France). *Contrib. Mineral. Petrol.* **96**, 406–413.

Pin, C. and Lancelot, J. (1978) Un example de magmatisme Cambrien dans le Massif Central. les métadiorites quartziques intrusives dans la série du Lot. *Bull. Soc. Géol. France.* (7) XX, 203–8.

Pin, C. and Lancelot, J. (1982) U-Pb dating of an early palaeozoic bimodal magmatism in the French Massif Central and of its further metamorphic evolution. *Contrib. Mineral. Petrol.* **79**, 1–12.

Pin, C., Majerowicz, A. and Wojciechowska, I. (1988) Upper Palaeozoic oceanic crust in the Polish Sudetes, Nd–Sr isotope and trace element evidence. *Lithos* **21**, 195–209.

Pin, C. and Peucat, J.-J. (1986) Ages des épisodes de métamorphisme Paléozoique dans le Massif Central et le Massif Armoriquain. *Bull. Soc. gèol. Fr.* (8) II, 461–469.

Pin, C. and Vielzeuf, D. (1983) Granulites and related rocks in Variscan Median Europe: A dualistic interpretation. *Tectonophysics* **93**, 47–74.

Pin, C. and Vielzeuf, D. (1988) Les granulites de haute-pression d'Europe moyenne témoins d'une subduction éo-hercynienne. Implications sur l'origine des groupes Leptyno-amphiboliques. *Bull. Soc. Géol. Fr.* (8) IV, 13–20.

Platt, J.P. (1986) Dynamics of orogenic wedges and the uplift of high-pressure metamorphic rocks. *Geol. Soc. Am. Bull.* 97, 1037–1053.

Platt, J.P. (1987) The uplift of high-pressure–low-temperature metamorphic rocks. *Phil. Trans, R. Soc. London A* **321**, 87–103.

Podvin, P. (1988) Ni–Mg partitioning between synthetic olivines and orthopyroxenes: application to geothermometry. *Am. Mineral.* **73**, 274–280.

Pognante, U. (1984) Eclogitic versus blueschist metamorphism in the internal western Alps along the Susa valley traverse. *Sci. Geol. Bull. Strasbourg* **37**, 29–36.

Pognante, U. (1985) Coronitic reactions and ductile shear zones in eclogitised ophilite metagabbro, western Alps, north Italy, *Chem. Geol.* **50**, 99–109.

Pognante, U. (1988) Early Alpine eclogitisation in talc/chloritoid bearing Mg–metagabbros and in jadeite–Fe-omphacite bearing metatrondhjemites from the ophiolites of the western Alps. *Rend. Soc. Ital. Mineral. Petrol.* **43**, (in press).

Pognante, U. (1989a) Lawsonite, blueschist and eclogite formation in the southern Sesia zone (western Alps, Italy). *European J. Mineral.* **1**, 89–104.

Pognante, U. (1988b) Tectonic implications of lawsonite formation in the Sesia zone (Western Italian Alps). *Tectonophysics* **162**, 219–227.

Pognante, U., Compagnoni, R. and Gosso, G. (1980) Micro-mesostructural relationships in the continental eclogitic rocks of the Sesia-Lanzo Zone (Italian Western Alps). *Rend. Soc. Ital., Mineral. Petrol.*, **36**, 169–186.

Pognante, U. and Kienast, J.R. (1987) Blueschist and eclogite transformations in Fe–Ti gabbros: a case from the western Alps ophiolites. *J. Petrol.* **28**, 271–292.

Pognante, U., Lombardo, B. and Venturelli, G. (1982) Petrology and geochemistry of Fe–Ti gabbros and plagiogranites from the western Alps ophiolites, *Schweiz. Mineral. Petrogr. Mitt.* **62**, 457–472.

Pognante, U., Perotto, A., Salino, C. and Toscani, L. (1986) The ophiolitic peridotites of the western Alps: record of the evolution of a small oceanic-type basin in the Mesozoic Tethys. *Tschermaks Mineral. Petrogr. Mitt.* **35**, 47–65.

Pognante, U., Roesli, U. and Toscani, L. (1985) Petrology of ultramafic and mafic rocks from the Lanzo peridotite body (western Alps), *Lithos* **18**, 201–214.

Pognante, U. and Sandrone, R. (1989) Eclogites in the northern Dora-Maira nappe (Western Alps, Italy). *Mineralogy and Petrology* **40**, 57–92.

Pognante, U., Talarico, F. and Benna, P. (1988) Incomplete blueschist re-crystallization in high-grade metamorphics from the Sesia-Lanzo unit (Vasario-Sparone subunit, western Alps): a case history of metastability. *Lithos* **21**, 129–142.

Pognante, U., Talarico, F., Rastelli, N. and Ferrati, N. (1987) High pressure metamorphism in the nappes of the valle dell'Orco traverse (western Alps), *J. metamorphic Geol.* **5**, 397–414.

Polanski, A. (1955) On the metamorphism of crystalline formations of the Sowie Mts (Middle Sudeten). (Polish with English summary). *Arch. Mineral.* **18**, 285–292.

Pollack, H.N. and Chapman, D.S. (1977) On the regional variation of heat flow, geotherms and lithospheric thickness. *Tectonophysics* **38**, 279–296.

Ponomarenko, A.I. (1975) Alkremite—a new variety of aluminous hyperbasite (xenoliths from the kimberlite of the Udachnaya pipe). *Dokl. Akad. Nauk SSSR* **225**, 928–931 (in Russian).

Ponomarenko, A.I., Sobolev, N.V., Pokhilenko, N.P., Lavrentyev, Yu. G. and Sobolev, V.S. (1976). Diamond-bearing grospidite and diamond-bearing kyanite eclogite from the Udachnaya kimberlite pipe, Yakutia. *Dokl. Akad Nauk. SSSR, Earth Sci. Sect.* **226**, 158–161.

Ponomarenko, A.I. Spetsius, Z.V. and Sobolev, N.V. (1980) A new type of diamond-bearing rock—garnet pyroxenite *Dokl. Akad. Nauk SSSR* **251**, 438–441 (in Russian).

Ponomarenko, A.I. and Spetsius, Z.V. (1978) Coesite of eclogites from the Udachnaya kimberlite pipe, Yakutia. Abstracts XI Gen. Meeting Internat. Mineral. Assoc. Novosibirsk **2**, 23.

Pouba, Z., Padera, K. and Fiala, J. (1985) Omphacite granulite from the NE marginal area of the Bohemian Massif (Rychleby Mts.). *Neues Jahrb. Mineral. Abh.* **151**, 29–52.

Powell, R. (1985a) Geothermometry and geobarometry: a discussion. *J. Geol. Soc. London* **142**, 29–38.

Powell, R. (1985b) Regression diagnostics and robust regression in geothermometer/geobarometer calibration: the garnet–clinopyroxene geothermometer revisited. *J. Metamorphic Geol.* **3**, 327–342.

Powell, R. and Holland, T.J.B. (1985) An internally consistent thermodynamic dataset with uncertainties and correlations: 1. Methods and a worked example. *J. Metamorphic Geol.* **3**, 327–342.

Powell, D. and Holland, T.J.B. (1988) An internally consistent dataset with uncertainties and correlations: 3 Applications to geobarometry, worked examples and a computer program. *J. Metamorphic Geol.* **6**, 173–204.

Prinz, M., Manson, V.R., Hlava, P.F. and Keil, K. (1975) Inclusions in diamonds: garnet lherzolite and eclogite assemblages. *Phys. Chem Earth* **9**, 797–816.

Puchelt, H., Matthes, S. and Richter, P. (1978) Aussagemöglichkeiten der Seltenen Erden und anderer Spurenelemente für die Eklogite des Münchberger Gneisgebiets. *Fortschr. Mineral.* **56**, 106–108.

Quadt, A. and Gebauer, D. (1988) Sm/Nd-, U–Pb- and Rb–Sr dating of high-pressure ultramafic to felsic rocks from the Moldanubian area of NE Bavaria (FRG) and the Saxonian Granulite Massif (GDR). *Chem. Geol.* **70**, 15.

Qvale, H. and Stigh, J. (1985) Ultramafic rocks in the Scandinavian Caledonides. In *The Caledonide Orogen—Scandinavia and Related Areas* (eds Gee, D.G. and Sturt, B.A.) pp. 693–715, Wiley.

Raase, P. (1974) Al and Ti contents of hornblende, indicators of pressure and temperature of regional metamorphism. *Contrib. Mineral. Petrol.* **45**, 231–236.

Råheim, A. and Green, D.H. (1974a) Experimental determination of the temperature and pressure dependence of the Fe–Mg partition coefficient for coexisting garnet and clinopyroxene. *Contrib. Mineral. Petrol.* **48**, 179–203.

Råheim, A. and Green, D.H. (1974b) Experimental petrology of lunar highland basalt composition and applications to models for the lunar interior. *J. Geol.* **82**, 607–622.

Råheim, A. and Green, D.H. (1974c) Talc-garnet-kyanite-quartz schist from an eclogite-bearing terrane, western Tasmania. *Contrib. Mineral. Petrol.* **43**, 223–231.

Råheim, A. and Green, D.H. (1975) P, T paths of natural eclogites during metamorphism—a record of subduction. *Lithos* **8**, 317–328.

Raith, M., Hörmann, P.K. and Abraham, K. (1977) Petrology and metamorphic evolution of the Penninic ophiolites in the western Tauern Window (Austria). *Schweiz. Mineral. Petrogr. Mitt.* **57**, 187–232.

Raith, M., Mehrens, C. and Thoele, W. (1980) Gliederung, tektonischer Bau und metamorphe Entwicklung der penninischen Serien im südlichen Venediger Gebiet, Östtirol. *Jahrb. Geol. Bundesanstalt, Wien* **123**, 1–37.

Rajlich, P. and Synek, J. (1987) A cross section through the Moldanubian of the Bohemian Massif and the structural development in its ductile domains. *Neues Jahrb. Geol. Paläont. Monatsh.* **11**, 689–698.

Rao, B.B. and Johannes, W. (1979) Further data on the stability of staurolite + quartz and related assemblages. *Neues Jahrb. Mineral. Abh.* **10**, 437–447.

Ravasz-Baranyai, L. (1969) Eclogite from the Meczek Mountains, Hungary. *Acta geol. Hung.* **13**, 315–322.

Reale Ufficio Geologico (1908) *Carta Geologica della Alpi Occidentali alla scala 1:400,000.*

Reid, A.M., Brown, R.W., Dawson, J.B., Whitfield, G.G. and Siebert, J.C. (1976) Garnet and pyroxene compositions in some diamondiferous eclogites. *Contrib. Mineral. Petrol.* **58**, 203–220.

Reid, A.M. and Dawson, J.B. (1972) Olivine–garnet reaction in peridotites from Tanzania. *Lithos*, **5**, 115–124.

Reinecke, T. (1983) Mineralogie und Petrologie der Mangan- und Eisenreichen Metasedimente von Andros/Kykladen/Griechenland. Dissertation Braunschweig, 256 pp.

Reinsch, D. (1977) High pressure rocks from Val Chuisella (Sesia-Lanzo Zone, Italian Alps). *Neues Jahrb. Mineral. Abh.* **130**, 89–102.

Reinsch, D. (1979) Glaucophanites and eclogites from Val Chiusella (Sesia–Lanzo-Zone, Italian Alps). *Contrib. Mineral. Petrol.* **70**, 257–266.

Reynard, B. and Ballèvre, M. (1988) Coexisting amphiboles in an eclogite from the Western Alps: new constraints on the miscibility gap between sodic and calcic amphiboles. *J. metamorphic Geol.* **6**, 333–350.

Ribiero, A. (1974) Contribution à l'étude tectonique de Trás-os-Montes Occidental. *Mem. Serv. Geol. Portugal* **24**, 1–168.

Richardson, S.H. (1986) Latter-day origin of diamonds of eclogite paragenesis. *Nature*, **322**, 623–626.

Richardson, S.H., Gurney, J.J., Erlank, A.J. and Harris, J.W. (1984) Origin of diamonds in old enriched mantle. *Nature* **310**, 198–202.

Richter, W. (1973) Vergleichende Untersuchungen an ostalpinen Eklogiten. *Tschermaks Mineral. Petrogr. Mitt.* **19**, 1–50.

Rickard, R.S., Harris, J.W., Gurney, J.J. and Cardoso, P. (1989) Mineral inclusions in diamonds from Koffiefontein Mine. In *Kimberlites and related rocks, Vol. 2. Their mantle/crust setting, diamonds and diamond exploration.* Geol. Soc. Austr. Spec. Publn. 14, Blackwells, Carlton, 1054–1062.

Ridley, J. (1984) Evidence of a temperature-dependent 'blueschist' to 'eclogite' transformation in high-pressure metamorphism of metabasic rocks. *J. Petrol.* **25**, 852–870.

Ridley, J. and Dixon, J.E. (1984) Reaction pathways during the progressive deformation of a blueschist metabasite: the role of chemical disequilibrium and restricted range equilibrium. *J. Metamorphic Geol.* **2**, 115–128.

Ries, A.C. and Shackleton, R.M. (1971) Catazonal complexes of North-West Spain and North Portugal, remnants of a Hercynian thrust plate. *Nature Phys. Sci.* **234**, 65–68.

Ringwood, A.E. (1975) *Composition and petrology of the Earth's Mantle.* McGraw-Hill, New York 618.

Ringwood, A.E. (1977) Synthesis of the pyrope-knorringite solid-solution series. *Earth Planet. Sci. Letts.* **36**, 443–448.

Ringwood, A.E. (1982) Phase transformations and differentiation in subducted lithosphere: implications for mantle dynamics, basalt petrogenesis, and crustal evolution. *J. Geol.* **90**, 611–643.

Ringwood, A.E. and Green, D.H. (1966) An experimental investigation of gabbro–eclogite transformation and some geophysical consequences. *Tectonophysics* **3**, 383–427.

Robert, C., Javoy, M. and Kienast, J.R. (1985) Coefficients de distribution et mesures isotopiques des éclogites et micaschistes de la zone Sesia-Lanzo (Alpes italiennes). *Bull. Soc. Fr. Mineral. Cristall.* **108**, 699–711.

Roberts, D. and Gee, D.G. (1985) An introduction to the structure of the Scandinavian Caledonides. In *The Caledonide Orogen—Scandinavia and Related Areas* (eds Gee, D.G. and Sturt, B.A.) pp. 55–68, John Wiley, Chichester.

Robertson, E.C.F., Birch, F. and Macdonald, G.L.F. (1957) Experimental determination of jadeite stability relations to 25,000 bars. *Am. J. Sci.* **255**, 115–137.

Robinson, D.N. (1979) Diamond and graphite in eclogite xenoliths from kimberlite. In *The Mantle Sample: Inclusions in Kimberlite and Other Volcanics* (eds Boyd, F.R. and Meyer, H.O.A.) Am. Geophys. Union, Washington, D.C. 50–58.

Robinson, D.N., Gurney, J.J. and Shee, S.R. (1984) Diamond eclogite and graphite eclogite xenoliths from Orapa, Botswana. In *Kimberlites II. The Mantle and Crust-Mantle Relationships* (ed. Kornprobst, J.) Elsevier, Amsterdam 11–24

Roermund, H. van (1985) Eclogites of the Seve Nappe. Central Scandinavian Caledonides. In *The Caledonide Orogen—Scandinavia and Related Areas* (eds Gee, D.G. and Sturt, B.A.) pp. 873–866, John Wiley, Chichester.

Rogers, N.W. (1977) Granulite xenoliths from Lesotho kimberlites and their composition of the lower crust. *Nature* **270**, 681–684.

Rogers, N.W. and Hawkesworth, C.J. (1982) Proterozoic age and cumulate origin for granulite xenoliths, Lesotho. *Nature* **299**, 409–413.

Roques, M. (1971) Structure géologique du Massif Central. In *Symp. J. Jung. Géologie, Géomorphologie et structure profonde du Massif Central Français.* Plein Air Service. Clermont Ferrand, 17–32.

Rossi, G. (1982) The crystal-chemistry of high-pressure clinopyroxenes. *Terra Cognita* **2**, 304.

Rost, F. and Brenneis, P. (1978) Die Ultramafitite im Bergzug sudlich des Ultentales, Provinz Alto Adige (Oberitalien). *Tschermaks Mineral. Petrogr. Mitt.* **25**, 257–286.

Rost, F., Wannemacher, J. and Anton, J. (1979) Garnet ultramafites in the Ivrea Zone and other crystalline zones along the Periadriatic Lineament. *Mem. Sci. Geol. Padova* **33**, 183–191.

Rost, F., Wannemacher, J. and Grigel, W. (1974) Die Ultramafitite der Alpe Arami und Croveggio Tessin (Schweiz.). *Schweiz. Mineral. Petrogr. Mitt.* **54**, 353–369.

Rozen, O.M., Zorin, Yu.M. and Zayachkovskiy, A.A. (1972) A diamond find in Precambrian eclogite of the Kokchetav block. *Doklady Akademii Nauk SSSR* **203**, 163–165, (English translation).

Rozen, O.M., Zorin, Yu.M. and Zayachkovskii, A.A. (1972) the discovery of diamond in connection with eclogites in Precambrian deposits of the Kokchetav massif. *Dokl. Akad. Nauk. SSSR* **203**, 674–676 (in Russian).

Rubie, D.C. (1983) Reaction-enhanced ductility: the role of solid–solid univariant reactions in deformation of the crust and mantle. *Tectonophysics* **96**, 331–352.

Rubie, D.C. (1984) A thermal–tectonic model for high-pressure metamorphism and deformation in the Sesia Zone, Western Alps. *J. Geol.* **92**, 21–36.

Rubie, D.C. (1986) The catalysis of mineral reactions by water and restrictions on the presence of aqueous fluid during metamorphism. *Mineral. Mag.* **50**, 399–415.

Rubie, D.C. (1990) Mechanisms of reaction-enhanced deformability in minerals and rocks. In *Deformation Processes in Minerals, Ceramics and Rocks*, (eds Barber, D.J. and Meredith, P.G.) Unwin Hyman, London (in press).

Rubie, D.C. and Brearley, A.J. (1987) Metastable melting during the breakdown of muscovite + quartz at 1 kbar. *Bull. Minéral.* **110**, 533–549.

Rubie, D.C. and Brearley, A.J. (1988) The application of non-equilibrium thermodynamics to partial melting and rates of multicomponent diffusion in granitic melts. *Trans. Am. Geophys. Union* **69**, 1475–1476.

Rubie, D.C. and Thompson, A.B. (1985) Kinetics of metamorphic reactions at elevated temperatures and pressures: an appraisal of available experimental data. In *Metamorphic Reactions, Advances in Physical Geochemistry 4*, (eds Thompson, A.B. and Rubie, D.C.) Springer-Verlag, New York, 27–79.

Rudnick, R.L., McDonough, W.F., McCulloch, M.T. and Taylor, S.R. (1986) Lower crust xenoliths from Queensland, Australia: evidence for deep crustal assimilation and fractionation of continental basalts. *Geochim. Cosmochim. Acta* **50**, 1099–1115.

Rutter, E.H. and Brodie, K.H. (1985) The permeation of water into hydrating shear zones. In *Metamorphic Reactions, Advances in Physical Geochemistry 4*, (eds Thompson, A.B. and Rubie, D.C.) Springer-Verlag, New York, 242–250.

Safronov, A.F. Nikishov, N.K., Botkunov, A.I., Gotovtev, V.V. and Makhoto, V.F. (1980) Composition of garnets and pyroxenes from diamond-bearing eclogites in the pipes "Mir" and "Udachnaya". *Geol. i. Geofiz* (9), 76–82 (in Russian).

Saliot, P. (1979) La jadeite dans les Alpes française. *Bull. Soc. Fr. Mineral. Cristall.* **102**, 391–401.

Salje, E. (1986) Heat capacities and entropies of andalusite and sillimanite. The influence of fibrolitization on the phase diagram of the Al_2SiO_5 polymorphs. *Am. Mineral.* **71**, 1366–1371.

Sanc, I. and Rieder, M. (1983) Lamellar pyroxenes and their petrogenetic significance: three examples from the Czech Massif. *Contrib. Mineral. Petrol.* **84**, 73–83.

Sanders, I.S. (1982) Eclogite and cofacial gneiss from Glenelg, NW Scotland (abstr.) *Terra Cognita* **2**, 317.

Sanders, I.S., van Calsteren, P.W.C. and Hawkesworth, C.J. (1984) A Grenville Sm-Nd age for the Glenelg eclogite in north-west Scotland. *Nature* **312**, 439–440.

Sandrone, R., Leardi, L., Rossetti, P. and Compagnoni, R. (1986) P–T conditions for the eclogitic re-equilibration of the metaophiolites from Val d'Ala di Lanzo (internal Piemontese zone, western Alps). *J. metamorphic Geol.* **4**, 161–178.

Santallier, D.S. (1988) Mineralogy and crystallisation of the Seve eclogites in the Vuoggatjalme area, Swedish Caledonides of Norbotten. *Geol. Fören. Stockholm Förh.* **110**, 89–98.

Santallier, D., Briand, B., Ménot, R-P. and Piboule, M. (1988) Les complexes leptyno-amphiboliques (C.L.A.): revue critique et suggestions pour un meilleur emploi de ce terme. *Bull. Soc. géol. Fr.* (8) IV, 3–12.

Santallier, D., Floc'h, J.P. and Guillot. P.L. (1978) Quelques aspects du metamorphisme Dévonien en Bas-Limousin (Massif Central, France). *Bull. Minéral.* **101**, 77–88.

Sass, J.H. and Lachenbruch, A.H. (1979) Thermal regime of the Australian continental crust. In *The Earth, its Origin, Structure and Evolution.* (ed. McElhinny, M.W.) Academic Press, London, 391–352.

Sautter, V. and Harte, (1988) Diffusion gradients in an eclogite xenolith from the Roberts Victor kimberlite pipe: (1) Mechanism and evolution of garnet exsolution in Al_2O_3-rich clinopyroxene. *J. Petrol.* **29**, 1325–1352.

Saxena, S.K. (1979) Garnet–clinopyroxene geothermometer. *Contrib. Mineral. Petrol.* **70**, 229–235.

Saxena, S.K. and Eriksson, G. (1983) Theoretical computation of mineral assemblages in pyrolite and lherzolite. *J. Petrol.*, **24**, 538–555.

Scharbert, H. (1954) Die eklogitischen Gesteine des südlichen Grossvenedigergebietes (Osttirol). *Jahrb. Geol. Bundesanstalt, Wien* **97**, 39–63.

Scharbert, H.G. (1973a) Garnets from Moldanubian garnet pyroxenites and associated pyriclasites, Bohemian Massif of Lower Austria, Austria. *Earth Planet. Sci. Lett*, **20**, 415–418.

Scharbert, H.G. (1973b) Pyrope-rich garnets from Moldanubian garnet pyroxenites, Bohemian Massif of Lower Austria, Austria. *Neues Jahrb. Mineral. Monatsch.* **2**, 89–93.

Scharbert, H.G. and Carswell, D.A. (1983) Petrology of garnet-clinopyroxene rocks in a granulite facies environment, Bohemian Massif of Lower Austria. *Bull. Mineral*, **106**, 761–774.

Scharbert, H.G. and Kurat, G. (1974) Distribution of some elements between coexisting ferromagnesian minerals in Moldanubian granulite facies rocks, Lower Austria, Austria. *Tschermaks Mineral. Petrogr. Mitt.* **21**, 110–134.

Schellhorn, O., Gebauer, D. and Grünenfelder, M. (1982) Eclogites of the Limousin area: a key for the geodynamic reconstruction of the Western part of the French Massif Central. *Terra Cognita* **2**, 326.

Schermerhorn, L.J.G. and Kotsch, S. (1984) First occurrence of lawsonite in Portugal and tectonic implications. *Comun. Serv. geol. Portugal* **70**, 23–29.

Schleicher, H. and Kramm, U. (1987) U/Pb and Rb/Sr dating on eclogitogenic polymetamorphic rocks from the Black Forest, FRG. *Terra Cognita* **7**, 168.

Schliestedt, M. (1980) *Phasengleichgewichte in Hochdruckgesteinen von Sifnos, Griechenland.* PhD Thesis, Technische Universität Braunschweig.

Schliestedt, M. (1986) Eclogite–blueschist relationships as evidenced by mineral equilibria in the high-pressure metabasic rocks of Sifnos (Cycladic Islands), Greece, *J. Petrol.* **27**, 1437–1459.

Schliestedt, M., Altherr, R. and Matthews, A. (1987) Evolution of the Cycladic crystalline complex: petrology, isotope geochemistry and geochronology. In *Chemical Transport in Metasomatic Processes* (ed. Helgeson, H.C.) Reidel (Dordrecht) 389–428.

Schliestedt, M. and Matthews, A. (1987) Transformation of blueschist facies rocks as a consequence of fluid infiltration, Sifnos (Cyclades), Greece. *Contrib. Mineral. Petrol.* **97**, 237–250.

Schmidt, K. (1967) Die Eklogitvorkommen des kristallinen Grundgebirges in NE-Bayern. I. Der Granat der Eklogite Eklogitamphibolite und Granatamphibolite des Münchberger Gneisgebietes. *Neues Jahrb. Mineral. Abh.* **107**, 74–112.

Schreyer, W. (1977) Whiteschists: their compositions and pressure temperature regimes based on experimental, field, and petrographic evidence. *Tectonophysics* **43**, 127–144.

Schreyer, W. (1985) Metamorphism of crustal rocks at mantle depths: high-pressure minerals and mineral assemblages in metapelites. *Fortschr. Mineral.* **63**, 227–261.

Schreyer, W. (1988a) Experimental studies on metamorphism of crustal rocks under mantle pressures. *Mineral. Mag.* **52**, 1–26.

Schreyer, W. (1988b) Subduction of continental crust to mantle depths: petrological evidence. *Episodes* **11**, 97–104.

Schreyer, W., Abraham, K. and Behr, H.J. (1975) Sapphirine and associated minerals from the kornerupine rock of Waldheim, Saxony. *Neues Jahrb. Mineral. Geol. Paläont. Abh.* **126**, 1–27.

Schreyer, W., Baller, T. and Chopin, C. (1985) High-pressure synthesis of ellenbergerite in the system MgO-Al_2O_3-SiO_2-H_2O. *Terra Cognita* **5**, 327.

Schreyer, W. Horrocks, P.C. and Abraham, K. (1984) High-magnesium staurolite in a sapphirine-garnet rock from the Limpopo belt, southern Africa. *Contrib. Mineral. Petrol.* **86**, 200–207.

Schreyer, W., Massonne, H.-J. and Chopin, C. (1987) Continental crust subducted to depths near 100 km: implications for magma and fluid genesis in collision zones. In: *Magmatic Processes: Physiochemical Principles.* (ed. Mysen, B.O.) Geochemical Society Spec. Publ. **1**, 155–163.

Schreÿer, W. and Seifert, F. (1969) High-pressure phases in the system MgO–Al_2O_3–SiO_2–H_2O. *Am. J. Sci.* **267A**, 407–443.

Schulze, D.J. (1986) Calcium anomalies in the mantle and a subducted metaserpentinite origin for diamonds. *Nature* **319**, 483–485.

Schulze, D.J. and Helmstaedt, H. (1979) Garnet pyroxenite and eclogite xenoliths from the Sullivan Buttes latite, Chino Valley, Arizona. In *The Mantle Samples: Inclusions in Kimberlites and other Volcanics.* (eds Boyd. F.R. and Mayer, H.O.A.) *Proc. 2nd Int. Kimberlite Conference* **2**, American Geophysical Union, Washington, D.C., 318–329.

Schulze, D.J. and Helmstaedt, H. (1988) Coesite-sanidine eclogites from kimberlite: products of mantle fractionation or subduction? *J. Geol.* **96**, 435–443.

Schüssler, U., Richter, P. and Okrusch, M. (1987) Amphibolites of the KTB target Oberpfalz, Bavaria. *Terra Cognita* **7**, 164.

Scott Smith, B.H., Danchin, R.V., Harris, J.W. and Stracke, K.J. (1984) Kimberlites near Orrorroo, South Australia. In *Kimberlites: II The Mantle and Crust–Mantle Relationships.* (ed, Kornprobst, J.) Developments in Petrology 11B, Elsevier, Amsterdam, 112–142.

Selverstone, J. (1986) Possible consequences of CO_2–H_2O immiscibility on blueschist-facies metamorphism. *Geol. Soc. Amer.Abstr.* **18**, 745.

Selverstone, J., Spear, F.S., Franz, G. and Morteani, G. (1984) High-pressure metamorphism in the S.W. Tauern Window, Austria: P–T paths from hornblende-kyanite-staurolite–schists. *J. Petrol.* **25**, 501–531.

Sen, S.K. and Bhattacharya, A. (1984) An orthopyroxene–garnet thermometer and its application to Madras Charnockites. *Contrib. Mineral. Petrol.* **88**, 64–71.

Seranne, M. and Seguret, M. (1987) The Devonian basins of western Norway: tectonics and kinematics of an extending crust. In *Continental Extensional Tectonics* (eds Coward, M.P., Dewey, J.F. and Handcock, P.L.) Geol. Soc. London Spec. Publ. No. 28, 537–548.

Shee, S.R. and Gurney, J.J. (1979) The mineralogy of xenoliths from Orapa. Botswana. In *The Mantle Sample: Inclusions in Kimberlites and Other Volcanics*, (eds Boyd, F.R. and Meyer, H.O.A.) Am. Geophys. Union, Washington, D.C., 37–49.

Shervais, J.V., Taylor, L.A., Lugmair, G.W., Clayton, R.N., Mayeda, T.K. and Korotev, R.L. (1988) Early Proterozoic oceanic crust and the evolution of subcontinental mantle: eclogites and related rocks from southern Africa. *Bull. Geol. Soc. Am.* **100**, 411–423.

Shilo, N.A., Kaminskii, F.V. Palandzhyan, S.A., Tilman, S.M., Tkachenko, Lavrova, L.D. and Shepeleva, K.A. (1978) First diamond finds in alpine-type ultramafic rocks of the northeastern U.S.S.R. *Dokl, Akad. Nauk SSSR* **241**, 179–182 (in Russian).

Sibson, R.H., Moore, J. McM. and Rankin, A.H. (1975) Seismic pumping—a hydrothermal fluid transport mechanism. *J. Geol. Soc. London* **131**, 653–659.

Sills, J.D., Ackermand, O., Herd, R.K. and Windley, B.F. (1983) Bulk composition and mineral parageneses of sapphirine-bearing rocks along a gabbro-lherzolite contact at Finero, Ivrea Zone, N. Italy. *J. Metamorph. Geol.* **1**, 337–351.

Silver, L.T. and Deutsch, S. (1963) Uranium lead variations in zircon: a case study. *J. Geol.* **71**, 747–758.

Skjerlie, F.J. (1969) The Pre-Devonian rocks in the Askvoll–Gaular area and adjacent districts. Western Norway. *Nor. Geol. Unders.* **258**, 325–359.

Škvor, V. and Zeman, J. (1969) The Variscides of the Bohemian Massif. *Tectonophysics* **8**, 145–155.

Slodkevich, V.V. (1982) Graphite octahedra at Beni Bousera, N. Morocco, associated with a garnet–clinopyroxene layered complex. *Zap. Vses. Mineral. Obshch.* 1982, 13–33 (in Russian).

Smellie, J.L. and Stone, P. (1984) 'Eclogite' in the Ballantrae Complex: a garnet–clinopyroxenite segregation in mantle harzburgite? *Scott. J. Geol.* **20**, 315–327.

Smith, B.K., (1985) The influence of defect crystallography on some properties of orthosilicates. In *Metamorphic Reactions, Advances in Physical Geochemistry 4*, (eds Thompson, A.B. and Rubie, D.C.) Springer-Verlag, New York, 98–117.

Smith, C.B., Allsopp, H.L., Kramers, J.D., Gurney, J.J. and Jagoutz, E. (1986) Isotopic and geochemical studies of kimberlite and included xenoliths. *Fourth Int. Kimberlite Conf., Geol. Soc. Australia Abstracts* **16**, 329–331.

Smith, C.B., Gurney, J.J., Harris, J.W., Robinson, D.N., Shee, S.R. and Jagoutz, E. (1989) Sr and Nd isotopic systematics of diamond-bearing eclogite xenoliths and eclogitic inclusions in diamond from Southern Africa. In *Kimberlites and related rocks, Vol 2. Their mantle/crust setting, diamonds and diamond exploration.* Geol. Soc. Australia Spec. Publn, Blackwells, Carlton, 853–863.

Smith, D.C. and Wilson, C.R. (1985) Garnet-olivine equilibration during cooling in the mantle. *Am. Mineral.* **70**, 30–39.

Smith, D.C. (1971) A tourmaline-bearing eclogite from Sunnmöre, Norway, *Nor. Geol. Tidsskr.* **51**, 141–147.

Smith, D.C. (1980a) Highly aluminous sphene (titanite) in natural high-pressure hydrous-eclogite-facies parageneses in the Precambrian suite of eclogitic lenses in the Selje and Vartdalsfjorden districts of the basal gneiss region, S.W. Norway. *26th International Geological Congress Abstracts* **1**, 145.

Smith, D.C. (1980b) A tectonic mélange of foreign eclogites and ultramafics in west Norway. *Nature* **287**, 366–367.

Smith, D.C. (1981) A reappraisal of factual and mythical evidence concerning the metamorphic and tectonic evolution of eclogite-bearing terrain in the Caledonides. *Terra Cognita* **1**, 73 (abstract).

Smith, D.C. (1983) Nomenclature, fluids and deformation, geodynamics and nodules. *Terra Cognita* **3**, 329–334.

Smith, D.C. (1984) Coesite in clinopyroxene in the Caledonides and its implications for geodynamics, *Nature* **310**, 641–644.

Smith, D.C. (1985) The coesite-eclogite facies: from microtextures to megatectonics. *Terra Cognita* **5**, 418–419.

Smith, D.C. (1987) Appendix for Excursion Guide: Eclogite field symposium in Western Norway, 1987. Uppsala Caledonides Symposium.

Smith, D.C., Kienast, J.-R., Kornprobst, J. and Lasnier, B. (1982) Eclogites and their problems: an introduction to the First International Eclogite Conference (F.I.E.C.). *Terra Cognita* **2**, 283–295.

Smith, J.V. and Dawson, J.B. (1975) Chemistry of Ti-poor spinels, ilmenites and rutiles from peridotite and eclogite xenoliths. *Phys. Chem. Earth* **9**, 309–322.

Smulikowski, K. (1960) Comments on eclogite facies in regional metamorphism. *Rep. Int. Geol. Congr.* **XXI** *Copenhagen*, Pt. XIII, 372–382.

Smulikowski, K. (1964) An attempt at eclogite classification. *Bulletin de l'Académie Polonaise des Sciences: Série géologique et géographique* 12, 27–33.

Smulikowski, K. (1965) Chemical differentiation of garnets and clinopyroxenes in eclogites. *Bulletin de l'Academie Polonaise des Sciences: Série géologique et géographique* **13**, 11–18.

Smulikowski, K. (1967) Eclogites of the Snieznik Mts. in the Sudetes. *Geologia Sudetica* **3**, 7–180 (Polish, with English extended abstract).

Smulikowski, K. (1968) Differentiation of eclogites and its possible causes. *Lithos* **1**, 89–101.

Smulikowski, K. (1972) Classification of eclogites and allied rocks. *Krystallinikum* **9**, 107–130.

Smulikowski, K. (1980) Interrelations between eclogites and mafitic rocks of the granulite facies. *Archiwum. Mineral.* **36**, 5–21.

Smulikowski, K. and Bakun-Czubarow, N. (1969) Corundum-bearing eclogite amphibolite forming a loaf-shaped inclusion in the granulites of Bystrzyca Gorna (Sowie Gory, Middle Sudetes, Poland). *Bulletin de I'Academie Polonaise des Sciences serie de sciences géologiques et géographiques* **17**, 1–6.

Smulikowski, K. and Bakun-Czubarow, N. (1973) New data concerning the granulite–eclogite rock series of Stary Gieraltów, East Sudetes, Poland. *Bull. Acad. Pol. Sci. Sér. Sci. Terre* **21**, 25–34.

Smulikowski, K. and Smulikowski, W. (1985) On the Porphyroblastic Eclogites of the Śnieźnik Mountains in the Polish Sudetes. In *Chemistry and Petrology of Eclogites* (eds Smith, D.C. Franz, G. and Gebauer, D.) *Chem. Geol.* **50**, 201–222.

Smyth, J.R. (1980) Cation vacancies and the crystal chemistry of breakdown reactions in kimberlite omphacites. *Am. Mineral.* **65**, 1185–1191.

Smyth, J.R. and Caporuscio, F.A. (1984) Petrology of a suite of eclogite inclusions from the Bobbejaan kimberlite: II Primary phase compositions and origin. In *Kimberlites II: The Mantle and Crust–Mantle Relationships* (ed. Kornprobst, J.) Elsevier, Amsterdam, 121–131.

Smyth, J.R. and Hatton, C.J. (1977) A coesite-sanidine grospydite from the Roberts Victor kimberlite. *Earth Planet. Sci. Letts.* **34**, 284–290.

Smyth, J.R., McCormick, T.C. and Caporuscio, F.A. (1984) Petrology of a suite of eclogite inclusions from the Bobbejaan kimberlite: I. Two unusual corundum-bearing kyanite eclogites. In *Kimberlites II: The Mantle and Crust–Mantle Relationships* (ed. Kornprobst, J.) Elsevier, Amsterdam, 109–119.

Sobolev, N.V. (1977) *Deep-seated inclusions in kimberlite and the problem of the composition of the mantle.* Am. Geophys. Union, Washington, D.C.

Sobolev, N.V. (1984) Crystalline inclusions in diamonds from New South Wales, Australia. In *Kimberlite occurrence and origin: a basis for conceptual models in exploration.* (eds Glover, J.E. and Harris, P.G.) Univ. W. Australia, Perth, 213–226.

Sobolev, N.V. Bothkunov, A.I. and Kuznetsova, A.I. (1970) A diamond-bearing eclogite with calcium-rich garnet from the 'Mir' pipe, Yakutia. *Int. Geol. Rev.* **12**, 1357–1358.

Sobolev, N.V., Dobretsov, N.L. Bakirov, A.B. and Shatsky, V.S. (1986) Eclogites from various types of metamorphic complexes in the USSR and the problem of their origin. *Geol. Soc. Amer. Mem.* **164**, 349–363.

Sobolev, N.V., Galimov, E.M., Ivanovskaya, I.N. and Yefimova, E.S. (1979) Isotope composition of carbon of diamonds containing crystalline inclusions. *Dokl. Akad. Nauk SSSR* **249**, 1217–1220 (In Russian).

Sobolev, N.V. and Kuznetsova, I.K. (1965) More facts on the mineralogy of eclogite from Yakutian kimberlite pipes. *Dokl. Akad. Nauk. SSSR, Earth Sci. Sect.* **163**, 137–140.

Sobolev, N.V. Kuznetsova, I.K. and Zyuzin, N.I. (1968) The petrology of grosydite xenoliths from the Zagadochnaya kimberlite pipe in Yakutia. *J. Petrol.* **9**, 253–280.

Sobolev, N.V. and Lavrentyev, Yu.G. (1971) Isomorphic sodium admixture in garnets formed at high pressure. *Contrib. Mineral. Petrol.* **31**, 1–12.

Sobolev, N.V., Lavrentyev, Yu. G. and Usova, L.V. (1972) Minor elements in rutile from eclogites. *Geol. i. Geofiz.* (11), 108–112 (in Russian).

Sobolev, N.V. and Shatskii, V.S. (1987) Carbon mineral inclusions in garnets of metamorphic rocks. *Geol. i. Geofiz.* 77–80.

Sobolev, N.V. and Shatskii, V.S. (1988) Accessory diamonds in high pressures metamorphic rocks of the Earth's crust. Abstr. Internat. Symp. *Composition and processes of deep-seated zones of the continental lithosphere,* Novosibirsk 1988, 182–183.

Sobolev, N.V., Sobolev, A.V., Claui-Long, D.S., Shatsky, V.S. and Foster, L.D. (1988) Caledonian age of crustal metamorphism of superhigh pressures; U-Pb dating of zircon contained accessory diamond inclusions. Abstr. Internat. Symp. *Composition and processes of deep-seated zones of the continental lithosphere.* Novisibirsk 1988, 184.

Sobolev, V.S., Sobolev, N.V. and Lavrentyev, Yu.G. (1972) Inclusions in diamond from diamondferious eclogite. *Dokl. Akad. Nauk. SSSR., Earth Sci. Sect.* **207**, 121–123.

Söllner, F., Köhler, H. and Müller-Sohnius, D. (1981a) Rb/Sr Altersbestimmungen an Gesteinen der Münchberger Gneismasse (MM), NE Bayern. Teil 1. Gesamtgesteinsdatierungen. *Neues Jahrb. Mineral. Abh.* **141**, 90–112.

Söllner, F., Köhler, H. and Müller-Sohnius, D. (1981b) Rb/Sr Altersbestimmungen an Gesteinen der Münchberger Gneismasse (MM), NE Bayern. Teil 2, Mineraldaterungen. *Neues Jahrb. Mineral. Abh.* **142**, 178–198.

Spalla, M.I., De Marial, L., Gosso, G., Miletto, M. and Pognante, U. (1983). Deformazione e metamorfismo della zona Sesia-Lanzo meridionale al contatto con la Falda Piemontese e con il Massiccio di Lanzo, Alpi occidentali. *Mem. Soc. Geol. Ital.* **26**, 499–514.

Spear, F.S. and Franz, G. (1986) *P–T* evolution of metasediments from the Eclogite Zone, south-central Tauern Window, Austria, *Lithos* **19**, 219–234.

Spear, F.S. and Selverstone, J. (1983) Quantitative *P–T* paths from zoned minerals: theory and tectonic applications. *Contrib. Mineral. Petrol.* **83**, 348–357.

Spear, F.S., Selverstone, J. and Hickmott, D. (1984) *P–T* paths from garnet zoning: a new technique for deciphering tectonic processes in crystalline terrains. *Geology* **12**, 87–90.

Spetsius, Z.V., Nikishov, K.N. and Makhoto, V.F. (1984) Kyanite eclogite with sanidine from the Udachnaya kimberlite pipe. *Dokl. Akad. Nauk SSSR Earth Sci. Sect.* **279**, 138–141.

Steel, R.J., Siedlecka, A. and Roberts, D. (1985) The Old Red Sandstone basins of Norway and their deformation: a review. In *The Caledonide Orogen—Scandinavia and Related Areas* (eds Gee, D.G. and Sturt, B.A.) pp. 293–316. John Wiley, Chichester.

Stenger, R., Baatz, K., Klein, H. and Wimmenauer, W. (1987) Metamorphic evolution of the pre-Hercynian basement of the Schwarzwald (FRG). *Terra Cognita* **7**, 167.

Stephens, M.B. (1988) The Scandinavian Caledonides: a complexity of collisions. *Geology Today*, Jan–Feb. 1988, 20–26.

Stephens, M.B. and Gee, D.G. (1985) A tectonic model for the evolution of the eugeoclinal terranes in the central Scandinavian Caledonides. In *The Caledonide Orogen—Scandinavia and Related Areas* (eds Gee, D.G. and Sturt, B.A.) pp. 953–978. John Wiley and Sons, Chichester.

Stephens, M.B., Gustavson, M., Ramberg, I.B. and Zachrisson, E. (1985) The Caledonides of central-north Scandinavia—a tectonostratigraphic overview. In *The Caledonide Orogen—Scandinavia and Related Areas* (eds Gee, D.G. and Sturt, B.A.) pp. 135–162. John Wiley, Chichester.

Stephens, M.B. and Roermund, H. van (1984) Occurrence of glaucophane and crossite in eclogites of the Seve Nappes, southern Norbotten Caledonides, Sweden. *Norsk Geol. Tidsskr.* **64**, 155–163.

Stephens, W.E. and Dawson, J.B. (1977) Statistical comparison between pyroxenes from kimberlites and their associated xenoliths. *J. Geol.* **85**, 433–449.

Stettner, G. (1960) Erläuterungen zur Geologischen Karte von Bayern 1:25000, Blatt Nr. 5836, Münchberg. Bayer, Geol. Landesamt.

Stettner, G. (1960) Über Bau und Entwicklung der Münchberger Gneismasse. *Geol. Rdsch.* **49**, 350–375.

Stille, P. and Oberhänsli, R. (1987) Petrology and Alpine metamorphic evolution of the metabasic and ultrabasic rocks in the Berisal crystalline complex (Switzerland) *Lithos.* **20**, 169–180.

Stöckert, B. (1985) Compositional control on the polymorphism (2M1–3T) of phengite white mica from high pressure parageneses of the Sesia Zone (lower Aosta valley, Western Alps; Italy). *Contrib. Mineral. Petrol.* **89**, 52–58.

Stöckert, B., Jäger, E. and Voll, G. (1986) K-Ar age determinations on phengites from the internal part of the Sesia Zone, Western Alps. *Contrib. Mineral. Petrol.* **92**, 456–470.

Storre, B., Johannes, W. and Nitsch, K.-H. (1982) The stability of zoisite in H_2O–CO_2 mixtures. *Neues Jahrb. Mineral. Mtsh.* **9**, 395–406.

Stosch, H.-G. and Lugmair, G.W. (1987) Geochronology and Geochemistry of eclogites from the Münchberg Gneiss Massif, FRG. *Terra Cognita* **7**, 163.

Stracke, K.J., Ferguson, J. and Black, L.P. (1979) Structural setting of kimberlites in southeastern Australia. In *Kimberlites, Diatremes and Diamonds: their Geology, Petrology and Geochemistry.* (eds Boyd F.R. and Meyer, H.O.A.) *Proc. 3rd Int. Kimberlite Conference.* **1**, American Geophysical Union, Washington, D.C., 71–91.

Subramanian, A.P. (1956) Mineralogy and petrology of the Sittampundi Complex, Salem District, Madras State, India. *Bull. Geol. Soc. Am.* **67**, 317–390.

Suen, C.J. and Frey, F.A. (1987) Origins of the mafic and ultramafic rocks in the Ronda peridotite. *Earth Planet. Sci. Lett.* **85**, 183–202.

Suk, M. (1969) Mapping of metamorphic zonation in the Bohemian Massif. *Geol. Survey Prague Bull.* **44**, 209–218.

Suk, M. (1983) *Petrology of Metamorphic Rocks.* Developments in Petrology 9 Elsevier, 322 pp.

Suk, M. *et al* (1984) *Geological history of the territory of the Czech Socialist Republic,* Geological Survey of Czechoslovakia, Publishing House of Czechoslovak Academy of Sciences, Prague, 396 pp.

Suppe, J. and Armstrong, R.L. (1972) Potassium-Argon dating of Franciscan metamorphic rocks. *Am. J. Sci.* **272**, 217–233.

Svoboda, J. (1966) *Regional geology of Czechoslovakia, Part I The Bohemian Massif.* Geological Survey of Czechoslovakia, Publishing House of Czechoslovak Academy of Sciences, Prague, 668 pp.

Switzer, G. and Melson, W.G. (1969) Partially melted kyanite eclogite from the Roberts Victor mine, South Africa. *Smithsonian Contr. Earth Sci.* **1**, 1–9.

Takasu, A. (1984) Prograde and retrograde eclogites in the Sambagawa metamorphic belt, Besshi district, Japan. *J. Petrol.* **25**, 619–643.

Takayama, M. (1986) Mode of occurrence and significance of jadeite in the Kamuikotan metamorphic rocks, Hokkaido, Japan. *J. Metamorphic Geol.* **4**, 445–454.

Takeda, H. (1972) Crystallographic studies of coexisting aluminan orthopyroxene and augite of high-pressure origin. *J. Geophys. Res.* **77**, 5798–5811.

Tarney, J. and Windley, B.F. (1977) Chemistry, thermal gradients and evolution of the lower continental crust. *J. Geol. Soc. London* **134**, 153–172.

Taylor, H.P. and Coleman, R.G. (1968) O^{18}/O^{16} ratios of coexisting minerals in glaucophane-bearing metamorphic rocks. *Geol. Soc. Amer. Bull.* **79**, 1727–1756.

Tera, F. and Wasserburg, G.J. (1972) U, Th, Pb systematics in the three Apollo 14 basalts and the problem of initial Rb in lunar rocks. *Earth Planet. Sci. Lett.* **14**, 281–304.

Teufel, S. (1988) Vergleichende U–Pb– and Rb–Sr–Altersbestimmungen an Gesteinen des Übergangsbereiches Saxothuringikum/Moldanubikum, NE Bayern. *Göttinger Arb. Geol. Paläont.* **35**, 1–87.

Theye, T. (1988) Aufsteigende Hochdruckmetamorphose in Sedimenten der Phyllit-Quartzit-Einheit Kretas und des Peloponnes. *Dissertation, Univ. Braunschweig*, 224 pp.

Thierrin, J. (1983) Les éclogites et le complexe gabbroique du Val Sarsura (Silvretta). *Schweiz. Mineral. Petrogr. Mitt.* **63**, 479–496.

Thompson, A.B. (1983) Fluid-absent metamorphism. *J. Geol. Soc. London* **140**, 533–547.

Thompson, A.B. and England, P.C. (1984) Pressure–temperature–time paths of regional metamorphism II. Their inference and interpretation using mineral assemblages in metamorphic rocks. *J. Petrol.* **25**, 929–955.

Thompson, A.B. and Ridley, J.R. (1987) Pressure–temperature–time (P–T–t) histories of orogenic belts. *Phil. Trans. R. Soc. London. A* **321**, 27–45.

Thompson, J.B. Jr. (1981) An introduction to the mineralogy and petrology of the biopyriboles. In *Amphiboles and other Hydrous Pyriboles—Mineralogy* (ed. Veblen, D.R.) Mineral. Soc. Am. Reviews in Mineralogy, 9A, 141–188.

Thompson, R.N. (1975) Is upper-mantle phosphorus contained in sodic garnet? *Earth Planet. Sci. Lett.* **26**, 417–424.

Thurston, S.P. (1985) Structure, petrology and metamorphic history of the Nome Group blueschist terrane, Salmon Lake area, Seward Peninsula, Alaska. *Geol. Soc. Amer. Bull.* **96**, 600–617.

Tingle, T.N. and Green, H.W. (1987) Carbon solubility in olivine: implications for upper mantle evolution. *Geology* **15**, 324–326.

Toft, P.B., Hills, D.V. and Haggerty, S.E. (1988) Crustal evolution and the granulite to eclogite transition in xenoliths from kimberlites in the West African craton. *Tectonophysics* (in press).

Tollmann, A. (1982) Grossräumiger variszischer Deckenbau im Moldanubikum und neue Gedanken zum Variszikum Europas. *Geotekton. Forsch. Stuttgart* **64** (I–II), 1–91.

Triboulet, C. (1974) Les glaucophanites et roches associées de L'île de Groix (Morbihan, France): étude minéralogique et petrogénétique. *Contrib. Mineral. Petrol.* **45**, 65–90.

Triboulet, C. (1978) Co-existing blue and blue-green amphiboles from Ile de Groix (Morbihan, France). *J. Petrol.* **19**, 653–668.

Triboulet, C. and Audren, C. (1985) Evolution des amphiboles et de leurs associations au cours d'un metamorphisme progressif polyphasé dans des métabasites de la Vilaine (Bretagne méridionale). *Schweiz. Mineral. Petrol. Mitt.* **65**, 279–298.

Triboulet, C. and Audren, C. (1988) Controls on P–T–t deformation path from amphibole zonation during progressive metamorphism of basic rocks (estuary of the river Vilaine, South Brittany, France). *J. Metamorphic Geol.* **6**, 117–133.

Troll, G. (1978) The 'Altkristallin' of eastern Tyrol between Tauern Window and Periadriatic Lineament. In Inter-Union Commission for Geodynamics, Scientific Report 38 *Alps, Apennines, Hellenides.* 149–154.

Trommsdorff, V., Evans, B.W. and Richter, W. (1975) Eklogit/Rodingit-Übergänge in Ultramafiten der Cima-Lunga-Serie. *Schweiz. Mineral. Petrogr. Mitt.* **55**, 572–574.

Tsai, H.M., Meyer, H.O.A., Moreau, J. and Milledge, H.J. (1979a) Mineral inclusions in diamond: Premier, Jagersfontein and Finsch kimberlites, South Africa, and Williamson Mine, Tanzania. In *Kimberlites, Diatremes and Diamonds.* (eds Boyd, F.R. and Meyer, H.O.A.) Am. Geophys. Union, Washington, D.C., 16–26.

Tsai, H.M., Shieh, Y.N. and Meyer, H.O.A. (1979b), Mineralogy and S^{34}/S^{32} ratios of sulfides associated with kimberlite, xenoliths and diamonds. In *The Mantle Sample: Inclusions in Kimberlites and Other Volcanics.* (eds Boyd, F.R. and Meyer, H.O.A.) Am. Geophys. Union, Washington, D.C., 87–103.

Tsong, I.S.T., Knipping, U., Loxton, C.M., Magee, C.W. and Arnold, G. (1985) Carbon on surfaces of magnesium oxide and olivine single crystals—diffusion from the bulk or surface contamination? *Phys. Chem. Minerals* **15**, 261–270.

Tual, E., Jahn, B.M. and Bougault, H. (1983) REE geochemistry of basalts from DSDP Leg 83: Hole 504 B. *Terra Cognita* **3**, 145.

Turner, F.J. (1968) *Metamorphic Petrology: Mineral. Field and Tectonic Aspects.* 1st Edn, McGraw-Hill, New York 403 pp.

Turner, F.J. (1981) *Metamorphic Petrology: Mineralogical Field and Tectonic Aspects.* 2nd edition, McGraw-Hill 525 pp.

Udovkina, N.G., Muravitskaya, G.N. and Laputina, I.P. (1978) Phase equilibria in the talc–garnet–kyanite rocks of the Kokchetav block, northern Kazakhstan. *Izvestiya Akademii Nauk SSSR Seriya Geologicheskaya* **7**, 55–64 (in Russian).

Ungaretti, L., Lombardo, B., Domeneghetti, C. and Rossi, G. (1983) Crystal-chemical evolution of amphiboles from eclogitised rocks of the Sesia-Lanzo Zone, Italian western Alps. *Bulletin Minéralogie* **106**, 645–672.

Ungaretti, L., Smith, D.C. and Rossi, G. (1981) Crystal-chemistry by X-ray structure refinement and electron microprobe analysis of a series of sodic-calcic to alkali amphiboles from the Nybö eclogite pod, Norway, *Bull. Minéral.* **104**, 400–412.

Van Breemen, O., Aftalion, M., Bowes, D.R., Dudek, A., Mísař, Z., Povondra, P. and Vrána, S. (1982). Geochronological studies of the Bohemian Massif, Czechoslovakia, and their significance in the evolution of Central Europe. *Trans. R. Soc. Edinburgh Earth Sci.* **73**, 89–108.

Van Breemen, O. and Hawkesworth, C.J. (1980) Sm–Nd isotopic study of garnets and their metamorphic host rocks. *Trans. Roy. Soc. Edinburgh: Earth Sciences* **71**, 97–102.

Van Calsteren, P.W.C., Buelrijk, N.A.I.M., Hebeda, E.H., Priem, H.N.A., Den Tex, E., Verdurmen, E.A.T. and Vershure, R.H. (1979) Isotopic dating of older elements (including the Cabo Ortegal Mafic-ultramafic Complex) in the Hercynian orogen of N.W. Spain: manifestations of a presumed early Palaeozoic mantle plume. *Chem. Geol.* **24**, 35–56.

Van der Molen, I. and Van Roermund, H.L.M. (1986) The pressure path of solid inclusions in minerals: the retention of coesite inclusions during uplift. *Lithos* **19**, 317–324.

Van Roermund, H., Kullerud, K., Mørk, M.B.E. and Stephens, M.B. (1985) Allochthonous high pressure metamorphic rocks in the Swedish Caledonides. *Terra Cognita* **5**, 437.

Vearncombe, J.R. (1983) High pressure–low temperature metamorphism in the Gran Paradiso basement, Western Alps. *J. metamorphic Geol.* **1**, 103–115.

Vejnar, Z. (1965) Bemerkungen zur lithostratigraphischen Beziehung zwischen mittelbohmischen Algokium und dem Moldanubikum. *Neues Jahrb. Geol. Paläont. Monatsh.* **2**, 102–111.

Vejnar, Z. (1966) Peridotites and serpentinites of the Český Les Mountains. *Krystalinikum* **4**, 163–169.

Vejnar, Z. (1971) Grundfragen des Moldanubikums und seine Stellung in der Bohmischen Masse. *Geol. Rdsch* **60**, 1455–65.

Velde, B. (1965) Phengite micas: synthesis, stability, and natural occurrence. *Am. J. Sci.* **263**, 886–913.

Velde, B. (1967) Si^{4+} content of natural phengites. *Contrib. Mineral. Petrol.* **14**, 250–258.

Velde, B., Harvé, F. and Kornprobst, J. (1970) The eclogite–amphibolite transition at 650°C and 6.5 kbar presssure, as exemplified by basic rocks of the Uzerche area, Central France. *Am. Mineral.* **55**, 933–974.

Vernon, R.H. (1976) *Metamorphic Processes.* Allen and Unwin, London.

Vialon, P. (1966) Etude géologique du Massif cristallin Dora-Maira (Alpes cottiennes internes-Italie), *Trav. Lab. Geol. Grenoble* **4**, 293 pp.

Viterbo-Bassani, C. and Blackburn, C. (1968) The eclogitic rocks of the 'eclogitic micaschist formation', Sesia-Lanzo Zone (western Alps, Italy), *Memorie Istituti Geologia e Mineralogia Università Padova* **27**, 1–45.

Vivier, G., Ménot, R.-P. and Giraud, P. (1987) Magmatismes et structuration orogénique paléozoïque de la chaîne de Belledonne (massifs cristallins externes alpins). Le domaine nord-oriental. *Géol. Alp. Grenoble* **63**, 25–53.

Vogel, D.E. (1967) Petrology of an eclogite- and pyrigarnite-bearing polymetamorphic rock complex at Cabo Ortegal, NW Spain. *Leidse Geol. Meded.* **40**, 121–213.

Vogel, D.E. (1984). Cabo Ortegal, Mangle Plume or Double Klippe? In *Ophiolites and ultramafic rocks—a tribute to Emile den Tex.* (eds Zwart, H.J., Hartman, P. and Tobi, A.C.) Geol. Mijnbouw 63, 131–140.

Voll, G. (1960) Stoff, Bau und Alter in der Grenzzone Moldanubikum/Saxothuringikum in Bayern unter besonderer Berücksichtigung gabbroider amphibolitischer und kalksilikatführender Gesteine. *Beih. Geol. Jahrb. Hannover* **42**, 1–382.

Von Frese, R.R.B., Hinze, W.J., Olivier, R. and Bentley, C.R. (1987) Satellite magnetic anomalies and continental reconstructions. *Gondwana six: structure, tectonics, and geophysics, Geophysical Monograph*, **40**, 9–15.

Von Gehlen, K. and Schmitt, G.E. (1987) Ophiolitic serpentinites within the Moldanubian of NE-Bavaria, FRG, identified by means of relictic primary minerals. *Terra Cognita* **7**, 164–5.

Von Raumer, J.F. (1974) Zur Metamorphose amphibolitischer Gesteine im Altkristallin des Mont–Blanc– und Aiguilles–Rouges–Massivs. *Schweiz. Mineral. Petrogr. Mitt.* **54**, 471–488.

Vrána, S. and Jakes P. (1982) Orthopyroxene and two-pyroxene granulites from a segment of charnockitic crust in southern Bohemia. *Věsn. Ůstřed Ůstavu Geol. CAV.* **5**, 129–143.

Vuichard, J.P. and Ballèvre, M. (1988) Garnet–chloritoid equilibria in eclogitic pelitic rocks from the Sesia zone (Western Alps): their bearing on phase relations in high pressure metapelites. *J. metamorphic Geol.* **6**, 135–158.

Wasilewski, P. and Warner, R.D. (1988) Magnetic petrology of deep crustal rocks—Ivrea Zone, Italy. *Earth Planet. Sci. Lett.* **87**, 347–361.

Wasilewski, P. and Weidner, G. (1989) Redox in the continental lithosphere: a magnetic approach. *Geology* (submitted).

Wass, S.Y. and Hollis, J.D. (1983) Crustal growth in southeastern Australia—evidence from lower crustal eclogitic and granulitic xenoliths. *J. metamorphic Geol.* **1**, 25–45.

Waters, D.J. (1976) *Structural, metamorphic and geochronological studies in the south-east Tauern.* Unpublished D.Phil. thesis, University of Oxford.

Watson, K.D. and Morton, D.M. (1969) Eclogite inclusions in kimberlite pipes at Garnet Ridge, northeastern Arizona. *Am. Mineral*, **54**, 267–285.

Wayte, G.J. Worden, R.H., Rubie, D.C. and Droop, G.T.R. (1989) A TEM study of disequilibrium plagioclase breakdown at high pressure: the role of infiltrating fluid. *Contrib. Mineral. Petrol.* **101**, 426–437.

Webb, S.A.C. and Wood, B.J. (1986) Spinel–pyroxene–garnet relationships and their dependence on Cr/Al ratio. *Contrib. Mineral. Petrol.* **92**, 471–480.

Weber, K. and Behr, H.-J. (1983) Geodynamic Interpretation of the Mid-European Variscides. In *Intracontinental Fold Belts* (eds Martin, H. and Eder, W.) Springer-Verlag, Heidelberg, 427–469.

Weissenbach, N. (1965) Die geologische Neuaufnahme des saualpenkristallins. *Carinthia II* **73**, 5–22.

Welch, M. (1987) *Experimental studies of selected amphiboles in the system $Na_2O–CaO–MgO–Al_2O_3–SiO_2–H_2O–F_2$ and its subsystems.* University of Edinburgh unpubl. Ph.D. thesis. 200 pp.

Welch, M. (1989) An experimental study of glaucophane-type amphiboles in the system $Na_2O–MgO – Al_2O_3–SiO_2–SiF_4$ (NMASF), with implications for natural systems. *Contrib. Mineral. Petrol.* (Submitted).

Wells, P.R.A. (1977) Pyroxene thermometry in simple and complex systems. *Contrib. Mineral. Petrol.* **62**, 129–139.

White, A.J.R. (1964) Clinopyroxenes from eclogites and basic granulites. *Am. Mineral.* **49**, 883–888.

White, S.H. and Knipe, R.J. (1978) Transformation- and reaction-enhanced ductility in rocks. *J. Geol. Soc. London* **135**, 513–516.

Whitfield, G.G. (1971) A petrological and mineralogical study of peridotite and eclogite xenoliths from certain kimberlite pipes. Unpubl. M.Sc thesis. Rhodes University.

Wieseneder, H. (1934) Beiträge zur Kenntnis der ostalpinen Eklogite. *Tschermaks Mineral. Petrogr. Mitt.* **46**, 174–211.

Wijbrans, J.R. York, D. and Schliestedt, M. (1988) *Laser probe $^{40}Ar/^{39}Ar$ spectra of single crystals of phengite from the Cycladic blueschist belt, Greece, Abstract AGU*, Baltimore.

Williams, P.F. and Compagnoni, R. (1983) Deformation and metamorphism in the Bard area of the Sesia–Lanzo zone, western Alps, during subduction and uplift. *J. Metamorphic Geol.* **1**, 117–140.

Wilson, C.R. and Smith D.C. (1984) Cooling rate estimates from mineral zonation: resolving power and applications. In *Kimberlites: II. The Mantle and Crust-Mantle Relationships* (ed. Kornprobst, J.) Developments in Petrology 11B, Elsevier, Amsterdam, 265–275.

Wilson, C.R. and Smith, D.C. (1985) A numerical study of mineral zoning formed during the cooling of finite volume grains. *J. Geophys. Res.* **90**, 10, 351–10, 358.

Wimmenauer, W. and Adam, A. (1985) High pressure phenomena in acid rocks accompanying eclogites in the Schwarzwald, SW Germany. *Terra Cognita* **5**, 426.

Wimmenauer, W. and Stenger, R. (1987) Acid and intermediate HP-metamorphic rocks in the Schwarzwald (FRG). *Terra Cognita* **7**, 167–8.

Winchester, J.A. and Floyd, P.A. (1976) Geochemical magma type discrimination: application to altered and metamorphosed basic igneous rocks. *Earth Planet. Sci. Lett.* **28**, 459–469.

Windom, K.E. and Boettcher, A.L. (1980) Mantle metasomatism and the kimberlite–lamprophyre association: evidence from an eclogite nodule from the Roberts Victor Mine. *J. Geol.* **88**, 705–712.

Windom, K.E. and Unger, C.P. (1988) Stability of the assemblage albite plus forsterite at high temperatures and pressures with petrologic implications. *Contrib. Mineral. Petrol.* **98**, 390–400.

Winkler, H.G.F. (1974) *Petrogenesis of Metamorphic Rocks.* 3rd edn, Springer-Verlag, Berlin 320 pp.

Wood, B.J. (1974) The solubility of alumina in orthopyroxene coexisting with garnet. *Contrib. Mineral. Petrol.* **46**, 1–15.

Wood, B.J. (1978) Reactions involving anorthite and $CaAl_2SiO_6$ pyroxene at high pressures and temperatures. *Am. J. Sci.* **278**, 930–942.

Wood, B.J. (1979) Activity-composition relationships in $Ca(Mg,Fe)Si_2O_6$–$CaAl_2SiO_6$ clinopyroxene solid solutions. *Am. J. Sci.* **279**, 854–875.

Wood, B.J. (1987) Thermodynamics of multicomponent systems containing several solid solutions. In *Thermodynamic Modeling of Geological Materials: Minerals, Fluids and Melts.* (eds Carmichael, I.S.E. and Eugster, H.P.) Reviews in Mineralogy, 17, Min. Soc. America, Michigan, 71–96.

Wood, B.J. and Banno, S. (1973) Garnet–orthopyroxene and orthopyroxene–clinopyroxene relationships in simple and complex systems. *Contrib. Mineral. Petrol.* **42**, 109–124.

Wood, B.J. and Henderson, C.M.B. (1978). Composition and unit cell parameters of synthetic non-stoichiometric Tschermakitic clinopyroxenes. *Am. Mineral.* **63**, 66–72.

Wood, B.J., Holland, T.J.B. and Kleppa, O.J. (1980) Thermochemistry of jadeite–diopside pyroxenes. *Geochim. Cosmochim. Acta* **44**, 1363–1371.

Wood, B.J., Holland, T.J.B., Newton, R.C. and Kleppa, O.J. (1980) Thermochemistry of jadeite–diopside pyroxenes. *Geochim. Cosmochim. Acta* **44**, 1363–1371.

Wood, B.J. and Holloway, J.R. (1984) A thermodynamic model for subsolidus equilibria in the system CaO–MgO–Al_2O_3–SiO_2. *Geochim. Cosmochim. Acta* **48**, 159–176.

Wyllie, P.J. (1979) Magmas and volatile components. *Am. Mineral.* **64**, 489–500.

Yamada, H. and Takahashi, E. (1984) Subsolidus phase relations between coexisting garnet and two pyroxenes at 50 to 100 kbar in the system CaO–MgO–Al_2O_3–SiO_2. In *Kimberlites II: The Mantle and Crust-Mantle Relationships.* (ed. Kornprobst, J.) *Developments in Petrology.* **11B**, 247–255, Elsevier.

Yoder, H.S. and Tilley, C.E. (1962) Origin of basalt magmas: an experimental study of natural synthetic rock systems. *J. Petrol.* **3**, 342–532.

Yokoyama, K., Brothers, R.N. and Black, P.M. (1982) Regional eclogite facies in the high pressure metamorphic belt of New Caledonia. *Geol. Soc. Amer. Mem.* **164**, 407–423.

Yokoyama, K., Brothers, R.N. and Black, P.M. (1986) Regional eclogite facies in the high-pressure metamorphic belt of New Caledonia. In *Blueschists and Eclogites.* (eds Evans, B.W. and Brown, E.H.) *Geol. Soc. Amer. Mem.* **164**, 407–423.

Yund, R.A. (1986) Interdiffusion of NaSi-CaAl in peristerite. *Phys. Chem. Minerals* **13**, 11–16.

Yund, R.A., Smith, B.M. and Tullis, J. (1981) Dislocation-assisted diffusion of oxygen in albite. *Phys. Chem. Minerals* **7**, 185–189.

Yund, R.A. and Tullis, J. (1980) The effect of water, pressure, and strain on Al/Si order–disorder kinetics in feldspar. *Contrib. Mineral. Petrol.* **72**, 297–302.

Zakrutkin, V.V. (1968) The evolution of amphiboles during metamorphism. *Zap. Vses. Mineral. Ova.* **96**, 13–23.

Ziegler, P.A. (1984) Caledonian and Hercynian crustal consolidation of Western and Central Europe—a working hypothesis. *Geol. Mijnbouw* **63**, 93–108.

Zindler, A., Staudigel, H., Hart, S.R. Endres, R. and Goldstein, S. (1983) Nd and Sr isotopic study of a mafic layer from Ronda ultramafic complex. *Nature* **304**, 226–230.

Zoubek, V. (1969) Age relations in metamorphic terrains of the Bohemian Massif, some methods and results. In *Age Relations in High Grade Metamorphic Terrains* (ed. Wynne-Edwards, H.R.) Geol. Assoc. Canada Spec. Paper 5, 73–81.

Zwart, H.J. (1969) Metamorphic facies series in the European orogenic belts and their bearing on the causes of orogeny. In *Age Relations in High Grade Metamorphic Terrains.* (ed. Wynne-Edwards, H.R.) Geol. Assoc. Canada Spec. Paper 5, 7–16.

Index

References to tables are printed in italics, references to illustrations in bold face type. Localities are listed by country and/or region. Only selected mineral references are given.

FSC
www.fsc.org
MIX
Papier | Fördert gute Waldnutzung
FSC® C083411

Kolibri 360 GmbH
Ferdinand-Jühlke-Straße 7
99095 Erfurt, Deutschland
produktsicherheit@kolibri360.de